PARTICULATE CARBON
Formation During Combustion

PUBLISHED SYMPOSIA
Held at
General Motors Research Laboratories
Warren, Michigan

1980 R. L. Klimisch, G. T. Wolff, eds., *Particulate carbon: Atmospheric life cycle*, Plenum Press, New York. In preparation.

1980 D. C. Siegla, G. W. Smith, eds., *Particulate carbon: Formation during combustion.* Plenum Press, New York, 1981.

1979 R. C. Schwing, W. A. Albers, Jr., eds., *Societal risk assessment: How safe is safe enough?* Plenum Press, New York, 1980.

1978 J. N. Mattavi, C. A. Amann, eds., *Combustion modeling in reciprocating engines*, Plenum Press, New York, 1980.

1978 G. G. Dodd, L. Rossol, eds., *Computer vision and sensor-based robots*, Plenum Press, New York, 1979.

1977 D. P. Koistinen, N.-M. Wang, eds., *Mechanics of sheet metal forming: Material behavior and deformation analysis*, Plenum Press, New York, 1978.

1976 G. Sovran, T. A. Morel, W. T. Mason, eds., *Aerodynamic drag mechanisms of bluff bodies and road vehicles*. Plenum Press, New York, 1978.

1975 J. M. Colucci, N. E. Gallopoulos, eds., *Future automotive fuels: Prospects, performance, perspective*, Plenum Press, New York, 1977.

1974 R. L. Klimisch, J. G. Larson, eds., *The catalytic chemistry of nitrogen oxides*, Plenum Press, New York, 1975.

1973 D. F. Hays, A. L. Browne, eds., *The physics of tire traction*. Plenum Press, New York, 1974.

1972 W. F. King, H. J. Mertz, eds., *Human impact response*. Plenum Press, New York, 1973.

1971 W. Cornelius, W. G. Agnew, eds., *Emissions from continuous combustion systems*, Plenum Press, New York, 1972.

1970 W. A. Albers, ed., *The physics of opto-electronic materials*. Plenum Press, New York, 1971.

1969 C. S. Tuesday, ed., *Chemical reactions in urban atmospheres*. American Elsevier, New York, 1971.

1968 E. L. Jacks, ed., *Associative information techniques*. American Elsevier, New York, 1971.

1967 P. Weiss, G. D. Cheever, eds., *Interface conversion for polymer coatings*. American Elsevier, New York, 1968.

1966 E. F. Weller, ed., *Ferroelectricity*, Elsevier, New York, 1967.

1965 G. Sovran, ed., *Fluid mechanics of internal flow*. Elsevier, New York, 1967.

1964 H. L. Garabedian, ed., *Approximation of functions*. Elsevier, New York, 1965.

1963 T. J. Hughel, ed., *Liquids: Structure, properties, solid interactions*. Elsevier, New York, 1965.

1962 R. Davies, ed., *Cavitation in real liquids*. Elsevier, New York, 1964.

1961 P. Weiss, ed., *Adhesion and cohesion*. Elsevier, New York, 1962.

1960 J. B. Bidwell, ed., *Rolling contact phenomena*, Elsevier, New York, 1962.

1959 R. C. Herman, ed., *Theory of traffic flow*. Elsevier, New York, 1961.

1958 G. M. Rassweiler, W. L. Grube, eds., *Internal stresses and fatigue in metal*, Elsevier, New York, 1959.

1957 R. Davies, ed., *Friction and wear*, Elsevier, New York, 1959.

PARTICULATE CARBON
Formation During Combustion

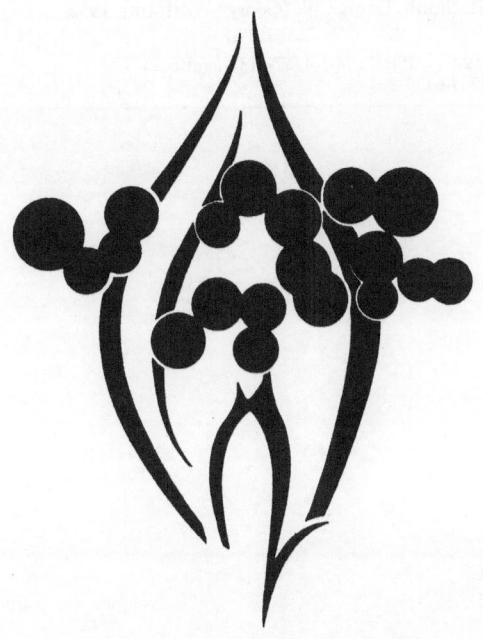

Edited by
DONALD C. SIEGLA and GEORGE W. SMITH
General Motors Research Laboratories

SPRINGER SCIENCE+BUSINESS MEDIA, LLC 1981

Library of Congress Cataloging in Publication Data

Main entry under title:

Particulate carbon, formation during combustion.

"Proceedings of an international symposium on particulate carbon, formation during combustion, held October 15-16, 1980, at the General Motors Research Laboratories, Warren, Michigan" — Verso t.p.
Includes bibliographies and index.
1. Combustion — Congresses. 2. Soot — Congresses. 3. Carbon — Congresses. I. Siegla, Donald C. II. Smith, George W. (George Wolfram), 1932-

QD516.P36	541.3'61	81-15363
ISBN 978-1-4757-6139-9 ISBN 978-1-4757-6137-5 (eBook)		AACR2
DOI 10.1007/978-1-4757-6137-5		

Proceedings of an international symposium on Particulate Carbon:
Formation During Combustion, held October 15–16, 1980, at the
General Motors Research Laboratories, Warren, Michigan

© 1981 Springer Science+Business Media New York
Originally published by Plenum Press, New York in 1981

PREFACE

The goal of the symposium, "Particulate Carbon: Formation During Combustion", held at the General Motors Research Laboratories on October 15 and 16, 1980, was to discuss fundamental aspects of soot formation and oxidation in combustion systems and to stimulate new research by extensive interactions among the participants. This book contains the papers and discussions of that symposium, the 26th in an annual series covering many different disciplines which are timely and of interest to both General Motors and the technical community at large.

The subject of this symposium has considerable relevance for man in his effort to control and preserve his environment. Emission of particulate carbon into the atmosphere from combustion sources is of concern to scientists and laymen alike. The hope of reducing this emission clearly requires an understanding of its formation during the combustion process, itself an area of considerable long-term research interest. It is our hope that this symposium has served to summarize what is known so that what remains to be learned can be pursued with greater vigor.

The scientists and engineers who participated in this symposium came from seven countries and were chosen because of their expertise in physical and chemical characterization, formation and oxidation kinetics, premixed and diffusion combustion, turbulent mixing, spray combustion, and combustion in engines. Attendance was limited in order to enable a closer contact among the participants, the intent being to foster not only discussion among experts in related fields but also interaction among those conducting fundamental research and those responsible for the design of practical combustion systems. Hence, new channels of communication were established in order to create an awareness of the concerns and objectives of others and thus enhance the productivity of future research.

The organization of the papers in this volume follows that of the symposium itself. The chapters are arranged in five groups: an overview of the field, characterization of combustion-formed particulate carbon, basic formation and oxidation mechanisms and kinetics, formation during continuous combustion, and formation during intermittent combustion. Thus the subject matter of the book progresses from the fundamental to the practical. The volume, like the symposium, concludes with a summary and look to the future by Prof. Palmer, one of the pioneers in the field.

One confusing aspect of the literature of particulate carbon became apparent during the symposium: the diversity of nomenclature. Various participants used different terminology in referring to the same entity or concept. (Dr. Medalia addresses this problem in his remarks on pg. 89.) For instance, several different names have been used to describe the quasi-spherical units of which a soot agglomerate is composed: *primary particles, elementary particles, spherical particles,* and *spherical units.* The simple term *spherule* (which makes no implications concerning the elementary or primary nature of the quasi-spherical units) was used by some conferees to avoid confusion. Prof. Hilliard has pointed out that *spherule* possesses the added merit of priority, having been used by Lord Rayleigh [Proc. Roy. Soc., Vol. 29 (1879), p. 71] over 100 years ago to describe colliding droplets. Names applied to describe the total soot structural entity have ranged from *particulate* (usually an adjective) to *particle, agglomerate, aggregate, aciniform carbon* (see pg. 89), and *cluster* (all nouns). The terms *agglomeration, coagulation, clustering,* and *coalescence* have been used to describe aspects of the aggregation process. No confusion arises if the usage is carefully and properly defined, which, we feel, is the case with the papers in this volume. Our purpose in this paragraph has been to call attention to a need which all expanding fields of endeavor share: terminology which is accurate, concise, consistent, and capable of growth.

The organization of the symposium and the publication of these proceedings resulted from the collaboration of many people. We are extremely grateful to our symposium advisors, John B. Heywood, Jack B. Howard, and Howard B. Palmer, for helping us to select topics and speakers for a stimulating and worthwhile meeting. To our session chairmen, Klaus H. Homann, Irvin Glassman, John P. Longwell, and John B. Heywood, we express our indebtedness for participating in and guiding the progress of the discussions. And, of course, we are grateful to the speakers and participants of the symposium who contributed the essential ingredient of a successful conference: free and enthusiastic exchange of ideas. We thank John D. Caplan, William G. Agnew, Nils L. Muench, Charles A. Amann, Frank E. Jamerson, and Leroy R. Buzan, for their support and guidance. Advice on technical and organizational matters from Walter A. Albers, Jr., Richard J. Blint, Donald P. Koistinen, James A. Mattavi, and Richard C. Schwing was greatly appreciated. The assistance of R. Thomas Beaman with physical arrangements was invaluable. We are grateful to Tanvir Ahmad, James H. Bechtel, Richard J. Blint, James P. Myers, Steven L. Plee, David M. Roessler, Mary Lee Smith, Robin Stevenson, and Richard E. Teets for aiding in the transcription and editing processes. We appreciate the contributions of Jane K. Fiebelkorn, Elizabeth L. Kalata, Linda D. Sroda, and Mildred D. Woodhouse who expertly assisted in typing and the administrative details of the symposium. Finally, we particularly thank David N. Havelock for supervising the conversion of the edited manuscripts into this volume.

Donald C. Siegla
Engine Research Department
General Motors Research Laboratories
Symposium Cochairman

George W. Smith
Physics Department
General Motors Research Laboratories
Symposium Cochairman

CONTENTS

- 1. Howard Palmer - Pennsylvania State University
- 2. Heinz Wagner - Universität Göttingen
- 3. Bjorn Magnussen - Universitetet I Trondheim
- 4. Toshi Kadota - Wayne State University
- 5. Koon Neoh - Massachusetts Institute of Technology
- 6. Gilles Prado - Massachusetts Institute of Technology
- 7. Arthur Mellor - Purdue University
- 8. David Roessler - General Motors Research Laboratories
- 9. John Heywood - Massachusetts Institute of Technology
- 10. Donald Siegla* - General Motors Research Laboratories
- 11. Adel Sarofim - Massachusetts Institute of Technology
- 12. Naeim Henein - Wayne State University
- 13. Douglas Olson - AeroChem Research Laboratories
- 14. Jack Longwell - Massachusetts Institute of Technology
- 15. Klaus Homann - Technische Hochschule Darmstadt
- 16. Michal Dyer - Sandia National Laboratories
- 17. Jim Myers - General Motors Research Laboratories
- 18. Bill Flower - Sandia National Laboratories
- 19. Jack Howard - Massachusetts Institute of Technology
- 20. Jim Bittner - Massachusetts Institute of Technology
- 21. Ralph Kowalik - Exxon Research and Engineering Company
- 22. Ray Edelman - Science Applications, Inc.
- 23. Milton Lee - Brigham Young University
- 24. Tanvir Ahmad - General Motors Research Laboratories
- 25. Richard Farmer - Science Applications, Inc.
- 26. Antonio D'Alessio - University of Naples
- 27. Irwin Glassman - Princeton University
- 28. George Smith* - General Motors Research Laboratories
  *Symposium Co-chairman

Not Pictured
- Jacques Lahaye
- Robin Stevenson
- Steve Plee
- L. A. Ruth

Co-authors Not in Attendance:
- Fred Faxvog
- William Blazowski
- Hartwell F. Calcote
- E. Wong
- Keith Bartle

Organizers, Session Chairmen, and Authors
of the 1980 GMR Symposium
PARTICULATE CARBON: Formation During Combustion

SOOT FORMATION — AN OVERVIEW

H. Gg. WAGNER

Institut für Physikalische Chemie der Universität
Göttingen, West Germany

ABSTRACT

Formation of soot can take place in all kinds of practical combustion systems, especially in systems which operate on a diffusion flame concept. The influence of macroscopic system parameters on the amount of soot formed is rather well investigated for different systems. Even though these results often look unrelated to each other, the primary soot particles formed in various combustion processes are rather similar. Their mean diameter is usually a few hundred Angstroms. These particles often stick together and form long branched or straight chain aggregates and their number densities are very similar even under widely varied conditions, a consequence of their formation mechanism.

What comes out of a chimney or an exhaust pipe may look quite different. The formation of soot is usually accompanied by the formation of heavy hydrocarbons. Soot particles are very good absorbers. It depends on their history (time, temperature, concentrations), how much of other substances they absorb until they reach the open atmosphere.

"Particle history" is also important for the formation process of the soot particles. Even though they experience different surroundings during their growth in premixed combustion or in various types of diffusion flames, the results of the formation processes are rather similar.

Soot particles formed during combustion can be removed only by oxidation, by reactions with OH, O atoms or O_2. Experiments show that this process is slow compared to the oxidation of hydrocarbon molecules and the necessary residence times for particle removal strongly increase with particle size.

INTRODUCTION

Stationary combustion devices like furnaces or combustors operate on a diffusion flame concept. The gaseous or liquid fuel is injected into the air and forms a diffusion flame, which has a large stability range and can be rather easily adjusted to the conditions required. These favorable properties are largely due to the fact that there exist concentration fields in which the flame can find its optimum positions. In reciprocating engines with fuel injection into the cylinder, combustion also takes place in a mixture with varying fuel-air ratios, because there is not enough time to form homogeneous gaseous mixtures prior to combustion. Even when the fuel/air mixture is formed outside of the cylinder by a carburetor or some injection system, the burning mixture is not completely homogeneous. This need not be a disadvantage for all operating conditions of the engine of e.g., a passenger car, but it can give rise to the intermediate formation of soot.

In technical combustion processes, soot usually forms at temperatures from 1000 to 2500°C. Palmer [1] recognized that the elementary soot particles do not differ much in size, no matter whether they come from a furnace flame, a piston engine, a combustion chamber or a premixed flame ($d \sim 20$-50 nm). The total amount of soot formed in technical combustion processes is always very small compared to the amount of carbon present in the fuel consumed, and the time available for the soot formation process in engines is very short indeed. (These are important differences compared to the processes applied for carbon black fabrication [2]). The formation of soot at high temperatures is usually accompanied or preceded by the formation of unsaturated hydrocarbons, especially acetylenes and condensed unsaturated ring systems, polycyclic aromatics and other compounds. These hydrocarbons are thermodynamically relatively stable at temperatures of interest here [3-5], and kinetically stable as compared to paraffins or even olefins. Even in the presence of solid carbon, a certain amount of some of these hydrocarbons will remain in the gas phase in thermal equilibrium [4]. When such burned gases are cooled down, these hydrocarbons can adsorb on soot particles so that the "particulate carbon" leaving an exhaust pipe may look rather different from the soot particles passing through the exhaust valve of an engine.

The detailed investigation of the soot formation process in combustion devices is either practically impossible or very complicated. Therefore, research work on the kinetics of soot formation, performed in recent years, has followed three lines:

  i. soot formation in premixed flames,
 ii. soot formation by high temperature pyrolysis and
iii. soot formation in diffusion flames, which contains elements of both processes mentioned.

APPEARANCE OF SOOT

Soot generated in combustion processes is not a uniquely defined substance. (For details, see [1, 2, 6].) It normally looks lustre-less black and is different from graphite. The main constituents are carbon atoms. It contains up to 10 mole percent hydrogen and even more if it is young, as well as traces of other elements. A good deal of that hydrogen can be extracted in organic solvents, where it appears mostly in condensed aromatic ring compounds. Sometimes, materials are emitted which, when cooled, look like tar or a glassy substance, either black, brown or even yellow. The inspection of soot coming, for example, from a premixed flame or the exhaust valve of a diesel engine shows under the electron microscope that the building bricks of that soot are spherical or near-spherical elementary soot particles with mean diameters around 20 to 30 nm, corresponding to about 10^6 carbon atoms. These elementary particles adhere to each other to form straight or branched chains. These chains agglomerate and form the visible macroscopic soot flocculates, generally as a fluffy substance.

For the investigation of the internal structure of the elementary particles, x-ray studies have been very useful [7-10]. Phase contrast electron micrographs [11, 12] show that carbon black particles can be considered as an arrangement of bent carbon layers which follow the shape of the particle surface [11-13]. Many dislocations and lattice defects are present. The density of these particles seems to be less than 2 g/cm^3, due to large interplanar spacing.

Electron diffraction indicates the presence of single C-C bonds in soot [14]. Soot particles collected during their growth show much stronger ESR signals [6], than those of fully developed particles [15-17]. Heat treatment improves the degree of ordering [18] and reduces the hydrogen content.

PATHWAYS FOR SOOT FORMATION

The time available for soot formation in combustion processes is of the order of milliseconds. During this time a solid phase of soot particles is formed from the fuel molecules via their oxidation and/or pyrolysis products.

The possible paths for soot forming processes can be seen in Fig. 1, where the logarithm of the molar mass (g/mol) of a species is plotted against its hydrogen content (as mole fraction in the molecule) [19]. Various fuel molecules are shown in the lower left corner, with $X_H > 0.5$, the final soot particles at the right top corner. Also indicated is the growth line of polyacetylenes. It is immediately obvious that soot is not formed along the growth line of the acetylenes, or the polycyclic aromatic compounds alone. In order for the fuel molecules to evolve to soot, there must take place a (chemical!) condensation of species with the right hydrogen content or a (chemical!) condensation of species with higher hydrogen content and consecutive dehydrogenation, or a combination of these two limiting pathways from the lower left to the upper right corner in Fig. 1 (which is supported by the experimental data available up to now).

LIMITS OF SOOT FORMATION IN PREMIXED SYSTEMS

Pure hydrocarbons exposed to high temperature will, after sufficient time, reach their equilibrium state, which is graphite, and hydrogen with some stable hydrocarbons like acetylenes, large polycyclic aromatics, etc. in the gas phase [4, 5].

For premixed hydrocarbon/air systems soot is expected to appear when there is just not enough oxygen to transform the hydrocarbon into CO and H_2, i.e., when C/O approaches one [20]. Experimentally determined limits of soot formation [21-25], however, do not occur at C/O = 1 but usually close to C/O $\approx$ 0.5.

Three different methods, Bunsen burner [21], stirred reactor [25] and flat flame [23-25] give quite similar results even though the formation of polyhedral flames influences the limits to some extent [26] (Table 1).

TABLE 1
C/O at Limit of Soot Formation

	Bunsen flame [+)	Stirred reactor	Flat flame
C_2H_6-air	0.48		0.47
C_3H_8-air	0.47[++)		0.53[++)
C_4H_{10}-air	0.46[++)		0.54[++)
C_2H_4-air	0.60[+)	0.60	0.62[++)
C_4H_8-air	0.48[++)	0.48	
C_6H_6-air	0.57[++)	0.57	0.62[++)
Methylnaphthalene-air	0.42[++)	0.62	

[+) *Except for small fuel molecules, the Bunsen flame starts to emit soot at the tip [21].*
[++) *Polyhedral flames.*

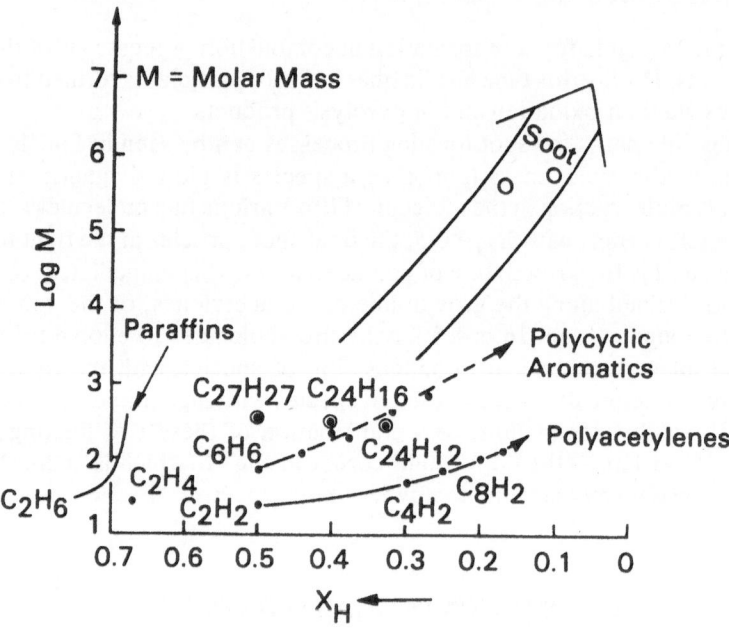

Fig. 1. Paths for soot formation. Molar mass of species plotted as a function of their hydrogen mole fraction X_H[19].

The critical C/O ratio is only weakly, if at all, dependent on pressure or dilution with inert gas at constant temperature [6]. Increasing temperature generally allows richer mixtures to be burned without the onset of sooting, as long as the flame front remains smooth [23, 24]. When the flames become polyhedral, the soot concentration in the filaments starts to increase with increasing temperature: the situation is more related to that for a diffusion flame and the gas jet forming the filament has a C/O ratio above the mean value in the unburned gas [26].

THE SOOT YIELD

Beyond the soot limit the yield of soot in premixed systems initially increases very rapidly indeed (Fig. 2) with increasing C/O ratio. This rapid rise of soot volume fraction, f_v, with C/O is to a certain extent typical for the fuel (closer to the upper limit of inflammability problems of the flame stability start to interfere). This increase in the amount of soot is not so much due to an increase in particle number but to an increase of the size of the elementary spherical soot particles with increasing C/O ratio.

The influence of pressure on the soot yield seems to be very strong; it is strongly enhanced by increasing pressure. In their study of turbulent premixed flames McFarlane et al. [27] found that the soot yield varied as p^{2-3} for fuels like hexane and hexene. However, in the case of benzene the effect of pressure was only weak.

In keeping with the influence of pressure, back mixing brings about a profound reduction in the amount of soot produced at a given C/O ratio [25, 28].

In the burned gases of premixed flames the water-gas components CO_2 and H_2O can be found (in equilibrium with CO and H_2 if C/O is not too large) as well as CO

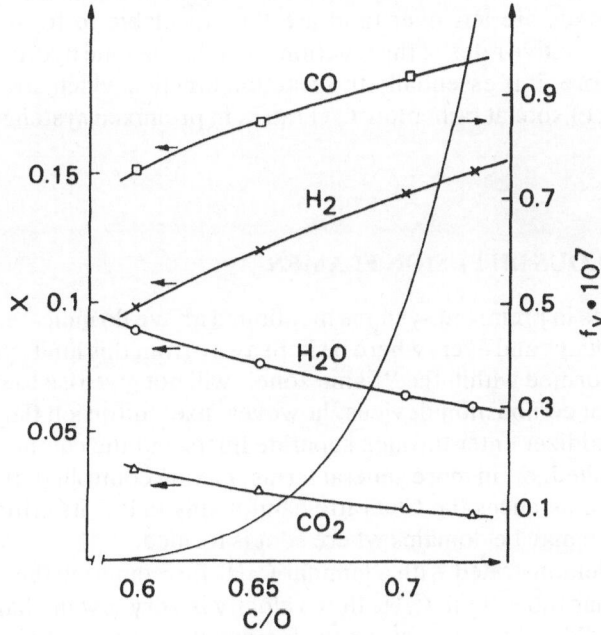

Fig. 2. Soot volume fraction f_v (in cm³/cm³) as a function of C/O for C_2H_4/air flames burning at 1 bar. Also shown are the approximate mole fraction X of the water-gas components. The maximum C_2H_2 mole fraction in the reaction zone of the flame at C/O $\sim$ 0.6 is X $\approx$ 0.03.

and H_2 (and other hydrocarbons not shown in Fig. 2) in the presence of soot.

This indicates, that the amount of soot which can be formed in a given flame is determined in the main oxidation zone. The oxygen contained in H_2O and in CO_2 is, under these air flame conditions, tied up for the time available. The reactions of H atoms, which could reduce H_2O or CO_2 via

$$CO_2 + H \rightarrow CO + OH$$

$$H_2O + H \rightarrow H_2 + OH$$

have rather high activation energies of 108 and 86 kJ/mol respectively and the direct reduction of CO_2 and H_2O at soot particles is likewise too slow.

The main oxygen consuming branching reaction

$$H + O_2 \rightarrow OH + O$$

has a higher activation energy than the reaction of hydrocarbons with the radicals H, O and OH. These radicals can attack the fuel early in the reaction zone, forming hydrocarbon radicals and unsaturated species, H_2, H_2O and CO. The hydrocarbons left over are mostly the thermodynamically rather stable ones like acetylenes and a large variety of condensed ring systems [87, 88].

How much of the oxygen is stored in CO_2 and H_2O and how much of the hydrocarbon compounds are left over (and are thus available to form larger species) depends on the relative rates of the reactions which consume hydrocarbons and CO and H_2. Therefore, it is essentially the reaction kinetics which are responsible for the appearance of soot at rather low C/O ratios in premixed systems.

SOOT IN GASEOUS DIFFUSION FLAMES

The soot limits in premixed systems mentioned above do indicate a borderline. If a system can always and everywhere be kept away from this limit, the intermediate hydrocarbons formed within the "flame zone" will not give rise to soot emission.

Most practical combustion devices, however, use "diffusion flame type" burning. Fuel and oxidizer enter through separate inlets and the combustion process is diffusion controlled, or, in more general terms, mixing controlled. It is obvious that under such circumstances the C/O ratio cannot stay below its critical limit everywhere, and there may be domains where soot is formed.

This can be demonstrated with a laminar C_2H_4 flow entering the surrounding air through a circular tube. If the C_2H_4 flow velocity is very low the flame looks like a Bunsen burner flame. Only a faint soot luminosity is visible below the reaction zone. A slight increase of the flow velocity increases the yellow luminosity and a further increase of the flow velocity generates the usual picture of a laminar diffusion flame. Still further increase in the flow velocity increases the height of the flame, until at a certain height it starts to emit soot, to smoke [see e.g., 29].

It is obvious that formation and emission of soot from diffusion flames is influenced, among other things, by the flow conditions. Whether a flame smokes or not depends on whether the soot, once formed, has time enough to burn out before radiation losses and diffusion of fresh cold air quench its oxidation.

The reciprocal value of the height at which a flame on a given burner starts to smoke and flame luminosity measurements have been used to gauge the tendency of different fuels to smoke. This tendency increases from paraffin to mono-and di-olefins, benzenes and naphthalenes [30-32] and is different from the order observed for the critical C/O ratio in premixed flames, thus pointing to some different mechanism. Glassman *et al* [33] have recently confirmed the importance of fuel structure in determining the smoking tendency and shown that the C/H ratio of the fuel is not of itself significant. They demonstrated the importance of temperature in determining the smoke point: at the same stoichiometric flame temperature acetylene is actually more resistant to soot formation than ethylene. This influence of temperature is also demonstrated in Fig. 3 where the effect of various additives to the fuel in suppressing soot, as determined by the increase in fuel flow at the smoke point is shown: all additives fall in one line when the data are plotted against the total sensible heat content of the fuel plus additive stream at the average flame temperature.

Increase of pressure usually promotes soot formation [34, 35] in diffusion flames. The influence of the oxidant composition [36] and of additives on soot formation in diffusion flames are rather complicated because they can interfere in many ways.

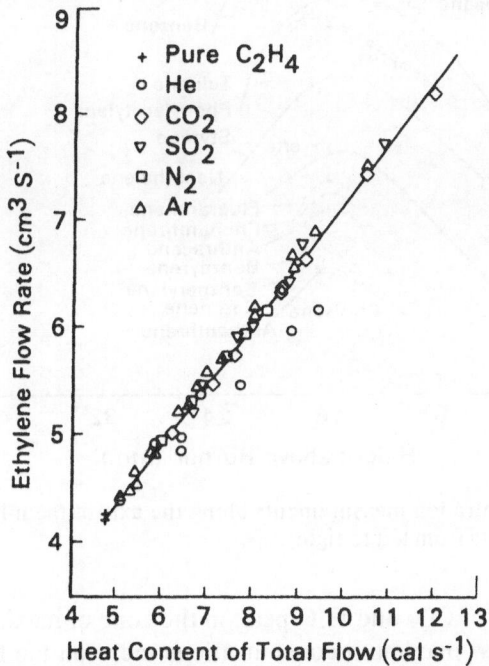

Fig. 3. Diffusion flame: Effect of additives to the fuel in suppressing soot as determined by the increase in fuel flow at the smoke point plotted against the total sensible heat content of the fuel + additive stream at average flame temperature [33].

THE SOOTING STRUCTURE OF LAMINAR DIFFUSION FLAMES

Before following the growth and oxidation of soot particles in a diffusion flame we should have a brief look at the concentration distribution along the axis of the flame. Fig. 4 shows as an example a n-hexane/air diffusion flame: the parent fuel disappears rapidly along the flame axis while temperature increases and a variety of hydrocarbon products arise. The main "pyrolysis products" are (besides hydrogen) alkanes, C_2H_4 and other olefins, and towards larger heights C_2H_2, C_6H_6 and a large number of different polycyclic aromatics and related compounds, albeit each in low concentrations.

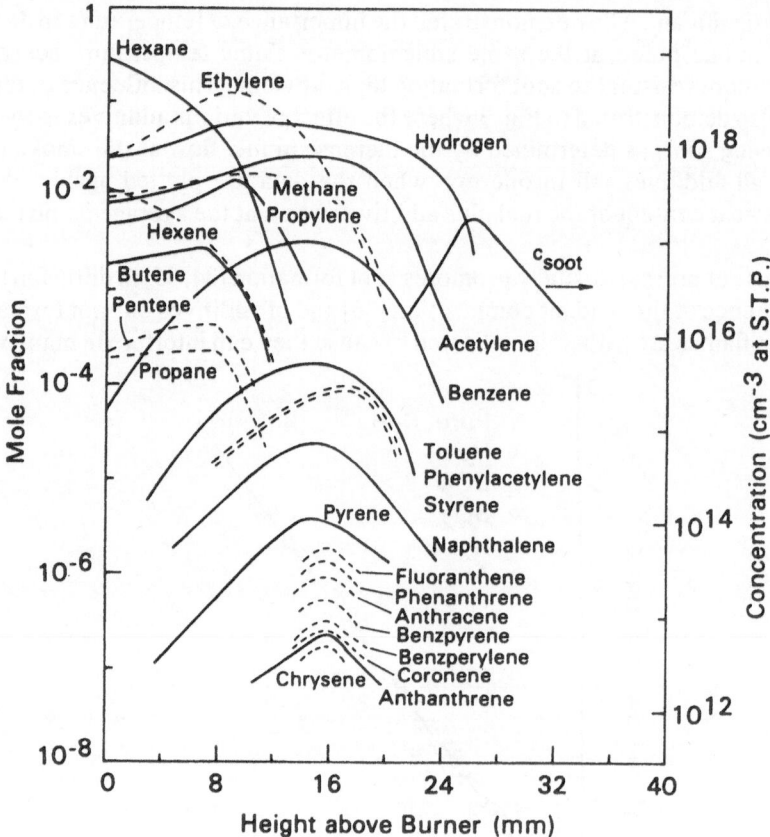

Fig. 4. Species concentration measurements along the axis of the n-hexane/air diffusion flame [37] (gas flow shown from left to right).

The concentrations of CO_2 and H_2O peak in the zone of maximum temperature while CO and H_2 show maximum concentrations more on the fuel side. Oxygen penetrates into the unburned fuel region already close to the burner mouth and radicals may also diffuse into the fuel flow, thus generating a very special situation for "pyrolysis" of the fuel, for the growth conditions of soot.

The "soot field" of a flat C_2H_4/air diffusion flame is shown for a height of 40 mm above the burner in Fig. 5. The soot is contained in a region a few millimeters wide to the fuel side of the flame zone, the zone of maximum temperature. The particle number density is highest near the flame zone ($> 10^{12}$ cm^{-3}) and decreases sharply into the fuel side while the particle radius increases. The soot volume fraction f_V shows a maximum in an intermediate position. Particle generation (high N) and soot formation (high f_V) take place at different locations.

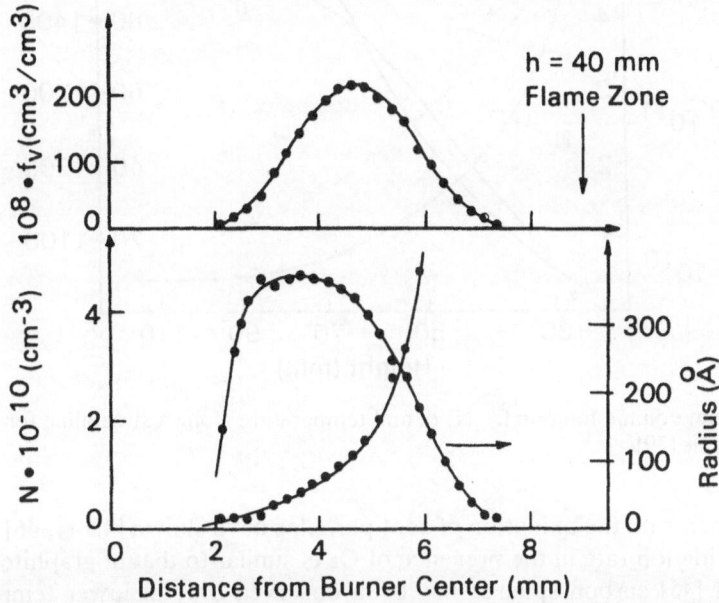

Fig. 5. Soot loading (f_V), particle mean size, and number density (N) profiles across the sooting region in the early stages of a laminar ethylene/air diffusion flame established on a Wolfhard-Parker burner at 40 mm above the burner. (The fuel-air partition in the burner occurs at 3 mm from the centre. The arrow marked "flame zone" indicates the position of the stoichiometric fuel-air interface at this height of 40 mm.) [38].

The flow patterns of these hydrocarbon-air diffusion flames are such that streamlines from the air side passing upwards through the flame zone will initially see increasing fuel concentrations and lower temperatures (compared to temperature in the reaction zone). The particles formed near the reaction zone try to follow the streamlines into the fuel rich interior of the flame where surface growth increases soot loading rapidly (f_V growth). At the same time, particles are coagulating and their number density N is falling along the streamline. Fig. 6 shows f_V, N, particle diameter d and temperature along a streamline in a C_2H_4/diffusion flame. These results are very typical for the kinetics of soot formation, not only in diffusion flames, but also in premixed flames or in high temperature pyrolysis.

For overventilated flames, the structure described pertains only to the lower region of the flame. At some height, particle oxidation will start. Because particle burnout time is significantly longer for larger particles, the smoking of a flame will also depend on the soot forming process in the early stage of the flame. Roper [40,

41] has shown that flames start to smoke when the ratio of soot oxidation length to diffusion flame height exceeds about one.

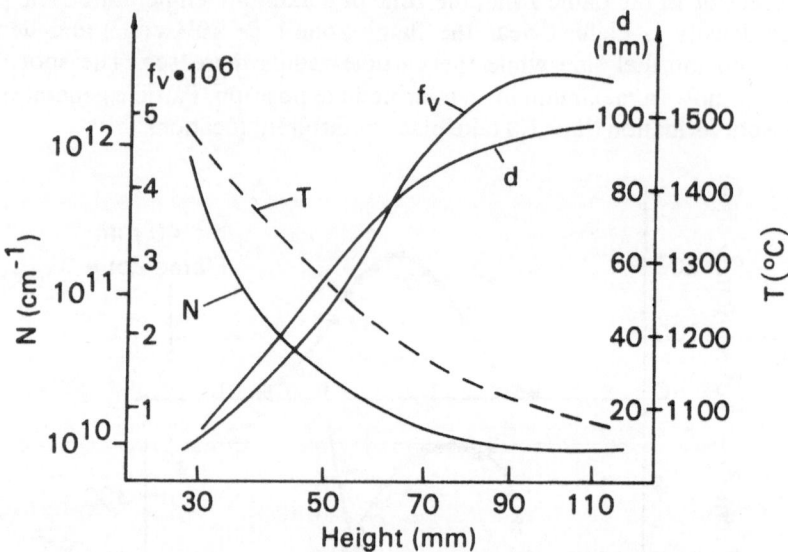

Fig. 6. Soot volume fraction f_v, N, d, and temperature along a streamline for a C_2H_4/air diffusion flame [39].

Experiments on the oxidation of soot particles have shown [42-44, 46] that their surface oxidation rate in the presence of O_2 is similar to that of graphite. Internal burning of [45] carbon spheres has been observed only at lower temperatures. Therefore data on graphite oxidation might be used together with those on soot oxidation [2] to give the order of magnitude of the lifetime of 100 $\overset{\circ}{A}$ soot particles as a function of O_2 partial pressure and temperature. (See Fig. 7.) For temperatures below 1700 K, the lifetime increases exponentially with decreasing temperature (activation energy 167kJ/mol). As soon as the O_2 pressure rises above a certain limit, it plays only a minor role. Therefore addition of too much cool air only reduces the temperature and increases the lifetime of the particles. Above 1800 K, P_{O_2} becomes important for all partial pressures.

TURBULENT DIFFUSION FLAMES

In their pioneering work on soot distribution in free turbulent diffusion flames, Dalzell, Williams and Hottel [47] found that the soot mass concentration increases rapidly along the axis to 10^{-6} g/cm^3, 10^8 clusters/cm^3 at a height of 80 to 100 burner diameters from the burner mouth and then decreases rapidly by oxidation. This location of the maximum soot concentration in turbulent flames has also been found e.g., by Magnussen [48] and Becker [49] for much higher Reynolds numbers.

Becker's study [49] on propane and C_2H_2 flames with Reynolds numbers up to 30000 demonstrates that the soot concentration field is essentially mixing con-

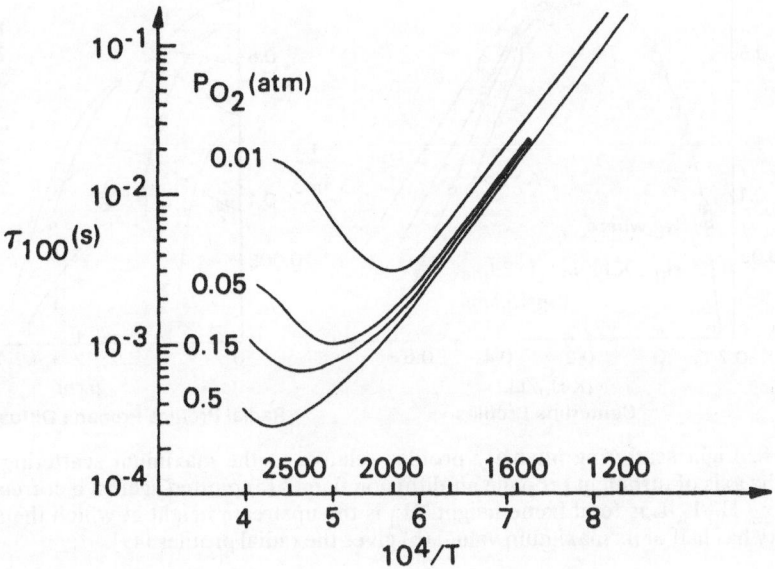

Fig. 7. Lifetime τ for 100 Å soot particles as a function of temperature for various partial pressures of oxygen, based on measurements of Appleton, Fenimore and Jones, Kahn, Lee, Beer and Thring and the model of Strickland-Constable. See Ref. [100].

trolled. No pronounced effect of the Reynolds number on soot formation in these turbulent flames could be observed. The principle aerodynamic parameter, which allows good correlation of various properties of these flames is the Richardson number Ri_L, the ratio of buoyant to momentum forces.

Light scattering intensities I (assumed to be proportional to soot concentration) for orders of magnitude variations of the Richardson number are shown in Fig. 8. The right side gives the radial profiles at various heights; the left side gives the axial profiles normalized to height and approximate position of maximum intensity. On the upstream diffusion-flame side, the measurements agree excellently. On the downstream side, where the carbon particles burn, agreement should not be and is not quite as good.

The positions of maximum temperatures, maximum scattering intensity and other properties correlate well with $(Ri_L)^{1/3}$. The maximum amount of soot present increases steeply with Ri from the forced convection limit (Ri = 0) to a maximum of 0.2×10^{-6} g/cm^3 and decreases slightly toward the natural convection limit (R $\rightarrow \infty$). Somewhat higher absolute soot concentrations are obtained in confined turbulent gas diffusion flames [50].

Although these descriptions of time-averaged quantities are important, it is the unsteady nature of turbulent flames which is of great fundamental significance. Short exposure photographs of turbulent flames show the strong fluctuation of luminous domains, particularly in the burnout zone where the soot-containing pockets appear to be separated by large zones free of soot. On the basis of the high intermittency factor, Magnussen, like Hottel [47], assumes that soot is formed and

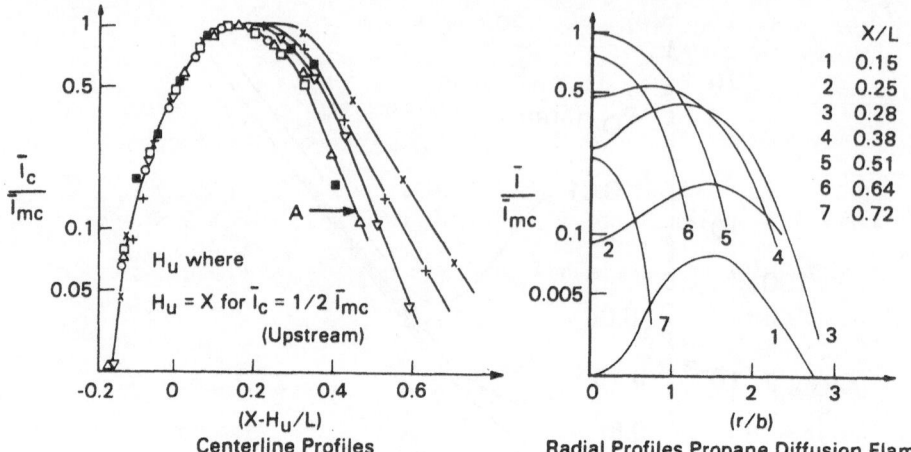

Fig. 8. Light scattering intensity profiles relative to the maximum scattering intensity along the axis of turbulent propane air diffusion flames (a) plotted against a corrected flame height x - H_u/L. L is total flame height. H_u is the upstream height at which the scattering intensity has half of its maximum value. (b) gives the radial profiles [49].

contained in eddies separated by regions with little or no soot. The combustion proceeds mainly by the break up of bigger eddies into smaller ones which burn away (largest particles in largest eddies). Further experiments will help to clarify this situation.

FUEL SPRAY FLAMES

Among the most important diffusion flames today are fuel spray flames. The range of conditions under which fuel spray flames must operate in diesel engines, continuous combustion chambers, and furnaces, with fuels from kerosene to heavy oil, is very wide indeed. Fuel and air enter the burner separately with different velocities and the fuel droplets are not distributed uniformly through the combustion zone. Size distribution, trajectories, velocities and penetration depths of the droplets, as well as their evaporation, play an important role. When the fuel spray approaches the flame front, droplets start to evaporate and the fuel vapor burns in a more or less premixed flame. There is increasing evidence that individual droplet burning, as such, does not play an important role in the dense sprays typical of very many practical combustion systems (see Chigier [51]). There seems to be some overall analogy with gaseous diffusion flames as has been emphasized by the work of Onuma *et al.* [52] who found that the shape of and species concentration profiles in their turbulent kerosene-spray flames are indistinguishable from those obtained in propane diffusion flames. In situ measurements of soot formation in combustors are still rare. Let us consider as one example the measurements of Prado *et al.* [53] of the soot distribution in a cylindrical turbulent continuous flow combustor (Fig. 9). Kerosene supplied through an air-assist atomizer was burned under overall stoichiometric conditions.

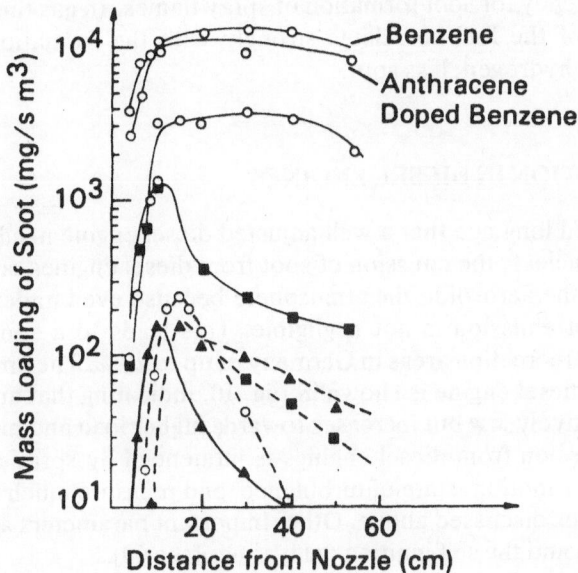

Fig. 9. Influence of atomizing pressure on axial profiles of soot mass loading in a continuous kerosene combustor for fuel equivalence ratio = 1 [53]. Atomizing pressure, ○ = 12 psig; □ = 15 psig; △ = 20 psig; cold gas velocity — 0.96 m/s; ---- 2.67 m/s.

For low cold air velocity (solid lines), soot formation occurs early in the flame within 1/2 to 1 burner diameter, well within the recirculation zone.

Maximum as well as exhaust soot concentrations decrease with increasing nozzle pressure due to improved atomization and better mixing.

At high flow velocities, the maximum soot concentration appears outside the recirculation zone and is much lower, due to better local fuel/air ratio. The soot burnout rate, though little kerosene is still present, is always higher than for the low flow velocity (residence time is about 1/3 of that for low velocity). In contrast to the case for low velocity, it decreases with increasing atomizing pressure due to increased oxygen consumption in the main flame zone [54].

Soot collected (again about 10^{-6} g/cm^3!) downstream has the usual appearance (diameter of the elementary particles d $\approx$ 200Å, which form chains; close to nozzle partially coalesced units).

These results point to the importance of mixing and atomization in controlling soot formation in spray flames, and they indicate that measurements of exhaust concentrations do not necessarily provide real insight.

In the following we will consider some findings on the influence of various parameters: (for details see: [56-58]). An overall stoichiometric or lean combustion is important in suppressing soot emission. Fuel pockets may lead to significant soot emission. A decrease in droplet size (without reduced penetration depth) leads to more rapid evaporation of the fuel. An increase in spray cone angle can reduce soot emission. Some degree of swirl of the combustion air may also reduce soot loading. Increasing pressure favors soot formation. Recirculation of burned gas can strongly reduce the tendency for soot formation of spray flames. In gas turbine combustors the H/C ratio of the fuel correlates strongly with the formation of soot in the direction: more hydrogen, less soot.

SOOT FORMATION IN DIESEL ENGINES

Ricardo stated long ago that a well adjusted diesel engine need not emit visible smoke. Nevertheless, the emission of soot from diesel engines contributes an important part of the aerosol in the atmosphere because even under good operating conditions, soot emission is not negligible. This leads to a soot content in the aerosol in some recreation areas in Germany of up to 20 %. The emission of a 2.4L, 4-cylinder [59] diesel engine is shown in Fig. 10, indicating that around 2500 r/min emission is relatively low but increases towards higher load and higher speed.

The soot emission from diesel engines is influenced by spray atomization and configuration, air motion, state of turbulence, and pressure much as in the continuous spray flames discussed above. Other important parameters are injection timing and duration and the self ignition delay times [see 60].

Direct injection, prechamber and swirl chamber engines differ somewhat in their sooting characteristics and the way sooting depends on their operation parameters. It is not immediately obvious from the data reported in the literature which engine type is superior for general applications. (At the exit of the prechamber soot concentrations seem to be rather high because of the relatively rich mixtures formed there, but the conditions in the main chamber in most cases lead to good burnout [60].)

Inspection of soot particles collected from diesel engines at the exhaust valve show the usual log normal (elementary soot) particle size distributions which depend very little on the mode of operation and the type of engine used. The mean

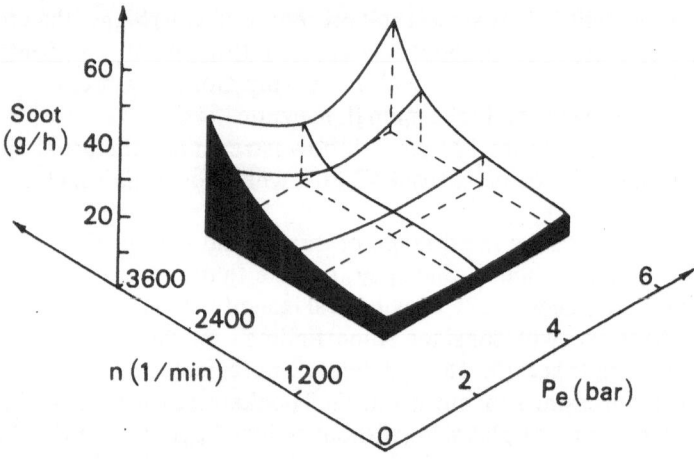

Fig. 10. Soot emission of a 2.4 L four-cylinder diesel engine [59].

diameters are around 200 to 300 Å. They are contained in clusters [61], the size of which increases slightly with the fuel/air ratio. Together with the soot, large hydrocarbon molecules are formed, as already mentioned for flames and combustors.

If soot from a diesel engine is collected in the exhaust system or at low exhaust temperature, these heavy hydrocarbons condense on the agglomerates, change their appearance into that of tarry material and increases the H/C ratio towards one. Time resolved measurements made using UV absorption show that soot formation commences in the region of the spray cone immediately after ignition. The soot concentration increases rapidly to a maximum ($>10^{-7}$ g/cm^3) just 5° (crank angle) later, or 14° after the start of injection. At 21° a nearly even soot distribution is reached [63].

The injection timing is, as mentioned above, important. Early injection [63] allows more fuel to be injected, vaporized and mixed before ignition. A high rate of fuel injection may also reduce sooting. Increasing inlet air temperature cuts ignition delays and reduces spray penetration and therefore promotes soot formation [60].

If, due to other requirements, it is not possible to prepare the mixtures in a way that they are below the critical C/O ratio, prior to combustion, their real values should be kept as close to the critical C/O as possible because soot volume fraction and particle size grow strongly with increasing C/O. (see Fig. 2)

SOOT FORMATION KINETICS

Up to now, we have considered largely phenomenological aspects of soot formation in various combustion environments. We have seen that soot formation and the combustion of soot particles are largely determined by the local effective C/O ratio, the local hydrocarbon concentrations, by the temperature, and by the residence time of the "particles" both in the soot-forming region and the burning region. As mentioned at the beginning there are two comparatively simple systems for the investigation of the kinetics of "homogeneous" soot formation: premixed flat flames and shock wave pyrolysis, which we will consider now. (For flow tube pyrolysis see Lahaye and Prado [64].)

For simplicity the process shall be considered from the "burned gas" side, where one finds soot particles, and thermodynamically relatively stable hydrocarbons, and where the soot particles have reached their "final size" (within the time available of the order of milliseconds).

Coagulation — It was noted in early studies of low pressure acetylene flames [73, 74] that a significant fraction of the observed particle growth of the elementary soot particles is due to coagulation, whereby particles collide and fuse, thus reducing their overall number concentration and increasing the average size. This is shown for a typical premixed flame in Fig. 11 (See Fig. 6 for comparison). Often it has also been observed that the final number of soot particles per cm^3 is very similar under widely different conditions.

This coagulation of particles has also been observed in shock wave pyrolysis experiments. Graham et al. [66, 67] investigated it there in detail, using laser light scattering and absorption technique.

References pp. 25-28.

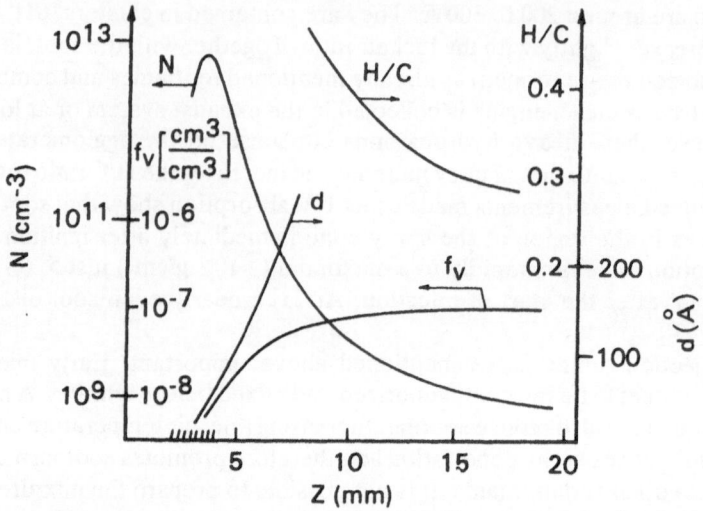

Fig. 11. Particle number density N, mean diameter d and soot volume faction f_v for a CH_4/O_2 flame (equivalence ratio 1.27) at atmospheric pressure as a function of the height above the burner (after D'Alessio [65]).

When particles coagulate, meet and adhere, the decrease in particle number density N can be expected to occur according to the Smoluchowski equation [e.g., 68]

$$\frac{dN}{dt} = - k_{th} (d) \cdot N^2$$

where the rate constant k_{th} depends on the particle diameter or the mean particle volume $\bar{v}$. For small particles compared with ℓ the mean free path, k_{th} can be described in terms of free molecule theory; if $d > \ell$, continuum approach is used: important in that the values of k_{th} can be calculated (also in cases where shear flow or turbulent shear cause coagulation). For the conditions we consider here ($d < \ell$) one can write [68]

$$\frac{dN}{dt} = - \frac{6}{5} k_{th} f_v^{1/6} N^{11/16}$$

with f_v, the soot volume fraction ($f_v = N \cdot \bar{v}$) and

$$k_{th} = \frac{5}{12} \left(\frac{3}{4\pi}\right)^{1/6} \left(\frac{6kT}{\rho}\right)^{1/2} G' \alpha$$

ρ = particle density, k = Boltzmann constant. G' is a factor which takes care of the increase of collision cross section compared to hard sphere (due to dispersion or electric forces); its value is around 2. The factor α (≈ 6.55) is related to the particle size distribution function [71]. (It is 5.66 for monodisperse systems.)

The growth of the mean particle volume is given by

$$\frac{d\bar{v}}{dt} = \frac{6}{5} k_{th} f_v \bar{v}^{1/6}$$

and for f_v = const. one should obtain

$$(\bar{v})^{5/6} \sim k_{th} \cdot f_v \cdot t.$$

This relation has been found to be applicable in the experiments of Graham *et al.* [66, 67, 72] for the description of the later stage of the pyrolysis of a large number of hydrocarbons, mostly ring systems. The observed rate constants k_{coag} values were close to the theoretically determined values.

Measurements of the soot particle coagulation in flames of several fuels (C_2H_4/air and O_2 enriched air, C_6H_6/air, CH_4/O_2, at different C/O ratios) at normal pressure confirmed their results, the form of the rate law, and the approximate values of the rate constants [75-77].

Some examples for measured k_{coag} values in C_2H_4 flames are given in Table 2.

Variation of the C/O ratio has hardly any influence on k_{coag} and on the final particle number, except in case of some metal additives (see Table 2) which do influence k_{coag} and the final particle number [77].

Integration of dN/dt shows that for long times ($N_0 \cdot f_v^{5/6} \cdot k_{th}^{1/6} \cdot t \gg 1$; $t \geq 1$ ms for Graham's conditions, N_0 the initial particle number density at $t = 0$), the relation.

$$N \approx (k_{theo}^{1/6} t)^{-6/5}$$

holds. During the time available in combustion processes, the particle number

TABLE 2

Effect of Additives on the Soot Particle Coagulation Rate Constant in Premixed Atmospheric C_2H_4/Air Flames for C/O = 0.76.*

Additive	Concentration	$k_{coag} 10^9$ $cm^{2.5}$/ms	$\dfrac{k_{coag}}{k_{th}}$
—	—	11	4
H_2	1%	10.2	4
NO	1.2 %	13.4	5
SO_2	1.1%	12.6	5
KCl	0.01 m		0.38
KCl	0.1 m		0.1

*(k_{th} theoretical coagulation rate constant for hard spheres. These data are based on Sarofim's value of the soot refractive index n. If other values are used the absolute vaues of k_{coag} change slightly).

tends towards a "constant" value. The dependence of that "final particle number" on f_V, is very weak (the particle volume increases nearly proportional to f_V). This is in fair agreement with what has been observed in technical processes. It has also been confirmed quantitatively for laminar flames over a wide range of conditions. This expression for the "final particle number density" is independent of the initial particle number density N_0. Therefore, the information about the formation of the initial particles is lost during the soot formation process, and it is unimportant how many of these initial particles are formed, as long as their number is large enough. Close to the critical C/O the situation becomes somewhat different.

These results establish that coagulation is a property of the soot itself and does not depend, as a process, on the nature of the parent molecules; and it may well, as already mentioned, obscure the details of the processes taking place at the beginning of soot formation.

It is expected that when particles are charged, their coagulation rate will be changed. The expected magnitude of the effect has been considered by Brock *et al.* [78] and by Howard [79]. A demonstration of the effect has been provided recently by Haynes [77] who seeded sooting C_2H_4/air flames with metal additives (see Table 2). Some of these additives could strongly reduce coagulation rate because, as shown directly by Homann [80], the presence of metal promotes sufficient charging of the small soot particles which subsequently resist coagulation by Coulombic repulsion.

Electric fields do have a strong influence on the soot formation process and on the appearance of sooting flames (see the papers of Weinberg [55]).

Growth of Soot Mass in Shock Waves — The growth of the soot volume fraction f_V in a diffusion flame and in a premixed flame is shown in Fig. 6 and 11. They show that the amount of soot detectable in particles starts to rise with a high rate and grows towards a constant level in premixed flames (or a maximum in diffusion flames where the soot particles are oxidized afterwards). A similar shape of the growth curve of f_V is observed in high temperature pyrolysis experiments (see Fig. 12. The transmission curve is a measure of the f_V profile). When a hydrocarbon/argon mixture is shock heated, a certain induction period is required before soot particle coagulation and an increase of f_V can be observed [81-85]. This induction time is characteristic of the fuel; it depends on fuel concentration, temperature and pressure. Apparent activation energies of these induction periods for soot formation are about 30 to 50 kcal/mol [82, 84, 85]. They are close to those for the pyrolysis of the hydrocarbon at higher concentrations.

If these pyrolysis rate data are used to calculate the consumption of the shock heated hydrocarbon, most of it should have disappeared when soot particles become visible in the pyrolysis process.

If hydrocarbons like alkanes, olefins, benzene, etc. are exposed to high temperatures at low concentrations, the molecules decompose in a unimolecular reaction by breaking at first the weakest bond and forming radicals (in most cases).

Towards higher concentrations or lower temperatures, the formal rate laws change, the apparent energies of activation decrease and consecutive reactions become more important. Under these conditions the number of different species increases and the number of possible reactions, including ring closure or addition

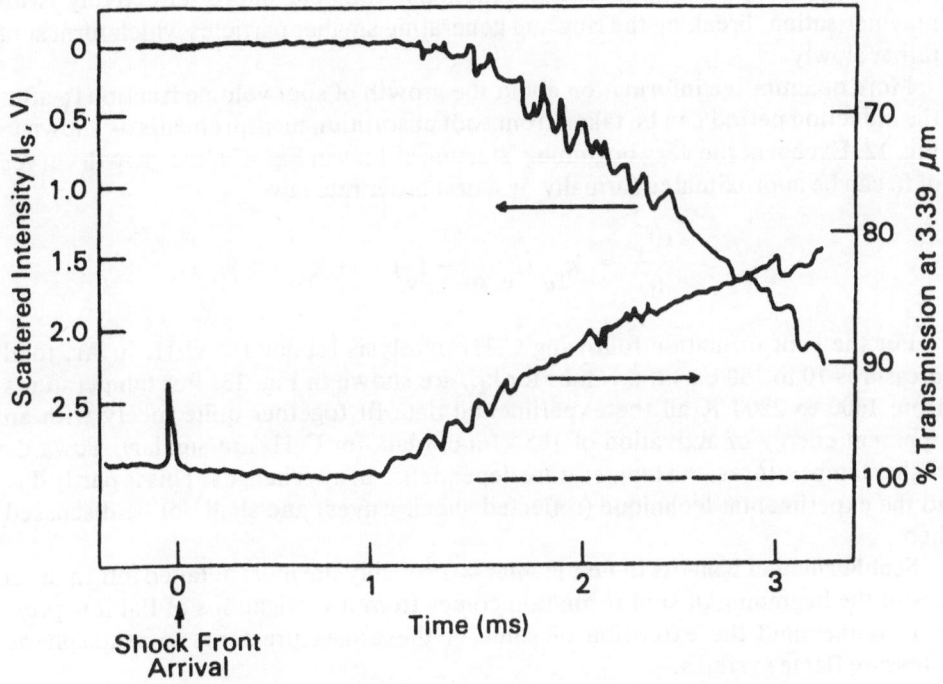

Absorption at 3.39 μm and Scattering at 488 nm

Fig. 12. Transmission and scattering of light during shock tube pyrolysis of ethyl benzene in argon at 1750 K. Adapted from [72].

reactions of the Diels-Alder type, may become more important. (An activation energy of 100 kJ/mol means only a handicap of 10^3 in the rate constant.) As at low hydrocarbon concentrations there is a tendency to form hydrogen and thermo-dynamically stable compounds like acetylene, etc. This fact is used in the Bartho-lome-Sachsse process for technical C_2H_2 production.

Quantitative studies of the rates of soot formation using absorption at various wavelength and its dispersion, light scattering and depolarization, as well as emis-sion, have shown the characteristic properties of various fuels concerning the growth of f_v, d and N. Graham *et al.* [72] characterize the fuels by stating the fractional conversion of the hydrocarbon (at a fixed amount of C per cm^3) into soot after a fixed time of 2.5 ms. (These data include induction period as well as some part of the f_v growth curve.) In case of C_2H_2, C_2H_4, etc. this fractional conversion of the hydrocarbon into soot increases with temperature in the temperature range considered (up to 2300 K). For hydrocarbons, like aromatics and ring compounds, the fractional conversion shows a pronounced maximum around 1750 K. This "maximum rate of soot formation" around 1750 K is interpreted by Graham *et al.* [72] as arising from the competition between two pathways for pyrolysis — at lower temperature, the molecules do not fragment but retain their underlying aromatic structure. These species rapidly and effectively lead to soot. Towards higher temperatures, thermal fragmentation reactions compete more effectively with polymerisation, breaking the ring and generating smaller particles which form soot rather slowly.

More quantitative information about the growth of soot volume fraction f_v after the induction period can be taken from soot absorption measurements as shown in Fig. 12. Except at the very beginning (starting at 1 ms in Fig. 12), the growth curves of f_v can be approximated formally by a first order rate law

$$\frac{df_v}{dt} = k_{f_v} (f_{v,\infty} - f_v) = -k_\infty \cdot [S]$$

For the soot formation following C_2H_4 pyrolysis (about 1% C_2H_4 in Ar, total pressures 10 to 150 bar) the values for k_{f_v} are shown in Fig. 13. For temperatures from 1600 to 2200 K all the experimental data fit together quite nicely with an apparent energy of activation of 185 kJ/mol (data for C_2H_2 are similar). Towards higher temperatures, the temperature dependence of k_{f_v} changes. This is partly due to the experimental technique (reflected shock waves) and shall not be discussed here.

Soot Formation Kinetics in Flat Flames — Probably the most detailed information about the beginning of soot formation comes from investigations of flat low pressure flames and the extension of some of these measurements to atmospheric pressure flame systems.

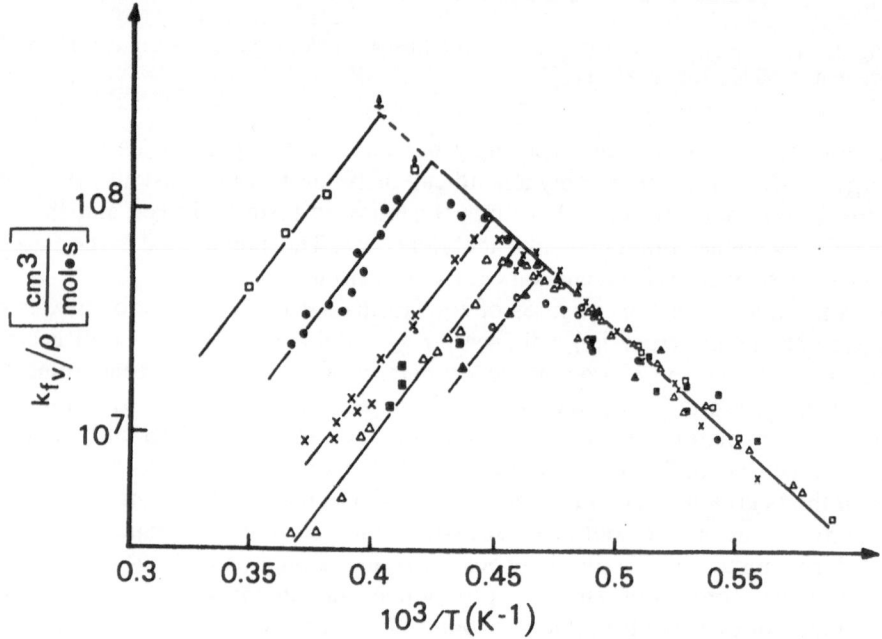

Fig. 13. Rate constant k_{f_v} for the growth of f_v. About 1% C_2H_4 in Ar; ρ = total density. The different symbols are for different pressures from 10 to 150 bar [84].

For low pressure C_2H_2/O_2 flames, measurements of Howard [86], Long [87], and ourselves [88] agree quite well, even if there exist some quantitative differences due to slightly different conditions.

Parameters for the standard flame used by various authors are: $C/O = 1.4$, $p = 2.63$ kPa, flow velocity of the unburned gas $v = 50$ cm/s. Some of its properties, which are to a certain degree typical, are shown in Fig. 14. Most of the oxygen is consumed at about 12 mm above the burner where the water-gas components H_2, H_2O, CO_2, CO have nearly reached equilibrium [88].

Some C_2H_2 is left over and stays in the burned gases. Polyacetylenes pass through a maximum around 15 mm above the burner. A large variety of polycyclic compounds such as pyrene and acenaphthylene and many others seem to originate from the maximum. Some proceeds with time like $C_{14}H_8$ in Fig. 14. A great many other hydrocarbons including various ring systems with and without side chains, mainly those having masses > 250, begin to appear and disappear in the hot gases behind the oxidation zone [88]. At 3.5 to 4 cm above the burner, their concentration passed below the limit of detectability of the mass spectrometer used, which has a mass range up to 500, so that particles larger than 500 mass units could not be measured. Similar results have been obtained by D'Alessio et al. [89] in their studies of sooting CH_4/O_2 flames at atmospheric pressure.

The formation of these higher hydrocarbons must again proceed via fast reactions, which are not easy to follow experimentally because so many different species are involved. In C_2H_2 flames their formation starts with reactions of acetylenes and radicals by way of a fast radical polymerization.

Even though the early products are similar to those obtained in pyrolysis experiments, an initiation reaction as in the pyrolysis experiments is not required here

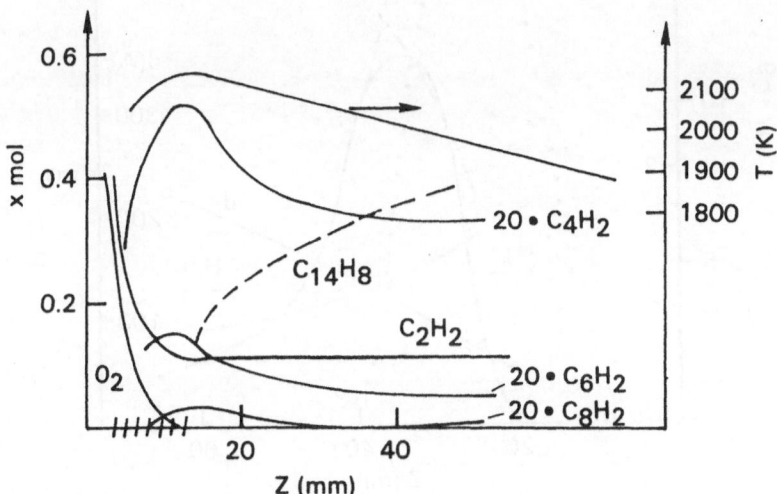

Fig. 14. Temperature and concentration profiles for C_2H_2, C_4H_2, C_6H_2, C_8H_2 and O_2 as a function of temperature for a C_2H_2-O_2 flame burning at 20 torr with 50 cm/s and $C/O = 1.4$. The hatched range on the abscissa indicates the luminous reaction zone [88]. See Ref. [100].

because the radical concentration is high (C_2 reaches 5×10^{11} particles/cm^3 [98]). When the species become large enough, they show branching and will undergo cyclisation in uni- or bimolecular reactions. Acetylenes and other highly unsaturated species are known to be capable of undergoing cyclisation, for example by Diels-Alder reactions. Such products are always found in flames. Some of these, especially the larger ones, are at least as stable thermodynamically as C_2H_2 under conditions applied here [5]. An important feature of these early steps is that, once the molecules are large enough and sufficiently compact, radical addition complexes will have long lifetimes, thus preserving the radical character of the large species. Unimolecular reactions acting at internal bonds will not change the substance, as in oxidation reactions only outer atoms or small groups will be taken off.

Measurements for young soot particles do point to a high H/C ratio (see Fig. 15) and strong radical character of these early particles. These young soot particles are surrounded in the gas phase by the species described above and those shown in Fig. 14, including polycyclic aromatics. The results of Long [87], showing a maximum of the polymeric material concentration at about 22 mm above the burner which is higher than the final amount of soot, do fit well into this picture if some reaction is allowed during isokinetic sampling, especially in the radical-rich highly reactive domain where soot formation starts to become visible.

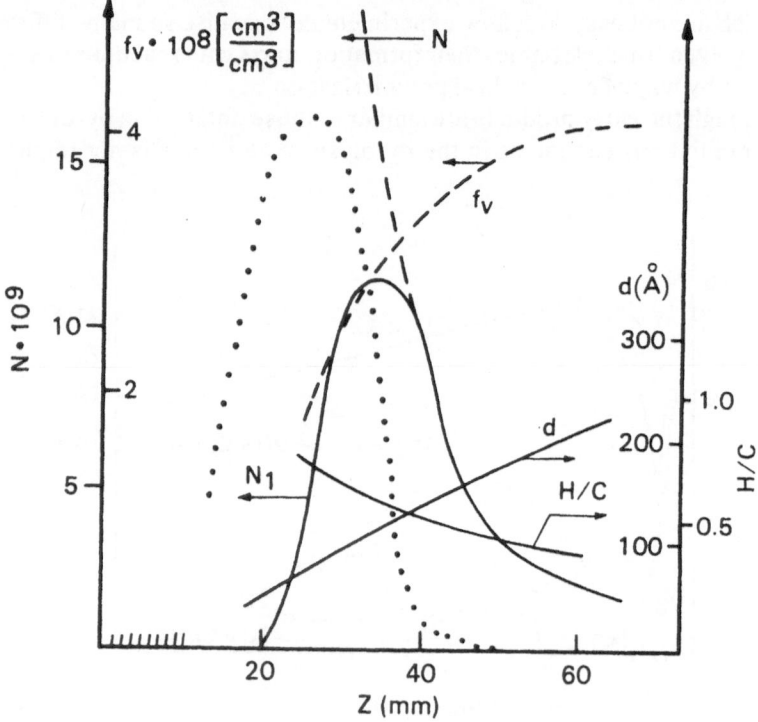

Fig. 15. Flame as in Fig. 14. Volume fraction f_v in cm^3/cm^3; particle diameter d (in Å) and particle number density in cm^{-3} and H/C ratio. Dotted curve gives a measure of absorbing polycyclic material in the flame zone. This and the N_1 curve are taken from Howard [86]. See Ref. [100].

It is a well-known feature that ions are also formed in these flames. Ion-molecule reactions can be extremely fast [91]. Unsaturated species are formed during oxidation, therefore the formation of larger species could proceed via ionic polymerization. Whether this will be a significant contribution depends on the ion concentration.

For conditions similar to those of the $C/O = 1.4$ standard flame, Homann [80, 92] found at the end of the oxidation zone two maxima in the mass distribution of the ions. One belongs to the chemiions, the concentration of which increases towards the fresh gas; they grow very little in the oxidation zone. The other stronger maximum belongs to ions whose mass increases away from the burner to positively charged soot particles (up to about 6000 mass units, closing the gap between mass-spectrometric and electron microscopic results). This non-equilibrium ionization is essentially concentrated in the region where the volume fraction f_V grows. Homann [80] attributes this to an ionization mechanism related to the mass growth of soot particles. The absolute concentrations of the ions in these flames are still the subject of discussion [86, 92, 93]. Two independent groups who determined absolute ion concentrations came to the conclusion that the ion concentrations are well below the soot particle concentration in the main soot-forming region.

The growth of soot volume fraction f_V and the decay of particle number N in the standard flame are shown in Fig. 15.

The N and the f_V curves exhibit the typical behavior discussed above. Two curves for the particle number are shown. If one counts as soot particles those above a certain size, say 50 Å diameter, their number density will first increase and then decrease. For another size limit, the curve will look different on the burner side. This apparent difference is due to the experimental methods (see Fig. 15) applied and to the question: at what size is a species a soot particle? A similar situation arises with the particle diameter during the later phase of the process. During the early phase of soot formation the particles coagulate and become spherical again. In the later phase, when the growth of f_V becomes very slow, the elementary soot particles continue to coagulate, not forming spheres anymore, but the particle chains mentioned. Optical methods see these particle chains as a whole while under an electron microscope one can still count the elementary particles: the results are apparently different mean diameters.

In the early phase of soot formation an absorption signal (Fig. 15) appears. Howard [86] attributes it to heavy hydrocarbons whose volume fraction seems to be close to that of the soot $f_{V, \infty}$ finally present.

It is in this region where the differences in the soot formation tendencies and soot formation rates of various fuels like C_2H_2, C_2H_4, paraffins and C_6H_6 are most important[88] due to the different size and reactivity of the building bricks formed.

These observations in C_2H_2-O_2 low pressure flames are to a certain extent typical, and propane or ethylene flames will be rather similar. There is no question, however, that benzene flames form soot more readily than acetylene flames [88, 94]. Mass spectrometric analysis of the main reaction zone shows a composition rather different from that which is found for aliphatic fuels [88]. Acetylenes are still major products, albeit in reduced amounts. The concentration of aromatic species is about two orders of magnitude higher than for the C_2H_2 flames.

References pp. 25-28.

A very detailed investigation of intermediate products of a near sooting benzene-oxygen flame has been performed by Bittner and Howard [95]. This shows, on the one hand, that the chemistry of these systems is complex indeed; on the other hand it is obvious that there are many more comparatively large molecules with intact rings which may act as "foundations" or "building blocks" for a more rapid growth of the soot volume fraction than in C_2H_2 flames. This is in agreement with the finding of Homann [80] that in sooting low pressure benzene-oxygen flames there are no positively charged soot particles with masses > 1400 mass units; f_v reaches its final values while the particles are still comparatively small. These measurements close a gap in the experimental observations of the various phases of soot formation.

The rate determining steps for the formation of soot in combustion processes, the growth of f_v and particle coagulation have been realized and the differences compared to the normal condensation process have been established. It is now necessary to improve the quantitative information about the growth of soot. Here, non-intrusive measurements with lasers offer a unique possibility to reach sufficient accuracy within a reasonable time for the measurements of the various rates involved in the soot growth.

SOOT PARTICLES

The coagulation of soot particles poses some problems which I should mention. Electron micrographs show that elementary soot particles look spherical — the large ones and the small ones. One way to explain this would be to consider the particles as liquid drops [96, 97] which collide and form a larger liquid drop which later on pyrolyses. This concept has been evaluated quantitatively by Lahaye [98] for "low temperature" pyrolysis. While around 900 K the mass of the condensing macromolecules must be about 750 atomic mass units, under combustion conditions, at T = 1500K the mass would have to be more than 1800 atomic mass units (at 1800 K more than 2700 a.m.u.) [99]. Even if the young soot particles are not considered to be liquid drops, they must have a rather high internal mobility (which is often a property of large macromolecules). They are very reactive so that part of the empty space between two particles after collision may readily be filled from the gas phase. The mass required for filling the gaps is proportional to the volume of the particles. Therefore, not much substrate is required for small particles to come close to spherical after coagulation again. (Some caution is also required with respect to the particle sampling procedures.)

This overview covers a selection of experimental results on soot formation in combustion [100]. Many more details including literature are given in a review article by Dr. Haynes and myself which will appear soon in "Progress in Energy and Combustion Science".

REFERENCES

1. H. B. Palmer, C. F. Cullis, in "Chemistry and Physics of Carbon," ed. P. L. Walker, Marcel Dekker, New York, (1965), p. 265; H. B. Palmer, D. Seery, Ann. Rev. Phys. Chem., Vol. 24 (1973), p. 253.
2. J. B. Donnet, A. Voet, "Carbon Black", Marcel Dekker, New York, (1976).
3. Ullmann 10, (1958), p. 7.
4. M. Cowperthwaite, S. H. Bauer, J. Chem. Phys. Vol. 36, No. 7 (1962), p. 1743; R. E. Duff, S. H. Bauer, J. Chem. Phys. Vol. 36, No. 7 (1962), p. 1754.
5. S. E. Stein, J. Phys. Chem., Vol. 82 (1978), p. 566.
6. K. H. Homann, Comb. and Flame, Vol. 11 (1967), p. 265; K. H. Homann, Angew. Chemie Vol. 80 (1968), p. 425.
7. B. E. Warren, Phys. Rev. Vol. 59 (1941), p. 693; B. E. Warren, P. Bodenstein, Acta Cryst. Vol. 18 (1965), p. 289.
8. A. E. Austin, Proc. 3rd. Conf. on Carbon, (1958), p. 389.
9. R. E. Franklin, Acta Cryst., Vol. 3 (1950), p. 107; R. E. Franklin, Acta Cryst. Vol. 4 (1951), p. 253; R. E. Franklin, Proc. Roy. Soc. A, Vol. 209 (1950), p. 196.
10. F. A. Heckmann, Rubber Chem. Technol., Vol. 37 (1964), p. 1245; F. A. Heckmann, D. E. Harlin, Rubber Chem. Technol., Vol. 39 (1966), p. 1.
11. L. L. Ban, "Surface and Defect Properties of Solids", The Chemical Society, London, Vol. 1 (1972); L. L. Ban, W.M. Hess, 9th Bienn Conf. on Carbon (1963), p. 162; L. L. Ban, W. M. Hess, Noreleco Reporter, Vol. 13 (1966), p. 4.
12. P. A. Marsh, A. Voet, T. J. Mullem, L. D. Price, Rubber Chem. Technol., Vol. 43 (1970), p. 470; P. A. Marsh, A. Voet, T. J. Mullem, L. D. Price, Carbon, Vol. 9 (1971), p. 797.
13. S. Ergun, 10th Bienn. Conf. on Carbon, Bethlehem, Pa. (1971), p. 164.
14. S. H. Bauer, 10th Symposium (International) on Combustion (1965), p. 511.
15. C. R. Marchand, Acad. Sci. Paris, Vol. 238 (1954), p. 1645; C. R. Marchand, Acad. Sci. Paris, Vol. 239 (1954), p. 1609.
16. D. J. Ingram, J. G. Tapley, R. Jackson, R. L. Bon, A. R. Murnagham, Nature, Vol. 174 (1954), p. 797.
17. A. Pacault, Chemistry and Physics of Carbon, New York, Vol. 107 (1977).
18. E. Fitzer, K. Müller, W. Schäfer, Carbon, Vol. 7 (1971), p. 237.
19. K. H. Homann, FVM Frankfurt, Vol. 327 (1978), p. 137.
20. Landolt Bornstein IV Band, Vol. 4b (1972), p. 364.
21. J. C. Street, A. Thomas, Fuel, Vol. 34 (1955), p. 4; A. Thomas, Comb. and Flame, Vol. 6 (1962), p. 46.
22. C. P. Fenimore, G. W. Jones, G. E. Moore, 6th Symposium (International) on Combustion (1956), p. 242.
23. R. G. Millikan, J. Phys. Chem., Vol. 66 (1962), p. 794; R. G. Millikan, J. Opt. Soc. Am., Vol. 51 (1961), p. 535, p. 698; "Temperature, Its Measurement and Control in Science and Industry", Reinhold, New York, London, Vol. III₂ (1962), p. 497; R. G. Millkan, W. I. Foss, Comb. and Flame, Vol. 6 (1962), p. 210.
24. J. Flossdorf, H. Gg. Wagner, Z. Phys. Chem., Vol. NF 54 (1967), p. 8.
25. F. J. Wright, 12th Symposium (International) on Combustion (1969), p. 867; F. J. Wright, Comb. and Flame, Vol. 15 (1970), p. 217.
26. J. Flossdorf, W. Jost, H. Gg. Wagner, Ber. Bunsenges. Phys. Chem., Vol. 78 (1974), p. 378.
27. J. J. McFarlane, F. H. Holderness, F. S. E. Whitcher, Comb. and Flame, Vol. 8 (1964), p. 215.
28. W. S. Blazowski, Combust. Sci. Tech., Vol. 21 (1980), p. 87.
29. A. G. Gaydon, H. G. Wolfhard, "Flames, Their Structure, Radiation and Temperature", 3rd ed. Chapman and Hall, London (1970).
30. A. G. Gaydon, G. Wittingham, Proc. Roy. Soc., Vol. A179 (1947), p. 303.
31. J. R. Arthur, D. H. Napier, 5th Symposium (International) on Combustion (1955), p. 303.

32. *A. E. Clark, T. G. Hunter, F. R. Garner, J. Inst. Petrol., Vol. 32 (1946), p. 627; A. E. Clark, J. Odgers, P. Rayan, 8th Symposium (International) on Combustion (1962), p. 982; A. E. Clark, F. W. Stringer, A. J. Harrison, 10th Symposium (International) on Combustion (1965), p. 1151.*

33. *K. P. Schug, Y. Manheimer-Timnat, P. Yaccarino, I. Glassman, Combust. Sci. Tech., (to appear).*

34. *W. G. Parker, H. G. Wolfhard, J. Chem. Soc. (1950), p. 2038.*

35. *R. L. Schalla, G. E. McDonald, 5th Symposium (International) on Combustion (1955), p. 316.*

36. *I. S. McLintock, Comb. and Flame, Vol. 12 (1968), p. 217.*

37. *J. Kern, G. Spengler, Erdöl-Kohle-Erdgas-Petrochem., Vol. 23 (1970), p. 813.*

38. *B. S. Haynes, H. Gg. Wagner, Ber. Bunsenges. Phys. Chem., Vol. 84 (1980), p. 499.*

39. *J. H. Kent, H. Jander, H. Gg., Wagner, 18th Symposium (International) on Combustion, (to appear).*

40. *F. G. Roper, Comb. and Flame, Vol. 29 (1977), p. 219.*

41. *F. G. Roper, C. Smith, A. C. Cummingham, Comb. and Flame, Vol. 29 (1977), p. 227.*

42. *S. W. Radcliff, J. P. Appleton, Comb. Science and Technol., Vol. 4 (1971), p. 171.*

43. *K. B. Lee, M. W. Thring, J. M. Beer, Comb. and Flame, Vol. 6 (1962), p. 137.*

44. *C. P. Fenimore, G. W. Jones, J. Phys. Chem., Vol. 71 (1967), p. 593.*

45. *L. Kurylko, R. M. Essenhigh, 14th Symposium (International) on Combustion (1973), p. 1375.*

46. *I. M. Kahn, C. H. T. Wang, B. E. Langridge, Comb. and Flame, Vol. 17 (1971), p. 409.*

47. *W. M. Dalzell, G. C. Williams, H. C. Hottel, Comb. and Flame, Vol. 14 (1970), p. 161.*

48. *B. F. Magnussen, 15th Symposium (International) on Combustion (1975), p. 1415; B. F. Magnussen, 16th Symposium (International) on Combustion (1977), p. 719.*

49. *H. A. Becker, S. Yamazaki, 16th Symposium (International) on Combustion (1977), p. 681.*

50. *K. Hein, Com. Sci. and Technol., Vol. 5 (1972), p. 195.*

51. *N. A. Chigier, Prog. Energy Combust. Sci., Vol. 2 (1976), p. 97.*

52. *Y. Onuma, M. Ogasawara, 15th Symposium (International) on Combustion (1975), p. 453; Y. Onuma, M. Ogasawara, T. Inoue, 16th Symposium (International) on Combustion (1977), p. 561.*

53. *G. P. Prado, M. L. Lee, R. A. Hites, D. P. Hoult, J. B. Howard, 16th Symposium (International) on Combustion (1977), p. 649.*

54. *J. P. Appleton, J. B. Heywood, 14th Symposium (International) on Combustion (1973), p. 777.*

55. *F. J. Weinberg, Symposium of the Faraday Society, Vol. 7 (1973), pp. 120-132.*

56. *W. S. Blazowski, Progr. Energy Combust. Sci., Vol. 4 (1978), p. 177.*

57. *A. H. Lefebvre, 15th Symposium (International) on Combustion (1975), p. 1169.*

58. *A. M. Mellor, Progr. Energy Combust. Sci., Vol. 1 (1976), p. 111.*

59. *W. Hilburger, Diplomarbeit, Stuttgart (1976).*

60. *N. A. Henein, Progr. Energy Combust. Sci., Vol. 1 (1976), p. 165.*

61. *C. V. Vuk, M. A. Jones, J. H. Johnson, SAE paper 760131 (1976).*

62. *G. Greeves, J. O. Meehan, Paper C88/75 presented to Institution of Mechanical Engineers Conference on "Combustion in Engines," Cranfield, (1975); V. K. Duggal, T. Priede, I. M. Khan, SAE Technical Paper 780227 (1978).*

63. *I. M. Khan, Inst. Mech. Eng. Proc., Vol. 184 (1969), p. 35.*

64. *J. Lahaye, G. Prado, Chemistry and Physics of Carbon, (P. L. Walker and P. A. Thrower, eds.), Marcel Dekker, New York, Vol. 14 (1978), pp. 168-294.*

65. *A. D'Alessio, A. Di Lorenzo, A. Borghese, F. Beretta, S. Masi, 16th Symposium (International) on Combustion (1977), p. 695.*

66. *S. C. Graham, 16th Symposium (International) on Combustion (1977), p. 663.*

67. *S. C. Graham, J. B. Homer, J. L. J. Rosenfeld, Proc. Roy. Soc. A, Vol. 344 (1975), p. 259.*

68. *F. S. Lai, S. K. Friedländer, J. Pich, C. M. Hidy, J. Colloid and Interface Sci., Vol. 39 (1972), p. 395.*

69. C. M. Hidy, J. R. Brock, The Dynamics of Aerocolloidal Systems, Pergamon (1970).
70. G. D. Ulrich, Comb. Science Technol., Vol. 4 (1971), p. 47.
71. S. C. Graham, A. Robinson, J. Aerosol, Vol. 7 (1976), p. 261.
72. S. C. Graham, J. B. Homer, J. L. H. Rosenfeld, 10th International Shock Tube Symposium, Kyoto (1975), p. 621.
·73. U. Bonne, H. Gg. Wagner, Ber. Bunsenges. Phys. Chem., Vol. 69 (1965), p. 35.
74. K. H. Homann, H. Gg. Wagner, 11th Symposium (International) on Combustion (1967), p. 371.
75. A. D'Alessio, A. Di Lorenzo, A. F. Sarofim, F. Beretta, S. Masi, C. Venitozzi, 15th Symposium (International) on Combustion (1975), p. 1427.
76. B. S. Haynes, H. Jander, H. Gg. Wagner, Ber. Bunsenges. Phys. Chem., Vol. 84 (1980), p. 585.
77. B. S. Haynes, H. Jander, H. Gg. Wagner, 17th Symposium (International) on Combustion (1979), p. 1365.
78. G. M. Hidy, J. R. Brock, "The Dynamics of Aerocolloidal Systems", Pergamon, Oxford (1970).
79. J. B. Howard, 12th Symposium (International) on Combustion (1969), p.877.
80. K. H. Homann, (private communication).
81. W. J. Hooker, 7th Symposium (International) on Combustion (1959), p. 949.
82. W. Buckendahl, Diplomarbeit Göttingen (1970).
83. I. L. Mar'yasin, Z. A. Nabutowski, Kinetics and Catalysis, Vol. 10 (1969), p. 983; I. L. Mar'yasin, Z. A. Nabutowski, Kinetics and Catalysis, Vol. 11 (1970), p. 706; I. L. Mar'yasin, Z. A. Nabutowski, Kinetics and Catalysis, Vol. 14 (1973), p. 139.
84. C. C. Geck, Diplomarbeit Göttingen (1975).
85. R. B. Cundall, D. E. Fussey, A. J. Harrison, D. Lampard, 11th International Shock Tube Symposium, Seattle (1977); R. B. Cundall, D. E. Fussey, A. J. Harrison, and D. Lampard, J. C. S., Faraday Trans. I, Vol. 74 (1978), p. 1403; R. B. Cundall, D. E. Fussey, A. J. Harrison, D. Lampard, J. C. S., Faraday Trans. I, Vol. 75 (1979), p. 1390.
86. J. B. Howard, 12th Symposium (International) on Combustion (1969), p. 877; R. T. Ball, J. B. Howard, 13th Symposium (International) on Combustion (1971), p. 353; B. L. Wersborg, J. B. Howard, G. C. Williams, 14th Symposium (International) on Combustion (1973), p. 929; J. B. Howard, B. L. Wersborg, G. C. Williams, Faraday Symposia of the Chemical Society, Vol. 7 (1973), p. 109; B. L. Wersborg, A. C. Yeung, J. B. Howard, 15th Symposium (International) on Combustion (1974), p. 1439; B. L. Wersborg, L. K. Fox, J.B. Howard, Comb. and Flame, Vol. 24 (1975), p. 1.
87. E. E. Tompkins, R. Long, 12th Symposium (International) on Combustion (1969), p. 625; R. Long, E. E. Tompkins, Nature, Vol. 213 (1967), p. 1011.
88. K. H. Homann, H. Gg. Wagner, Ber. Bunsenges. Phys. Chem., Vol. 69 (1965), p. 20; U. Bonne, H. Gg. Wagner, Ber. Bunsenges. Phys. Chem., Vol. 69 (1965), p. 35; K. H. Homann, H. Gg. Wagner, 11th Symposium (International) on Combustion (1967), p. 371; K. H. Homann, M. Mochizuki, H. Gg. Wagner, Z. Phys. Chem. NF, Vol. 37 (1963), p. 299; K. H. Homann, H. Gg. Wagner, Proc. Roy. Soc. A, Vol. 307 (1968), p. 141.
89. A. D'Alessio, A. Di Lorenzo, F. Beretta, C. Venitozzi, 14th Symposium (International) on Combustion (1973), p. 941; A. D'Alessio, A. Di Lorenzo, A. F. Sarofim, F. Beretta, S. Masi, C. Venitozzi, 15th Symposium (International) on Combustion (1975), p. 1427.
90. Th. Just, (private communication).
91. H. F. Calcote, "Ion-Molecule Reactions", (J. L. Franklin, ed.), Plenum Press, New York, Vol. 2 (1972), pp. 673-706.
92. K. H. Homann, Ber. Bunsenges. Phys. Chem., Vol. 83 (1979), p. 738.
93. R. S. Tse, P. Michaud, J. L. Delfau, Nature, Vol. 272 (1978), p. 153; J. L. Delfau, P. Michaud, A. Barassin, Combust. Sci. Tech., Vol. 20 (1979), p. 165.
94. W. Morgeneyer, Dissertation, Göttingen (1968).
95. J. P. Bittner, J. B. Howard, 18th Symposium (International) on Combustion (to appear).

96. C. W. Sweitzer, G. L. Heller, Rubber World, Vol. 134 (1956), p. 855.
97. W. G. Parker, H. G. Wolfhard, J. Chem. Soc. London (1950), p. 2038.
98. G. Prado, J. Lahaye, J. de Chim. Physique, Vol. 72 (1975), p. 483; G. Prado, J. Lahaye, Water, Air and Soil Pollution, Vol. 3 (1974), p. 473.
99. K. H. Homann, 16th Symposium (International) on Combustion (1976), p. 717.
100. H. Gg. Wagner, 17th Symposium (International) on Combustion (1979), p. 3; B. S. Haynes, H. Gg. Wagner, Progress in Energy and Comb. Sci. (to appear).

DISCUSSION

A. S. Gordon (University of California, San Diego)

Many years ago I was probing flames and observed that the amount of acetylene present in a benzene diffusion flame (a very big sooter) was less than was present in a methane diffusion flame. Furthermore, the temperature distribution inside the flame was generally somewhat lower for the benzene flame than for the acetylene flame. This suggests to me that acetylene does not have to be the only precursor of soot. I would like your comment.

Wagner

Yes, this is true, acetylene is not the only precursor of soot. We have to keep in mind that at high temperatures acetylene is a fairly stable substance. Going from some type of fuel molecule by various types of reactions at the right temperature towards molecules (or species) of large mass, one can hardly avoid producing some of the thermodynamically stable components like acetylene, polyacetylenes [4] or polycyclic aromatics [3, 5]. I do know your measurements on diffusion flames. Of course, any system has its special problems and a comparison of the concentrations of certain species in diffusion flames requires what one may call "similar conditions". Here, I would like to refer to the experiments of Glassman and co-workers and to the soot profile measurements of Dr. Haynes and Dr. Kent in our laboratory.

N. A. Chigier (University of Sheffield)

You mentioned the importance of atomization and drop size distribution in practical systems. One of the surprising things which is becoming apparent from examination of sprays in diesel engines is that they appear to be very poor atomizers: in practice, we appear to have a great deal of bulk liquid coming out through the system. The atomization and formation of droplets seems to be very poor. Can you make a comment on the question of atomization in practical systems in the light of what you said about trying to produce small droplets?

Wagner

You very probably know much more about that problem than I do. There are definitely problems with the quality of the injection and atomization. (Indeed some engine types intentionally spray the fuel at the wall.) One essential problem is that the injection time is very short so that the duration of "start" and "stop" is not extremely different from the total injection duration. It is known that, when injection stops, even bigger drops may be released from the nozzle. (For details refer to ref. [100].) Nevertheless, the systems work and the observations mentioned give us hope that the soot emission from diesel engines can be further reduced even if the cars are driven by people like me.

I. Glassman *(Princeton University)*

I think we'd better be careful when we say good atomization and poor atomization, because if you go to the poorest atomization — where there is complete droplet separation — you get a perfectly spherical, symmetric flame around each droplet, and you don't get any soot. If we look at the early work of Goldsmith and Penner — where they did single hanging droplets — on one single droplet where there were low enough Grashof numbers for the gravimetric system to work, they formed a complete flame. In my examination of the literature, I couldn't find anybody saying they observed soot formation. And so it's the flowfield/droplet interaction which determines whether one soots, not necessarily the type of atomization.

Wagner

That's definitely true (see ref. [100]).

SESSION I
CHARACTERIZATION OF COMBUSTION-FORMED PARTICULATE CARBON

Session Chairman
K. H. HOMANN
Technische Hochschule Darmstadt
Darmstadt, West Germany

MORPHOLOGY AND INTERNAL STRUCTURE
OF SOOT AND CARBON BLACKS

J. LAHAYE

*Centre de Recherches sur la Physico-Chimie des Surfaces Solides, C.N.R.S.
Mulhouse, France*

G. PRADO*

Massachusetts Institute of Technology, Cambridge, Massachusetts

ABSTRACT

From the viewpoint of morphology and internal structure soot collected in exhaust gases of engines and carbon blacks are identical materials. They are made of aggregates of pseudospherical particles; the size of individual particles goes from ca. 10 to 500 nm; a single aggregate can include several tens of particles.

The morphology (size and shape) of aggregates can be characterized by electron microscopy, scattering and absorption of light, ultracentrifugation and also adsorption of gases and liquids (pure liquids and solutions).

Electron microscopy has been used to give a two dimensional description of aggregates; the development of Quantimets permitted a quantitative description of soot. The third dimension of carbon black aggregates can be investigated by using electron microscopy combined with stereoscopic methods (observation under two different angles).

Scattering and absorption of light allow one to compute equivalent diameters of aggregated particles.

Ultracentrifugation, carried out on suspensions in water, gives equivalent Stokes diameter distributions of carbon blacks.

A convenient technique has become standard; it is based on the absorption of oil or dibutyl-phthalate in the voids between the particles constituting the aggregates.

Specific surface areas and porosity of soot are currently determined by

* *Present address; Centre de Recherches sur la Physico-Chimie des Surfaces Solides – C.N.R.S., Mulhouse, France*

References pp. 50-51.

gas adsorption (essentially nitrogen at the temperature of liquid nitrogen). Iodine adsorption from iodine solution in water has also become standard practice.

The internal structure of carbon blacks (i.e. the crystalline organization of carbon inside the material) has been systematically studied by X-ray diffraction, electron microscopy of oxidized particles, dark field and phase contrast electron microscopy.

The crystalline structure is essentially graphitic (turbostratic). Layer planes of carbon are the basic building blocks of carbon black; they are partially oriented parallel to the surface but also around centers inside the particles.

External layer planes are continuous from one particle to its neighbors. Therefore, in the final material, individual particles have no real existence; the aggregate is the constituent unit of soot.

INTRODUCTION

While carbon is a given element of the periodic classification, the word "carbon" covers a large variety of materials containing mainly the carbon element but also heteroelements such as hydrogen, oxygen, sulfur, etc.

Carbon blacks are carbonaceous materials obtained *from the gas phase* generally by thermal decomposition or incomplete combustion of carbonaceous gases. The morphology of such materials appears in Fig. 1. They are made of aggregates of up to several tens of pseudospherical individual particles. For particles formed in flames, the diameter of the primary spheres depends little upon the conditions of formation and is usually in the range 10 to 40 nm. In pyrolytic systems (thermal

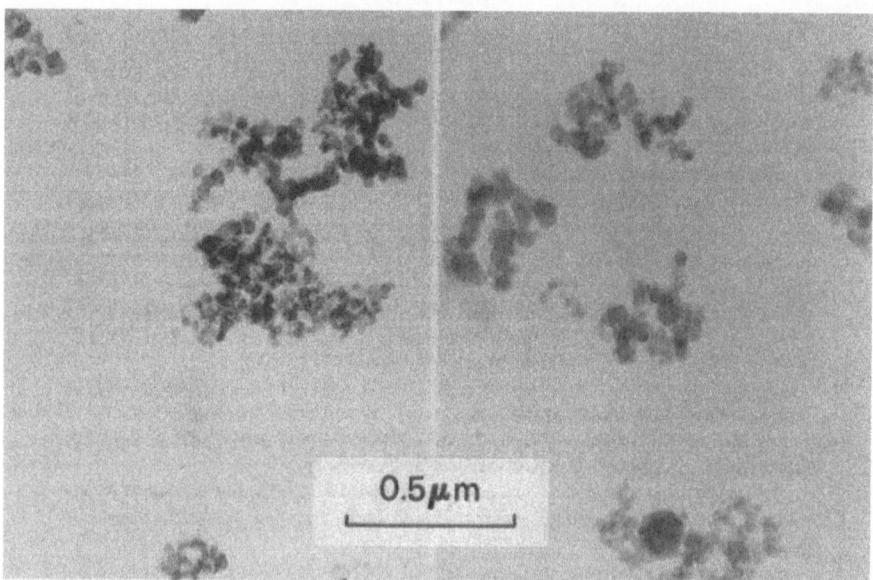

Fig. 1. Transmission electron micrograph. (Left): Soot collected from a diesel engine. (Right): HAF N 330 carbon black.

black), considerably larger spherical particles (up to 500 nm) have been observed. It will be seen later that an aggregate is a single entity, made up of carbon layers which are continuous from one particle to the next. From the stand-point of the final material, the concept of individual spherical particles for most of the blacks appears to be irrelevant.

The objective of the present review is to summarize for combustion specialists the information obtained on the morphology and internal structure of carbon blacks in recent years. In addition a brief description of the techniques used to obtain this information is included. A more extensive coverage of carbon black formation, properties and practical applications is available in the recent book of Donnet and Voet [1] and two recent review articles [2, 3] dealing with mechanism of soot formation and properties, respectively.

Most of the information presented was obtained on commercial carbon blacks. The relevance of such information to soot particles formed in practical combustion devices has often been questioned. However, from the point of view of morphology and internal structure on which is focussed the present review, there appear to be no differences between soot and carbon black, as illustrated in Figs. 1 and 2 where soot collected at the exhaust pipe of a diesel engine and a commercial carbon black are compared. Morphology, as observed by transmission electron microscopy (Fig. 1), and internal structure, as revealed by phase contrast electron microscopy, (Fig. 2) are identical.

MORPHOLOGY OF SOOT AGGREGATES

The degree of aggregation of soot particles is often referred to by professionals of the carbon black industry as "structure". This word is confusing as it does not refer

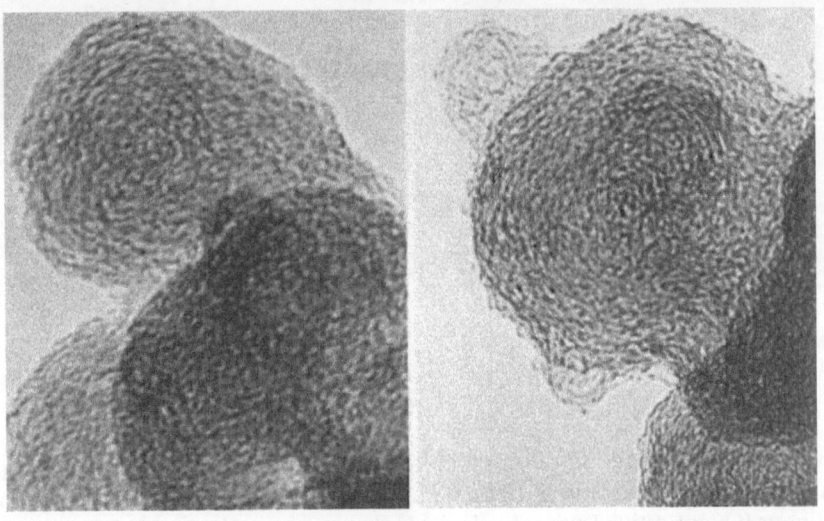

10 nm

Fig. 2. Phase contrast electron micrograph. Left: Soot collected from a diesel engine. Right: XC 72 carbon black (Cabot Corp.)

References pp. 50-51.

to internal structure (crystalline organization) of the material. It will not be used in the present review but it can be useful to keep this in mind when consulting the references.

The morphology (size and shape) of aggregates can be characterized by:
- electron microscopy
- scattering and absorption of light
- ultracentrifugation
- adsorption of gases and liquids (pure liquids or solutions).

Observation of Aggregate by Electron Microscopy — *Two Dimensional Characterization* – It is relatively easy for a trained observer to differentiate qualitatively, upon electron micrograph observations, amongst carbon blacks having different degrees of aggregation. For example, in Fig. 3, it is clear that aggregates in sample A are much more developed than those in sample B.

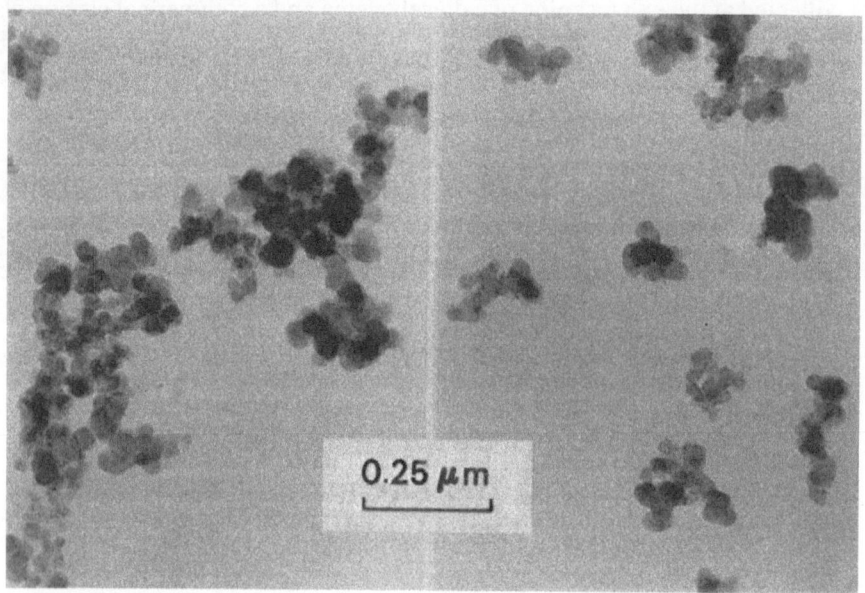

0.25 μm

Fig. 3. Transmission electron micrograph. (Left) A: High degree of aggregation. (Right) B: Low degree of aggregation.

Quantitative classification of carbon blacks according to their aggregation was first attempted by Cohan and Watson [4] and by Medalia and Heckman [5] before Quantimets (Quantitative Image Analysis) became commonly used in laboratories. In their method, the silhouette of the aggregate is compared to an ellipse [5] (Figs. 4a and 4b) and the following parameters are defined:

A   :   Area of silhouette

$$Q = \frac{K_A}{K_B} \quad : \quad \text{Anisotropy of ellipse}$$

B   :   Bulkiness $= \dfrac{\text{Area of ellipse}}{\text{Area of silhouette}} = \dfrac{4\pi K_A K_B}{A}$

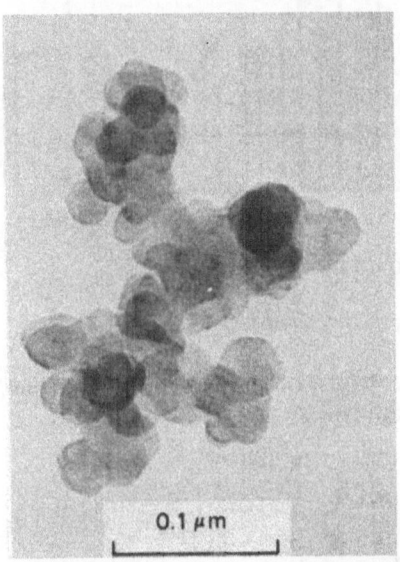
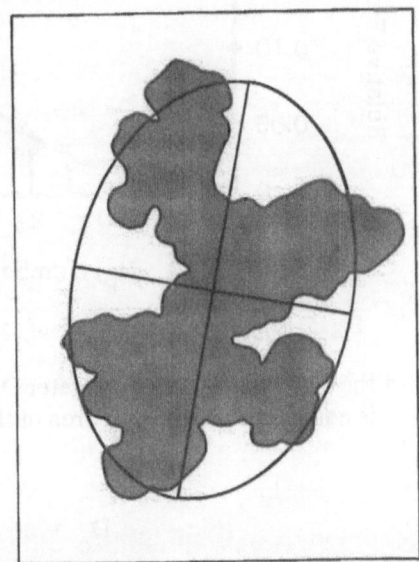

Fig. 4. Carbon black aggregate. (a) Electron micrograph. (b) Elliptic silhouette [5].

These parameters permit one to characterize quantitatively the aggregate. For example [5], Fig. 5 gives for a furnace black the relative frequency of aggregates as function of aggregate area and number of individual particles per aggregate. This method, however, has to rely somewhat on the operator's judgement. Also, it is time consuming and therefore not appropriate for systematic characterization of soot.

Human operators can be replaced by an electro-optical system and a computer [6, 7]. The microscope and specimen preparation are kept unchanged; the micrographs are analyzed using a scanning technique.

Automated image analysis does not have these limitations and is now being increasingly used for aggregate characterization. This technique was first systematically applied to electron micrographs of carbon black by Hess *et al.* [8]. The black units are analyzed individually for their projected area, perimeter, total projected length (number of chord intercepts) and chord length distribution [9, 10]. From these experimental values, the two following parameters are computed for each aggregate [10]:

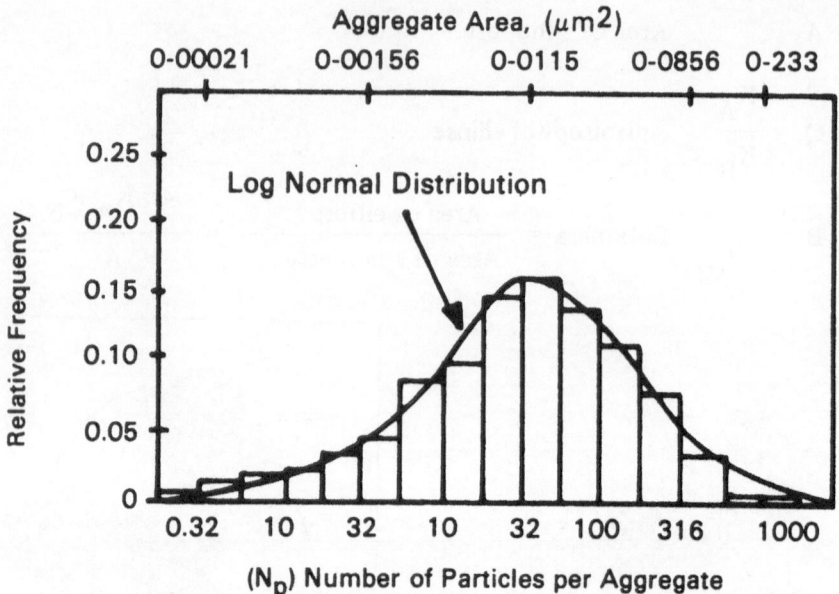

Fig. 5. Frequency distribution of aggregates of Vulcan 3 (sum of 7 runs) [5].

- the equivalent sphere diameter D_e, diameter of a sphere whose projected area is equal to the projected area of the aggregate A.

$$D_e = \left(\frac{4A}{\pi}\right)^{1/2}$$

- the Circularity Shape Factor (CSF) [11] derived from measurements of aggregate area A and perimeter P.

$$CSF = \frac{P^2}{4\pi A}$$

For a sphere CSF = 1; its value increases with the complexity of the aggregate. Fig. 6 [10] is a typical example of what can be obtained by Quantitative Image Analysis.

These methods have been intensively used by the carbon black industry, but one has to notice, however, that they give a *two-dimensional* description of three dimensional features. An important effort has been devoted to the development of stereo characterization of carbon blacks.

The Third Dimension of Carbon Black Aggregates – It is well known that the size of an object to each eye is slightly different in a direction perpendicular to a line joining the eyes. From this information the brain gives a three-dimensional image. The third dimension of carbon black aggregates can be reached by using the scanning electron microscope combined with stereoscopic methods [12].

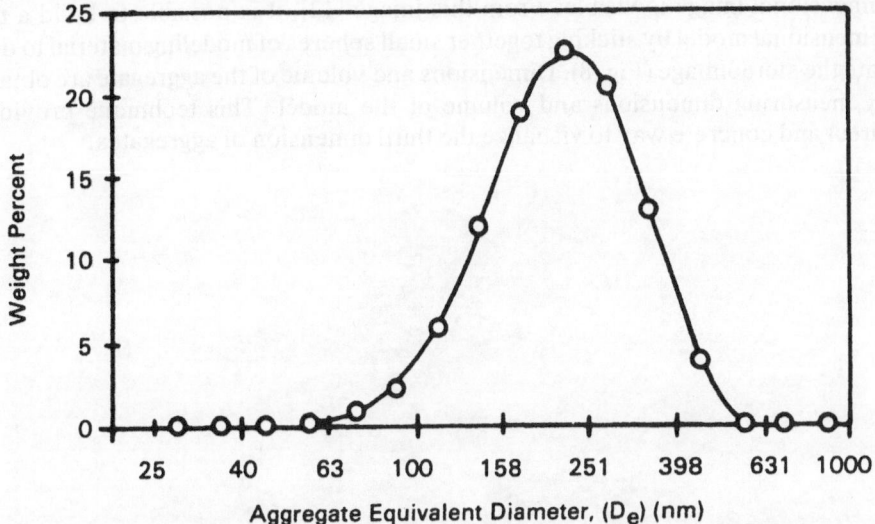

Fig. 6. Quantitative image analysis of Vulcan 6 N 220 conventional technology. Weight percent as function of aggregate equivalent diameter (D_e) (nm) [10].

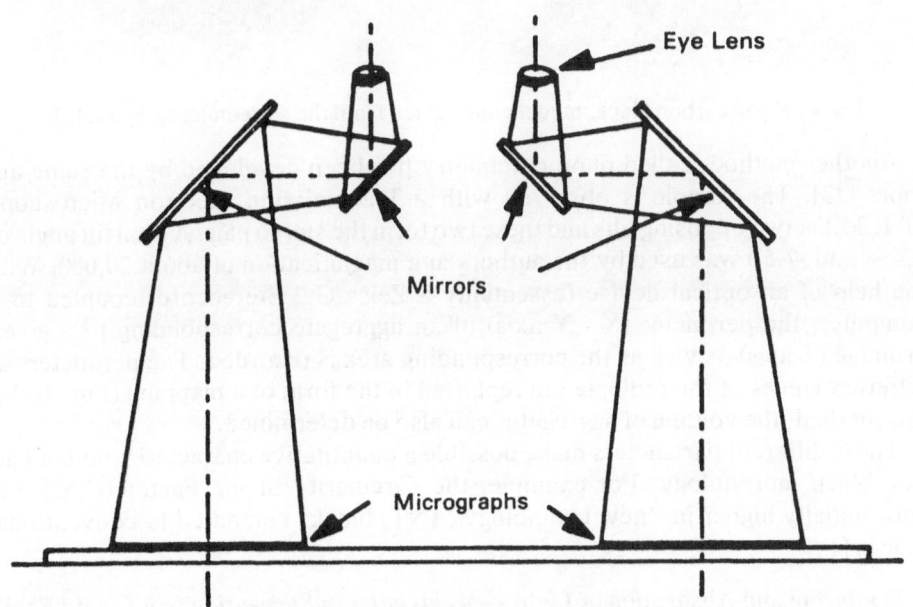

Fig. 7. Folding mirror stereoscope [12].

A carbon black sample is placed on the stage of an electron microscope. The stage can be tilted at $\pm$ 5° with respect to the electron beam. Viewing and analysis of the two stereoimages can be accomplished with the aid of a folding mirror stereoscope (Fig. 7). This instrument is constructed so that each eye sees a separate micrograph. The micrographs are brought into register so that only one three-dimensional image is visible. From this image [12], it is possible to build a three dimensional model by sticking together small spheres of modeling material to duplicate the stereoimage (Fig. 8). Dimensions and volume of the aggregate are obtained by measuring dimensions and volume of the model. This technique provides a direct and concrete way to visualize the third dimension of aggregates.

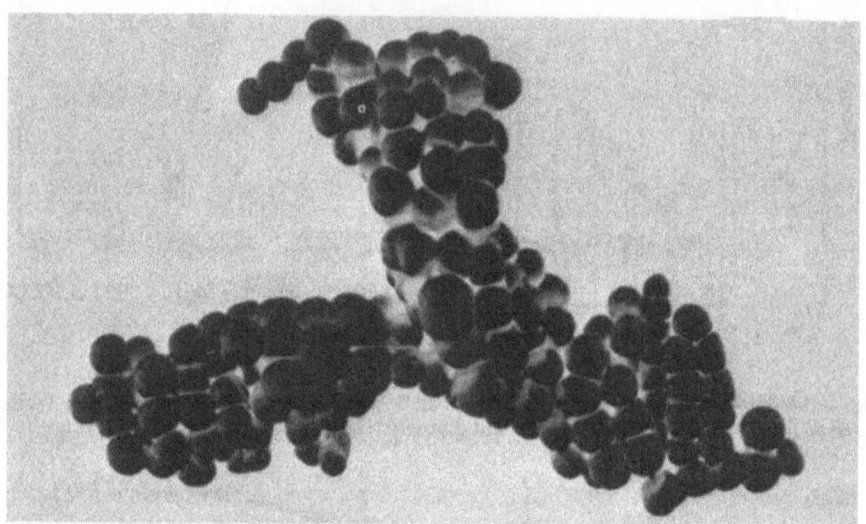

Fig. 8. N 550 carbon black: model constructed from the stereomicrographs [12].

Another method, called photogrammetry has been developed by the same authors [13]. The sample is observed with a Transmission Electron Microscope (T.E.M.) at two opposing tilts and these two form the stereo pair. A total tilt angle of 15° + and -7.5°) was used by the authors at a magnification of about 20,000. With the help of an optical device (essentially a Zeiss G.2 Stereocord) coupled to a computer, the perimeter (X - Y axis) of an aggregate corresponding to a given ordinate (Z axis) as well as the corresponding area is recorded. The perimeters at different values of the ordinate are replotted in the form of a mapping (Fig. 9). By this method, the volume of aggregates can also be determined.

These different parameters make possible a quantitative characterization of carbon black morphology. For example, the Circularity Shape Factor (C.S.F.) is substantially higher in "new technology" (NT) blacks compared to conventional blacks [14].

Scattering and Absorption of Light — *Aggregates in Suspension in a Liquid Phase* – Light scattering technique is a classical way of characterizing macromolecules in solution (Zimm Plot). The same technique can be used for the characterization of a carbon black suspension. For a series of slightly oxidized rubber blacks in suspension in water. Donnet *et al.* [15] determined Z, the ratio of intensities of scattered light at 45° and 135° with respect to incident light. The higher is Z, the higher the degree of aggregation [16]. The values of Z allow a classification of blacks according to their degree of agglomeration. The authors found the same classification for the

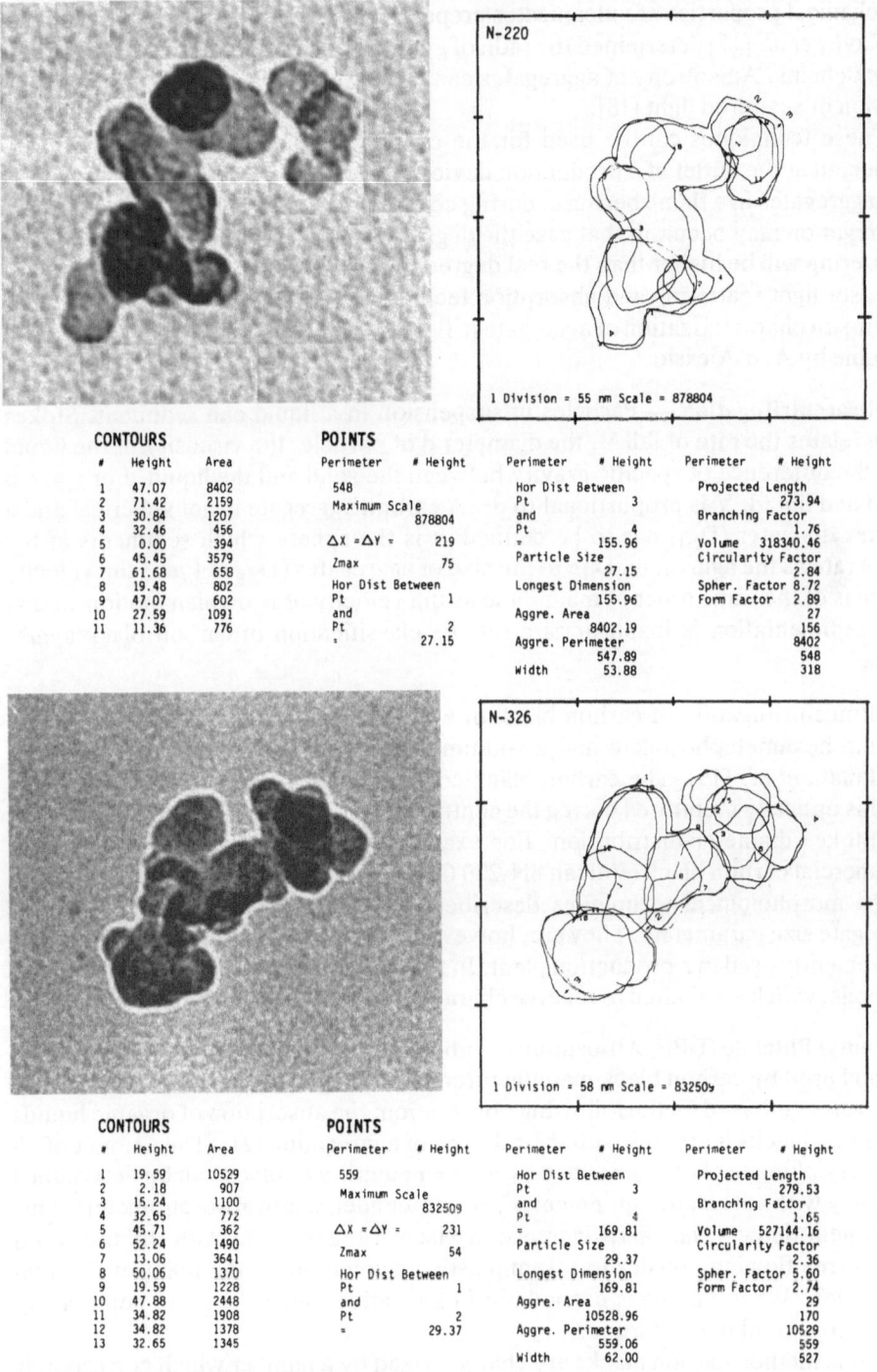

Fig. 9. Micrograph and finished plot of N 220 (top) and N 326 (bottom) aggregates [13].

References pp. 50-51.

mechanical properties of vulcanizates prepared with the corresponding blacks.

Ravey *et al.* [17] determined the radii of gyration and molecular masses of carbon black chains. Anisotropy of aggregates can be obtained by analysis of the depolarization of scattered light [18].

These techniques can be used for the characterization of final carbon blacks collected at the outlet of a production device. They are not adapted to characterizing aggregates in a flame because, during collection as well as in the solution, more aggregation may occur; in that case the degree of aggregation as measured by light scattering will be higher than the real degree of aggregation in the flame.

Laser light scattering and absorption techniques are more and more often used for in-situ characterization of aggregate in flames. They are described in the present volume by A. d'Alessio.

Ultracentrifugation — Particles in suspension in a liquid can sediment. Stokes Law relates the rate of fall V, the diameter d of particle, the viscosity of the liquid and the difference of specific gravity between the solid and the liquid. For a given solid and liquid, V is proportional to d^2. A carbon aggregate is not spherical and a Stokes diameter (D_{St}) has to be defined: it is the sphere which sediments at the same rate as the aggregate. Due to the size of aggregates ($D_{St} < 1 \mu m$) the velocity of fall is of the same order of magnitude as the velocity of Brownian motion so that free sedimentation is inappropriate for the classification of carbon black aggregates.

Ultracentrifugation of carbon black in a water suspension (the dispersants are sodium hexametaphosphate and a sodium lignosulfonate) has been achieved by Heckman, *et al.* [19]. The carbon black concentration at the external part of the rotor is optically monitored during the centrifugation, allowing the determination of the Stokes diameter distribution. For example, Fig. 10 shows the results for a commercial carbon black (Vulcan 6N-220 CT).

The morphological techniques described thus far provide absolute values of aggregate size parameters. They are, however, mainly research tools and cannot be conveniently used in a production plant. In the next two sections faster and simpler methods, which are limited to relative characterization of aggregates, are reviewed.

Dibutyl Phtalate (DBP) Absorption Number of Carbon Blacks [20] — A universal method used by carbon black manufacturers for the determination of the degree of aggregation is based on the following observation: the absorption of organic liquids by carbon blacks increases with their degree of aggregation [21]. The amount of oil or dibutylphtalate (DBP) absorbed by a given quantity of black can be determined precisely with automatic equipment (C.W. Brabender/Cabot absorptiometer). This instrument detects the sharp increase in viscosity associated with the transition from a free-flowing powder to a semiplastic agglomerate at the point of maxium absorption. When mixing is carried out in a rotative mixer, the end point corresponds to a rapid rise in torque.

By this method carbon blacks are characterized by a number which corresponds to the volume (in cm^3) of DBP absorbed per 100 g of sample. A correlation has been developed by Medalia [22] for the void volume between aggregates calculated from

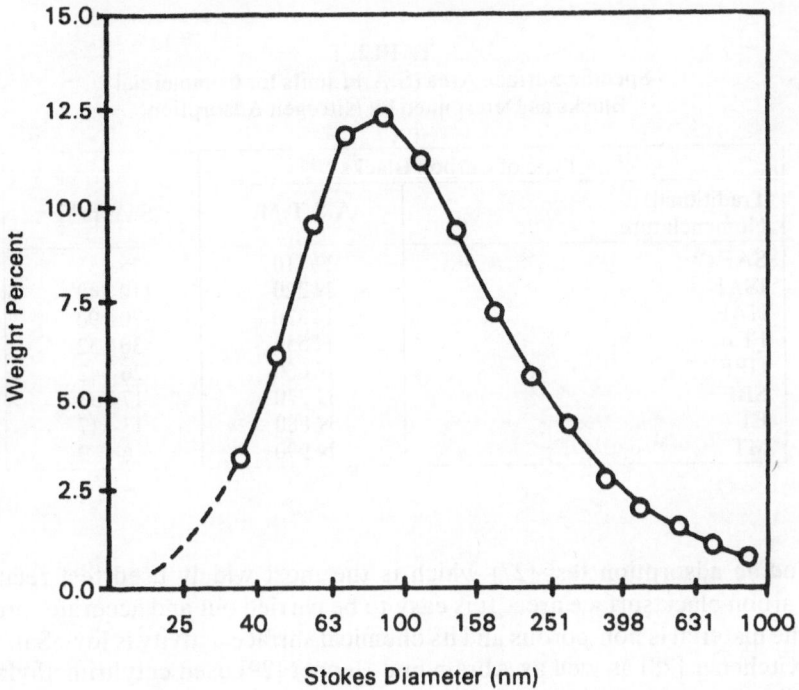

Fig. 10. Stokes diameter distribution plot. Vulcan 6N — 220 conventional technology [19].

electron microscopic data and DBP absorption numbers. Using some adjusting parameters, the author obtained a good agreement for eight carbon blacks of various degrees of aggregation and particle sizes.

Specific Surface Area and Porosity of Soot — The specific surface area and porosity of soot, as of any solid, can be determined from adsorption and desorption isotherms of gases. Systematic determination began with the development of the B.E.T. method. Since that time, extensive measurements have been published [23]. Nitrogen adsorption isotherms on carbon blacks at the temperature of liquid nitrogen are routinely used to measure surface areas. Table 1 [1] lists values obtained for several types of commercial blacks. Other gases such as krypton may also be used. In general carbon blacks are not porous and the "BET" surface is equal to the external surface. This is not the case for oxidized blacks which may possess an important microporosity.

Adsorption-desorption isotherms as well as other methods such as the measurement of heat of adsorption of hydrocarbon vapors [24, 25] or immersion in liquids [26] are mainly used in research studies.

For industrial applications [3] two main types of methods have been developed:
- Gas adsorption using one single-point of the isotherm. This is legitimate provided the samples are non-porous and the parameter c of the BET equation [23] is nearly constant, as is the case for nitrogen on most carbon blacks.

References pp. 50-51.

TABLE 1
Specific Surface Area (S.A.) Limits for Commercial
Blacks as Determined by Nitrogen Adsorption.

Type of Carbon Blacks		S.A. m²/g
Traditional Nomenclature	A.S.T.M.	
SAF	N 110	125-155
ISAF	N 220	110-140
HAF	N 330	70- 90
FEF	N 550	36- 52
GPF	N 660	26- 42
SRF	N 770	17- 33
FT	N 880	13- 17
MT	N 990	6- 9

- Iodine adsorption test [27] which is the most widely used test relating to carbon black surface area. It is easy to be carried out and accurate, provided the material is non-porous and its chemical surface activity is low. Saleeb and Kitchener [28] as well as Abram and Bennet [29] used cetyltrimethylammonium bromide (CTAB) and sodium di-(2-ethyl-hexyl) sulfosuccinate (aerosol OT) as adsorbates on carbon blacks for measurements of the external surface areas.

Morphology of Individual Particles — As pointed out in the introduction, individual particles have no real existence from the standpoint of the final material. In other words, soot is made of irreversible aggregates of pseudo-spherical particles. However, characterization of the spherical units constituting the aggregates is useful for the understanding of the mechanism of soot formation as they are the precursors of the aggregates. Moreover, carbon blacks obtained by thermal decomposition of hydrocarbons in the absence of air (thermal blacks) are mainly constituted of single spherical particles.

Electron microscopy is the major technique used to characterize individual carbon black particles (Figs. 1 and 2). By measuring on electron micrographs the diameter of a large number of particles (1,000 or more), the size distribution of a sample can be accurately established. For that purpose a semi-automatic particle size analyser (TGZ, Zeiss) is often employed [1, 2, 30, 31]. Size distributions of commercial carbon black or soot are usually log-Gaussian. From the distribution curve, average parameters (diameter, surface or volume) are readily computed. In Table 2 are indicated the arithmetic mean diameters D_A of several carbon blacks used for the reinforcement of rubber.

Accurate determinations of size distribution of carbon blacks may also bring valuable information on mechanisms of particle growth (coalescence and/or surface growth). This is discussed in detail in another article of the present volume by Prado and Lahaye.

TABLE 2
Mean Arithmetic Particle Diameters D_A as Determined
by Electron Microscopy.

Type of Carbon Blacks		D_A (nm).
Traditional Nomenclature	A.S.T.M.	
SAF	N 110	11- 19
ISAF	N 220	20- 25
HAF	N 330	26- 30
FEF	N 550	40- 48
GPF	N 660	49- 60
SRF	N 770	61-100
FT	N 880	101-200
MT	N 990	201-500

MICROSTRUCTURE

The first part of the review has been concerned with the morphology of carbon blacks or soot without any consideration to their internal structure, that is to say the crystalline organization of carbon inside the material.

A systematic study of carbon microstructure began with the development of X-ray crystallography. Complementary information was obtained by combining chemical oxidation of particles and transmission electron microscopy. More recently, an important improvement in the description of internal structure of carbon was achieved by employing dark field and phase contrast electron microscopy.

In graphite, carbon atoms are located in planes of symmetrical hexagons (Fig. 11). Each plane of hexagons can be deduced from the previous one by displacing over a distance d equal to the length of the carbon-carbon bond. The crystallographic unit cell can be defined by the three vectors $\vec{a}, \vec{b}$ (in the plane of hexagons), and $\vec{c}$ perpendicular to $\vec{a}, \vec{b}$.

$$|\vec{a}| = |\vec{b}| = 0.2461 \text{ nm}$$

$$|\vec{c}| = 0.6708 \text{ nm}$$

The distance ($\vec{c}$ direction) between two consecutive planes is half of $|\vec{c}|$ i.e. 0.3354 nm.

X-ray Diffraction — A good survey of X-ray characterization of carbon black particles is given by Donnet and Voet [1]. The diffraction patterns show two or three diffuse rings at about the same position as the most intense graphitic diffractions.

In 1934, Warren [32] indicated that carbon black presented a degenerated graphitic structure: carbon atoms are located in planes of hexagons but the layers are disoriented by random rotations in their planes ("turbostratic structure").

A series of investigations was carried out between 1950 and 1965 [33]. To sum up these works, soot particles can be described as made of small crystallites presenting

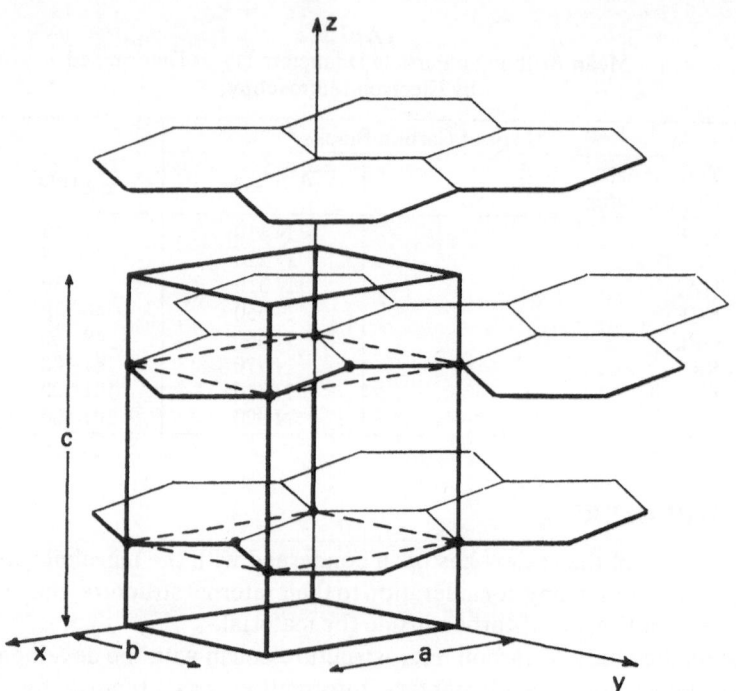

Fig. 11. Schematic of graphite structure.

a turbostratic graphitic structure, randomly oriented and connected to one another by single planes or by amorphous carbon. Austin [34] measured the values of crystalline parameters of a large variety of carbon black; crystallites consist, on average, of three to four turbostratic layers at interplanar distances of 0.34 to 0.36 nm. In analyzing the (002) reflection of carbon blacks, Ergun [35] questioned the existence of crystallites and suggested that the layer of carbon atoms is the basic building block of the carbon black particles. Kmetko [36] studying the graphitization of channel blacks, proposed that the layer planes of the crystallites are oriented parallel to the surface; Boehm [37], Akamatu, *et al.* [38] and Kasatotshkin, *et al.* [39] came to the same conclusion. A model for carbon black microstructure was proposed by Heckmann and Harling in 1966 [40] and is reproduced in Fig. 12.

Transmission Electron Microscopy of Carbon Blacks — Donnet and Bouland [41] conducted in 1964 a series of electron microscopy observations of thermal carbon blacks oxidized with nitric acid. They were able to show that the oxidized particles are composed of numerous concentric shells (Fig. 13). The existence of carbon layers parallel to the surface confirms the X-ray data. Particle coalescence can be noticed in the micrographs. External layer planes are continuous from one particle to its neighbors.

Dark Field and Phase Contrast Electron Microscopy — Among the different publications concerned with dark field and phase contrast electron microscopy, one can mention those of A. Oberlin on the application of dark field electron microscopy

Fig. 12. Schematic of Heckman and Harling's 1966 model of carbon black microstructure [1], by courtesy of Marcel Dekker, Inc.

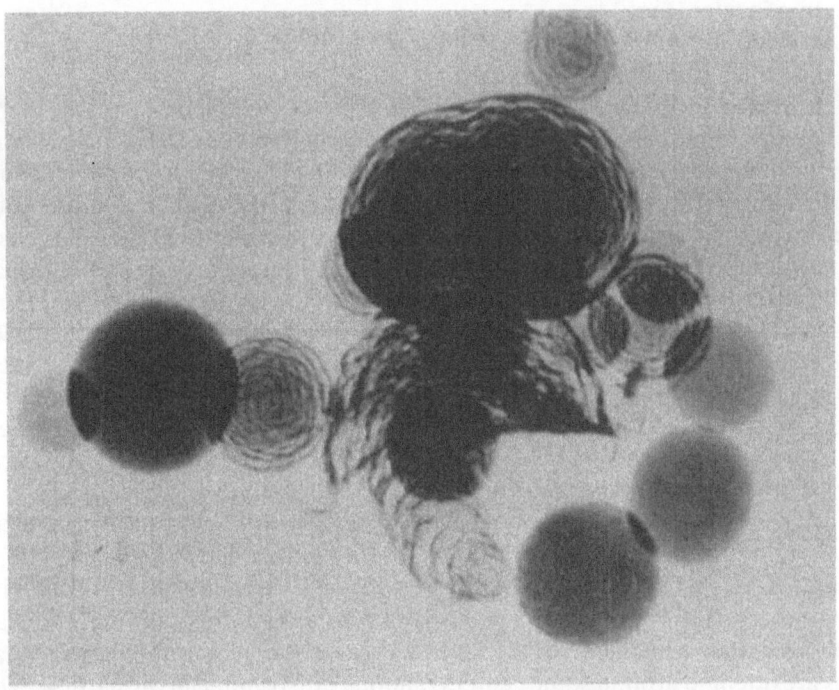

Fig. 13. FT carbon black oxidized with nitric acid [1], by courtesy of Marcel Dekker, Inc.

References pp. 50-51.

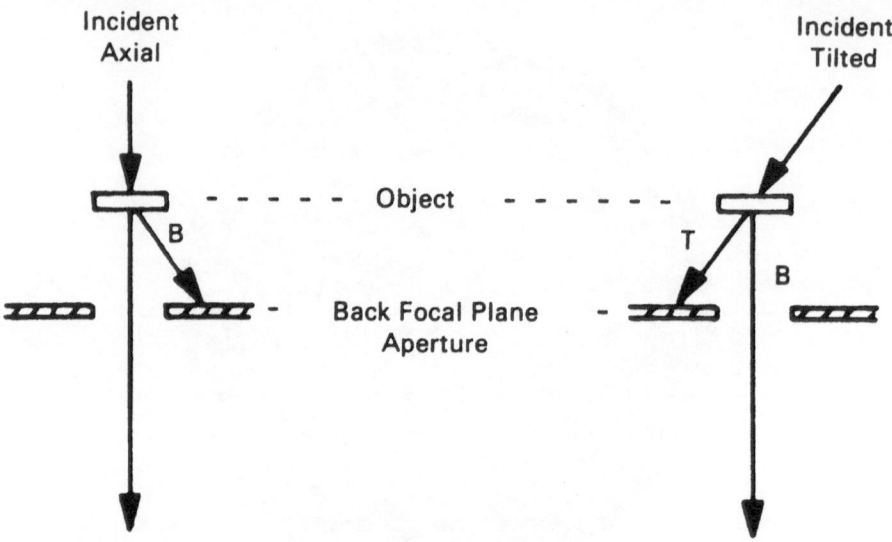

Fig. 14. Schematic diagram illustrating bright (left) and dark field (right) diffraction contrast with an objective aperture; T: transmitted beam, B: Bragg beam.

[42] and those of Heidenreich *et al.* [43]. The two techniques will be described in a very schematic way.

Let us consider (Fig. 14):
  i. an electron beam reaching the object to be observed,
  ii. the transmitted beam and
  iii. the Bragg diffraction beam [(002) diffraction for example].

If an aperture removes the diffracted beam allowing only the transmitted beam to reach the image plane, a bright field image is produced on which the object appears dark (classical transmission electron microscopy). If the objective aperture is placed so as to intercept the transmitted beam and accept the Bragg beam, a dark field image is produced. By using a part of the (002) ring, only the graphitic planes which satisfy the Bragg relation corresponding to the (002) reflection yield bright spots on a dark background. In removing the aperture and allowing Bragg diffracted image and transmitted image to superimpose, an interference pattern is obtained; it gives a visualization of the reticular planes (carbon layer planes in the case of carbon). A dark field image of FT thermal black (002) reflection is represented in Fig. 15 [44]. The wedge-shape zones indicate concentric rings of surface-parallel layer groups which are analogous to the layers of an onion.

The development of phase contrast electron microscopy led to the direct imaging of layer planes in carbon blacks. Fig. 16 shows a phase contrast micrograph of a graphitized I.S.A.F. black [45]. Interlayer spacings, as measured on the micrograph, are c = 0.344 nm. There appears to be a parallel orientation of the planes along the particle periphery. In Fig. 2, however, in the bulk of the particles, layer planes are oriented around some centers randomly distributed in the particle. These centers would correspond to the primary nuclei mentioned in another article of the present volume by G. Prado and J. Lahaye. In the same figure the continuity of

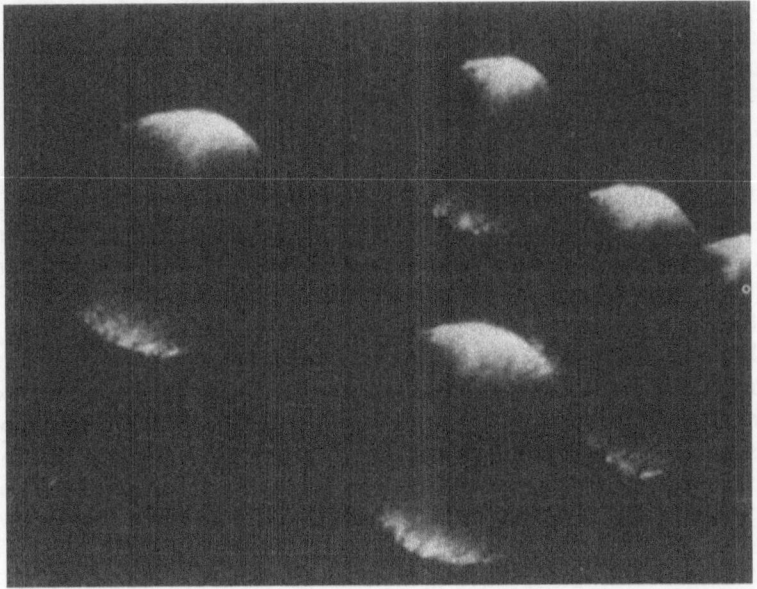

Fig. 15. Dark field ((002) reflection) electron micrograph of FT (N 880) black [1], by courtesy of Marcel Dekker, Inc.

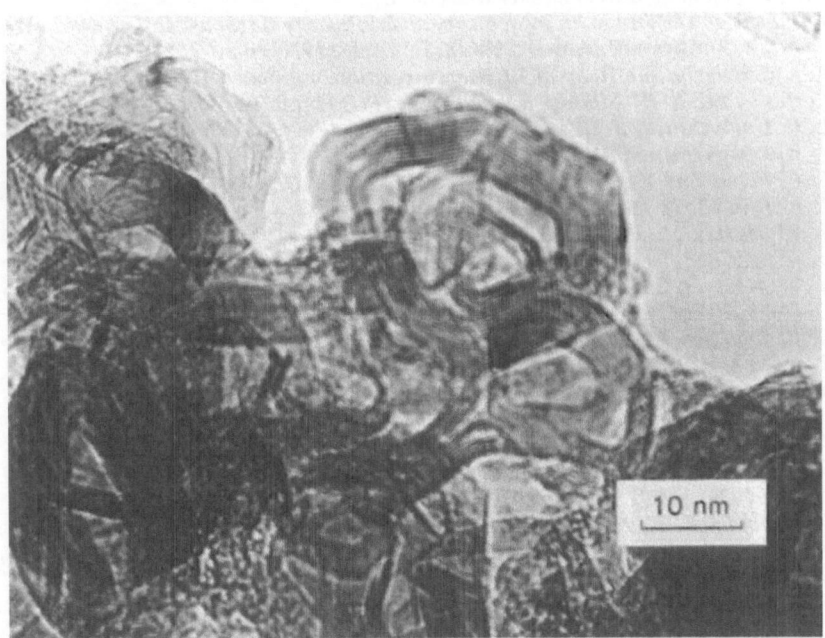

Fig. 16. High resolution phase contrast electron micrograph of graphitized ISAF (N 220) black [1], by courtesy of Marcel Dekker, Inc.

References pp. 50-51.

carbon network through the aggregate appears clearly. From the point of view of the final material, individual particles have no existence. Aggregates constitute the real entity.

CONCLUSIONS

From the standpoint of morphology and microstructure, soot and carbon blacks appear identical, and the large amount of information obtained on commercial carbon blacks is most probably valid for soot particles emitted from combustion equipment.

For the other properties, comparison should be drawn with much caution. For example, the amount of hydrocarbons condensed on a particle surface is usually very low (<1%) for carbon black, but can reach very high levels in soot from combustion equipment (up to 50%). This amount of extractible depends not only on combustion operating conditions, but also on sampling procedures. In most cases, it will be necessary to remove the condensible fraction from the solid particulate before applying the different methods described in the present review to soot particles.

REFERENCES

1. J. B. Donnet and A. Voet, "Carbon Black – Physics, Chemistry and Elastomer Reinforcement", Marcel Dekker Inc. (1976).
2. J. Lahaye and G. Prado, in "Chemistry and Physics of Carbon", Edit. by P. L. Walker and P. A. Thrower, Marcel Dekker Inc., Vol. 14 (1978), pp. 168-294.
3. A. I. Medalia and D. Rivin, in "Characterization of Powder Surfaces", Edit. by G. D. Parfitt and K. S. W. Sing, Academic Press (1976) pp. 279-351.
4. L. H. Cohan and J. H. L. Watson, Rubber Age, Vol. 68 (1951), p. 687.
5. A. I. Medalia and F. A. Heckman, Carbon, Vol. 7 (1969), p. 562.
6. C. Fisher and M. Cole, The Microscope, Vol. 16 (2), (1968), p. 81.
7. C. Fisher, The Microscope, Vol. 19 (1), (1971), p. 1.
8. W. M. Hess, L. L. Ban and G. C. Mc Donald, Rubber Chem. Technol., Vol. 42 (1969), p. 1209.
9. W. M. Hess, G. C. Mc Donald and E. Urban, Rubber Chem. and Technol., Vol. 46 (1), (1973), p. 204.
10. E. Redman, F. A. Heckman and J. E. Connolly, "Particle size analysis conference Proceedings", 3d, (1978), p. 51.
11. M. Rink, J. Microsc. Oxford, Vol. 107 (1976), p. 267.
12. H. N. Mercer, A. H. Boyer, P. L. Brusky and M. L. Deviney, Rubber Chem. and Technol., Vol. 49 (4), (1976), p. 1068.
13. H. N. Mercer, A. H. Boyer and M. L. Deviney, Rubber Chem. and Technol. Vol. 52 (2), (1979), p. 377.
14. E. Redman, F. A. Heckman and J. E. Connoly, Meeting of the Rubber Division, American Chemical Society, Chicago, Ill. 1977; Abstract in Rubber Chem. Technol., Vol. 50 (1977), p. 1000.
15. J. B. Donnet, C. Eckhardt and A. Voet, Rev. Gen. Gaoutchouc Plast. Vol. 44 (5), (1967), p. 627, Vol. 44 (12), (1967), p. 1505.
16. S. Premilat and P. Horn, J. Chim. Phys., Vol. 63 (3), (1966), p. 463.
17. J. C. Ravey and S. Premilat, J. Chim. Phys., Vol. 67 (1), (1970), p. 147.
18. J. C. Ravey and S. Premilat, J. Chim. Phys., Vol. 67 (1), (1970), p. 157.

19. F. A. Heckman, E. Redman and J. E. Conolly, Complementary Studies of Carbon Black Aggregate Morphology by Analytical Centrifugation and Quantitative Image Analysis – Technical Service Report Cabot Corporation, Spring (1977).
20. ASTM D 2414-65, Standard Method of Testing, Carbon Black – Dibutyl Phthalate Absorption Number, American Society for Testing and Materials, Philadelphia, Pa., (1966).
21. B. Schubert, F. P. Ford and F. Lyon, Analysis of Carbon Black, Encyclopedia of Industrial Chemical Analysis John Wiley and Sons, Inc., Vol. 8 (1969), p. 225.
22. A. I. Medalia, J. Colloid Interf. Sci., Vol. 32 (1970), p. 115.
23. M. Bastick, P. Chiche and J. Rappeneau, Les Carbones – Tome II (Masson et Cie), (1965), pp. 24-160.
24. W. R. Smith, Rev. Gen. Caoutchouc, Vol. 41 (1964), p. 367.
25. N. N. Avgul and A. V. Kiselev, "Chemistry and Physics of Carbon", P. L. Walker Jr., Marcel Dekker, New York, Vol. 6 (1970).
26. W. H. Wade, M. L. Deviney, W. A. Brown, M. H. Knoosch, and D. R. Wallace, Rubber Chem. Technol., Vol. 45 (1972), p. 117.
27. ASTM D 1510-65, Method of Tests for Iodine Adsorption Number of Carbon Black, American Society for Testing and Materials, Philadelphia, Pa, (1966).
28. F. Z. Saleeb and V. A. Kitchener, J. Chem. Soc. (1965), p. 911.
29. V. C. Abram and M. C. Bennett, J. Colloid Interf. Sci., Vol. 27 (1968), p. 1.
30. G. Prado, PhD Dissertation Thesis, Strasbourg University, (1972).
31. G. Prado, and J. Lahaye, J. Chim, Phys., Vol. 4 (1975), p. 483.
32. B. E. Waren, J. Chim. Phys. Vol. 2 (1934), p. 551.
33. J. Mering et J. Maire, Les Carbones, Tome I (Masson & Cie), (1965), p. 162.
34. A. E. Austin, Proceeding 3rd Conf. on Carbon, (1958), p. 389.
35. S. Ergun, Carbon, Vol. 6 (1968), p. 141.
36. E. A. Kmetko, Proc. 1st and 2nd Conf. on Carbon, (1956), p. 21
37. H. P. Boehm, Z. Anorg. Allgem. Chem., Vol. 297 (1958), p. 315.
38. H. Akamatu and H. Kuroda, Proc. 4th Conf. on Carbon, (1960), p. 363.
39. V. L. Kasatotshkin, V. M. Lukianovitch, N. M. Popov and K. V. Tschmutov, J. Chim. Phys., Vol. 52 (1964), p. 822.
40. F. A. Heckman and D. E. Harling, Rubber Chem. Technol. Vol. 39 (1966).
41. J. B. Donnet and J. C. Bouland, Rev. Gen. Caoutchouc, Vol. 41 (1964), p. 407.
42. A. Oberlin, Carbon, Vol. 17 (1979), p. 7.
43. R. D. Heidenreich, W. M. Hess and L. L. Ban, J. Appl. Cryst., Vol. 1 (1968), p. 1.
44. W. M. Hess, L. L. Ban, F. J. Eckert and V. Chirico, Rubber Chem. Technol., Vol. 41 (1968), p. 356.
45. P. A. Marsh, A. Voet, T. J. Mullens and L. D. Price, Carbon, Vol. 9 (1971), p. 797.

DISCUSSION

L. C. Yang (*Jet Propulsion Laboratory*)

We have some experience on other properties of diesel particles, as compared to the carbon blacks. We found a considerable difference in electrical conductivity. Also we're interested in thermal conductivity differences. These physical properties are not necessarily the same for the two cases, and they are very important to several techniques currently being pursued in this country as possible methods for diesel particle destruction.

Lahaye

I thank you for your comment on electrical conductivity of soot and carbon blacks. It is an important parameter which may contribute to the properties of carbon black/polymer mixtures and, in the case of conductive blacks used in battery cells, to the performance of batteries. Absolute conductivity cannot be reached, but comparative values can be obtained for comparable blacks. When referring to electrical conductivity and to thermal conductivity it is important to mention the methods used for their determination.

Yang

Since, as you mentioned, the morphologies are quite similar, one can compress to a standard density and then compare microscopic resistivity. That's a way to do it from an engineering standpoint.

A. F. Sarofim *(Massachusetts Institute of Technology)*

Can you tell from any of your phase contrast electron microscopies if soot particles are formed as aggregates of smaller particles, each of which would presumably have structure parallel to the surface? You'd expect a series of concentric spheres within a larger sphere.

Lahaye

Phase contrast electron microscopy, as well as electron microscopy of oxidized soot indicate indeed that spherical soot units (spherules) contain a series of concentric spheres. In a pure coagulation model involving only collisions between spherical particles of carbon, one would expect particles with internal structure independent of the structure of their neighbors. In fact external carbon layers are common to several particles. Two explanations can be proposed, and probably both are right, according to the system one is referring to. The first is surface growth: material deposits from the gas phase onto the surface of the aggregates. The other one occurs when two particles collide: mobility of the molecules is sufficiently high that rearrangement of carbon layers at the surface of aggregate is possible. At low temperature, in a pyrolytic system, important surface growth is expected to be significant. In a flame, at high temperature, surface growth is probably very limited.

M. Brown *(Exxon Research)*

Have any measurements been made on oxidized carbon black to see if it burns from surface erosion versus internal particle combustion?

Lahaye

Yes, many investigations were carried out in the sixties. For example, if a thermal black is introduced onto the stage of an electron microscope in the presence of a very small amount of oxygen, what is noticed is very interesting: particles remain essentially unoxidized, but when a particle does begin to burn, the rate is very fast and the internal part of the particle is burnt out. An onion shell structure is obtained. It appears that crystalline organization of carbon inside the particle is lower than outside and, provided oxygen is able to diffuse, there is oxidation of the internal part of the particle. A systematic study of thermal black oxidation by air has been carried out [1]. Pores are formed in the case of furnace blacks. These experiments refer to relatively slow oxidation at low temperature. At higher temperatures, such as those encountered in a flame, the mechanism might be different.

R. Stevenson *(General Motors Research Laboratories)*

With regard to the question of external versus internal organization of the particle, have you ever done an optical diffraction scan across a phase contrast micrograph of one of these spherical particles to look at the optical diffraction pattern as a function of distance from the center?

Lahaye

By fracturing a thermal black particle with an ultramicrotome and imaging parts from the periphery and from the core of the same particle, it has been possible to show* that interlayer distances are identical.

* *P. A. Marsh, A. Voet, T. J. Mullens and L. D. Price, Rubber Chem. Technol. Vol 43, p. 470 (1970).*

A. Snelson *(IIT Research Institute)*

Have you tried looking at carbon blacks to expose them to something like bromine, which, in the case of graphite, will intercalate and swell the layers dramatically? If you did it with carbon blacks, do you see a sort of expansion or not? Or has it not been done?

Lahaye

Bromine is able to intercalate into graphite*. At low temperature a slight intercalation may occur with some carbon blacks. It is possible, however, to obtain a significant intercalation with other reagents, such as silver bichromate and potassium permanganate in sulfuric medium**. These reagents intercalate into graphitic carbon structure and after intercalation, well organized parts of particles are oxidized with a much higher rate than disorganized parts. It is an analytical method to determine the degree of organization of carbon blacks.

* *M. C. Robert, M. Oberlin and J. Mering, Chemistry and Physics of Carbon – Ed. P. L. Walker and P. A. Thrower (M. Dekker Inc.), Vol. 10, p 141 (1973).*
** *P. Ehburger and J. B. Donnet, Carbon Vol. 11, pp. 309-16 (1973).*

S. H. Bauer *(Cornell University)*

In considering morphology of these particles we should perhaps ask the questions: 1) Why are they so much like spheres? and 2) Why do these spheres tend to form chains i.e., the sort of a bead-like structure? I wanted to point out that this is not a property of carbon black alone. A large number of particles including metallic particles and metallic oxides, have the same morphology and external appearance. That is, they tend to form spheres when they're first generated, and the spheres tend to form little bead-like chains. To me this suggests very strongly that surface free energies (I don't want to use the term surface tension) play a very important role in the initial agglomeration of the elementary units. The fact that you have continuity between the planes, I think, must result from a secondary effect — perhaps a rearrangment after they come together — otherwise it would be very difficult to see how you get the spheres.

Lahaye

Yes, we shall hear more about this later on (Dr. Prado's presentation), but we can summarize that part which is related to your question. Individual pseudospherical particles can be formed by collision of species or small nuclei whose surface free energy is sufficiently low so that, as in liquids, spherical particles are formed. The particles are carbonized, and when two of them collide, they are no longer able to fuse: aggregates of small pseudospherical particles are obtained. It can be noticed, however, that this mechanism involves a modification of the surface free energy of the material by chemical reactions (carbonization) which is not the case for metallic particles.

J. B. Howard *(Massachusetts Institute of Technology)*

If you measure the surface area per unit mass of carbon black or soot with the B.E.T. technique, can you account for the observed area in terms of the superficial area of these units within the aggregates, or is the area too large to be accounted for by the superficial area?

Lahaye

Are you asking me whether there is a possibility of microporosity and capillary condensation?

Howard

Well, that could follow from the answer.

Lahaye

Usually, carbon blacks are not porous, unless they have been postoxidized; they exhibit no hysteresis during desorption of gases. It is possible to use the B.E.T.

method to determine the surface area of carbon black aggregates; there is no major problem, providing the surface area is high enough to use nitrogen as an adsorbent. For large particles of soot (low specific surface area), krypton has to be used. Because of the uncertainties in the value of the projected area of a krypton molecule at the temperature of liquid nitrogen, the accuracy is poor, but comparative values can be reached.

Howard

And the nitrogen area observed is no larger than you could calculate by taking the diameter of these roughly spherical units?

Lahaye

B.E.T. surface area is smaller than the one obtained by electron microscopy. Indeed, by the latter method, surface area is computed from average diameter of particles. Therefore, in aggregates, the surface common to two contiguous particles is counted twice. When the surface areas as determined by B.E.T. is higher than those obtained by electron microscopy, porosity can be expected.

OPTICAL PROPERTIES AND MORPHOLOGY
OF PARTICULATE CARBON:
VARIATION WITH AIR/FUEL RATIO

D. M. ROESSLER, F. R. FAXVOG*, R. STEVENSON and G. W. SMITH

General Motors Research Laboratories
Warren, Michigan

ABSTRACT

The specific optical extinction, composition, and morphology of diesel particulate emissions vary strongly with engine operating conditions such as air/fuel ratio. Particles produced at low air/fuel ratios are highly agglomerated, consist primarily of elemental carbon, and have large values of specific optical extinction; at high air/fuel ratios particles are less agglomerated, contain considerable volatile organic material, and have decreased specific extinction. Calculations based on Mie theory and the Rayleigh approximation for spheroids show that the variation in optical properties depends primarily on particle shape (i.e., morphology), but that composition also plays a role.

We have developed a simple nucleation/depletion model for the size of the fundamental spherical subunits (spherules) which agglomerate to form diesel particles. The model correctly estimates the magnitude of the spherule radius (~10 nm) and predicts that the spherule size decreases with formation temperature.

INTRODUCTION

Despite keen interest in the particulate emissions from diesel engines and a rich miscellany of research on their generation and character, many fundamental questions remain unanswered. For example, it is known that such particles have variable chemical composition and physical morphology, but there is only meager information on how such changes are related to the combustion process. The

---

Present address: Honeywell Systems & Research Center, 2600 Ridgeway Parkway, Minneapolis, Minnesota 55413.

References pp. 82-84.

present work was undertaken to see whether systematic variation of engine operating conditions changes the nature of the particulate emissions sufficiently to affect their fundamental optical properties.

The papers by Vuk *et al.* [1, 2] provide a useful introduction to the literature on diesel particulate emissions. Although there are legitimate concerns about the influence of the sampling technique [3, 4] on morphology and chemical composition, it is generally accepted that the particles range in shape and size from single spheres of about 0.03 μm (30 nm) diameter to clusters, often chain-like, extending over 1 μm in length. Furthermore, the composition may range from almost entirely carbon to dominantly hydrocarbon. In spite of these variations, there have been substantial efforts to use optical parameters, such as smoke opacity, as a measure of particulate mass concentration [for example, references 5-7]. Carey [8] noted that one could not assume that a particular engine load-speed condition would always produce the same amount of smoke. Calibration of smoke meters against one another therefore required their simultaneously monitoring the same particulate emissions. Bryzik and Smith [9] reported an extensive study of the opacity of diesel smoke as a function of several engine parameters and observed significant trends. Similar systematic patterns were derived by Laresgoiti *et al.* [10], who found that both the particulate concentrations and smoke opacity were highly dependent on engine speed and load.

In the present work we have used opacity data as a basis to obtain a more fundamental optical parameter, the specific extinction A_E. Unlike opacity, A_E depends only on the nature of the particles, not their mass concentration. We have observed large changes in A_E as a function of engine operating conditions. These have been correlated with variations in the morphology and chemistry of the particles. Following a description of the experimental determinations of particle optical properties, chemistry, and morphology, we shall present a model for the optical properties of diesel particles. Although the model is approximate, we shall show how it can give useful insights as to the way in which the size, shape, and chemistry of the particles affect their optical properties. Finally we offer an explanation for the characteristic size of the fundamental spherical subunits (spherules) comprising typical particles emitted by diesel engines.

MEASUREMENT OF OPTICAL PROPERTIES

Consider a beam of light passing through a smoke formed by particles of mass concentration M. The amount of light transmitted will depend on M, the distance through the smoke L, and the fundamental optical properties of the particles. The particulate opacity, θ, can be written as

$$\theta = 1 - \exp(-A_E ML) \tag{1}$$

where A_E is the specific optical extinction of the particles and is dependent on their refractive index, size and shape. The product $A_E M$ is the extinction coefficient, a measure of the attenuation per unit distance. The quantity A_E, however, is independent of the amount of material present and varies only if the character, rather than the number, of particles varies. Clearly the success of any attempt to calibrate opacity against mass concentration depends on the constancy of A_E. We have shown elsewhere [11] that A_E for black smokes of different particle sizes and refractive indices can indeed vary widely (from less than 5 $m^2 \cdot g^{-1}$ to more than 10 $m^2 \cdot g^{-1}$).

The diesel particles of the present study were generated by a variety of engines run over a range of operating conditions. Table 1 lists these engines and conditions.* The 2.1 L engine was operated on only one cylinder to preclude averaging effects of multi-cylinder operation. The other engines, B-E, were operated on all 8 cylinders.

TABLE 1
Engines and Operating Condition Ranges

Engine	Speed (r/min)	Air/Fuel Ratio
A*(2.1L)	1250-1750	15-65
B (5.7L)	1000-1500	18-87
C (5.7L)	640-1860	17-83
D (5.7L)	630-1750	19-88
E (5.7L)	1500-2000	22-100

*Engine A was operated on only one cylinder. The remaining engines, B. through E, were all operated on 8 cylinders but had slight variations in combustion and prechamber design.

Fig. 1 is a schematic diagram of the sampling arrangement for the optical experiments. A conventional commercial opacity meter (Celesco #107) was used to sample the raw exhaust. Following dilution, a particulate sample was collected on a filter for gravimetric determination of the mass concentration. In some experiments, only part of the raw exhaust was diluted, permitting use of a smaller dilution tunnel.

Additional optical data could be obtained by sampling the diluted exhaust both in a photoacoustic cell [12] and in a long path transmission cell. The latter was a simplified laboratory version of the commercial opacity meter, but having a path length of nearly 2 m rather then the nominal 0.15 m of the Celesco #107. These transmission measurements were made using a helium-neon laser, emitting at 0.6328 μm wavelength. The Celesco #107 meter employs a light-emitting diode giving green rather than orange radiation. Since the optical properties of carbon vary little with wavelength in the range 0.55-0.65 μm, any differences in the optical properties of the particles as determined by the opacity meter and the long tube could thus be attributed primarily to effects of dilution ratio and temperature differences between the two units.

*We use the term air/fuel ratio to denote the overall average value for the combustion chamber (measured by metering the air and fuel flow rates), while recognizing the heterogeneous nature of the mixture.

References pp. 82-84.

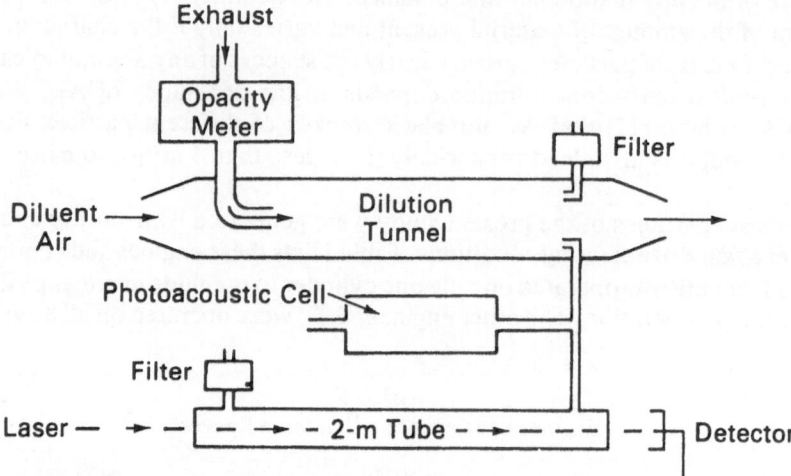

Fig. 1. Schematic of apparatus of optical measurements. Mass measurements were made at the filter positions.

The photoacoustic cell, which has been described elsewhere [12], measures the optical absorption at a wavelength, λ, of 10.6 μm, thus providing infrared data to complement the opacity results in the visible region. Infrared measurements of diesel particulate opacity are possible, but the low absorption makes them much more difficult than at visible wavelengths. The greater sensitivity of the photoacoustic device renders it more useful in the infrared, despite its greater instrumental complexity.

Fig. 2 shows the opacity of the particulate emissions from engine A as a function of mass concentration M. Mass data, determined gravimetrically from material collected on filters following dilution, have been corrected for dilution ratio (typically 10:1) and temperature to give the apparent mass concentration in the opacity meter. This procedure overestimates the true mass concentration to the extent that additional material condenses or adsorbs on the particles between the opacity meter and the filter.

As expected from equation (1) the opacity values shown in Fig. 2 increase with M. Despite considerable experimental scatter, the data suggest a dependence of the calibration on engine speed. The broken line is a fit to equation (1) with a value of $A_E = 9.6 \text{ m}^2 \cdot \text{g}^{-1}$.

Values of A_E derived for engine A vary between 3 and 14 $\text{m}^2 \cdot \text{g}^{-1}$ showing that significant differences exist between particles produced under different conditions. A_E was found to increase with mass concentration and with engine load but to decrease with overall air/fuel ratio.

Fig. 3 shows the variation of A_E with air/fuel ratio. The most significant feature is the increase in A_E as the overall mixture becomes richer, A_E doubles in value as the air/fuel ratio decreases from 25 to 15. The overall change in A_E is more than a factor of 3, representing substantial variations in the nature of the particles. It is evident

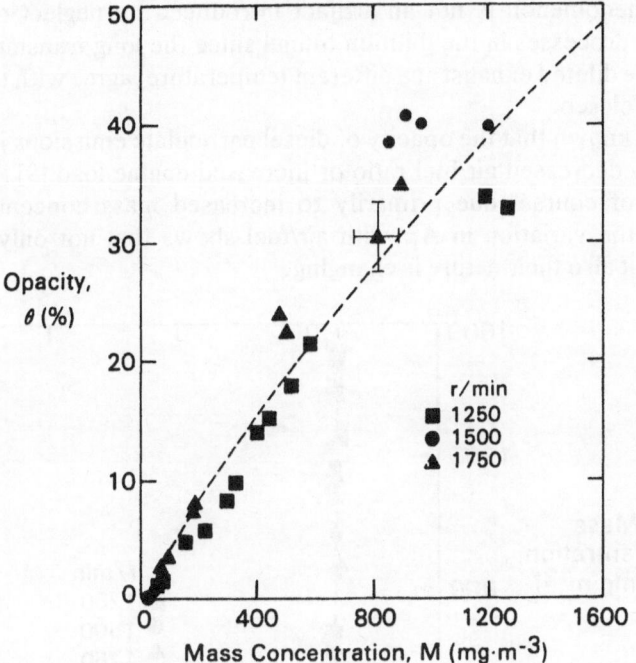

Fig. 2. Opacity versus mass concentration for engine A (2.1 L operated on 1 cylinder).

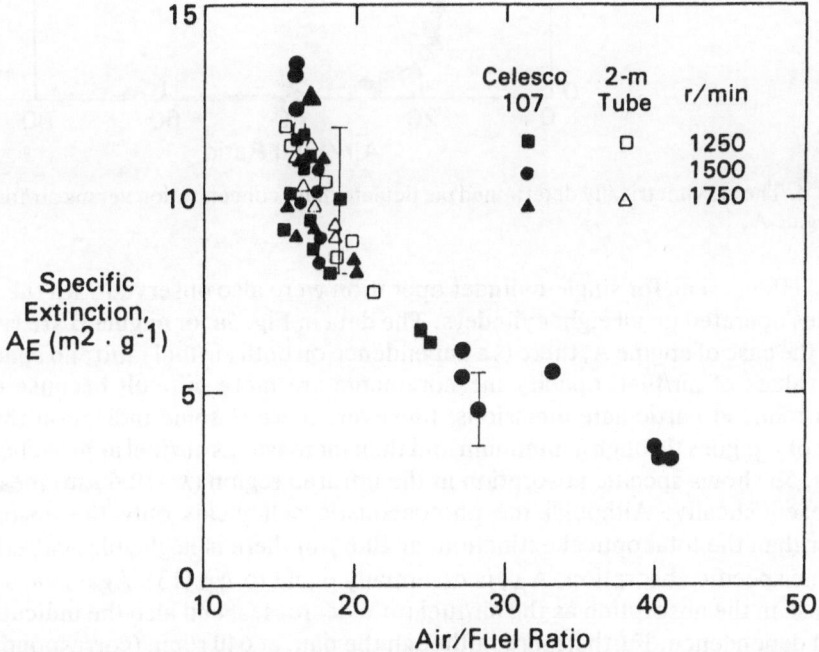

Fig. 3. Specific extinction versus overall air/fuel ratio for particulate emissions from engine A.

that the phenomenon is not an artifact introduced by neglect of condensation/ adsorption processes in the dilution tunnel since the long transmission tube data, taken on the diluted exhaust at a different temperature, agree with the data gathered using the Celesco.

It is well known that the opacity of diesel particulate emissions increases significantly with decreased air/fuel ratio or increased engine load [9]. This opacity increase is, of course, due primarily to increased mass concentration (Fig. 4). However, the variation in A_E with air/fuel shows that not only the number of particles but also their nature is changing.

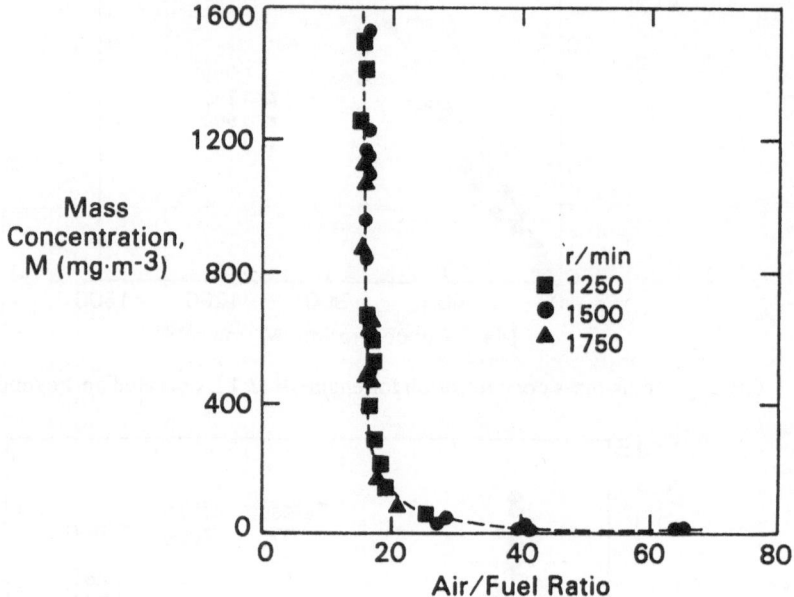

Fig. 4. The gravimetrically determined particulate mass concentration versus air/fuel ratio for engine A.

The effects seen for single-cylinder operation were also observed with the larger engines operated on all eight cylinders. The data in Fig. 5a for engine B are typical. As in the case of engine A, there is a dependence on both air/fuel ratio and speed. At high values of air/fuel, opacity measurements are more difficult because of the much reduced particulate emissions; however, there is some indication that the value of A_E goes through a minimum and then increases as air/fuel approaches 100.

Fig. 5b shows specific absorption in the infrared region ($\lambda = 10.6 \ \mu m$) measured photoacoustically. Although the photoacoustic cell yields only the absorption rather than the total optical extinction, at $10.6 \ \mu m$ there is negligible scattering so that the specific absorption, A_A, is essentially equal to A_E [13]. Again we see the increase in the absorption as the air/fuel ratio decreases, and also the indication of speed dependence. Furthermore, although the data at 640 r/min (corresponding to the engine at idle) show considerable scatter, the possibility that A_A may be increasing as air/fuel approaches 100 cannot be ruled out.

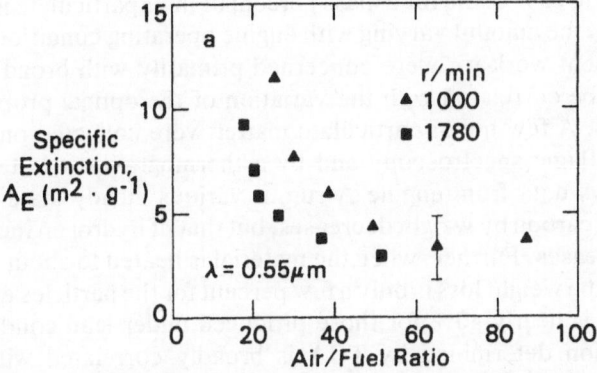

Fig. 5a. Specific extinction at 0.55 μm wavelength versus air/fuel ratio for engine B (5.7 L operated all 8 cylinders).

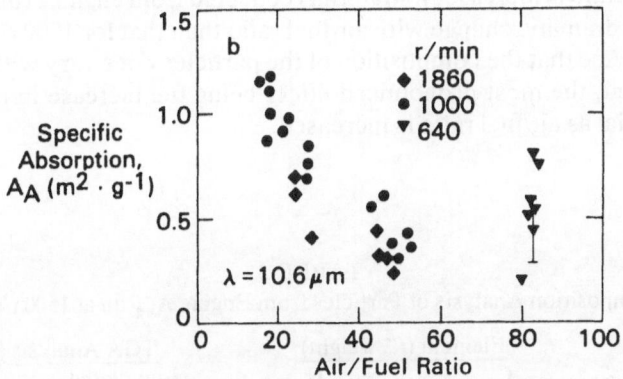

Fig. 5b. Specific absorption at 10.6 μm wavelength versus air/fuel ratio for engine C (5.7 L operated on all 8 cylinders). The triangles refer to the idle condition.

CHEMISTRY

The review by Lipkea et al. [2] points out some of the complexities of the chemistry of diesel particles. The primary constituents are elemental carbon and a variety of hydrocarbons, together with smaller amounts of sulfur, nitrogen and metallic compounds. Williams and Begeman [14] have recently described the results of several types of analyses, including gel permeation chromatography, elemental analysis, infrared absorption spectroscopy, NMR, thermogravimetry and gas chromatographic simulated distillation. Funkenbusch et al. [15] have reported a detailed analysis of the soluble organic fraction. The latter can vary from a few percent to nearly 80% of the total particulate mass; Mayer et al. [16], for example, found that up to 25% of the total mass percent of their particulate matter came from the engine oil, the amount varying with engine operating condition.

In the present work we were concerned primarily with broad chemical trends which might be correlated with the variation of the optical properties as air/fuel ratio changes. A few mg of particulate matter were collected on filter paper and analyzed by flame spectroscopy and by a thermogravimetric technique (TGA). Table 2 shows data from engine A run at various steady state conditions. The percentage of carbon by weight decreases, but that of hydrogen increases as the air/fuel ratio increases. Further, when the material is heated to about 550°C in an inert atmosphere, the weight loss is only a few percent for the particles emitted at low air/fuel ratio but is nearly 40% for those produced under lean conditions. The non-volatile fraction determined by TGA is broadly correlated with the inorganic fraction of the particulate matter while the volatile fraction is a crude measure of the organic fraction.

In Table 3 are given the results of TGA measurements of the non-volatile fraction for particles emitted by the 8-cylinder engines (B through E). The particulate chemistry shows the same trend as for engine A (i.e., an increase in the volatile fraction as the air/fuel ratio is increased). Materials collected from engines run at 2000 r/min shows a more dramatic change with air/fuel ratio than that for 1000 r/min.

Overall, we see that the composition of the particles does vary with both air/fuel ratio and speed, the most pronounced effect being the increase in the fraction of volatile material as air/fuel ratio is increased.

TABLE 2

Composition Analysis of Particles from Engine A, Run at 1500 r/min.

	Element (% weight)		TGA Analysis (% weight)	
Air/Fuel	C	H	Nonvolatile	Volatile
17.3	81.5	2.2	93.1	6.9
17.4	87.0	1.7	92.3	7.7
33	59.5	4.8	70.0	30.0
33	64.7	5.8	66.7	33.3
67	60.5	5.3	61.0	39.0
69	61.8	5.1	64.7	35.3

TABLE 3

Thermogravimetric Analysis for Engines B-E

	Non-volatile fraction (wt %)	
Air/Fuel	1000 r/min	2000 r/min
20	95 ± 4	95 ± 3
40	86 ± 8	73 ± 9
60	80 ± 13	60 ± 15

MORPHOLOGY

Particulate matter examined by electron microscopy showed significant variation in morphology as a function of air/fuel ratio. This variation was apparent even on very heavily loaded filters, but the differences can be more clearly noted by examining lightly loaded filters on which individual particles can be distinguished. Examples of particles collected on a Nuclepore® filter from engine D are shown in Fig. 6. These micrographs were taken using a JEOL 200C scanning transmission electron microscope operated in the secondary electron mode at an accelerating voltage of 200 kV. The instrumental resolution under these conditions is ~5 nm; however the useful resolution is less than this due to the fact that a ~10 nm thick Au/Pd coating was deposited to ensure that the sample was an electrical conductor (as required for the SEM technique).

The size of the particles can be estimated most conveniently by comparing them with the 200 nm diameter pores in the Nuclepore filter (which appear as dark regions). It is clear that the particles are (like carbon black) agglomerates of smaller, approximately spherical subunits (spherules). From a comparison of the morphology of particles formed under rich and lean conditions, two features are apparent:

1. On average, the number of spherules per particle is larger for rich mixtures than for lean mixtures.
2. Whereas particles formed during combustion of lean mixtures tend to be compact clusters, those formed from rich mixtures exhibit a much more elongated, chain-like appearance.

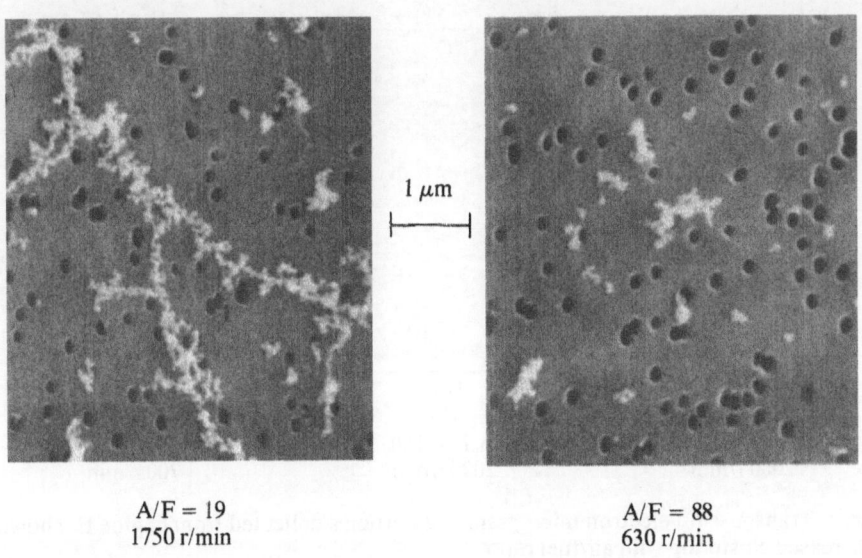

1 μm

A/F = 19   A/F = 88
1750 r/min   630 r/min

Fig. 6. Scanning electron micrographs of particles collected from engine D (5.7 L engine) at two operating conditions.

References pp. 82-84.

Similar features have also been observed for particles from engine A, but our most exhaustive study was conducted for engine E.* The engine was operated at 2000 r/min, with five air/fuel ratios ranging from 100 to 22, and also at idle (air/fuel ~85, 700 r/min). Particles were collected on Nuclepore filters to which were attached 3 mm copper grids with an amorphous carbon film for transmission electron microscopy (TEM). Particles collected on the copper grids were examined by TEM while regions of the Nuclepore filter immediately adjacent to the copper grids were examined by SEM. The resolution is higher for TEM than for SEM, both because of instrumental resolution and also because the TEM samples do not require the 10 nm Au/Pd coating. However, the improved TEM resolution revealed no significant differences in morphology from those observed by means of SEM: that is, the richer the mixture, the greater the number of spherules per particle and the greater the tendency to form a chain-like structure. This effect is very systematic: the number of spherules per particle increases progressively as the air/fuel ratio decreases. The extent of the chaining however is not as great as observed by SEM for engines A and D. This is illustrated in Figs. 7a and 7b which show particles collected at the extremes of air/fuel ratio examined.

It was noted earlier that some effect of engine speed on the optical properties is evident. This observation is supported by comparing particles collected at 2000 r/min (Figs. 7a and 7b) and those collected at 700 r/min (Fig. 7c). It may be noted that despite the high air/fuel ratio at 700 r/min (idle) the particles collected bear more resemblance to the low air/fuel particles at 2000 r/min.

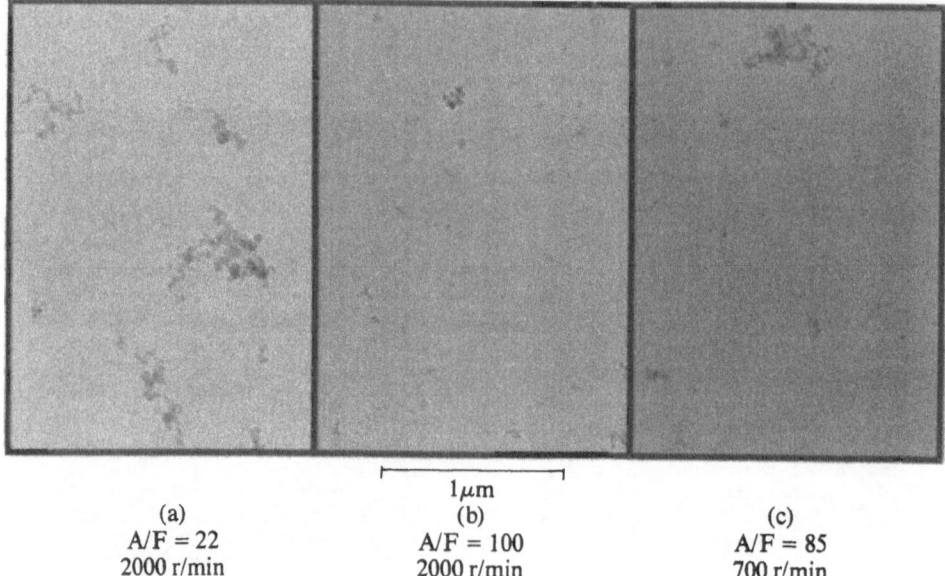

	1μm	
(a)	(b)	(c)
A/F = 22	A/F = 100	A/F = 85
2000 r/min	2000 r/min	700 r/min

Fig. 7. Transmission electron micrographs of particles collected from engine E, showing the increased clustering with air/fuel ratio.

*It should be recalled that engine E (operated on 8 cylinders) is subject to the average effects not expected for engine A (single cylinder).

BASIS OF ANALYSIS OF OPTICAL PROPERTIES

Since, as seen above, both chemistry and morphology of diesel particles change with engine operating conditions (primarily with air/fuel ratio), we must now determine how they are related to the observed changes in the specific optical extinction. The structural and chemical complexity of diesel particles precludes exact simulation of their optical properties by any simple model, but some elementary considerations furnish guidelines for further discussion.

Theoretical Considerations — The specific extinction, A_E, was introduced in equation (1) as the extinction coefficient per unit mass concentration. Alternatively we may express it in terms of the extinction cross section, C_E, per unit mass of material, i.e., [17]

$$A_E = \frac{C_E}{\rho V} \tag{2}$$

where ρ and V refer to the specific density and volume of a particle. The extinction efficiency, Q_E, is the ratio of the extinction cross section to the geometrical cross section, C_E/G. Furthermore, the average geometrical cross section of a convex particle with random orientation is $S/4$ where S is its surface area [17]. We thus have the relationship.

$$A_E = \frac{Q_E}{4\rho}\left(\frac{S}{V}\right) \tag{3}$$

describing the optical properties of particles in terms of Q_E and their density and shape.

For a sphere, A_E is simply $3Q_E/2\rho D$, where D is the diameter. For the special case of monodisperse isolated spheres, Q_E can be calculated from the Mie theory [18] for all values of refractive index m, diameter D, and wavelength λ. Fig. 8 shows the variation of Q_E with the size parameter D/λ for two materials. The broken curve is that for a non-absorbing material (such as a liquid hydrocarbon in the visible region) having a real refractive index, $m=1.5$. The extinction efficiency, Q_E, increases very rapidly with D/λ, exhibiting the well-known oscillatory structure (due to interference effects in monodisperse systems) before asymptotically approaching the value of 2 for large D/λ [17]. The solid curve is a plot of Q_E for a highly absorbing material (such as carbon) having a complex refractive index $m=2-1i$. The presence of absorption smooths out the oscillatory structure and increases the value of Q_E at low D/λ.

As far as optical properties are concerned, there are clearly three basic size regions: the Rayleigh region ($D \ll \lambda$), the Mie or resonance region ($D \sim \lambda$), and the geometric region ($D \gg \lambda$). It is apparent from Fig. 8 that chemistry is important for small particles where index differences appreciably affect Q_E. For particle dimensions close to the wavelength of light, the dependence on the refractive index is important only for materials which are both transparent and have well-defined

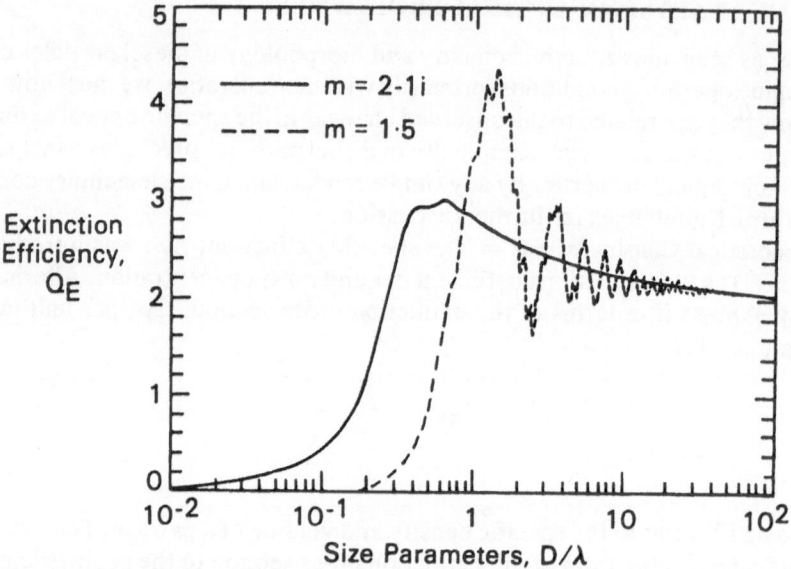

Fig. 8. Optical extinction efficiency of spheres calculated from the Mie theory as a function of the diameter/wavelength ratio. The complex index m=2-1i is typical of a material such as carbon in the visible region; the refractive index m=1.5 refers to a transparent material such as liquid hydrocarbons in the visible.

geometries (permitting the interference structure to appear). As either absorption increases, or as the shape becomes irregular, the oscillatory structure becomes smoothed out. For particles much larger than the wavelength, Q_E is independent of material composition (Fig. 8) and shape [17], approaching 2 as $D/\lambda \rightarrow \infty$.

Prediction of General Attributes of Specific Extinction — We are now in a position to predict some general attributes of the specific extinction coefficient for diesel particles:

1. Chemical composition can significantly affect A_E for particles with dimensions smaller than the wavelength. (This will therefore be of greater potential significance for infrared measurements than for those at visible wavelengths).
2. Particle shape is important throughout all size regimes because of the direct dependence of A_E on the surface to volume ratio (equation 3).
3. In the Rayleigh regime, the shape dependence of A_E is further enhanced since Q_E is also dependent on shape (see below).

Let us now turn to a more detailed consideration of effects of chemical composition and morphology on A_E.

EFFECTS OF CHEMICAL COMPOSITION ON SPECIFIC EXTINCTION

We begin our discussion by considering the simplest theoretical model — spherical particles for which the influence of chemical composition on optical properties manifests itself through the index of refraction. The most straightforward model compatible with the compositional data of Tables 2 and 3 is one which describes a particle in terms of two constituents, the primary one being elemental carbon (for which the refractive index should be similar to that of bulk carbon or graphite); the secondary being volatile material (presumably organic). Although a wide range of values has been reported [11, 19, 20], we take the refractive index of carbon in the visible region to be m=2-1i (as in Fig. 8). Furthermore we assume that the index of the organic material can be approximated by that for a transparent hydrocarbon: m=1.5 (also as in Fig. 8). We shall analyze the two-constituent spherical particles in two ways: by treating the components as uniformly mixed or as segregated, with the organic material coating a carbon core.

Mixed Constituent Model — Various methods of calculating an average refractive index for particles composed of two mixed constituents have been discussed by others [21-23], with effective-medium theories giving rise to differing estimates of the effective index m_E. However, for a carbon/hydrocarbon mixture (volume fraction f of hydrocarbon, hydrocarbon refractive index m_{HC}, carbon refractive index m_C), the simple linear expression

$$m_E = m_C(1-f) + m_{HC}f \tag{4}$$

gives values of m_E lying within the range calculated from the effective-medium theories. Although equation (4) is without strong theoretical justification, we have used it in the absence of satisfactory alternatives. Values of m_E can then be used in a straightforward application of Mie theory.

Coated Sphere Model — It is believed [24-26] that the more volatile component is condensed or adsorbed as a surface layer on the carbon core of the particle. Therefore, an optical model in which an absorbing carbon sphere is coated with a transparent hydrocarbon layer may be more realistic than one in which the constituents are uniformly mixed. The Mie theory has been adapted for calculations of the optical properties of coated spheres [27].

Results of Calculations — Fig. 9 shows the changes in the specific optical extinction of a sphere of diameter D as its composition is changed from pure carbon to half carbon/half hydrocarbon by weight. Tables 2 and 3 suggest that this represents a change as drastic as any encountered in the present work. The solid curve is that for monodisperse spheres of pure carbon. The dashed and dotted curves are for the two models for spherical particles containing 50% carbon, 50% hydrocarbon.

The three size regimes identified in discussing Q_E in Fig. 8 are also applicable to A_E. In the Rayleigh regime (D ≪ λ) A_E is independent of sphere size for absorbing material. The specific extinction exhibits a peak (and possibly further structure for less absorbing substances) at larger sizes where D is comparable to the optical wavelength. Finally, in the geometric regime (D ≪ λ), A_E varies inversely with D. Fig. 9 presents results of calculations only for monodisperse spheres. For polydisperse systems the main effect is a "smoothing" of the structural features [11, 19].

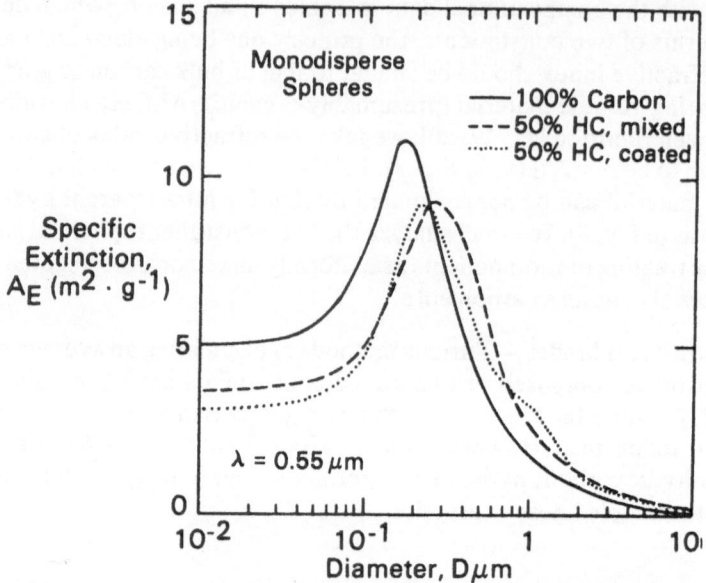

Fig. 9. The specific extinction of spheres as a function of their diameter D calculated at a wavelength of $\lambda=0.55\ \mu$m from the Mie theory. The solid curve represents carbon ($m=2$-li, $\rho=2$ g $\cdot$ cm^{-3}); the dashed curve represents homogeneous spheres containing half carbon, half hydrocarbon by weight ($m = 1.67$-0.33i, $\rho = 1.33$ g $\cdot$ cm^{-3}); the stippled curve represents a carbon core ($m = 2$-li, $\rho = 2$ g $\cdot$ cm^{-3}) coated with a hydrocarbon ($m = 1.5$, $\rho = 1$ g $\cdot$ cm^{-3}, the overall composition being half of each by weight.

It is apparent from Fig. 9 that the choice of the two-component model (i.e., mixed or coated) is not of major concern, and that the effect of changing composition is most pronounced for the smallest spheres. For unclustered spheres of less than 0.1 μm diameter, the calculated specific extinction drops about 30-40% for a compositional change from 100% carbon to 50% hydrocarbon. For larger particles the decrease is less and in fact, for particle dimensions larger than the wavelength of light, the specific extinction actually increases. This result is associated with the shift of the diameter for maximum light scattering to larger values as the material becomes more transparent [28]. The curve for the coated sphere exhibits a shoulder near 1 μm associated with the transparent coat, although the dominant feature is still the resonance for diameters near 0.2 μm.

In Fig. 10 is plotted the dependence of A_E on hydrocarbon content for three models, in the limit of small spheres. In all three the initial pure carbon sphere is taken to have a diameter of 0.03 μm, a value typical of the spherules which are the building blocks of diesel particles. Hydrocarbon has been added in three ways:

1. by uniformly mixing with the carbon while allowing diameter D to increase;
2. by uniform mixing while holding D constant;
3. by forming a coating of increasing thickness.

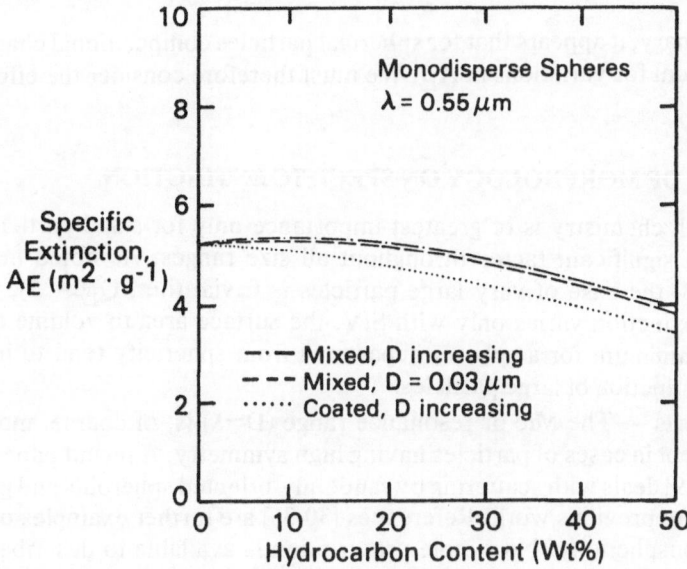

Fig. 10. The specific extinction of carbon hydrocarbon spheres at $\lambda = 0.55\ \mu m$ as a function of composition.

For the mixed systems there is remarkably little change with composition, at least for hydrocarbon contents up to 30% or so by weight. For the coated model there is a more pronounced and monotonic change. Extending the calculation to polydisperse size distributions or to particles of larger initial diameter results in even "flatter" curves. For polydisperse spheres having size distributions typical of those measured in practice for diesel particles, the Mie calculations show that increasing the weight percentage hydrocarbons from zero to 50%, decreases A_E by less than 20% for particles of mean diameter 0.03 μm and by less than 10% for particles of mean diameter 0.1 μm.

It will be recalled that in Figs. 3 and 5 A_E exhibited decreases by 50% or more in the visible region with increasing air/fuel ratio. The simple theoretical calculations based on spheres suggest that, even in the extreme case of small, unclustered monodisperse spheres, the observed changes in chemical composition with air/fuel ratio can account for only a portion of the observed change in A_E.

The situation becomes far more complicated for non-spherical particles. As noted previously, it is not clear that an optical model based on spheres is appropriate for the clusters and chains of Figs. 6 and 7, even though the particle subunits are essentially spherical. Unfortunately, extension of Mie theory to non-spherical shapes is non-trivial. For the special case where $D \ll \lambda$, the Rayleigh approximation may be used. As will be seen in the next section, for such particles the effects of composition are more dramatic than for spheres. In the extreme case where chain-like clusters are represented as Rayleigh cylinders, a 50% decrease in carbon content will reduce A_E by half.

References pp. 82-84.

In summary, it appears that for spherical particles compositional changes cannot fully account for variations in A_E. We must therefore consider the effects of morphology.

EFFECTS OF MORPHOLOGY ON SPECIFIC EXTINCTION

Although chemistry is of greatest importance only for small particles (Fig. 8), shape is a significant factor throughout all size ranges. As noted in discussing equation 3, the case of very large particles is trivial (i.e., $Q_E \approx 2$), so that the specific extinction varies only with S/V, the surface area to volume ratio. Since S/V is a minimum for a sphere, departures from sphericity tend to increase the specific extinction of large particles.

Size Effects — The Mie or resonance range ($D \sim \lambda$) is, of course, more complicated, except in cases of particles having high symmetry. A recent paper by Asano and Sato [29] deals with scattering by randomly-oriented spheroids and gives useful references to previous work. References [30-36] are further examples of the treatment of nonsphericity. No simple prescription is available to describe the shape effects in this size regime: they may be negligible, moderate, or severe, depending on the details of shape and material refractive indices.

For $D \ll \lambda$, it is possible to discuss shape effects more systematically, at least in cases of high symmetry. For spheres of diameter $D < 0.05$ μm, the size of diesel particle subunits (spherules), the Rayleigh region extends to wavelengths as short as the visible. Thus Fig. 9 tells us that small variations in the spherule diameter should have little effect on A_E. However, if clustering becomes significant, we might expect that A_E would increase (see, for example the solid curve of Fig. 9). A sphere of 0.2 μm diameter would have $A_E \approx 10$ m²g⁻¹, twice that calculated for a spherule. Examination of Figs. 9 and 3 may, at first glance, lead to the view that size variations with air/fuel ratio may be responsible for much of the observed changed in A_E.

There are, however, two further considerations. First, the variation in A_E with diameter is much reduced for polydisperse spheres [11, 19]. For typical diesel particle size distributions, the theoretical value of A_E would increase by only about 50% rather than doubling with increased diameter. Second, the magnitude of the variation of A_E with diameter depends sensitively on wavelength, as seen in Fig. 11. In the infrared ($\lambda = 10.6$ μm) the wavelength is much larger than the dimensions typical of diesel particles so that variations in particle size will not affect the value of A_E. However, we see from Figs. 5a and 5b that the experimental quantities vary in both the visible and infrared regions. Thus, although the data of Fig. 5a could, in principle, be attributed to size effects, a different explanation is required for those of Fig. 5b. Since clustering of particles alters their *shape* as well as size, we shall now consider the influence of shape effects on optical properties of particles.

Shape Effects — Size variations do indeed occur in diesel particulate emissions (e.g., the median diameter increases with engine load [37, 38] presumably due to agglomeration), but the arguments of the preceding paragraph suggest that shape effects must be considered as well. Jones [39, 40] has examined the effects of

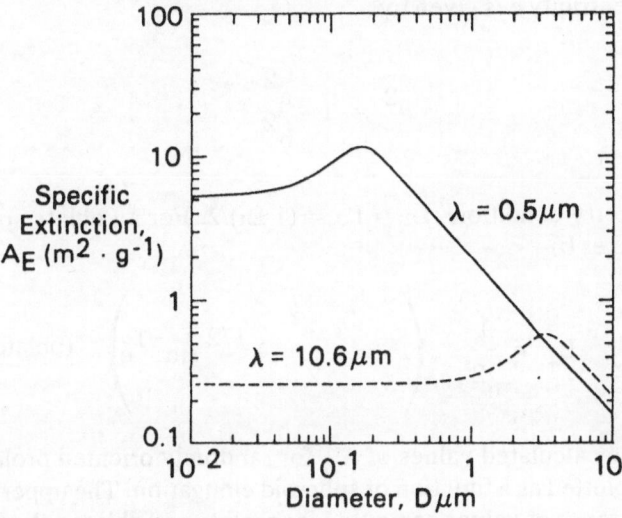

Fig. 11. The specific extinction of monodisperse carbon spheres at $\lambda = 0.5$ μm and $\lambda = 10.6$ μm, for $\rho = 2$ g $\cdot$ cm^{-3} and a constant refractive index m = 2-li.

chaining spheres together in the Rayleigh approximation, pointing out that to some extent such a chain may be approximately represented as an elongated spheroid. Both Jones [41] and Ravey [42, 43] have discussed optical properties of carbon aerosols in terms of spheroidal shapes. While some differences do exist in their treatments, they conclude that elongation tends to enhance the extinction.

Because of the complexity of shape effects, we, too, shall restrict our considerations to the Rayleigh regime and adopt a spheroidal model for clustering. It can be shown that, for randomly oriented Rayleigh ellipsoids [17],

$$A_A = \frac{2\pi}{3\rho\lambda} \text{ Re} \left(i \sum_{j=1}^{3} \frac{1}{L_j + \dfrac{1}{m^2 - 1}} \right) \tag{5}$$

where L_j is a shape factor for each of the three ellipsoid axes (j=1 to 3), such that $\sum_{j=1}^{3} L_j = 1$. For a prolate spheroid having minor axes a, major axis b, and elongation b/a,

$$L_1 = \frac{1-e^2}{e^2} \left(\frac{1}{2e} \ln \left(\frac{1+e}{1-e} \right) - 1 \right) \quad \text{(prolate spheroid)} \tag{6}$$

References pp. 82-84.

where the eccentricity e is given by

$$e^2 = 1 - \frac{a^2}{b^2}. \qquad (7)$$

From symmetry conditions $L_2 = L_3 = (1-L_1)/2$. For an oblate spheroid (minor axis a, major axes b)

$$L_1 = \frac{1}{e^2} \left(1 - (\frac{1-e^2}{e^2})^{1/2} \sin^{-1} e \right) \quad \text{(oblate spheroid)} \quad (8)$$

In Fig. 12 the calculated values of A_A for randomly oriented prolate and oblate spheroids are plotted as a function of spheroid elongation. The upper hatched band represents the range of values computed for prolate and oblate spheroidal approximations to carbon clusters.

As we have noted elsewhere [19] simplifications of this sort cannot take account of the "fluffiness" or porosity of the clusters. The presence of voids will reduce both the effective refractive index (decreasing A_A) and the density (increasing A_A). It thus appears that the effect of voids on A_A is, to some extent, mitigated. However, more quantitative understanding is needed (see for example [20] and [44]).

The upper band of Fig. 12 shows that the effects of elongation are most dramatic as b/a increases to about 10 or so and, thereafter, A_A increases more slowly. For the prolate spheroid, A_A almost doubles as b/a increases from unity to 10, but subsequently increases very little. For the limiting case of an infinite cylinder, A_A is 9.52 m²·g⁻¹, for the refractive index and specific density shown. To the extent that spheroids simulate the clustering of diesel particles we see that shape changes can in theory account for much, if not all, of the observed dependence of specific extinction on air/fuel ratio.

The lower band in Fig. 12 plots the calculated curves for prolate and oblate spheroids of a material having index and density approximating those for particles containing 50 w/o carbon, 50 w/o hydrocarbon. In this case the effects of shape are considerably less pronounced. However, from a comparison of the upper and lower bands it is evident that the effects of chemical composition (via the refractive index) are much stronger as particles become less spherical. Our diesel particle composition and morphology results suggest that the specific extinction lies to the left (low b/a) near the lower hatched band of Fig. 12 for particles produced at high air/fuel ratios, but moves to the right (high b/a) toward the upper hatched band as the mixture becomes richer.

Our model, though approximate, does indicate the probable significance of shape. The Rayleigh approximation for diesel particles is more nearly correct in the infrared than in the visible, especially in view of clustering effects on mean particle size. Fig. 13 presents the values of A_A calculated from the Rayleigh approximation for two proposed values of the infrared refractive index of carbon, m=3-2i and

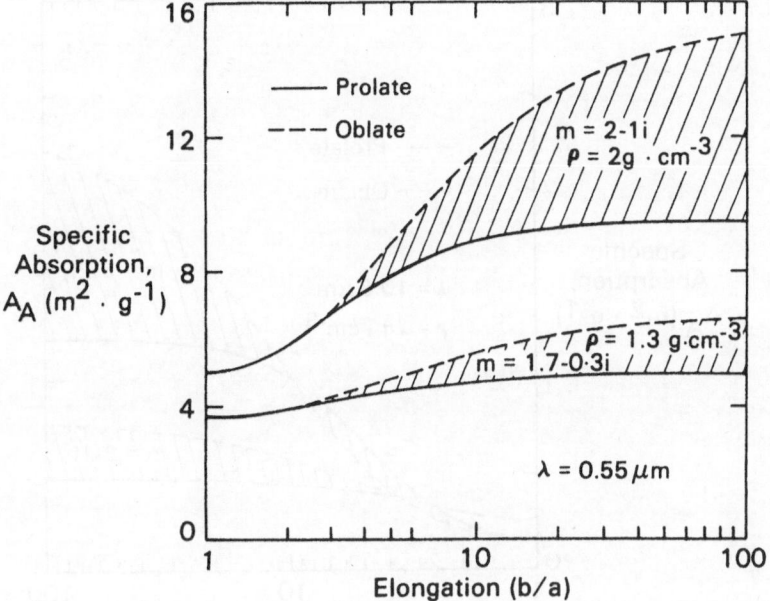

Fig. 12. The specific absorption at $\lambda = 0.55$ μm for Rayleigh spheroids. The upper hatched band refers to pure carbon, the lower to a mixture of 50% carbon, 50% hydrocarbon by weight.

$m = 2$-$1i$ [19]. The infrared index is even less well established than that for the visible region, and the large separation of the hatched bands at large elongations shows that the refractive index is a crucial parameter. Nonetheless, as in Fig. 12, we see that for both prolate (full curves) and oblate (broken curves) spheroids, the absorption is enhanced as the shape departs further from sphericity. It thus appears that particle shape and chemistry (as reflected in refractive index) both can contribute to changes in optical properties when particles become non-spherical.

It should be mentioned that, in spite of the approximation in our approach to the question of shape effects, the calculated magnitudes of A_A (Figs. 12 and 13) are fully consistent with the experimentally determined value (Figs. 3 and 5).*

A SIMPLE NUCLEATION/DEPLETION MODEL FOR THE SPHERULE SIZE

One of the interesting aspects of diesel particles, as well as other types of particulate carbon generated during combustion, is the near universality of size of the spherules which agglomerate to form the characteristic soot clusters (Figs. 6 and 7).

---

*It is true that we have calculated A_A rather than A_E. However, it should be noted that absorption is by far the most dominant mechanism for light attenuation in carbonaceous materials. Scattering can be neglected in the infrared regions and contributes only about 15% of the attenuation in the visible [13, 45].

References pp. 82-84.

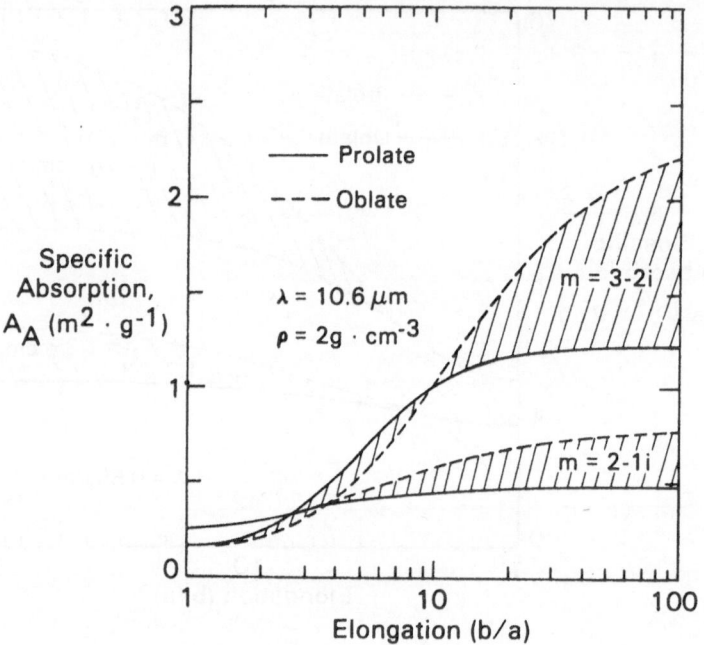

Fig. 13. The specific absorption at $\lambda = 10.6$ μm for Rayleigh spheroids. The two hatched bands refer to two different values for the complex refractive index of carbon.

That is, the spherule size is relatively insensitive (radii ~10 to 100 nm) to combustion pressure over a wide range of values (10^3 to 10^7 Pa) [46-48] and, in the case of diesel particles, to engine operating conditions (Fig. 7). In this section we describe briefly a simple nucleation/depletion model which enables us to calculate the size of the spherules. A more complete discussion will be given elsewhere [48].

As a starting point for the model we take the homogeneous nucleation theory which Lahaye et al., [49] applied to the formation of carbon black by thermal decomposition of benzene. In addition, we incorporate depletion as the spherule size-limiting mechanism for intermittent combustion such as that which occurs in diesel engines. From this simple model we obtain an estimate of the spherule radius. No attempt is made to address the question of agglomeration. It may be possible to extend the simple model to estimate spherule sizes for a variety of other intermittent processes: shock tube reactions, rapid evaporation/condensation, and gas-phase chemical reactions [50].

We assume, following Lahaye et al., that the combustion processes generate a highly supersaturated concentration of condensible soot precursor molecules (see Fig. 14). Subsequently formation of condensed phase nuclei capable of growth takes place. This nucleation eliminates supersaturation so that formation of additional nuclei becomes impossible [49]. The nuclei then grow by condensation to form the fundamental spherules (which may simultaneously and/or subsequently

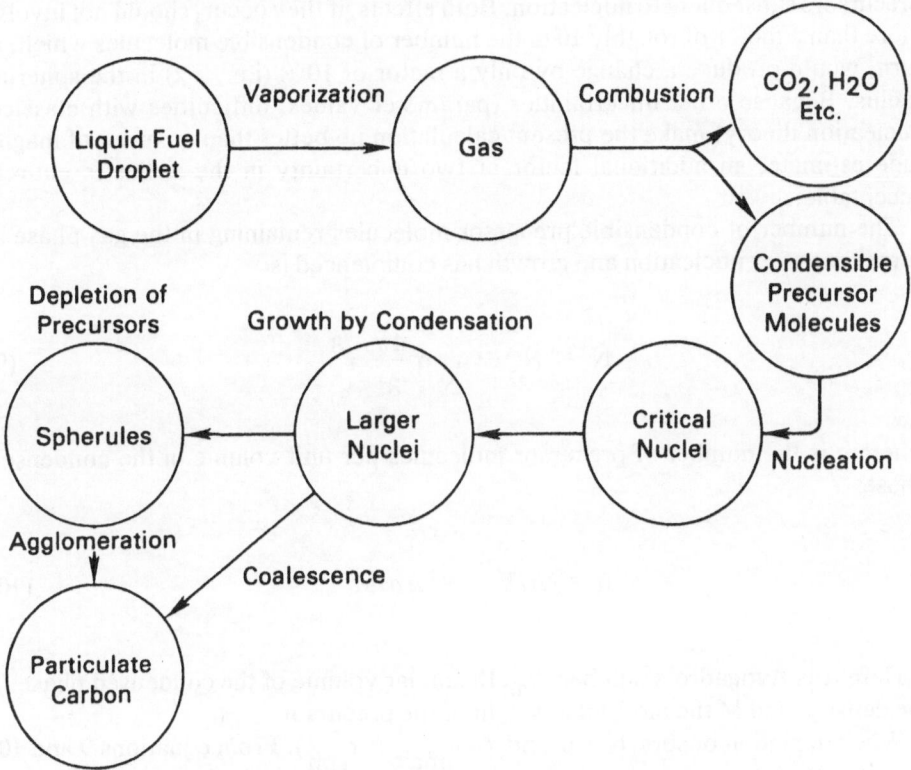

Fig. 14. Conceptual diagram for nucleation/depletion model.

agglomerate). We assume that during diesel combustion, generation of precursors occurs rapidly and then ceases (in times on the order of milliseconds); the maximum size attained by the spherules would then be limited by depletion of the gas-phase condensible precursors during condensation. Further, it is not unreasonable to suppose that few precursor molecules are generated subsequent to the nucleation and condensation processes.

Derivation of Expression for Spherule Radius — On the basis of these few assumptions it is a relatively simple matter to derive an expression for r_{sph}, the radius of a spherule. Let N_0 be the number per unit volume of condensible precursor molecules generated initially by the combustion reactions. Subsequently n_0 nuclei per unit volume are formed which are large enough to grow to produce the spherules. As growth proceeds, r, the nucleus radius, increases while N, the number of condensible precursor molecules per unit volume remaining in the gas phase, decreases. Ultimately N approaches zero (depletion) and growth ceases, yielding a maximum spherule size.

This simple picture ignores coalescence and agglomeration which may proceed simultaneously with growth. It also neglects formation of additional condensible precursors subsequent to nucleation. Both effects, if they occur, should not involve more than a factor of roughly 10 in the number of condensible molecules which, in turn, would produce a change by only a factor of $10^{1/3}$ (i.e., ~2) in the spherule radius. Because other uncertainties (parameter values, difficulties with classical nucleation theory) make the present calculation no better than an order of magnitude estimate, an additional factor of two uncertainty in the radius is entirely acceptable.

The number of condensible precursor molecules remaining in the gas phase at some time after nucleation and growth has commenced is

$$N = N_o - n_o q \frac{4\pi}{3} r^3 , \tag{9}$$

where q is the number of precursor molecules per unit volume in the condensed phase:

$$q = A/V_m = A\rho/M . \tag{10}$$

Here A is Avogadro's number, V_m the molar volume of the condensed phase, ρ the density, and M the molecular weight of the precursor.

When depletion occurs, $N \to 0$ and $r \to r_{max}$ ($=r_{sph}$). From equations 9 and 10, r_{sph} is given by

$$r_{sph} = \left(\frac{N_o}{n_o} \frac{M}{A\rho} \frac{3}{4\pi} \right)^{1/3} . \tag{11}$$

We note that the fundamental parameter in this equation is N_o/n_o, the ratio of the initial number of precursor molecules to the number of nuclei which grow (each per unit volume). (The ratio, of course may change from one local volume element to another, resulting in varying spherule sizes.) Under some circumstances nucleation might be heterogeneous, but we shall assume homogeneous nucleation to govern here [49, 51]. We shall therefore take n_o to be equal to n_c, the number of critical nuclei per unit volume which, according to classical nucleation theory [52-55], is approximately given by

$$n_c = N_o \exp(-\Delta G^*/kT) = N_o \exp(-4\pi r_c^2 \sigma/3kT) \tag{12}$$

where ΔG^* is the Gibbs free energy of formation of a critical nucleus, k the Boltzmann constant, T the absolute temperature, r_c the radius of a critical nucleus, and σ the surface energy (surface tension).

Substitution of equation 12 into equation 11 yields

$$r_{sph} = [\frac{3M}{4\pi A\rho} \exp (4\pi r_c^2 \sigma/3 \, kT)]^{1/3} \tag{13}$$

We now have a simple approximate expression for the spherule radius in terms of the precursor molecular weight M, the condensed phase density ρ, the radius of the critical nucleus r_c, the surface energy σ, and the formation temperature T. We particularly note that equation 13 is independent of N_0, as it should be for spherule size to be independent of initial precursor concentration and partial pressure (i.e., spherule size universality). This independence of N_0 is a consequence of our assumption of homogeneous nucleation in our depletion model, but the same result would hold for any depletion-limited system in which the number of nuclei is proportional to the number of precursor molecules (and presumably to the number of fuel molecules as well).

Uncertainties in the Calculation — A number of factors can influence the calculation of spherule size from equation 13. Among these are difficulties with classical homogeneous nucleation theory itself [53, 54] and uncertainties in the parameters needed for the computation. Uncertainties in the parameters of equation 13 constitute a serious problem. M, of course, must be estimated, as must ρ. Data for σ are available for few substances, and then only at temperatures considerably below those for diesel combustion. Estimation of the proper temperature for intermittent combustion processes is difficult; determination of r_c is also fraught with uncertainties. However, it is possible to obtain fairly good estimates of r_c from the Gibbs-Thomson equation [56], the Temperley method [57], or rules of thumb for the number of molecules in a critical nucleus [51, 53-55, 58] (for a fuller discussion see [48]).

Considerations involved in a choice of values for the surface energy σ are particularly troublesome. As mentioned above, very few data for organic precursors are available, none of it at diesel combustion temperatures ($\sim$1600 K). It is possible to correct for temperature by several empirical relations [59, 60] which predict reductions of up to 30-40% at temperatures approaching the boiling point. In addition Kirkwood and Buff [61] have shown that the surface energy of liquid droplets should decrease with their curvature according to the relation

$$\sigma_r = \sigma_\infty/(1 + 2\delta/r) \tag{14}$$

where σ_r is the value for a droplet of radius r, σ_∞ that for bulk liquid, and δ a distance on the order of a molecular diameter. For droplets the size of a critical nucleus, equation 14 predicts a further reduction in σ on the order of 20% or more. Numerous authors [53-55, 58, 62, 63] have pointed out the probable impropriety of using a macroscopic surface energy in calculations related to critical nuclei which contain only a few molecules. For the purposes of our estimates, we nevertheless correct for both temperature and curvature using the above techniques as the best available means for such compensation.

In Table 4 are summarized the parameters needed for calculation of r_{sph}, assuming two different types of precursor systems: carbon vapor condensing to graphitic spheres and an organic vapor — one heavier than, say, coronene (M=300) — condensing to liquid droplets. In the first case, carbon condensation to graphite, the surface energy value chosen is the mean of the range of values tabulated by Tabak *et al.* [64] for graphite basal surfaces. This choice is appropriate since high resolution phase contrast electron microscopy studies indicate that carbon blacks [65] and diesel particles [66] form primarily with their "graphitic" basal planes lying parallel to the spherule surfaces. The surface energy used for the organic precursor has been corrected for temperature and curvature as discussed above.

TABLE 4

Parameters for Calculation of r_{sph} from Equation 13

Precursor	$M \times 10^3$ (kg)	$\rho \times 10^3$ (kg/m³)	r_c (nm)	σ(N/m) Literature	σ(N/m) Corrected for Calc.
carbon/ graphite	12	~2 [a]	0.47	0.1 to 0.38 [b]	0.25
organic/ liquid	>300	~1.4 [c]	1.62	≤0.04 [d]	0.02 [e]

a "*Handbook of Chemistry and Physics*" (ref. 59), p. B-204.

b *Tabak et al.* (ref. 64).

c "*Handbook of Chemistry and Physics*" (ref. 59), p. C-65ff.

d *Assumed value, based on surface energies for organic liquids which are typically in range 0.02 to 0.04 N/m. [See M. P. Doss, "Physical Constants of the Principal Hydrocarbons," The Texas Co., N.Y., 1943.]*

e *Corrected for temperature and curvature.*

One assumption implicit in our estimates of spherule size is that r_{sph} does not change appreciably with chemical reactions which may continue in the condensed phase (e.g. dehydrogenation or oxidation). Although oxidation might be expected to have a considerable effect on size, it has been shown to occur for the most part internally, leaving a dense outer shell [67] and thus preserving the spherule size.

Result of Calculation — The calculated dependence of r_{sph} on formation temperature T is shown in Fig. 15 for the two precursors chosen. It is clear that the estimates do not differ by more than a factor of three over the entire temperature range, indicating the expected insensitivity to the nature of the precursor. The range of values calculated for r_{sph} (5 to 12 nm) for T=1600 K is in good agreement with the size found experimentally (~12 nm) (cf. Fig.7). Although we know of no published data for the temperature dependence of r_{sph} for diesel particles, it has been pointed out [68] that the particle radius of carbon black decreases with increasing temperature.

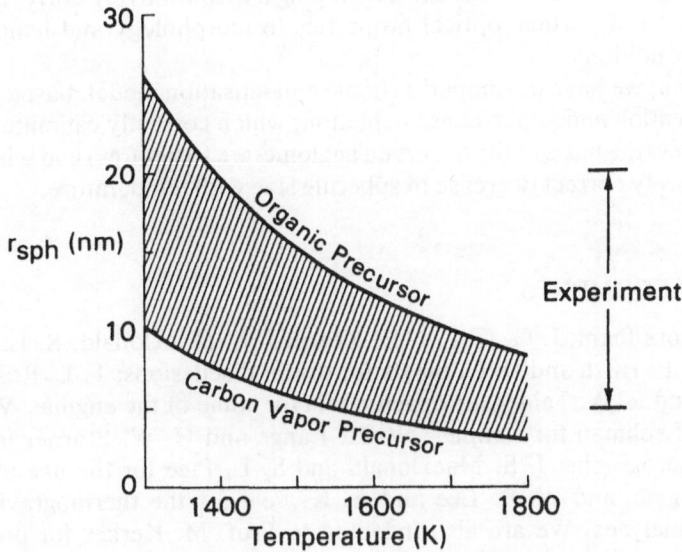

Fig. 15. Calculated spherule radius, r_{sph}, as a function of formation temperature for diesel particulate carbon, assuming two different precursors: a) carbon vapor condensation to graphite: b) organic vapor condensing to liquid nuclei. Calculation parameters from Table 1.

As a final cautionary note concerning the present calculation, we must point out the sensitivity of equation 13 to the parameters σ and r_c. Doubling σ can produce almost an order of magnitude change in r_{sph}; doubling r_c, almost two orders of magnitude change. However r_{sph} is relatively insensitive to M, because of the cube root dependence. We must also stress that the parameters for the calculation were chosen and corrected as judiciously as possible, so that we believe that our claim for order-of-magnitude validity is justified.

CONCLUDING REMARKS

We have seen that the systematic variations in the optical properties, chemical composition, and physical morphology with engine operating conditions provide useful information with which to develop greater understanding of diesel particulate carbon formation.

It is clear that the variations in optical properties with engine air/fuel ratio cannot be explained in a simple manner. It is necessary to take into account not only size and shape changes (as agglomeration occurs) but variations in particle chemistry, as well. Of these influences, however, that of shape appears to be dominant.

The difficulties in treating the optical properties are enhanced by the complexity of the Mie theory in the size range encountered in diesel particulate emissions. To render this problem tractable we have had to take advantage of simplifying assumptions, modeling the agglomerated particles as Rayleigh spheroids, in order to separate out and properly assess the effects of chemistry and shape. Nevertheless, these simplifications did not prevent our developing a quantitatively correct description of the relation of particle optical properties to morphology and hence to engine operating conditions.

In addition, we have developed a simple condensation model, based on homogeneous nucleation and vapor-phase depletion, which correctly estimates the size of the spherules (the basis of the observed agglomerate structures) and which predicts a qualitatitively correct decrease in spherule size with temperature.

ACKNOWLEDGMENTS

The authors thank J. G. Gay, J. E. Keem, J. S. MacDonald, S. L. Plee, J. B. Howard, J. P. Hirth and D. C. Siegla for useful discussions; T. L. Rosebrock, G. Shevzov, and R. W. Talder for opacity data for some of the engines; W. J. Mayer and D. C. Lechman for samples; W. H. Lange and H. W. Sturner for scanning electron micrographs; J. S. MacDonald and S. L. Plee for the use of the single cylinder engine; and W. R. Lee and A. K. Peat for the thermogravimetric and elemental analyses. We are also indebted to Prof. M. Kerker for providing the results of the coated spheroid calculations.

REFERENCES

1. C. T. Vuk, M. A. Jones and J. H. Johnson, SAE Paper 760131, Society of Automotive Engineers, Warrendale PA, February 1976.
2. W. H. Lipkea, J. H. Johnson and C. T. Vuk, SAE Paper 780108, Society of Automotive Engineers, Warrendale PA, February 1978.
3. K. Carpenter and J. H. Johnson, SAE Paper 790815, Society of Automotive Engineers, Warrendale PA, February 1979.
4. J. S. MacDonald, S. L. Plee, J. B. D'Arcy and R. M. Schreck, SAE Paper 800185, Society of Automotive Engineers, Warrendale PA, February 1980.
5. R. C. Bascom, W. S. Chiu and R. J. Padd, SAE Paper 730212, Society of Automotive Engineers, Warrendale PA, February 1973.
6. M. Vulliamy and J. Spiers, SAE Paper 670090, Society of Automotive Engineers, Warrendale PA, January 1967.
7. F. J. Hills, T. O. Wagner and D. K. Lawrence, SAE Paper 690493, Society of Automotive Engineers, Warrendale PA, February 1969.
8. A. W. Carey, Jr., SAE Paper 690492, Society of Automotive Engineers, Warrendale PA, February 1969.
9. W. Bryzik and C. O. Smith, SAE Paper 770718, Society of Automotive Engineers, Warrendale PA, February 1977.
10. A. Laresgoiti, A. C. Loos and G. S. Springer, Envir. Sci. Techn., Vol. 11 (1977), p. 973.
11. D. M. Roessler and F. R. Faxvog, Appl. Optics, Vol. 18 (1979), p. 1399.
12. F. R. Faxvog and D. M. Roessler, J. Appl. Phys., Vol. 50 (1979), p. 7880.
13. D. M. Roessler and F. R. Faxvog, J. Opt. Soc. Amer., Vol. 69 (1979), p. 1699.
14. R. L. Williams and C. R. Begeman, Proc. 17th Annual Purdue Air Quality Conference, Purdue University, May 1979.
15. E. F. Funkenbusch, D. G. Leddy and J. H. Johnson, SAE Paper 790418, Society of Automotive Engineers, Warrendale PA, February 1979.
16. W. J. Mayer, D. C. Lechman and D. L. Hilden, SAE Paper 800256, Society of Automotive Engineers, Warrendale PA, February 1980.

17. H. C. Van De Hulst, "Light Scattering by Small Particles," Wiley, N.Y., 1957, pp. 110-111.
18. G. Mie, Ann. Phys., Vol. 25 (1908) p. 337.
19. D. M. Roessler and F. R. Faxvog, J. Opt. Soc. Am., Vol. 70 (1980), p. 230.
20. J. Janzen, J. Colloid Interface Sci., Vol. 69 (1979), p. 436.
21. C. F. Bohren and N. C. Wickramasinghe, Astrophysics and Space Science, Vol. 50 (1977), p. 461.
22. C. G. Granqvist and O. Hunderi, Phys. Rev., Vol. B18 (1978), p. 2897.
23. D. E. Aspnes, J. B Theeten and F. Hottier, Phys. Rev., Vol. B20 (1979), p. 3292.
24. D. B. Kittelson and D. F. Dolan, Conf. on Carbonaceous Particles in the Atmosphere, Berkeley, California, March 1978.
25. N. J. Khatri, J. H. Johnson and D. G. Leddy, SAE Paper 780111, Society of Automotive Engineers, Warrendale PA, February 1978.
26. S. L. Plee and J. S. MacDonald, SAE Paper 800186, Society of Automotive Engineers, Warrendale PA, February 1980.
27. A. L. Aden and M. Kerker, J. Appl. Phys., Vol. 22 (1951), p. 1242.
28. F. R. Faxvog and D. M. Roessler, Appl. Optics, Vol. 17 (1978), p. 2612.
29. S. Asano and M. Sato, Appl. Optics, Vol. 19 (1980), p. 962.
30. B. W. Fowler and C. C. Sung, J. Opt. Soc. Am., 69 (1979), p. 756.
31. A. Cohen and P. Alpert, J. Appl. Phys., Vol. 50 (1979), p. 8262.
32. E. M. Purcell and C. R. Pennypacker, Astrophys. J., Vol. 186 (1973), p. 705.
33. R. M. Welch and S. K. Cox, Appl. Optics, Vol. 17 (1978), p. 3159.
34. J. M. Greenberg, J. Colloid Interface Sci., Vol. 39 (1972), p. 513.
35. R. H. Zerull, Beit. Phys. Atmos., Vol. 49 (1976), p. 168.
36. "Light Scattering by Irregularly Shaped Particles," D. W. Schuerman ed., Proc. of Int. Workshop, Albany, NY, 1979, Plenum Press, NY 1980.
37. P. J. Groblicki and C. R. Begeman, SAE Paper 790421, Society of Automotive Engineers, Warrendale PA, February 1979.
38. D. F. Dolan and D. B. Kittelson, SAE Paper 780110, Society of Automotive Engineers, Warrendale PA, February 1978.
39. A. R. Jones, Proc. Roy. Soc. Lond., A366 (1979), p. 111.
40. A. R. Jones, J. Phys. D: Appl. Phys., Vol. 12 (1979), p. 1661.
41. A. R. Jones, J. Phys. D: Appl. Phys., Vol. 5 (1972), p. L1.
42. J. C. Ravey, Colloid and Polymer Sci., Vol. 253 (1975), p. 292.
43. J. C. Ravey, S. Premilat and P. Horn, European Polymer Journal, Vol. 6 (1975), p. 1527.
44. S. C. Graham, Combustion Sci. and Techn., Vol. 9 (1974), p. 159.
45. D. M. Roessler and F. R. Faxvog, Appl. Optics, Vol. 19 (1980), p. 578.
46. J. B. Howard, B. L. Wersborg and G. C. Williams in "Faraday Symposium No. 7, Fogs and Smokes," Faraday Division, Chemical Society, London, 1973, p. 109 ff.
47. H. Gg. Wagner in "Seventeenth Symposium (International) on Combustion," The Combustion Institute, Pittsburgh, 1978, p. 3 ff.
48. G. W. Smith, to be submitted for publication.
49. J. Lahaye, G. Prado and J. B. Donnet, Carbon, Vol. 12 (1974), p. 27; J. Lahaye and G. Prado, Phys. Chem. of Carbon, Vol. 14 (1978), p. 167.
50. "Faraday Symposium No. 7, Fogs and Smoke," Faraday Division, Chemical Society, London, 1973.
51. N. A. Fuchs and A. G. Sutugin, "Highly Dispersed Aerosols," Ann Arbor Science Publishers, Ann Arbor MI, 1970, p. 8.
52. J. Frenkel, "Kinetic Theory of Liquids," Dover, NY, 1955, p. 376.
53. R. F. Strickland-Constable, "Kinetics and Mechanism of Crystallization," Academic Press, London, 1968, p. 44 ff.
54. J. P. Hirth and G. M. Pound, "Condensation and Evaporation," Pergamon Press, NY, 1963, p. 15ff.
55. D. Turnbull, Solid Sate Physics, Vol. 3 (1956), p. 225.
56. J. W. Mullin, "Crystallization," CRC Press, Cleveland, 1972, p. 139ff.

57. H. N. V. Temperley, "Changes of State," Cleaver-Hume, London, 1956, p. 89ff.
58. J. B. Homer and A. Prothero, Far. Trans. I (Phys. Chem.), Vol. 69 (1973), p. 673.
59. "Handbook of Chemistry and Physics," 61st Ed., CRC Press, Boca Raton, Fla., 1980, p. F-25ff.
60. "Handbook of Physics" (E. U. Condon and H. Odishaw, eds.) McGraw-Hill, NY, 1958, p. 5-94ff.
61. J. G. Kirkwood and F. P. Buff, J. Chem. Phys., Vol. 17 (1949), p. 338.
62. R. P. Andreas, in "Nucleation" (A. C. Zettlemoyer, ed.), Dekker, NY, 1969, p. 69ff.
63. G. D. Ulrich, Comb. Sci. and Tech., Vol. 4 (1971), 47.
64. R. G. Tabak, J. P. Hirth, G. Meyyick and T. P. Roark, Astrophys. J., Vol. 196 (1975), p. 457.
65. P. A. Marsh, A. Voet, T. J. Mullens and L. D. Price, Carbon, Vol. 9 (1971), p. 797.
66. R. Stevenson, to be submitted for publication.
67. J. B. Donnet and A. Voet, "Carbon Black," Dekker, New York, 1976, p. 84ff.
68. A. G. Amelin, Koll. Zh., Vol. 29 (1967), p. 16.

DISCUSSION

S. Japar (Ford Motor Company)

At Ford we have done a number of experiments using photoacoustic spectroscopy to analyze diesel vehicle exhaust — essentially the same experiment that Dave Roessler talked about, and we have found that if you normalize all the photoacoustic results to the mass concentration of carbon in the exhaust, you get rid of the variation: You get essentially a straight line relation between photoacoustic response and carbon mass loading in the exhaust. Our value of the mass specific absorption coefficient of carbon is then about 9 m²/g. We've done this now for four different cars running with and without EGR at various cruise speeds, FTP's, and the like. All the data seem to fall on one straight line.

Roessler

This is obviously still a controversial area. We have also done some photoacoustic work at various engine operating conditions and again find variation in the absorption. Only part of these changes can be attributed to particle composition changes. Let me emphasize that we see such changes only at very low air/fuel ratios. Under more normal conditions (or in measurements which average emissions over several driving cycles) one would not see such variations in the optical properties.

G. Prado (Massachusetts Institute of Technology)

You reported that for decreasing air/fuel ratio there is an increase of volatiles associated with the soot particles. These can be due to either increased production of volatiles in the combustion process or increased deposition due to the change of the structure of the soot particle — i.e. increased surface area. In recent experiments at MIT we found that it was essentially due to increased deposition and not increased production. Could you clarify your results in light of our observation? Did you make any measurement of total volatiles in the exhaust gas?

Roessler

We've not made such measurements.

H. Gg. Wagner *(Universität Göttingen)*

A general comment. I think that the fact that soot particles from diesel engines are fairly equal in size just tells us that they are formed under similar conditions.

G. W. Smith *(General Motors Research Laboratories)*

It is true that in our nucleation/depletion model we did assume that all the spherules were formed in the same way and therefore calculated a mean radius. The thing we were most interested in was an estimate of the spherule size as a test of depletion as the size limiting factor (starting from Prado and Lahaye's original nucleation model). The fact that we obtained the right order of magnitude suggests that depletion may, indeed, be the size limiting mechanism.

A. D'Alessio *(University of Naples)*

I have two comments: First, in order to take account of the effect of hydrocarbons on particles, we just consider them as a coating. One speculates that they may actually behave differently. A wavelength analysis of absorption may help to resolve this concern. Second: Polarized light scattering measurements should confirm whether your particle shapes are indeed following a trend with air/fuel ratio. Would it be possible to supplement your studies with other measurements in order to solve this problem?

Roessler

Yes. What we reported here are some initial findings using very simple techniques. These studies will form the basis of our future investigations.

I. Glassman *(Princeton University)*

Since the air/fuel ratio in the diesel engine is like that for a diffusion flame, the flame temperature is roughly the same irrespective of the air/fuel ratio in a generalized sense. What concerns me is that you reported all your data as a function of air/fuel ratio. One could idealize an experiment in which one could hold the oxygen partial pressure roughly constant, dilute with nitrogen or argon, and retain approximately the same temperature. It may be that the differences you find as a function of air/fuel ratio are due to the fact that you have different final temperatures in the actual system. If one could do a series of controlled experiments in which one controls not the air/fuel ratio, but the oxygen partial pressure and the temperature, then maybe when those two parameters were equal in different systems, one would find the same particle characteristics.

Roessler

Yes. As I mentioned, and as you realized, we have thus far considered only the external parameters controlling the engine. There are clearly many nice experiments that we would like to perform.

D. B. Kittelson (*University of Minnesota*)

We did some experiments using the integrating plate method for measuring light absorption cross sections, and we get results that are pretty much in agreement with what Steve Japar mentioned. We found that if we normalized our data with respect to the non-volatile component of the particles, we get a specific absorption coefficient of about 9 m^2/g, independent of particle morphology. We looked at a range of speed and load conditions for two engines, one a swirl chamber engine and one a direct injection engine. Our volatile fractions ranged from about 10% to about 75% and the particle size typically from 0.1 to 0.3 microns. We didn't see any systematic discernable effect that we could attribute to the influence of particle size or shape on the light absorption. However, when we used an integrating nephelometer to look at the light scattering component of extinction, we found that it was highly dependent on engine operating conditions. In fact, the data were very scattered. We couldn't correlate them with anything that we could identify. Perhaps some of the dependence you're seeing depends on the scattering component rather than on the absorption component.

Roessler

This talk was necessarily condensed, but we certainly have worried about variations in chemical composition. In fact we did some work with acetylene soot which we known to be almost entirely carbon. We found that in the infrared region, where there is neglible scattering, both extinction measurements and absorption measurements showed deviations from the Mie theory. In fact these data were our first clues that physical morphology might be a significant factor in the optical properties.

S. H. Bauer (*Cornell University*)

I would like to comment on the depletion model and perhaps add a subscript. The growth and depletion model is inherently reasonable based on the assumption that critical size nuclei are generated first, and that these grow by accretion of the surrounding material. Of course, the dimensions calculated by Dr. Smith are *mean* values; samples of condensates have a distribution of sizes. This is due to the fact that under homogeneous conditions, the critical nuclei appear *randomly spaced* throughout the volume occupied by the supersaturated vapor. The ultimate size attained by any spherule is therefore determined by its local volume which is free from other nuclei. We computed the distribution of sizes of such volumes (Voronoi polyhedra), which in turn determine the distribution of particle dimensions: it is log-normal, as is reported in the general literature for diverse preparations of fine powders. Furthermore the variance of the computed distributions range from 1.40 to 1.67, while the experimentally observed distributions range from 1.36 to 1.60.*

*C. F. Wilcox, S. Russo and S. H. Bauer, J. Phys. Chem. Vol. 83, p. 897 (1979).

B. S. Haynes *(Massachusetts Institute of Technology)*

That brings us to the point I wanted to make on the neglect of coagulation in the depletion model. I think it is an over-simplification to neglect it. As Professor Wagner discussed in his paper, the coagulating aerosol soon forgets where it started in terms of number density, that is, in terms of the size of the first particles. In some experiments we did at Göttingen, we were able to suppress coagulation by using metal additives.* We could decrease the size of the spherules (as you call them) or the primary spheres by a factor of 5 just by inhibiting coagulation. So I think the neglect of coagulation (and I'm not making any comments on my feelings about a physical condensation mechanism generating the first particles) does a disservice to your measurements since you're not taking account of one of the main physical processes.

*B. S. Haynes, H. Jander, and H. Gg. Wagner, Seventeenth Symposium (International) on Combustion, The Combustion Institute, Pittsburgh, Pa (1979), p 1365.

Smith

Our calculation was merely intended to be an order of magnitude estimate as a test of depletion as the size limiting process.

Haynes

The experimental data are not scattered anywhere near an order of magnitude. And an order of magnitude calculation is always going to bracket experimental data which hardly vary. In that case, why bother doing an order of magnitude calculation?

Smith

I don't know if we could do much better even if we took account of coagulation. There are sizable uncertainties in the data available to estimate surface energy and critical radius. Furthermore, the spherule radius relationship is insensitive to some of the parameters since they all occur inside a cube root. One interesting aspect of the depletion model is that it gives the right spherule size for particulate matter generated by other methods — such as rapid evaporation and condensation of metals.

A. T. Modak *(Northern Research)*

I would like clarification on the small particle limit of the Mie Theory, i.e., the Rayleigh limit. It's my understanding that, if scattering is negligible, all that the absorption coefficient would depend on would be the volume fraction which you have factored out. And then it would not depend upon the size of the particle. This is in contradiction to your conclusion that absorption coefficient depends on shape and size effects. Would you clarify that?

Roessler

For Rayleigh particles, size variations do not affect the specific absorption, but shape is important, as shown by our calculations for the spheroids, for example. Diesel particles are small enough to be Rayleigh-like in the infrared but are too big, at least when clustered, for the Rayleigh approximation to be valid in the visible region. Thus, size may affect the optical properties in the visible but not in the infrared region whereas the effects of shape are more important in the infrared than in the visible.

R. L. Bradow *(Environmental Protection Agency)*

We're very much interested in the application of the optical technique as a means of diagnosing the particle mass emission rate of a diesel engine undergoing cyclic testing. It appears that in your traces of air/fuel ratio, the real problems occur at air/fuel ratios of 20 to 1 or perhaps a bit smaller. With real diesel engines this is an extremely high fraction of the total load. After all, this low an air/fuel ratio is close to stoichiometry conditions that occur in premixed charges but rarely do in diesel engines, even at quite high loads. Do you think that there is a prospect that more nearly reasonable maximum load values might actually produce the kind of phenomenon that Steve Japar and Dave Kittelson talked about?

Roessler

As you know from smoke meter measurements cited in the literature, there is indeed a strong correlation between diesel opacity and mass concentration. Therefore people have been claiming, "We can calculate mass concentrations from opacity measurements." We were naturally interested because, if this is so, opacity would be useful for real-time quantitative mass measurements. However, we were always concerned about the scatter that appeared in such data (as shown in Fig. 2) even for a well-controlled engine. We asked ourselves if it was because the experiments were poorly done or if it indicated small but real variations in the optical properties. To determine the answer we therefore probed very extensively at extreme operating conditions. I think it's true that if you operate a diesel engine under most normal conditions, you would not observe the very pronounced trends that we saw. Once you get above an overall air/fuel ratio of about 30, you see relatively little further change in the optical properties. That means if you operate with the diesel under well-controlled conditions, an optical method may indeed provide an excellent monitor of mass concentration. However, by looking at the very low air/fuel conditions and seeing these variations, I think we now have a better understanding of the overall optical properties.

A. I. Medalia *(Cabot Corporation)*

Terminology is important in communicating within a field and between fields. Terminology has been a special problem in regard to soot formation because workers in this field come from diverse specialties. Confusion between "soot" and "carbon black" has been harmful to the carbon black industry because it has led to restrictive legislation based on the well-known carcinogenicity of certain types of soot.

The term "soot" actually covers a wide variety of natural and industrial by-products and laboratory preparations. Some of these, such as diesel soot, are morphologically rather similar to carbon black but contain considerably more soluble organic fraction. Others, such as chimney soots from domestic wood or coal fires, are quite different in appearance; they generally contain pieces of char, resinified or carbonized globules of tar, cenospheres, and other carbon particles which do not resemble carbon black in morphology. Since all of these can legitimately be called "particulate carbon," a new term is needed to describe the unique form of carbon formed in a flame.

A few suggested definitions are given below. The English word "aciniform" is of Latin origin and means "clustered like grapes." This list is not meant to be a comprehensive glossary of the field; such terms as "particulates", "primary particle", "spherule", and "aggregate" have been defined elsewhere. With this suggested terminology, it is perfectly proper to continue to refer to the product of a laboratory combustion experiment as "soot", while recognizing that this is only one type of soot, and that aciniform carbon may be only one of its constituents.

Aciniform Carbon (AC): Particulate carbon of turbostratic microstructure, formed in the gaseous phase, and generally composed of spheroidal particles fused together in aggregates of colloidal dimensions. Small amounts of other elements may be within the particles or bound to the surface.

Carbon Black: A manufactured product consisting almost entirely of AC.

Soluble Organic Fraction (SOF): Organic compounds which can be dissolved in a solvent; in particular, methylene chloride (MC).

Soot: Black powder containing carbon, SOF, ash, and other components, present in the smoke arising from incomplete combustion of organic matter.

Note: Carbon blacks have less than 0.2 % MC extract and less than 1 % ash, whereas soots generally exceed at least one of these limits by several fold.

THE CHEMICAL ANALYSIS OF PARTICULATE CARBON

M. L. LEE

Brigham Young University
Provo, Utah

K. D. BARTLE

University of Leeds
Leeds, United Kingdom

ABSTRACT

Considerable information concerning the formation of particulate carbon during combustion can be obtained by a detailed analysis of the organic compounds that are associated with the particles. As combustion products cool and are expelled from the combustion system, most organic vapors are adsorbed on the carbon particles. Characterization of these adsorbed materials is important in understanding the effects of different reaction conditions on the formation of particulate carbon.

Generally speaking, there are two approaches to the chemical analysis of particulate carbon. The particles can be probed directly by spectroscopic techniques or extracted with suitable solvents before analysis. The latter approach generally yields more useful information because the extracted sample can be resolved into a number of relatively pure components using modern chromatographic methods prior to qualitative and quantitative measurements.

High resolution gas chromatography represents the most powerful approach for resolving complex organic mixtures, and can be used effectively in the analysis of combustion products. The technique, however, is limited to the analysis of the more volatile sample components. High performance liquid chromatography can extend this range to higher molecular weight species. Polycyclic aromatic compounds containing up to ten and eleven rings have been successfully chromatographed using this technique. New approaches including supercritical fluid chromatography and field-flow fractionation offer the potential of extending chemical separations to even larger compounds.

This paper describes the different approaches to the chemical analysis of particulate carbon and compares the chemical compositions of particulate carbon formed in a variety of different combustion systems.

References pp. 100-104.

INTRODUCTION

The particulate matter associated with incomplete combustion of organic materials is of concern, in part because of potential health-related aspects. In addition to possible lung tissue irritation resulting from inhalation of respirable size soot particles, various organic chemicals, especially the polycyclic aromatic compounds (PAC), associated with the soot are known carcinogens [1].

The mechanism(s) of formation of particulate carbon is far from being completely understood, although it is generally accepted that the intact aromatic ring plays an important role as a building block in rapid soot formation [2]. In the formation of PAC, it is believed that two distinct reaction steps are involved, pyrolysis and pyrosynthesis. At high temperatures, organic materials are partially cracked to smaller, unstable molecules (pyrolysis). These fragments, mostly radicals, recombine to yield larger, relatively stable aromatic compounds (pyrosynthesis). The pyrosynthesis of PAC in combustion systems has recently been reviewed by Crittenden and Long [3]. In general, all organic materials containing carbon and hydrogen may serve as precursors of PAC. However, pyrolysis of a substance displaying chain branching or unsaturation results in an overall increase in the production of PAC. Furthermore, PAC are formed more easily from pyrolysis of compounds that already contain cyclic structures. When oxygen, nitrogen, or sulfur are present in the starting materials, oxygen-, nitrogen-, or sulfur-containing PAC can be expected.

The chemical characterization of combustion products can lead to valuable information useful in understanding the mechanisms of combustion and soot formation. The yields of soot and PAC, and the distributions of alkyl-substituted PAC can be quite different depending on conditions such as temperature, residence time in the combustion zone, and on the nature of the organic material burned.

The organic compounds produced during incomplete combustion are generally associated with the particulate matter which is also produced [4]. Natusch and Tomkins [5] have presented a theoretical model describing the quantitative adsorption of PAC onto fly ash owing to the sudden decrease in temperature as the fly ash exits from the stack. The particulate matter serves, therefore, as a convenient collection medium for these combustion products. The subsequent analysis and correlation, both qualitatively and quantitatively, of the organic material adsorbed on the particulate matter with the conditions of combustion are helpful in understanding soot formation.

In this paper, the modern analytical approaches and techniques for the analysis of PAC associated with combustion products are reviewed. Wherever possible, reference is made to the analysis of combustion products. In several cases, newer methods are described in light of their potential for the analysis of these materials. Finally, some general observations of the chemical compositions of soot formed in a variety of different combustion systems are given.

DIRECT ANALYSIS OF COMPOUNDS ADSORBED
ON PARTICULATE CARBON

While most analyses for organic compounds in particulate carbon currently begin with an extraction step, prospects for direct analysis (i.e. without a separation step) are now promising. For example, newer methods in infra-red (IR) spectroscopy have recently allowed the chemical characterization of the surfaces of a variety of solids; PAC adsorbed on various substrates may also be determined by direct fluorimetry and phosphorimetry at room temperature.

Fourier-transform (FT) IR offers advantages over "conventional" (dispersive) IR because of a pronounced increase in signal-to-noise ratio for the same measurement times [6]. The use of a computer for signal averaging also allows a full range of data treatment techniques, such as spectral subtraction etc., to be applied. Painter and Coleman [7] recently demonstrated the clear resolution of bands at 1695 cm^{-1} (aryl alkyl ketone) and 1575 cm^{-1} (COO$^-$) in the difference spectra of oxidized and unoxidized coal. IR photoacoustic spectroscopy may also afford advantages over the more usual methods for obtaining IR absorption spectra of solids and species adsorbed on surfaces [8-10]; as the sample absorbs photons, it loses energy (by thermal relaxation) which is transferred to the gas in contact with the sample. The consequent expansion of the gas yields pressure pulses, detected acoustically, if the light source is pulsed.

The high sensitivity possible with laser sources may make possible the direct observation of fluorescence spectra of PAC on particulate carbon surfaces. Moreover, phosphorescence spectra of PAC at very low concentrations may be recorded at room temperature by reflectance techniques if immobilized by adsorption so that non-radiative decay of the triplet state is prevented [11, 12]. Whether these approaches can be applied to the PAC adsorbed on particulate carbon is not yet clear.

EXTRACTION OF ORGANIC MATERIAL FROM PARTICULATE CARBON

PAC of molecular weights less than approximately 300 are soluble in numerous organic solvents, and there have been many recommendations for the best solvent for Soxhlet extraction from solid samples such as particulate carbon [13, 14]. Extraction with benzene rapidly gives high yields of PAC from atmospheric particulate matter [15], but dichloromethane has been preferred by a number of groups [16, 17] for extraction from carbon blacks. Benzene/methanol is an efficient solvent for PAC, and extracts even polar materials [18, 19], however methyl esters of polycarboxylic acids may be formed by esterification catalyzed by the highly acidic carbon surface [18]. Non-polar solvents were generally found to be more effective than polar solvents for extracting benzo[a]pyrene from carbon black [20].

The strong adsorption of PAC on certain types of particulate matter may make extraction difficult [21]. For the analogous case of fly-ash, low (<30%) recoveries of ^{14}C-benzo[a]pyrene have been recorded [22], and although higher recoveries were found for two- and three-ring PAC, larger compounds are likely to be incompletely extracted [22]. The use of naphthalene under reduced pressure as an extracting solvent has been shown [18] to yield more hexa- and heptacyclic PAC than benzene from carbon black, but the lower molecular weight compounds were lost and a number of artifacts were formed. Chlorine substituted benzenes, such as the dichloro- and trichlorobenzenes generally are more efficient in extracting the higher molecular weight PAC. An alternative to Soxhlet extraction is ultrasonication at room temperature [23], for which 95-98% recoveries of PAC from atmospheric particulate carbon has been shown [24]. The thermal desorption of PAC from carbon black into a gas chromatograph has been studied [25]. Excessive temperatures are required to desorb the higher molecular weight materials which limits the usefulness of this approach.

The variation of adsorptivity (demonstrated for ^{3}H-benzo[a]pyrene) with the origin of the particulate carbon is a further factor [19]; thus lampblack has low adsorptivity while the isotherm for active carbon shows very strong adsorption.

DIRECT ANALYSIS OF EXTRACTS

Luminescence Methods — A variety of analytical procedures are available for the analysis, without further separation, of the PAC mixture extracted from particulate carbon [26]. Although few have been applied in this context so far, the potential is very clear. Most prominent among these methods is fluorimetry at low temperatures for which there is a considerable improvement in selectivity and sensitivity. Thus, laser excitation of PAC in frozen-solution glasses at 4K yields sharp-line, characteristic spectra [27], but even more line-rich (Shpol'skii) spectra are produced if the PAC molecules are separated by large distances and embedded in a crystalline solvent (usually an n-alkane) lattice at low temperature [28]. Shpol'skii spectra can provide useful fingerprints of individual compounds [29]; thus PAC containing between three and ten rings were identified in extracts of coal and coal tar pitch [30]. Benzo[a]pyrene is well suited to determination by quasi-linear luminescence in, for example, Soxhlet extracts of aviation engine soots [31]. Selective laser excitation of a variety of other individual compounds in complex mixtures such as those from extracts of particulate carbon is certainly possible [32, 33].

Fluorimetry of PAC in nitrogen matrices is not subject to the inner-filter effects which affect quantitative measurements by the Shpol'skii method, although bandwidths are generally greater[34].

The selectivity of phosphorimetry is greater than that of fluorimetry, and the limits of detection are similar [26]; the method complements fluorescence measurements and is most applicable where the mixture contains strongly fluorescent, but weakly phosphorescent interfering species [35] (e.g. perylene in the determination of benzo[rst]pentaphene). Phosphorescence is generally observed at low temperature in frozen solutions, but as discussed above, may be observed at room temperature if collisional deactivation is reduced by adsorption [11, 12]. For example, strong room-temperature phosphorescence emission can be induced from nitrogen heterocycles adsorbed on silica gel plates containing a polymeric binder [12].

Quenchofluorimetry may be used to increase the selectivity of luminescence by making use of the property of a substitutent atom or group such as halogen, nitro, etc. in the solvent to enhance phosphorescence while quenching fluorescence [26]. The quenching of the flourescence of PAC containing only six-membered rings, but not of non-alternants, by electron acceptors such as nitro-methane [36] is a notable example; complementary quenchofluorimetry with electron donors suppresses the fluorescence of non-alternants and allows recognition of alternants in mixtures [37].

Modern approaches to luminescence analysis have been aimed at increasing the sensitivity and selectivity of the methods through the use of new excitation sources, especially lasers, and more sophisticated detectors and associated data-handling systems. Lasers allow detection in the sub part-per-trillion range [38], and if pulsed permit time resolution. Thus, benzo[a]pyrene and benzo[k]fluoranthene have sufficiently different fluorescence decay times in nitrogen matrices to produce excellent temporal resolution of their emission spectra [39]. Optoelectronic image detectors [40] allow the rapid gathering of the emission-excitation matrix, while derivative spectra [41] are accessible through computational procedures. Other new instrumental methods include modulation fluorescence [42] and synchronous luminescence [43] spectroscopy, in which both excitation and emission wavelengths are scanned with a constant interval; the spectrum of each PAC now consists of a single peak [44].

Mass Spectrometry — The simplicity of electron-impact mass spectra of PAC, consisting principally of intense molecular ions, makes this technique nearly ideal for group-type analysis of PAC mixtures. On the other hand, differentiation of PAC isomers by electron-impact mass spectra alone cannot be done in most cases because the mass spectra are indistinguishable. In as early as 1951, Brown [45] showed that compound types could be reliably determined in complex mixtures by selecting the appropriate masses which are characteristic of the compound types. Although early group-type analyses were made using conventional 70-eV electron-impact mass spectrometry, the use of low ionizing voltages has become more widespread [26]. Group-type mass spectral analysis has been used most extensively in the analysis of petroleum, but the techniques have recently been extended to the analysis of particulate matter [46-51]. The typical analytical approach is to introduce an aliquot of the PAC sample into the mass spectrometer through the direct introduction probe system and slowly vaporize the sample with increasing temperature. Plots can then be constructed that display relative abundance as a function of molecular weight.

A new technique of mass spectrometry that is growing rapidly is collision spectroscopy [52]. By introducing a collision gas in the field-free region, intermolecular processes such as charge-exchange and collision induced fragmentations can be studied. These reactions oftentimes provide unique spectra that can be used to determine specific compounds in complex mixtures. For example, Shushan *et al*. [53] recently described a method in which both the magnetic field (B) and the electric sector voltage (E) were scanned so that (B/E) was held at the constant value required to transmit stable parent ions of preselected m/e ratio; in this way fragment ions formed in the first field-free region, from the pre-selected parent ions, were successively transmitted. The resultant "daughter-ion" spectra of the four isomers, chrysene, triphenylene, benz[a]anthracene, and naphthacene showed considerable differences in the relative abundances of the M^+, $(M-H)^+$, and $(M-2H)^+$ ions, the latter two being the products of fragmentation of M^+ ions in the first field-free region.

Field ionization and field desorption mass spectrometry provide virtually single molecular ion spectra for PAC and, therefore, are attractive techniques for the determination of molecular weight distributions in mixtures. Although these techniques have not enjoyed very widespread use, the recent improvements in quantitative aspects have helped [54, 55].

Infrared Spectroscopy — The chief disadvantage of IR spectroscopy of PAC is the absence of sufficiently unique features in the spectra to allow identification. However, IR spectra of matrix-isolated PAC at low temperatures consist of very sharp vibrational lines [56], and it is possible to record highly resolved fingerprint spectra, as long as FT-IR is used to overcome the sensitivity limitations which apply to the dilute sample [34]. Analyses of moderately complex mixtures have been demonstrated for matrix-isolated FT-IR spectra obtained at 2 cm⁻¹ resolution [57].

HIGH RESOLUTION CHROMATOGRAPHIC METHODS

High Performance Liquid Chromatography — Until recently, sample preparation in the analysis of PAC has generally consisted of solvent extraction of the particulate matter and isolation of the PAC fraction by either solvent partitioning methods or column adsorption chromatography [58]. After the isolation of the PAC fraction(s), gas chromatographic or spectroscopic techniques have often been employed for the final analytical measurements. At present, high performance liquid chromatography (HPLC) does not approach the high separation efficiency of capillary column gas chromatography, but it does offer several advantages for the determination of PAC. Firstly, HPLC offers a variety of stationary phases capable of providing unique selectivity for the separation of PAC isomers that are often difficult to separate by gas chromatography. Secondly, UV-absorption and fluorescence spectroscopy provide extremely sensitive and selective detection for PAC. Thirdly, the analysis can be extended to higher molecular weight conpounds than can be analyzed by gas chromatography because of the temperature limitations of gas chromatographic columns and the low vapor pressures of large PAC. Finally, HPLC provides a useful technique for fractionation of PAC mixtures for subsequent analysis by other chromatographic and spectroscopic techniques.

Reverse-phase HPLC on chemically-bonded C_{18} (octadecyl) stationary phases is by far the most popular liquid chromatographic mode for the separation of PAC. Reverse-phase HPLC provides unique selectivity for the separation of PAC isomers and particularly alkyl-substituted compounds. The application of reverse-phase HPLC in the separation of PAC in atmospheric particulate matter [59] and automobile exhaust [60] has been described. The mechanism of retention of PAC on chemically-bonded C_{18} phases has not been established, but Wise *et al.* [61] has succeeded in relating retention to the shape of PAC solutes, particularly the length-to-breadth ratio.

Polar chemically-bonded stationary phases, used in conjunction with nonpolar mobile phases (normal phase HPLC) have also been employed for the separation of PAC. Wise *et al.* [59] have described the retention characteristics of over 90 unsubstituted and alkyl substituted PAC on a polar amine phase. On these polar columns, PAC separations are achieved similar to those obtained on the classical adsorbents such as silica and alumina. Blumer and Zander [62] studied the retention of a number of PAC on silica modified with polar nitrophenyl groups, and Lankmayr and Muller [63] compared nitrophenyl, amine, and C_{18} columns for the separation of PAC commonly found in dust samples. Chemielowiec and George [64] recently investigated the performance of several bonded phases (i.e., amine, nitrile, diol, ether, diamine, and quaternary ammonium) for normal phase separations of PAC, and advocated the use of the diamine column.

A major advantage of HPLC for the determination of PAC is the availability of extremely sensitive and selective detectors. Ultraviolet (UV) absorption and fluorescence detectors are ideally suited for the detection of PAC [26]. Fluorescence detection provides the best selectivity for individual PAC. Both filter fluorimeters and spectrofluorimeters have been used for the determination of PAC in diesel exhaust [65] and atmospheric particles [66]. Christensen and May [67] compared the sensitivity of several filter fluorimeters, a spectrofluorimeter, and UV detectors for PAC determinations.

Selective fluorescence quenching of certain PAC in the presence of nitromethane has been investigated as a selective detection system for HPLC [68, 69]. Konash *et al.* [69] found that the addition of 0.5 percent nitromethane to the mobile phase significantly quenches all the non-fluoranthenic PAC resulting in a selective chromatogram of only fluoranthenic PAC.

A recent development in spectroscopic detection of PAC in chromatographic effluents is the use of multichannel rapid scanning spectrometers [70, 71]. These detectors permit the recording of fluorescence spectra "on-the-fly."

The coupling of a liquid chromatograph to a mass spectrometer has been a difficult task. Several reviews [72-74] have described the construction, operating principles, and performance of various approaches.

Presently, HPLC offers the most advantageous technique for the separation of high molecular weight PAC (molecular weight >300). Recently, Blumer *et al.* [75, 76] reported the separation of approximately 100 constituents of coal tar pitch using a nitrophenyl bonded phase. In this application, approximately 70 percent of the tar could be eluted compared to only 30 percent by gas chromatography. Reverse-phase systems have also been shown to be effective in the separation of PAC up to a molecular weight of approximately 450 [77, 78].

Gas Chromatography — The complexities of PAC mixtures usually demand the greatest resolution possible for their analysis, and although excellent work has been done in the past using packed-columns [16, 79-82], capillary column gas chromatography is the method of choice. The use of capillary columns for the analysis of PAC has recently been reviewed [83]. Glass and, more recently, fused-silica columns are now used universally in this application. A variety of stationary phases have been used in capillary column separations of PAC, but because of slight selectivity effects and temperature stability, SE-52 (methylsilicone gum with 5% phenyl groups) and SE-54 (methylsilicone gum with 1% vinyl and 5% phenyl groups) have gained wide acceptance [83]. Liquid crystal phases have shown pronounced selectivity for PAC in packed column gas chromatography [84-86], but attempts to coat these phases onto glass capillaries have resulted in inefficient columns. Recent work [87] describes the use of mixed stationary phases composed of blends of a nematic liquid crystal with SE-52 which results in excellent efficiencies and selectivities for PAC in capillary column chromatography.

For most PAC analyses, capillary columns need be no longer than 10-25 meters with internal diameters of 0.2-0.3 mm and film thicknesses near 0.3 μm [83]. Analysis times can be shortened and elution temperatures lowered by the use of hydrogen as carrier gas at linear velocities up to 100 cm s^{-1}. Capillary columns have been used in the separation of PAC in atmospheric particulate matter [88-92], carbon black [16, 83] and particulate carbon from various combustion products [14, 50, 93].

The capillary column gas chromatograph is an excellent separation tool, but it is less effective for identification. Some information can be gained from chromatographic retention data, although it has been found that the Kovats retention indices of PAC are influenced by the stationary phase film thickness, the length of the column, the temperature programming rate, and carrier gas flow rate [94, 95]. For this reason, a new, reliable index system was defined by Lee *et al.* [95] based on the set of standards: naphthalene, phenanthrene, chrysene, and picene, and the retention indices of over 200 PAC were determined. The average 95% confidence limits for four measurements on each PAC were ± 0.25 index unit.

The most widely used gas chromatographic detector for PAC is the flame ionization detector (FID) because of its excellent response linearity, sensitivity, and reliability. FID response factors for a large number of PAC have been reported [79], but it is advisable to measure response factors for each chromatographic system if accurate quantitative results are desired.

In 1965, Cantuti *et al.* [96] showed that the response of the electron capture detector (ECD) for PAC was dependent on the structure of the compound. Bjørseth and Eklund [97] measured the ECD/FID response ratios for a number of PAC, and found that many isomers could be differentiated. In a more recent study, Grimsrud *et al.* [98] found that adding oxygen to the carrier gas greatly enhanced the ECD signal for certain PAC. This response enhancement was found to be related to the structure of the compound.

Driscoll *et al.* [99] found that by using both the photoionization detector (PID) with 10.2 eV lamp and the FID, one can differentiate between aromatic and aliphatic hydrocarbons. The relative molar response for the PID increases with unsaturation, and when the PID and FID responses are normalized to any alkane and the normalized PID/FID ratios calculated, those with a ratio of 5-10 are aromatics, 2-4 are alkenes, and < 2 are alkanes. Resonance enhanced 2-photon photoionization has recently been applied to the detection of PAC, and detection limits on the order of 10 picograms were obtained [100].

Gas-phase spectroscopic detectors for PAC include both UV [101] and fluorescence [102-104]. Spectrofluorimetric detectors offer more promise than UV detectors because of their inherent higher sensitivity and greater selectivity.

PAC containing nitrogen and/or sulfur heteroatoms can be detected using nitrogen-selective thermionic [105] and sulfur-selective flame photometric [106, 107] detectors. The selectivity of heterocycles over hydrocarbons for both detectors is $10^4 : 1$.

The most powerful approach available today for the analysis of complex PAC mixtures is capillary column gas chromatography/mass spectrometry (GC/MS) [83]. The maximum resolution of mixture components is of utmost importance in order to provide unambiguous mass spectral identifications of many PAC isomers. This approach was used in the identification of PAC in atmospheric particulates [88], carbon black [16], and particulates from various combustion products [50].

GC/MS is usually accomplished using electron impact ionization. As discussed earlier, the electron impact mass spectra of isomers are generally indistinguishable. The use of a mixed charge exchange-chemical ionization reagent gas that differentiates between isomers has recently been reported [108, 109]. This technique is sensitive to the ionization potentials and, therefore, structures of PAC isomers. Hunt et al. [110-112] have described the use of oxygen as reagent in both positive and negative chemical ionization mass spectrometry of PAC to differentiate isomers.

DISTRIBUTION OF PAC IN PARTICULATE CARBON

The distributions of PAC in various particulate carbon samples are dependent on combustion parameters such as fuel type, temperature, and fuel-to-air ratio. These distributions from different combustion sources differ considerably in type and concentration of species. This is illustrated by Lee et al. [50] in their comparison of chromatographic profiles of the PAC fractions obtained from coal, wood, and kerosene combustion products. The kerosene was burned in a turbulent diffusion flame, where mixing of the fuel with air was facilitated, and, hence, only the most stable PAC were formed. The absence of alkylated PAC is characteristic of more complete combustion. The uncontrolled combustion of coal and wood produced much more complex PAC mixtures with higher relative abundances of alkylated compounds.

Controlled combustion studies [113] of both kerosene and benzene fuels showed that larger amounts of PAC and particulate carbon were produced from the more aromatic fuel, benzene. Furthermore, the relative abundance of higher molecular weight PAC was greater for benzene. Individual compounds were identified and quantified by GC/MS.

The effect of combustion conditions on PAC production, irrespective of the fuel, can be dramatically demonstrated by comparison of the complexity of the capillary column gas chromatograms obtained from uncontrolled coal combustion [50] and combustion in a high temperature entrained flow gasifier [114]. While the former produced an extremely complex mixture, the latter was extremely simple, containing only very stable parent PAC.

The production of carbon blacks also involves high temperatures, and, therefore, the extractable PAC fractions are not as complex as other PAC mixtures [16]. A good review of the physical and chemical properties of various carbon blacks and related materials has been published [19]. Graphite was found to contain only trace quantities of naphthalene, but no other extractables. Likewise, activated carbon extracts contained only trace quantities of low molecular weight PAC.

There is considerable variation in the PAC content among the carbon blacks, themselves. Lampblacks (particle sizes ranging from 40 to 200 nm) contain large amounts of adsorbed PAC, and a variety of compounds including several sulfur-containing PAC have been identified by GC/MS [19]. Furnace blacks range in particle size from 20 to 50 nm with the larger particle sizes containing larger amounts of extractable PAC [16-19, 115]. Channel blacks (10-30 nm particle sizes) are strongly adsorptive for PAC, more so than activated carbon, and therefore yield only trace amounts of PAC [19].

Recent interest in the health effects of exposure to carbon blacks has led to the identification of a number of oxygen-containing and nitro-PAC in atmospheric particulate matter [116-120], auto and diesel exhausts [121], and carbon blacks [17-19, 122]. Several of the nitro-PAC have been shown to be potent frame shift mutagens [122]. It has been speculated that the oxidized PAC and nitro-derivatives are formed during high-temperature, air oxidation [19].

It was found in the analysis of different furnace blacks [16], that a particular carbon black which was produced from an ethylene tar contained some very high molecular weight compounds. Gas chromatographic methods permitted the analysis of PAC up to a molecular weight of 376 (8 rings). High performance liquid chromatography has recently extended that range to compounds of molecular weight 448 (11 rings) [78]. Lewis and Petro [123] have measured the molecular weight distribution in pitches by gel-permeation chromatography and mass spectrometry, and have evidence for PAC containing as many as 21 rings (molecular weight of 950). One of the major challenges facing analytical chemists today is the extension of high resolution chromatographic methods to higher molecular weight species. Even liquid chromatography, including its subclass of gel-permeation chromatography, weakens at very high molecular weights and eventually becomes inapplicable. Part of the solution may come from further development of new techniques such as supercritical fluid chromatography [124, 125] and field flow fractionation [126, 127].

REFERENCES

1. A. Dipple, in "Chemical Carcinogens," C. E. Searle (ed.), American Chemical Society, ACS Monograph No. 173, Washington, D.C., (1976), p. 245.
2. J. D. Bittner and J. B. Howard, Prog. Astronaut. Aeronaut., Vol. 62 (1978), p. 335.
3. B. D. Crittenden and R. Long, in "Carcinogenesis – A Comprehensive Survey: Polynuclear Aromatic Hydrocarbons," R. Freudenthal and P. W. Jones (eds.), Raven Press, New York, (1976), p. 209.
4. "Biologic Effects of Atmospheric Pollutants: Particulate Polycyclic Organic Matter," National Academy of Sciences, Washington, D.C., (1972).
5. D. F. S. Natusch and B. A. Tomkins, in "Carcinogenesis –A Comprehensive Survey: Polynuclear Aromatic Hydrocarbons," P. W. Jones and R. I. Freudenthal (eds.), Raven Press, New York, (1978), p. 145.

6. J. F. Ferraro and L. J. Basile (eds.), "Fourier Transform Infrared Spectroscopy: Applications to Chemical Systems," Academic Press, New York, (1978).

7. P. C. Painter and M. M. Coleman, International Laboratory, April (1980), p. 17.

8. M. G. Rockley, Appl. Spectroscop., Vol. 34 (1980), p. 405.

9. J. D. Low and G. A. Parodi, Appl. Spectroscop., Vol. 34 (1980), p. 76.

10. M. G. Rockley and J. P. Devlin, Appl. Spectroscop., Vol. 34 (1980), p. 407.

11. T. Vo-Dinh, F. Lue Yen, and J. D. Winefordner, Anal. Chem., 48 (1976), p. 1186; Talanta, Vol. 24 (1977), p. 146.

12. C. D. Ford and R. J. Hurtubise, Anal. Chem., Vol. 52 (1980), p. 656.

13. T. W. Stanley, J. E. Meeker, and M. J. Morgan, Environ. Sci. Technol., Vol. 1 (1967), p. 927.

14. F. S. C. Lee, T. J. Prater, and F. Ferris, in "Polynuclear Aromatic Hydrocarbons," P. W. Jones and P. Leber (eds.), Ann Arbor Science, Ann Arbor, (1979), p. 83.

15. R. C. Pierce and M. Katz, Anal. Chem., Vol. 47 (1975), p. 1743.

16. M. L. Lee and R. A. Hites, Anal. Chem., Vol. 48 (1976), p. 1890.

17. A. Gold, Anal. Chem., Vol. 47 (1975), p. 1469.

18. W. L. Fitch, E. T. Everhart, and D. H. Smith, Anal. Chem., Vol. 50 (1978), p. 2123.

19. W. L. Fitch and D. H. Smith, Environ. Sci. Technol., Vol. 13 (1979), p. 341.

20. K. Wettig, A. Ya. Khesina, A. B. Linnik, L. V. Krivosheva, and W. H. Doerre, Staub, Reinhaltung Luft, Vol. 29 (1969), p. 21.

21. H. L. Falk and P. E. Steiner, Cancer Res., Vol. 12 (1952), p. 30, 60.

22. W. H. Griest, J. E. Caton, M. R. Guerin, L. B. Yeatts, Jr., and C. E. Higgins, Anal. Chem., Vol. 52 (1980), p. 199.

23. G. Chalot, M. Castegnaro, J. L. Roche, R. Fontagnons, and P. Obaton, Anal. Chim. Acta, Vol. 53 (1971), p. 259.

24. C. Golden and E. Sawicki, Int. J. Environ. Anal. Chem., Vol. 4 (1975), p. 9.

25. G. T. Todd, Diss. Abs. Internat. (B), Vol. 31 (1971), p. 6671.

26. K. D. Bartle, M. L. Lee, and S. A. Wise, Chem. Soc. Rev., in press.

27. J. C. Brown, M. Edelson, and G. J. Small, Anal. Chem., Vol. 50 (1978), p. 1394.

28. E. V. Shpol'skii and T. N. Bolotnikova, Pure Appl. Chem., Vol. 37 (1974), p. 183.

29. A. Colmsjö and U. Stenberg, Anal. Chem., Vol. 51 (1979), p. 145.

30. J. A. G. Drake, D. W. Jones, B. S. Causey, and G. F. Kirkbright, Fuel, Vol. 57 (1978), p. 663.

31. J. Jager, Atmos. Environ., Vol. 2 (1968), p. 293.

32. J. R. Maple, E. L. Wehry, and G. Mamontov, Anal. Chem., Vol. 52 (1980), p. 920

33. Y. Yang, A. P. D'Silva, V. A. Fassel, and M. Iles, Anal. Chem., Vol. 52 (1980), p. 1350.

34. E. L. Wehry and G. Mamontov, Anal. Chem., Vol. 51 (1979), p. 643A.

35. L. V. S. Hood and J. D. Winefordner, Anal. Chim. Acta, Vol. 42 (1968), p. 199.

36. M. Zander, U. Breymann, H. Dreeskamp, and E. Koch, Z. Naturforsch., Vol. 32a (1977), p. 1561.

37. U. Breymann, H. Dreeskamp, E. Koch, and M. Zander, Fres. Z. Anal. Chem., Vol. 293 (1978), p. 208.

38. J. H. Richardson and M. E. Ando, Anal. Chem., Vol. 49 (1977), p. 955.

39. R. B. Dickinson, Jr., and E. L. Wehry, Anal. Chem., Vol. 51 (1979), p. 776.

40. Y. Talmi, D. C. Baker, J. R. Jadamec, and W. A. Saner, Anal. Chem., Vol. 50 (1978), p. 936A.

41. G. Talsky, L. Mayring, and H. Kreuzer, Angew. Chem. Internat. Ed., Vol. 17 (1978), p. 785.

42. T. C. O'Haver and W. M. Parks, Anal. Chem., Vol. 46 (1974), p. 1886.

43. J. B. F. Lloyd, Nature (London), Vol. 231 (1971), p. 64; Analyst, Vol. 99 (1974), p. 729.

44. T. Vo-Dinh, Anal. Chem., Vol. 50 (1978), p. 396.

45. R. A. Brown, Anal. Chem., Vol. 23 (1951), p. 430.

46. J. L. Shultz, A. G. Sharkey, Jr., and R. A. Friedel, Biomed. Mass Spectrom., Vol. 1 (1974), p. 137.

47. W. Giger and M. Blumer, Anal. Chem., Vol. 46 (1974), p. 1663.

48. D. Schuetzle, Biomed. Mass Spectrom., Vol. 2 (1975), p. 288.
49. R. A. Hites and W. G. Biemann, Advan. Chem. Ser., Vol. 147 (1975), p. 188.
50. M. L. Lee, G. P. Prado, J. B. Howard, and R. A. Hites, Biomed. Mass Spectrom., Vol. 4 (1977), p. 182.
51. M. Blumer, T. Dorsey, and J. Sass, Science, Vol. 195 (1977), p. 283.
52. R. G. Cooks (ed.), "Collision Spectroscopy," Plenum Press, New York, (1978).
53. B. Shushan, S. H. Safe, and R. K. Boyd, Anal. Chem., Vol. 51 (1979), p. 156.
54. S. E. Scheppele, P. L. Grizzle, G. J. Greenwood, T. D. Marriott, and N. B. Perreira, Anal. Chem., Vol. 48 (1976), p. 2105.
55. D. F. Barofsky, E. Barofsky, and R. Held-Aigner, Adv. Mass Spectrom., Vol. 7 (1978), p. 109.
56. I. R. Dunkin, Chem. Soc. Rev., Vol. 9 (1980), p. 1.
57. E. L. Wehry, G. Mamantov, R. R. Kemmerer, H. O. Brotherton, and R. C. Stroupe, in "Carcinogenesis-A Comprehensive Survey: Polynuclear Aromatic Hydrocarbons," R. Freudenthal and P. W. Jones (eds.), Raven Press, New York, (1976), p. 299.
58. M. L. Lee, M. Novotny, and K. D. Bartle, "Analytical Chemistry of Polycyclic Aromatic Compounds," Academic Press, in press.
59. S. A. Wise, W. J. Bonnett, and W. E. May, in "Polynuclear Aromatic Hydrocarbons: Chemistry and Biological Effects," A. Bjorseth and A. J. Dennis (eds.), Battelle Press, Columbus, Ohio, (1980), p. 791.
60. T. Nielson, J. Chromatogr., Vol. 170 (1979), p. 147.
61. S. A. Wise, W. J. Bonnett, F. R. Guenther, and W. E. May, J. Chromatogr. Sci., in press.
62. G.-P. Blumer and M. Zander, Z. Anal. Chem., Vol. 288 (1977), p. 277.
63. E. P. Lankmayr and K. Müller, J. Chromatogr., Vol. 170 (1979), p. 139.
64. J. Chmielowiec and A. E. George, Anal. Chem., Vol. 52 (1980), p. 1154.
65. D. E. Seizinger, Trends in Fluorescence, Vol. 1 (1978), p. 9.
66. D. Fechner and B. Seifert, Z. Anal. Chem., Vol. 292 (1978), p. 199.
67. R. G. Christensen and W. E. May, J. Liquid Chromatogr., Vol. 1 (1978), p. 385.
68. G.-P. Blumer and M. Zander, Z. Anal. Chem., Vol. 296 (1979), p. 409.
69. P. L. Konash, W. E. May, and S. A. Wise, Washington Chromatography Discussion Group Meeting, Gaithersburg, Maryland, May 21, (1980).
70. J. R. Jadamec, W. A. Saner, and Y. Talmi, Anal. Chem., Vol. 49 (1977), p. 316.
71. D. C. Shelly, W. A. Ilger, M. P. Fogarty, and I. M. Warner, Altex Chromatogram, Vol. 3(1) (1979), p. 4.
72. P. J. Arpino and G. Guiochon, Anal. Chem., Vol. 51 (1979), p. 682A.
73. D. E. Games, Proc. Anal. Soc., April, (1980), p. 110.
74. W. H. McFadden, J. Chromatogr. Sci., Vol. 18 (1980), p. 97.
75. G.-P. Blumer, R. Thoms, and M. Zander, Erdöl Kohle Erdgas Petrochem., Vol. 31 (1978), p. 197.
76. G.-P. Blumer and M. Zander, Compendium 78/79 Supplement to Erdöl Kohle Erdgas Petrochem., (1978), p. 1472.
77. D. Felscher and J. Stein, Z. Chem., Vol. 19 (1979), p. 303.
78. P. A. Peaden, M. L. Lee, Y. Hirata, and M. Novotny, Anal. Chem., Vol. 52 (1980), p. 2268.
79. R. C. Lao, R. S. Thomas, H. Oja, and L. Dubois, Anal. Chem., Vol. 45 (1973), p. 908.
80. R. C. Lao, R. S. Thomas, and J. L. Monkman, J. Chromatogr., Vol. 112 (1975), p. 681.
81. A. Hase, P. H. Lin, and R. A. Hites, in "Carcinogenesis-A Comprehensive Survey: Polynuclear Aromatic Hydrocarbons," R. Freudenthal and P. W. Jones (eds.), Raven Press, New York, (1976), p. 435.
82. G. Grimmer, H. Böhnke, and A. Hildebrandt, Z. Anal. Chem., Vol. 279 (1976), p. 139.
83. M. L. Lee and B. W. Wright, J. Chromatogr. Sci., Vol. 18 (1980), p. 345.
84. G. M. Janini, K. Johnston, and W. L. Zielinski, Jr., Anal. Chem., Vol. 47 (1975), p. 670.

85. *G. M. Janini, G. M. Muschik, and W. L. Zielinski, Jr., Anal. Chem., Vol. 48 (1976), p. 809.*

86. *G. M. Janini, G. M. Muschik, J. A. Schroer, and W. L. Zielinski, Jr., Anal. Chem., Vol. 48 (1976), p. 1974.*

87. *R. J. Laub, W. L. Roberts, and C. A. Smith, HRC & CC, Vol. 3 (1980), p. 355.*

88. *M. L. Lee, M. Novotny, and K. D. Bartle, Anal. Chem., Vol. 48 (1976), p. 1566.*

89. *A. Bjφrseth, Anal. Chim. Acta, Vol. 94 (1977), p. 21.*

90. *W. Giger and C. Schaffner, Anal. Chem., Vol. 50 (1978), p. 243.*

91. *A. Bjφrseth in "Carcinogenesis – A Comprehensive Survey: Polynuclear Aromatic Hydrocarbons," P. W. Jones and R. I. Freudenthal (eds.), Raven Press, New York, (1978), p. 75.*

92. *A. Bjφrseth and G. Eklund, Anal, Chim, Acta, Vol. 105 (1979), p. 119.*

93. *T. Doran and N. G. McTaggart, J. Chromatogr. Sci., Vol. 12 (1974), p. 715.*

94. *H. Beernaert, J. Chromatogr., Vol. 173 (1979), p. 109.*

95. *M. L. Lee, D. L. Vassilaros, C. M. White, and M. Novotny, Anal. Chem., Vol. 51 (1979), p. 768.*

96. *V. Cantuti, G. P. Cartoni, A. Liberti, and A. G. Torri, J. Chromatogr., Vol. 17 (1965), p. 60.*

97. *A. Bjφrseth and G. Eklund, HRC & CC, Vol. 2 (1979), p.22.*

98. *E. P. Grimsrud, D. A. Miller, R. G. Stebbins, and S. H. Kim, J. Chromatogr., Vol. 197 (1980), p. 51.*

99. *J. N. Driscoll, J. Ford, L. F. Jaramillo, and E. T. Gruber, J. Chromatogr., Vol. 158 (1978), p. 171.*

100. *C. M. Klimcak and J. E. Wessel, Anal. Chem., Vol. 52 (1980), p. 1233.*

101. *M. Novotny, F. J. Schwende, M. J. Hartigan, and J. E. Purcell, Anal. Chem., Vol. 52 (1980), p. 736.*

102. *H. P. Burchfield, E. E. Green, R. J. Wheeler, and S. M. Billedeau, J. Chromatogr., Vol. 99 (1974), p. 697.*

103. *D. J. Freed and L. R. Faulkner, Anal. Chem., Vol. 44 (1972), p. 1194.*

104. *R. P. Cooney and J. D. Winefordner, Anal. Chem., Vol. 49 (1977), p. 1057.*

105. *S. G. Wakeham, Env. Sci. Technol., Vol. 13 (1979), p. 1119.*

106. *B. Wenzel and R. Aiken, J. Chromatogr. Sci., Vol. 17 (1979), p. 503.*

107. *M. L. Lee, C. Willey, R. N. Castle, and C. M. White, in "Polynuclear Aromatic Hydrocarbons: Chemistry and Biological Effects," A. Bjφrseth and A. J. Dennis, (eds.) Battelle Press, Columbus, Ohio, (1980), p. 59.*

108. *M. L. Lee and R. A. Hites, J. Am. Chem. Soc., Vol. 99 (1977), p. 2008.*

109. *M. L. Lee, D. L. Vassilaros, W. S. Pipkin, and W. L. Sorensen, in "Trace Organic Analysis: A New Frontier in Analytical Chemistry," NBS Special Publication 519, U.S. Government Printing Office, Washington, D.C., (1979), p. 731.*

110. *D. F. Hunt, C. N. McEwen, and T. M. Harvey, Anal. Chem., Vol. 47 (1975), p. 1730.*

111. *D. F. Hunt, G. C. Stafford, Jr., F. W. Crow, and J. W. Russell, Anal. Chem., Vol. 48 (1976), p. 2098.*

112. *D. F. Hunt and S. K. Sethi, in "High Performance Mass Spectrometry: Chemical Applications," M. L. Gross (ed.), ACS Symp. Ser., 70, American Chemical Society, Washington, D.C., (1978), p. 150.*

113. *G. P. Prado, M. L. Lee, R. A. Hites, D. P. Hoult, and J. B. Howard, "Sixteenth Symposium (International) on Combustion" (1977), p. 649.*

114. *L. D. Hansen, L. R. Phillips, N. F. Mangelson, and M. L. Lee, Fuel, Vol. 59 (1980), p. 323.*

115. *L. Wallcave, D. L. Nagel, J. W. Smith, R. D. Waniska, Environ. Sci. Technol., Vol. 9 (1975), p. 143.*

116. *J. N. Pitts, Jr., K. A. Van Cauwenberghe, D. Grosjean, J. P. Schmid, D. Fitz, W. L. Belser, Jr., G. B. Knudson, and P. M. Hynds, Science, Vol. 202 (1978), p. 515.*

117. *R. C. Pierce and M. Katz, Environ. Sci. Technol., Vol. 9 (1975), p. 347.*

118. *R. C. Pierce and M. Katz, Environ. Sci. Technol., Vol. 10 (1976), p. 45.*

119. W. Cantreels and K. Van Cauwenberghe, J. Chromatogr., Vol. 131 (1977), p. 253.
120. J. N. Pitts, Jr., Philos. Trans. R. Soc. London Ser. A, Vol. 290 (1979), p. 551.
121. F. S.-C. Lee, T. M. Harvey, T. J. Prater, M. C. Paputa, and D. Schuetzle, Proc. ASTM Symp. on Sampling and Analysis of Toxic Organics in Source-Related Atmosphere, Boulder, Colorado, (1980), in press.
122. H. S. Rosenkranz, E. C. McCoy, D. R. Sanders, M.Butler, D. K. Kiriazides, and R. Mermelstein, Science, Vol. 209 (1980), p. 1039.
123. I. C. Lewis and B. A. Petro, J. Polymer Sci., Polym. Chem. Ed., Vol. 14 (1976), p. 1975.
124. R. E. Jentoft and T. H. Gouw, Anal. Chem., Vol. 48 (1976), p. 2195.
125. E. Klesper, Angew. Chem. Int. Ed. Engl., Vol. 17 (1978), p. 738.
126. J. C. Giddings, J. Chromatogr., Vol. 125 (1976), p. 3.
127. J. C. Giddings, S. R. Fisher, and M. N. Myers, Am. Lab., Vol. 10 (1978), p. 15.

DISCUSSION

J. Lahaye *(Centre National de la Recherche Scientique)*

Regarding the polyaromatic hydrocarbons formed in flames, some are by-products and some are intermediate species for the formation of soot. Probably the difference between both kinds of hydrocarbons depends on their reactivities. The reactivity depends quite a lot on the existence of side chains on the aromatic cyclic compound. For the compounds you obtain on soot which includes by-products, it would be interesting to know if there is or is not a side chain. Have you tried to do an examination by proton NMR?

Lee

We have not done a systematic analysis of the intermediates versus the products in the formation of soot. Recently we have been involved in developing techniques for the analysis of higher molecular weight species. But in the work that we have done, in looking at a wide variety of different combustion products, the parent aromatic system is the most stable. Those with alkyl side chains are less stable. Furthermore, we've noticed that compounds which have a 5-membered ring (such as acephenanthrylene, acenaphthylene, indenopyrenes, and cyclopenta[c, d]-pyrene), tend to be less stable. We have not analyzed combustion products by NMR to detect the presence of alkyl side chains, but mass spectral evidence suggests that the presence and length of alkyl side chains are dependent on many factors, such as the fuel and combustion conditions.

K. Smyth *(National Bureau of Standards)*

I would like to comment on using all of these techniques for actually learning something about soot formation. In a tar sample the soluble fraction can be quite high. But if the soot is collected in a flame in a region that is fairly hot, it is typical that less than 1% of the mass is soluble, and therefore, no matter what chemical characterization is performed, even if literally hundreds of compounds are identified, one cannot deduce from 1% of the mass fraction anything really meaningful about mechanisms. As Professor Lahaye pointed out, I think, when one goes down stream into cooler regions, one is collecting basically by-products and not anything that looks like a precursor.

Lee

Again, the amount and composition of the soluble fraction is dependent on the fuel and combustion conditions. Also, different kinds of particulate matter have different adsorptive properties and in some cases much of the organic material may not be extractable. Certainly the components in the soluble fraction, as well as the soot itself, are end products of combustion (or incomplete combustion). Whether or not they are also intermediates in soot formation, I don't think anyone really knows. It does seem reasonable, however, that knowledge concerning the composition of the end products of combustion should shed some light on the mechanisms of soot formation.

J. C. Hilliard *(University of Michigan)*

Is the retention index defined simply in terms of the number of rings in the molecule?

Lee

The number 200 was assigned for the two rings and 300 for the three rings. Yes, there was some reason behind assigning those numbers.

Hilliard

Based simply on the number of rings that you're considering in the molecule? Is it not possible that as the number of rings increases so does the number of possible ways in which you can arrange them? Isn't that necessarily going to alter what one might define as the aromatic double bond character of the large molecule? Isn't that going to affect the retention index from the column?

Lee

Yes, it does affect the retention index, and that's the reason why we obtained standards and measured their retention indices. The main purpose for developing the retention index is so that other people could use it if they don't have the standard compounds. The biggest problem in the exact structural identification of these species is comparison with standard compounds. So we have cataloged those retention index measurements for others to use.

J. B. Howard *(Massachusetts Institute of Technology)*

Going back to the previous comment, I think there may have been a little bit of misleading thrust there with regard to the pertinence of these kinds of measurements to the mechanisms of soot formation. It is true that some of the compounds reported on here may be by-products rather than intermediates to soot, but some of them may not be just by-products. I think that the complaint is really against the sampling techniques used in getting the material out of the flame and not against the techniques reported here by Dr. Lee. It would be very enlightening, in fact, if samples could be taken with techniques which give a good representation of what's in a soot-forming flame zone and then apply these techniques in the analysis.

A. Snelson *(ITT Research Institute)*

Have you made any effort to determine how successfully you can extract material from something like soot? It would seem to be intrinsically difficult.

Lee

That's a difficult question to answer. Some of these soot materials are extremely adsorptive; you can try about any organic solvent, and you still can't get them off completely. In other cases, the material comes off very rapidly. A lot of work has been done in looking at different solvents; the chlorinated benzenes are some of the best solvents for extracting these materials. But, again, it is difficult to determine if you get it all off. Recent work at Oak Ridge National Laboratories has shown that low recoveries (less than 30%) of ^{14}C-benzo[a]pyrene from spiked fly ash were obtained even by ultrasonic extraction. But, one doesn't even really know whether spiked samples are truly representative of real samples.

SESSION II
BASIC MECHANISMS
AND KINETICS

Session Chairman
I. GLASSMAN
Princeton University
Princeton, New Jersey

PRE-PARTICLE CHEMISTRY IN SOOT FORMATION

J. D. BITTNER and J. B. HOWARD

Massachusetts Institute of Technology
Cambridge, Massachusetts

ABSTRACT

Species and reactions involved in the early stages of soot formation prior to the formation of particles have been studied in flat low pressure flames of benzene using a molecular beam mass spectrometer system. In a near-sooting (equivalence ratio $\phi = 1.8$) benzene-oxygen-argon flame, the profiles of molar flux versus distance from the burner for several reactant, intermediate, and product species were calculated from the mole fraction profiles. The flux profiles indicate that CO, C_6H_6O and C_5H_6 are early intermediates in the ring destruction process but that C_2H_2 is not. The flux profiles of C_6H_6 and C_2H_2 are consistent with benzene consumption by OH addition and the OH + C_6H_6 reaction having a much larger rate coefficient than OH + C_2H_2. A mechanism is outlined for production of C_1, C_2, C_3, C_4 and C_5 species from benzene that is consistent with the flux profiles of intermediate species. The production of more hydrogenated species from those of low H/C ratio as they diffuse toward the burner surface is discussed.

Relative measurements of high molecular weight material (> 200 amu) as the fuel equivalence ratio is increased past the sooting limit ($\phi = 1.9$) in benzene flames suggest a sequential growth process from polycyclic aromatic hydrocarbons (PAH) in the mass range 120-210 amu to higher molecular weight hydrocarbons and then to soot. A preliminary analysis of reaction mechanisms suggests that the role of the aromatic hydrocarbon is to provide a structure capable of stabilizing by internal aromatic substitution reactions radicals formed from addition of non-aromatic species. The presence of PAH in the oxidation zone of sooting C_2H_2 flames was also observed and suggests that aromatics may play a similar role in flames of non-aromatic fuels.

INTRODUCTION

The large potential importance of diesel engines and of synthetic fuels has led to a surge of interest in the associated problem of soot formation. It is well known [1] that the formation of soot and polycyclic aromatic hydrocarbons (PAH) is strongly influenced by fuel type. Studies in gas turbine-type combustors [2-4] and simple laboratory systems such as laminar diffusion flames [5-9], premixed flames [10, 11] and well-stirred reactors [12, 13], have shown that aromatic fuels have a high propensity to form soot. Recent studies of soot formation in premixed flames using light scattering techniques [14, 15] have demonstrated the application of the laws of physical coagulation to the later stages of soot formation in flames of different fuels. These studies show that the important difference between fuels, and hence more generally a dominant factor in soot formation, is the mass of carbonaceous material that enters this coagulating system. Thus, understanding the early stages of the chemistry of the fuel oxidation and pyrolysis — or the pre-particle chemistry — is of primary importance in efforts to deal more effectively with fuel types and combustor designs to minimize particulate emissions. Since evidence suggests that the intact aromatic ring, not fragments thereof, is responsible for the propensity to form soot [16-19], the destruction of the aromatic ring when starting from an aromatic fuel and the formation of aromatic rings when starting from a non-aromatic fuel may be important processes in the pre-particle chemistry.

This paper reports results from a study of the role of the aromatic ring in the pre-particle chemistry of soot formation. Mole fraction and flux profiles of stable and free radical species in a benzene-oxygen-argon flame near the sooting limit are presented, and the production of C_1-C_5 hydrocarbons from the aromatic ring and the role of the aromatic ring in PAH and soot formation are discussed. Supplementary data from similar measurements in a sooting acetylene-oxygen-argon flame are also presented as a basis for discussion of whether aromatic species may also be important in soot formation from non-aromatic fuels.

EXPERIMENTAL ASPECTS

The molecular beam mass spectrometer system (Fig. 1) which has been developed to measure stable and radical species in fuel-rich and sooting flames, the calibration procedures and other experimental techniques have been described elsewhere [20, 21]. Both the benzene and acetylene flames were produced at a burner chamber pressure of 2.67 kPa and a cold gas velocity of 0.5 m s^{-1} at 298 K. The unburned gas composition of the benzene flame was 13.5 mol % C_6H_6, 56.5 mol % O_2 and 30.0 mol % Ar. This corresponds to a fuel equivalence ratio ϕ, of 1.8; the sooting limit is $\phi = 1.9$. The unburned gas composition of the acetylene flame was 52.9 mol % C_2H_2, 44.1 mol % O_2 and 3.0 mol % Ar which corresponds to $\phi = 3.0$. With the sooting limit at about $\phi = 2.45$, this is a moderately sooting flame.

PRODUCTION OF C_1-C_5 HYDROCARBONS FROM BENZENE

The apparent importance of the intact aromatic ring in promoting soot formation and the need for information on its destruction process were mentioned above. The distribution of non-aromatic hydrocarbons produced by the destruction of the aromatic ring may also be of primary importance in the soot and PAH formation processes. From structural considerations, many of the PAH observed in sooting flames cannot be formed by condensation reactions of intact aromatic rings. Results presented below suggest that even in flames of aromatic fuels, much of the carbon that ends up as soot may come from non-aromatic hydrocarbon intermediates.

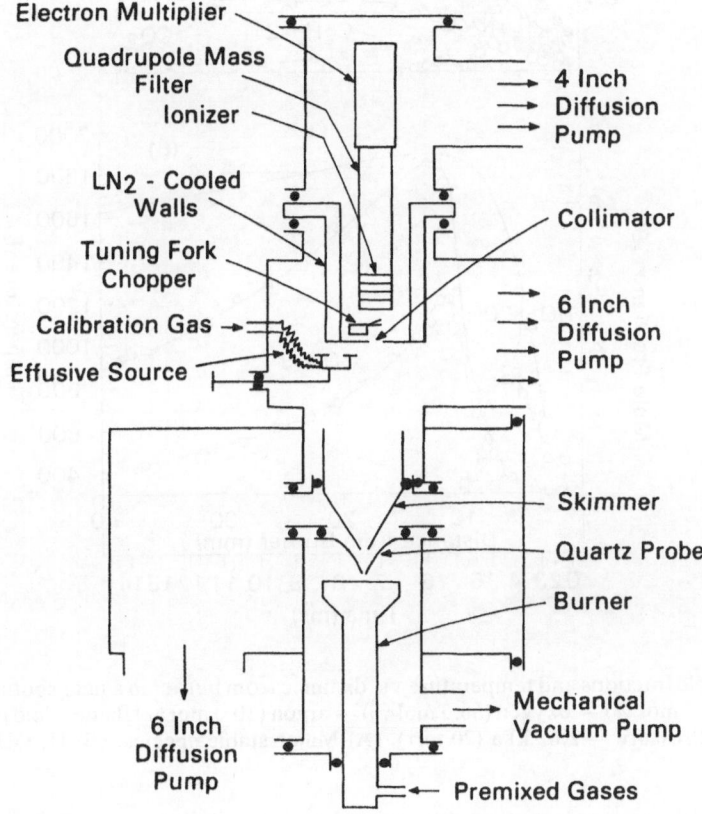

Fig. 1. Molecular beam mass spectrometer system for the study of flat low-pressure premixed laminar flames.

Mole fraction profiles of the major stable and radical species observed in the $\phi = 1.8$ benzene flame are shown in Fig. 2. Mole fraction profiles of forty other species with masses up to 202 amu have been presented elsewhere [21]. The follow-

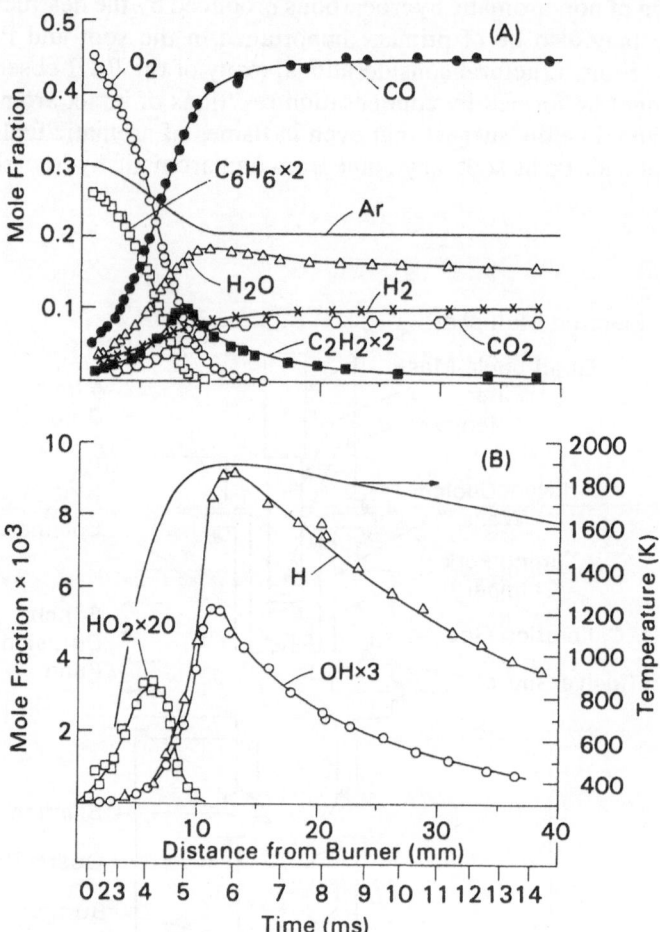

Fig. 2. Mole fractions and temperature vs. distance from burner in a near sooting ($\phi = 1.8$) benzene (13.5 mol %) — oxygen (56.5 mol %) — argon (30.0 mol %) flame. Cold gas velocity = 0.5 m/s. Pressure = 2.67 kPa (20 torr). (A) Major stable species. (B) H, OH, HO$_2$ and temperature.

ing discussion concentrates on the region between the burner surface and 16 mm above the burner, where the fuel is consumed and intermediate hydrocarbon species, PAH and soot are produced. To obtain insight into the chemical processes occurring in the flame, the diffusional processes that are of equal importance in determining the shapes of the mole fraction profiles were accounted for by calculating the molar fluxes (per unit area of the burner surface) F_i for eighteen species using the equation

$$F_i = X_i \frac{PA}{RT} \left(v - \frac{D_{i,\,mix}}{X_i} \frac{dX_i}{dz} \right) \tag{1}$$

where X_i is the mole fraction of species i, T is temperature, A is the area expansion ratio, v is the bulk gas velocity, $D_{i,mix}$ is the diffusion coefficient of species i in the mixture, z is the distance from the burner and R is the ideal gas constant [22].

The diffusion coefficients $D_{i,mix}$ were calculated from the binary coefficients D_{ij}, using the relationship of Fairbanks and Wilke [23] (Eq. 2).

$$D_{i,\ mix} = \frac{1-X_i}{\displaystyle\sum_{\substack{j=1 \\ j \neq i}}^{n} \frac{X_j}{D_{ij}}}$$
(2)

For all non-polar pairs the binary diffusion coefficients were calculated from Eqs. 3-5 [24]

$$D_{ij} = \frac{1.66 \times 10^{-3}\ [(M_i + M_j)/M_iM_j]^{1/2} T^{1.67}}{P\sigma_{ij}^{2}\ (\epsilon_{ij}/k)^{0.17}}$$
(3)

$$\sigma_{ij} = (\sigma_i + \sigma_j)/2$$
(4)

$$\epsilon_{ij}/k = [(\epsilon_i/k)(\epsilon_j/k)]^{1/2}$$
(5)

where D_{ij} is in cm²/s, P is pressure in atmospheres, M is molecular weight, T is temperature in K, and σ and ϵ/k are the Lennard-Jones (12-6) potential parameters in Å and K tabulated by Svehla [25] and Hirschfelder *et al.* [26]. When not available in references 25 or 26, the Lennard-Jones parameters were estimated from the critical pressure, critical temperature and acentric factor using the methods of Tee *et al.* [27] as described by Reid *et al.* [28]. For the only polar-polar pair considered, H_2O and C_6H_6O (assumed to be phenol), the method of Brokaw [29] also described by Reid *et al.* [28] has been used to estimate the binary diffusion coefficient. For the polar-nonpolar interactions of H_2O and C_6H_6O with other species, the binary diffusion coefficients were calculated from Eqs. 3-5 using the Lennard-Jones parameters of H_2O and C_6H_6O estimated by the method of Brokaw [29].

The smoothing and differentiation of the mole fraction profiles involved both pencil smoothing and numerical techniques. Smooth curves were drawn through the 30 to 50 experimental data points (ratios of signal of species i relative to the argon signal) taken for each profile. These pencil smoothed curves were entered into a computer program in the form of 151 evenly spaced points (0.1 mm) between 1 and 16 mm above the burner. The argon and other mole fraction profiles were calculated from the signal ratio profiles, calibration factors and an argon balance. These mole fraction profiles were further smoothed and differentiated by a simplified least squares procedure [30, 31].

The importance of diffusion in determining the shapes of the mole fraction profiles in this flame is apparent from the comparison of the mole fraction and flux profiles for the major stable species (Fig. 3). The inlet fluxes of O_2 and C_6H_6 are

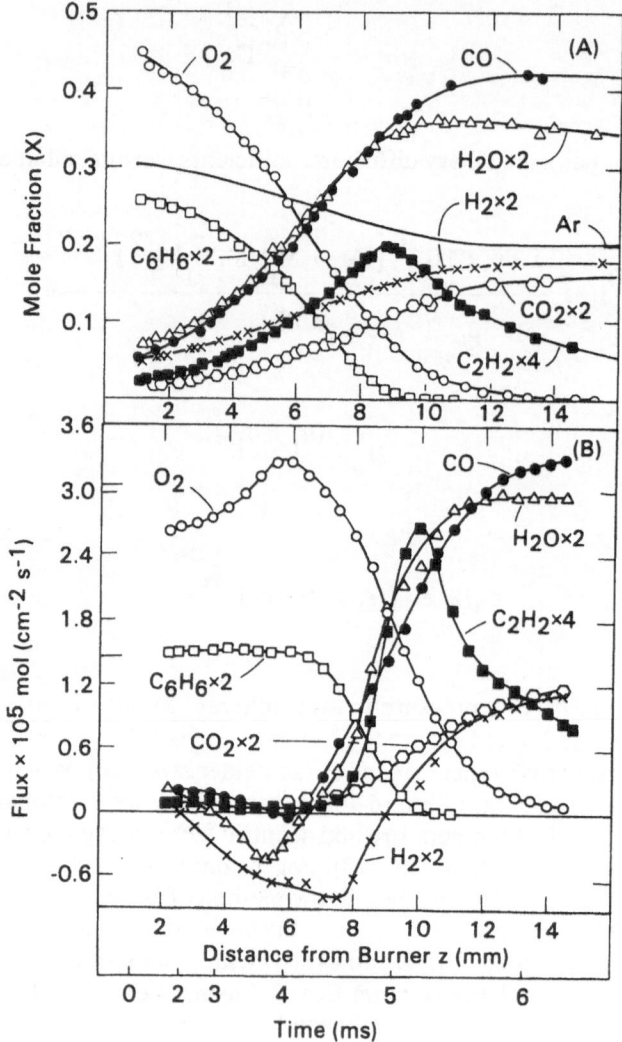

Fig. 3. Mole fractions (A) and molar fluxes (B) of major stable species vs. distance from burner in a near-sooting ($\phi = 1.8$) benzene (13.5 mol %) — oxygen (56.5 mol % — argon (30.0 mol %) flame. Cold gas velocity = 0.5 m/s. Pressure = 2.67 kPa (20 torr).

indicated by the hash marks on the ordinate. The constant C_6H_6 flux between 2 and 6 mm indicates that the 45% decrease in the C_6H_6 mole fraction in this region is caused by diffusion and not chemical reaction. The agreement between the inlet C_6H_6 flux and the C_6H_6 flux between 2 and 6 mm (difference of 6%) is good considering the uncertainties in the alignment of the temperature and mole fraction profiles and the uncertainty ($\pm$ 20%) in the diffusion coefficients. The shape of the O_2 flux profile between 2 and 6 mm suggests that the steep O_2 mole fraction gradient is not accurately reproduced by the 0.7 mm diameter orifice in the sampling probe.

The onset of significant benzene consumption between 6 and 7 mm is supported by the flux profiles of the products CO, CO_2 and H_2O and the major hydrocarbon intermediate C_2H_2 (Fig. 3b). The large mole fractions of these species present between 2 and 6 mm (Fig. 3a) are not produced by chemical reaction in this region, but they diffuse from the primary reaction zone which is clearly shown to be between 6 and 11 mm by the flux profiles of C_6H_6 (Fig. 3b) and nine other carbon-containing intermediate and product species (Figs. 4 and 5).

These flux profiles provide some insight into the production of lower molecular weight hydrocarbons from benzene. The C_2H_2 flux profile (Fig. 3b) exhibits an induction time between the onset of benzene decay at about 6 mm and the sharp increase in C_2H_2 flux at about 8 mm. On the other hand, the CO flux starts to increase at 6 mm, just as the benzene flux begins to decay. This behavior is not

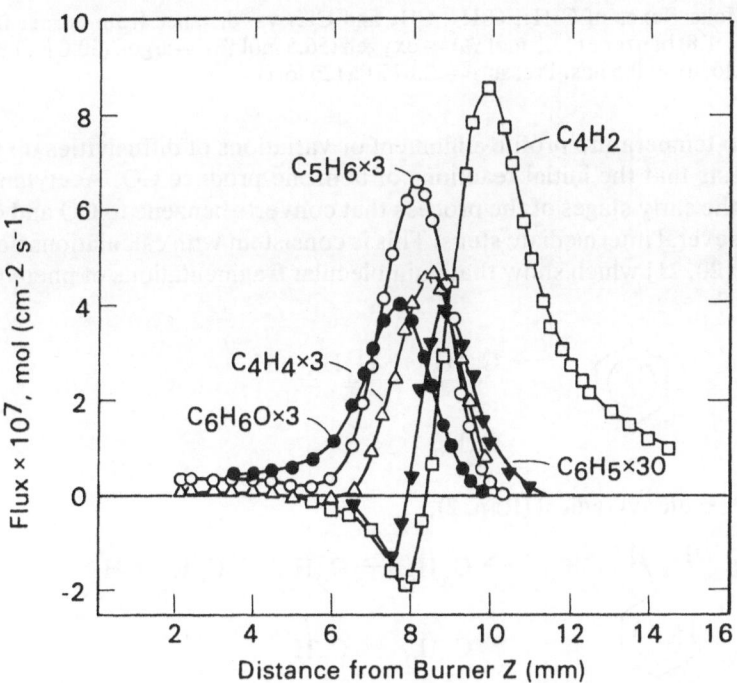

Fig. 4. Molar fluxes of C_6H_6O, C_6H_5, C_5H_6, C_4H_2 and C_4H_4 vs. distance from burner in a near-sooting ($\phi = 1.8$) benzene (13.5 mol %) — oxygen (56.5 mol %) — argon (30.0 mol %) flame. Cold gas velocity = 0.5 m/s. Pressure = 2.67 kpa (20 torr).

References pp. 135-137.

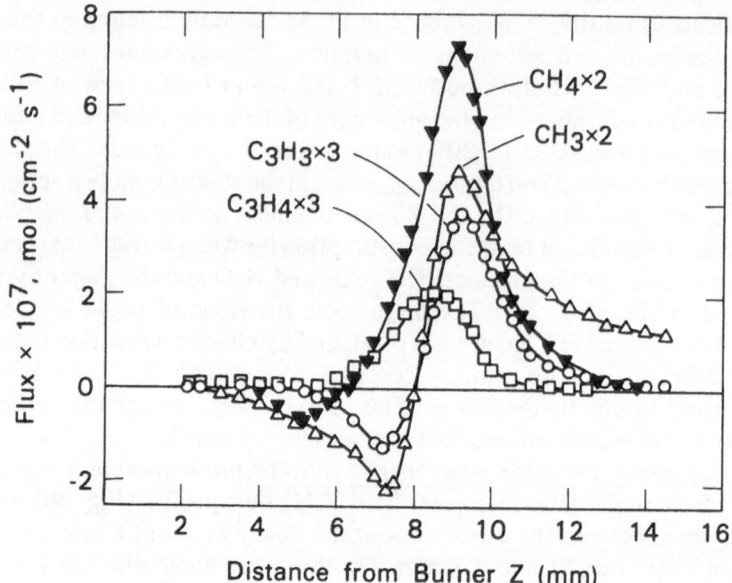

Fig. 5. Molar fluxes of C_3H_4, C_3H_3, CH_4 and CH_3 vs. distance from burner in a near-sooting ($\phi = 1.8$) benzene (13.5 mol %) — oxygen (56.5 mol %) — argon (30.0 mol %) flame. Cold gas velocity = 0.5 m/s. Pressure = 2.67 kPa (20 torr).

sensitive to temperature profile alignment or variations of diffusivities up to 20%. This suggests that the initial reactions of benzene produce CO. Acetylene is not formed in the early stages of the process that converts benzene to CO and CO_2 but only after several intermediate steps. This is consistent with calculations described elsewhere [20, 21] which show that unimolecular fragmentations of phenyl radical (Rxn. 1)

$$\rightarrow C_4H_3 \rightarrow C_4H_2 + H$$
$$+$$
$$C_2H_2$$
(Rxn. 1)

and cyclohexadienyl radical (Rxn. 2),

$$\rightarrow C_4H_5 \rightarrow C_2H_3 \rightarrow C_2H_2 + H$$
$$+ \qquad +$$
$$C_2H_2 \qquad C_2H_2$$
(Rxn. 2)

that produce C_2H_2 prior to CO, cannot account for the observed rate of benzene decay.

An alternative interpretation might be that C_2H_2 is consumed as fast as it is produced in the early stages of benzene disappearance. This implies that benzene decay is rate limiting and that the C_2H_2 produced is consumed very rapidly to give a near zero net C_2H_2 flux until about 8 mm. Since there is considerable C_2H_2 (0.035 mole fraction) at 7 mm (Fig. 3a), instantaneous conversion of C_2H_2 to CO should consume the C_2H_2 that has diffused upstream from the zone of net C_2H_2 production also. It is unlikely that C_2H_2 production would so nearly equal the rate of destruction in this zone as indicated by the C_2H_2 flux profile.

Although the mode of initial attack on the benzene ring is uncertain [21], the following reaction scheme, which is based on many studies of single reactions [32-46], is consistent with the early production of CO, the delayed production of C_2H_2 and the following sequence of positions (mm) of the flux profile maxima (Figs. 3-6):

$$C_6H_6O, 7.7; \quad C_5H_6, 8.1; \quad C_4H_4, 8.4; \quad C_3H_4, 8.5; \quad CH_4, 9.1;$$

$$CH_3, 9.1; \quad C_3H_3, 9.2; \quad C_2H_2, 9.8; \quad C_4H_2, 9.9$$

As discussed elsewhere [21], analysis of the mole fraction data for this flame led to the conjecture that benzene destruction was initiated by reaction with O-atoms. Because the O-atom signal was obscured by CH_4 in this rich flame, it has not been possible to confirm this hypothesis. However, the flux profiles for C_6H_6 and C_2H_2 indicate that the reaction of OH rather than O with benzene may be the primary source of C_6H_6O. If it is assumed, as postulated above, that the near zero C_2H_2 flux between 6 and 7.5 mm is due to a very low C_2H_2 production rate and a near zero consumption rate, rather than a delicate balance between high production and consumption rates, then a considerable fraction (22 mol %) of the benzene is consumed in a region (6-7.5 mm) where little or no C_2H_2 is being consumed. Since the mole fractions of C_2H_2 and C_6H_6 are comparable in this region ($X_{C_2H_2} = 0.026$, $X_{C_6H_6} = 0.0667$ at 6 mm; $X_{C_2H_2} = 0.0390$, $X_{C_6H_6} = 0.0292$ at 7.5 mm) the rate coefficient for the reaction that consumes C_6H_6 must be much greater than the rate coefficient for the reaction of C_2H_2 with the same radical. Thus, $k_{O+C_6H_6} \gg k_{O+C_2H_2} = 5.2 \times 10^{13} \exp(-1860/T)$ cm^3 mol^{-1}s^{-1} [47] or $k_{OH+C_6H_6} \gg K_{OH+C_2H_2} = 2.1 \times 10^{12} \exp(-890/T)$ cm^3 mol^{-1}s^{-1} [20]. Low temperature studies (298-462K) [48, 49] of the $O + C_6H_6$ reaction are in excellent agreement and give $k_{O+C_6H_6} = 1.1 \times 10^{13} \exp(-2060/T)$ cm^{3}mol^{-1}s^{-1}. Although $k_{OH+C_6H_6} \approx 5k_{O+C_6H_6}$ at room temperature [50], non-Arrhenius behavior of the rate coefficient for the OH + C_6H_6 reaction between 296 and 473 K due to occurrence of both addition and abstraction makes estimation of $k_{OH+C_6H_6}$ at higher temperatures impossible [37]. An analysis of the present flame data that assumes OH + C_6H_6 is the primary reaction for benzene consumption does not give reasonable Arrhenius parameters. Nevertheless, a mean value of $k_{OH+C_6H_6}$ of about 6×10^{13} cm^3 mol^{-1}s^{-1} can account for benzene decay between 7 and 9 mm. An activation energy of 13.2 kJ mol^{-1} for the OH + C_6H_6 reaction would account for the increase in $k_{OH+C_6H_6}$ between room temperature [37] and 1750 K. Hence, it appears more likely that $k_{OH + C_6H_6} \gg k_{OH + C_2H_2}$ rather than $k_{O + C_6H_6} \gg k_{O + C_2H_2}$. Therefore, from this analysis of the flux profiles, OH addition to benzene appears to be a more reasonable route for benzene consumption than O-atom addition.

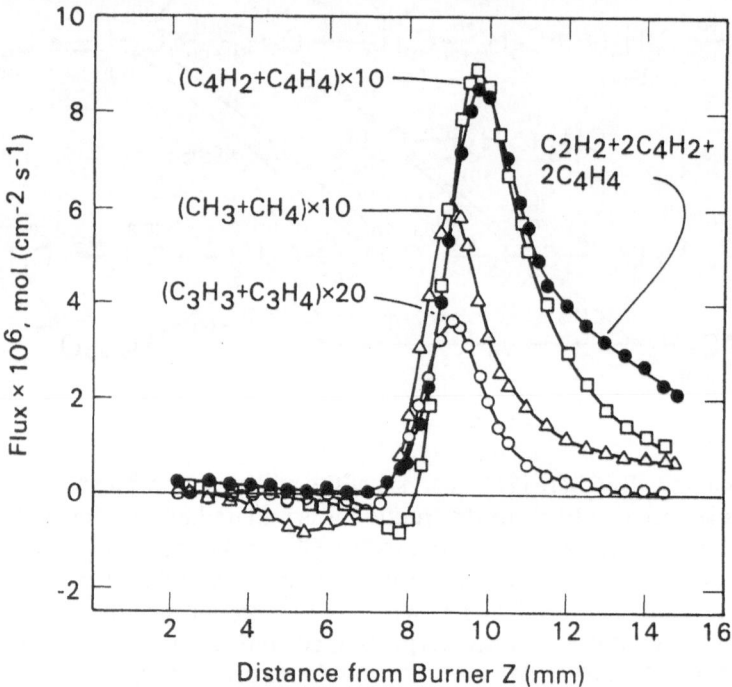

Fig. 6. Sums of fluxes of C_4H_2 and C_4H_4, CH_3 and CH_4, C_3H_3 and C_3H_4, and C_2H_2, 2 x C_4H_2 and 2 x C_4H_4 vs. distance from burner in a near-sooting ($\phi = 1.8$) benzene (13.5 mol %) — oxygen (56.5 mol %) — argon (30.0 mol %) flame. Cold gas velocity = 0.5 m/s. Pressure = 2.67 kPa (20 torr).

The early appearance of a mass 66 species (C_5H_6) along with the mass 94 species (C_6H_6O) (Fig. 4) at 6 mm where benzene decay and CO production begin (Fig. 3b) suggests that the initial attack by O or OH is followed by production of CO and a C_5H_6 hydrocarbon. Exclusion of a possible contribution to the mass 66 signal by a fragment of mass 94 is difficult since the appearance potential of the cyclo- $C_5H_6^+$ fragment ion from ground state phenol is low (estimated to be 10.2 eV from heats of formation available in the literature [51]). Fragmentation at lower electron energies of the energy-rich adduct formed by reaction of benzene with O or OH cannot be excluded even though the experimental ionization potential (IP) of mass 66, referenced to the experimental IP of mass 94 was 8.4 eV. However, two independent observations support the argument that the mass 66 signal is from a neutral species generated in the flame. Firstly, two C_5H_6 hydrocarbons have been detected by GC-MS analysis of batch samples collected via conventional microprobes from this flame [52]. Secondly, a natural flame ion unique to benzene flames has been observed at mass 67. This ion is presumably formed, as many hydrocarbon flame ions are, by protonation of an unsaturated neutral species at mass 66 [53].

An interesting feature of the proposed reaction scheme is the production of C_2H_2 from benzene by removal of one carbon atom at a time in the form of CO, H_2CO or CH_3, if the routes depicted by the dashed and dotted lines are neglected. The observation that the mass flux of C_2H_2 (1.78×10^{-4} g cm^{-2}s^{-1} maximum) approaches but never exceeds one-third the initial mass flux of benzene (5.67×10^{-4} g cm^{-2} s^{-1}) is consistent with this mechanism. The closeness of the approach of the C_2H_2 mass flux to one-third of the initial benzene mass flux suggests a two-stage mechanism whereby each mole of benzene produces a mole of C_2H_2 that is refractory relative to benzene and other hydrocarbon intermediates. This is consistent with the above hypothesis that OH is the major consumer of benzene and that $k_{OH+C_6H_6} \gg k_{OH+C_2H_2}$. Hence, prior to about 10 mm where benzene is almost depleted and the OH mole fraction rises dramatically, C_2H_2 is essentially only produced, i.e., its destruction rate is very slow. Even after 10 mm where OH mole fractions are high the C_2H_2 decay rate is slow enough that C_2H_2 persists beyond 40 mm (Fig. 2).

Another explanation for the close approach of the C_2H_2 mass flux to one-third of the initial C_6H_6 mass flux may be that a significant fraction of the C_2H_2 is produced from C_4 species via the routes depicted by the dotted and dashed lines. The observation, at 9-10 mm where the H-atom mole fraction rises rapidly, that H_2 and C_2H_2 become equilibrated with C_4H_2 through the overall reaction

$$2 \, C_2H_2 \rightleftharpoons C_4H_2 + H_2 \qquad \text{(Rxn. 3)}$$

indicates that a fast route for production of two moles of C_2H_2 from the C_4 hydrocarbons exists. The fact that mole fractions of C_2H consistent with equilibration of the route shown by the dotted and dashed lines were not observed in this flame although they were well within the limits of detectability of the apparatus [20, 21,

54] and the observation in similar acetylene flames of equilibria among hydrocarbon ions, C_2H_2, C_4H_2 and H_2 [55], suggest that ion-molecule reactions may be responsible for the observed equilibrium (Rxn. 3). This equilibrium is also responsible for the C_4H_2 flux maximizing near the C_2H_2 flux maxima rather than prior to that of C_3H_3.

The flux profiles of the C_1, C_3 and C_4 hydrocarbon intermediates (Fig. 4 and 5) demonstrate the importance of diffusion of even relatively large molecules in the main reaction zone (6-10 mm). The C_4H_4 flux (Fig. 4) increases from zero at about 6.5 mm. Between 6.5 and 8 mm where the net flux of C_4H_4 is increasing, the C_4H_2 flux and its derivative with respect to distance z are negative. Thus, C_4H_2 formed at z > 7.8 mm is diffusing towards the burner and being consumed at distances less than 7.8 mm. Some of this C_4H_2 may form C_4H_4 via the C_4H_3 radical as suggested in the reaction scheme discussed above. The sum of the C_4H_2 and C_4H_4 fluxes shown in Fig. 6 indicates that C_4H_4 formation almost accounts for the net consumption of C_4H_2 between 6 and 8 mm. Although this discrepancy may be within the experimental accuracy, the C_4H_6 and C_4H_3 mole fraction profiles (Fig. 7b) indicate that formation of C_4H_6 and C_4H_3 may account for some of the consumption of C_4H_2 and C_4H_4 between 6 and 8 mm.

Similar relationships exsts for the fluxes of C_3H_3 and C_3H_4, and CH_3 and CH_4. The C_3H_4 flux (Fig. 5) remains positive and maximizes closer to the burner than the C_3H_3 flux, which indicates net consumption of C_3H_3 at distances less than 7.2 mm. The sum of the C_3H_3 and C_3H_4 fluxes (Fig. 6) shows that the negative C_3H_3 flux is nearly offset by the positive C_3H_4 flux prior to 7.5 mm. This suggests that C_3H_4 is formed from C_3H_3 that has diffused from the region of net C_3H_3 production (z > 7.2 mm). The CH_3 and CH_4 flux profiles (Fig. 5) also indicate that the stable CH_4 is formed from the CH_3 radical as it diffuses towards the burner. The net consumption of CH_3 and CH_4 between 3 and 6 mm (Fig. 6) indicated by the negative flux with a negative slope may be due to reaction of CH_3 to form many of the methyl substituted aromatics observed in the flame or it may be caused by uncertainties in diffusion coefficients and alignment of temperature and mole fraction profiles.

Hence, these flux relationships suggest that species of the same carbon number exchange hydrogen rapidly via reactions with H and H_2. The more hydrogenated species are formed in the lower temperature region closer to the burner where their stability is more favorable thermodynamically. The more hydrogenated species appear to be formed from less hydrogenated species of the same carbon number diffusing towards the burner from the later stages of the primary reaction zone. As discussed below, the presence of these more hydrogenated species may be important in the formation of polycyclic aromatic material in both benzene and acetylene flames.

FORMATION OF PAH AND SOOT IN BENZENE FLAMES

Mole fraction profiles of several PAH observed in the $\phi = 1.8$ near-sooting benzene flame are shown in Fig. 8. Possible structures for these species and those at

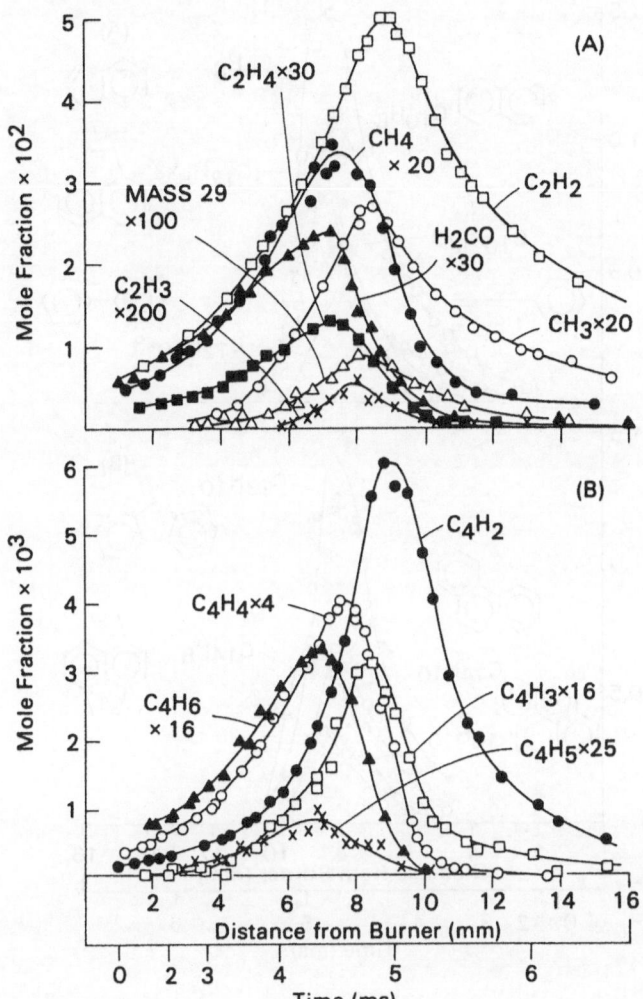

Fig. 7. Mole fractions of C_1 and C_2 (A) and C_4 (B) species vs. distance from burner in a benzene (13.5 mol %) — oxygen (56.5 mol %) — argon (30.0 mol %) flame. Cold gas velocity = 0.5 m/s. Pressure = 2.67 kPa (20 torr).

other masses at which signals were observed are shown in Table 1. The mole fractions of all PAH maximize between 8 and 10 mm above the burner. These relatively low mass PAH (< 228 amu) all disappear as does benzene by 10 or 11 mm. Much of the decrease in the PAH mole fractions is probably due to oxidation. However, some of these species continue to grow as shown in Fig. 9 by the high mass signals for the $\phi = 1.8$ flame. These signals were obtained by operating the quadrupole mass spectrometer as a high pass filter that transmits all ions larger than a specified mass. The signal from species of mass 200 and larger, $I_{M > 200}$, maximizes at about 8.5 mm, while $I_{M > 700}$ is much lower at 8.5 mm and maximizes at

References pp. 135-137.

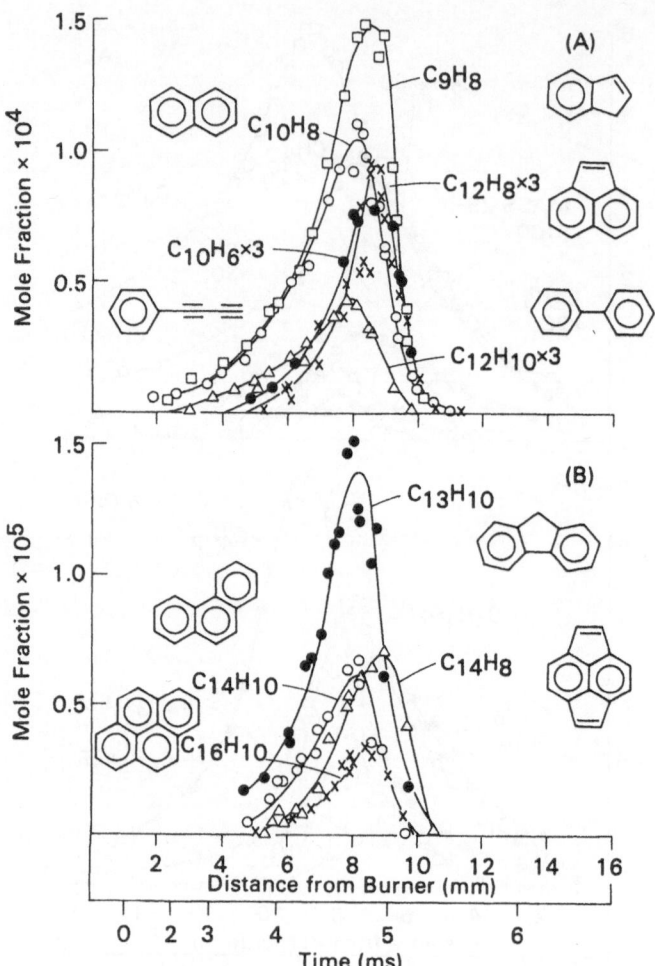

Fig. 8. Mole fractions of polycyclic aromatic hydrocarbons vs. distance from burner in a near-sooting ($\phi = 1.8$) benzene (13.5 mol %) — oxygen (56.5 mol %) — argon (30.0 mol %) flame. Cold gas velocity = 0.5 m/s. Pressure = 2.67 kPa (20 torr).

10.5 mm. This progression from PAH in the mass range 100–200 amu that maximize between 8 and 9 mm to material with mass greater than 700 amu that maximizes at 10.5 mm indicates that the PAH material continues to add carbon. Further indication of this growth is a shift of the mass distribution of the 200–700 amu material to higher masses from 8 to 10 mm. Since the mass numbers at which signals are observable between 200 and 300 amu correspond to aromatic structures it is believed that this high mass material is aromatic rather than polyacetylenic.

The correlation of the growth of this polycyclic material with the onset of soot formation was demonstrated by measurement of $I_{M>700}$ as ϕ was increased at

TABLE 1
Possible Structures of Several Hydrocarbons Observed Above Mass 90

MASS	FORMULA	NAME	STRUCTURE
92	C_7H_8	Toluene	
98	C_8H_2	Octatetrayne	
102	C_8H_6	Phenylacetylene	
104	C_8H_8	Styrene	
116	C_9H_8	Indene	
		Ethynyltoluene	
122	$C_{10}H_2$	Decapentayne	
126	$C_{10}H_6$	1-Phenylbutadiyne	
128	$C_{10}H_8$	Naphthalene	
140	$C_{11}H_8$	Butadiynyltoluene	
142	$C_{11}H_{10}$	Methylnaphthalene	
146	$C_{12}H_2$	Dodecahexayne	
152	$C_{12}H_8$	Biphenylene	
		Acenaphthalene	
154	$C_{12}H_{10}$	Acenaphthene	
		Biphenyl	
166	$C_{13}H_{10}$	Fluorene	
		Methylacenaphthalene	
		Phenalene	
168	$C_{13}H_{12}$	Methylbiphenyl	

MASS	FORMULA	NAME	STRUCTURE
170	$C_{14}H_2$	Tetradecaheptayne	
176	$C_{14}H_8$	Cyclopent[f,g]acenaphthylene	
		Cyclopent[b,c]acenaphthylene	
178	$C_{14}H_{10}$	Anthracene	
		Phenanthrene	
190	$C_{15}H_{10}$	4H-Cyclopenta[def]phenanthrene	
192	$C_{15}H_{12}$	Methylanthracene	
		Methylphenanthrene	
202	$C_{16}H_{10}$	Fluoranthene	
		Pyrene	
216	$C_{17}H_{12}$	Benzofluorene	
		Methylfluoranthene	
		Methylpyrene	
226	$C_{18}H_{10}$	Benzo[ghi]fluoranthene	
		Cyclopenta[cd]pyrene	
228	$C_{18}H_{12}$	Chrysene	
		Benz[a]anthracene	

References pp. 135-137.

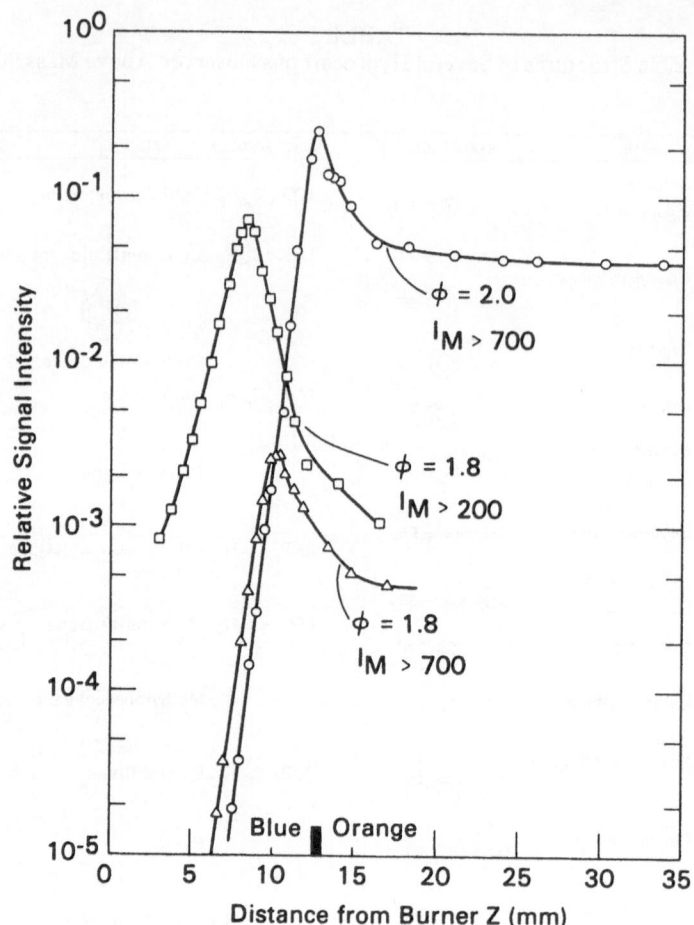

Fig. 9. Relative intensities of signals from high mass mode operation (see text) vs. distance from burner. (□) species larger than mass 200, $I_{M > 200}$; and (Δ) species larger than mass 700, $I_{M > 700}$, in near sooting ($\phi = 1.8$) benzene (13.5 mol %) — oxygen (56.5 mol %) — argon (30.0 mol %) flame. (○) species larger than mass 700, $I_{M > 700}$, in sooting ($\phi = 2.0$) benzene (14.7 mol %) — oxygen (55.3 mol %) — argon (30.0 mol %) flame. Pressure = 2.67 kPa (20 Torr), cold gas velocity = 0.5 m/s for both $\phi = 1.8$ and $\phi = 2.0$ flames. Black marker at 13 mm designates blue-orange boundary in sooting ($\phi = 2.0$) flame.

constant cold gas velocity, pressure and mole per cent argon. The orange luminosity characteristic of soot becomes visible at $\phi = 1.9$. The $I_{M>700}$ signal at $\phi = 2.0$, shown in Fig. 9 on the same basis as that at $\phi = 1.8$, rises sharply to a maximum and then declines rapidly at 13 mm where the orange luminosity becomes visible. The $I_{M>700}$ signal maximum at $\phi = 2.0$ is about one hundred times greater than that of the non-sooting $\phi = 1.8$ flame. This signal should be proportional to the number density of the high mass species. Fluorescence signals measured in atmospheric pressure C_6H_6-air flames show a similar behavior [14].

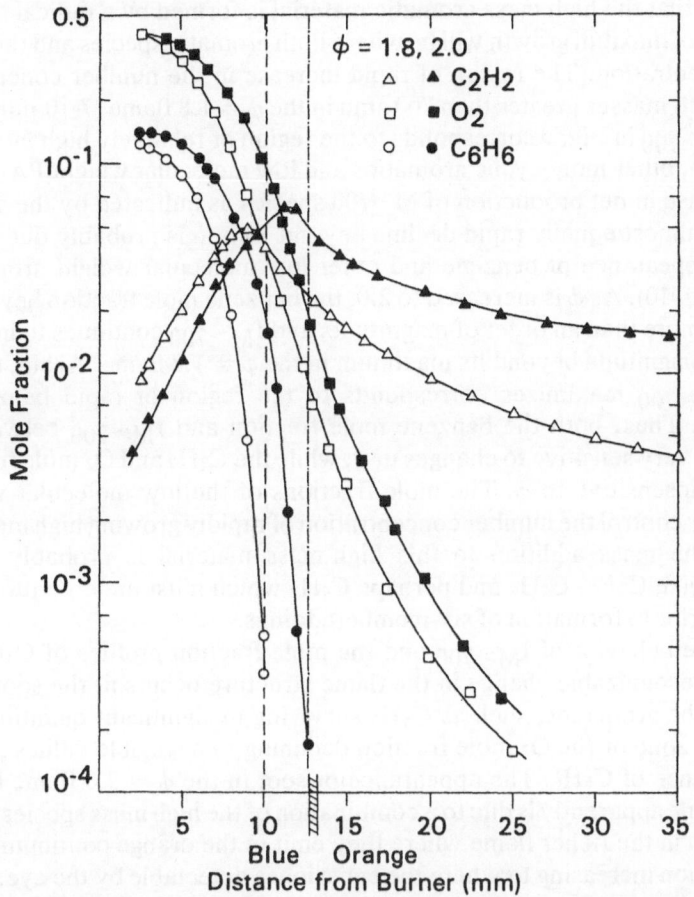

Fig. 10. Log of mole fraction of C_2H_2 (triangles), O_2 (squares) and benzene (circles) vs. distance from burner for near sooting ($\phi = 1.8$, open symbols) and sooting ($\phi = 2.0$, shaded symbols) flames. For $\phi = 1.8$: benzene = 13.5 mol %, oxygen = 56.5 mol %, argon = 30.0 mol %. For $\phi = 2.0$: benzene = 14.7 mol %, oxygen = 55.3 mol %, argon = 30.0 mol %. Pressure = 2.67 kPa (20 torr), v = 0.5 m/s for both flames. Dashed (---) and dotted (—·—·—) lines designate position of maxima of $I_{M \geq 700}$ in $\phi = 1.8$ and $\phi = 2.0$ flames, respectively. Shaded region at 13 mm designates blue-orange boundary in the sooting ($\phi = 2.0$) flame.

Comparison of the mole fraction profiles of C_6H_6, C_2H_2 and O_2 for the $\phi = 1.8$ and $\phi = 2.0$ flames (Fig. 10) with the $I_{M>700}$ profiles provides some insight into the possible role of aromatics in this rapid growth process. For both flames, the C_6H_6 mole fraction decreases very rapidly where the $I_{M>700}$ signals of Fig. 9 maximize [as indicated in Fig. 10 by the dashed (---) and dotted (—·—·—) lines for $\phi = 1.8$ and $\phi = 2.0$, respectively]. In the $\phi = 1.8$ flame, this point (10 mm) also corresponds to the region of the sharp increase in the mole fractions of H and OH (Fig. 2). Presumably, the rapid increase in H and OH occurs around 12.7 mm in the $\phi = 2.0$ flame.

References pp. 135-137.

Assuming that the high mass aromatic material is formed by a radical mechanism, the region of maxium growth will be where both aromatic species and radicals are in high concentration. The region of rapid increase in the number concentration of species with masses greater than 700 amu in the $\phi = 1.8$ flame (7-10 mm) as indicated by $I_{M>700}$ in Fig. 9 corresponds to the region of relatively high mole fractions of benzene, other monocyclic aromatics and low molecular weight PAH. The sudden decrease in net production of M>700 species as indicated by the sharp maximum and almost equally rapid decline beyond 10 mm is probably due to the very rapid disappearance of benzene and other low molecular weight aromatics near 10 mm (Fig. 10). As ϕ is increased to 2.0, the benzene mole fraction beyond 10 mm increases more than an order of magnitude, and $I_{M>700}$ continues to increase two orders of magnitude beyond its maximum in the $\phi = 1.8$ flame. Again, the point at which $I_{M>700}$ maximizes corresponds to the region of rapid benzene decay (12.7 mm). Thus, both the benzene mole fraction and $I_{M>700}$ between 10 and 13 mm are very sensitive to changes in ϕ, while the C_2H_2 and O_2 mole fractions are relatively insensitive to ϕ. The mole fractions of the low molecular weight aromatics may control the number concentration of rapidly growing high mass species. Most of the mass addition to this high-mass material is probably from non-aromatics like C_2H_2, C_4H_4 and perhaps C_4H_2 which must undergo further hydrogenation prior to formation of six-membered rings.

From the behavior of $I_{M>700}$ and the mole fraction profiles of C_2H_2, O_2 and C_6H_6, no recognizable change in the flame structure occurs at the soot limit with regard to the aromatics, such as C_6H_6 surviving in significant quantities into the post-flame zone or the O_2 mole fraction declining to negligible values prior to the disappearance of C_6H_6. The appearance of soot in the $\phi = 2.0$ flame but not the $\phi = 1.8$ flame apparently is due to a combination of the high mass species growing to larger sizes in the richer flame where they emit in the orange continuum and their concentration increasing to where the emission is detectable by the eye.

The structures of the lower molecular weight PAH formed in the early stages of this rapid growth process (Fig. 8, Table 1) indicate that they are products of reactions of aromatic species with C_2, C_3 and C_4 species. Furthermore, the region of rapid increase in the PAH mole fractions (7-8 mm) corresponds to the region where the mole fractions of more hydrogenated species, e.g., C_2H_4, C_4H_4 and C_4H_6 (Fig. 7), and phenyl and benzyl radicals [21] maximize. These observations are consistent with those of Smith [56, 57] in toluene pyrolysis between 1173 and 2173 K. He found the concentration of higher molecular weight hydrocarbons (up to $C_{14}H_{10}$) to maximize between 1573 and 1673 K where the mole fractions of phenyl radical, C_4H_4, C_3H_3, C_3H_4 and other more hydrogenated hydrocarbons were also near their maximum concentrations. Beyond 1673 K the concentration of these lower molecular weight species and the higher molecular weight hydrocarbons declined as C_2H_2 and C_4H_2 continued to increase, similar to the behavior observed in the present $\phi = 1.8$ flame. Thus, from the observations made in the present $\phi = 1.8$ benzene flame and Smith's toluene pyrolysis experiments, it appears that a proper mix of monocyclic aromatic species, e.g, benzene, phenyl radical, toluene, and benzyl radical, and non-aromatics such as C_2H_2, C_4H_4, C_4H_3, C_2H_4, etc. results in rapid production of higher molecular weight PAH. The role of the initial ring in this process may be to provide a structure that is capable of stabilizing the radical adduct formed by addition of non-aromatic species.

For rapid growth of aromatic structures by a free radical mechanism, the stabilization of the energy-rich adduct formed by addition to radicals to stable hydrocarbons may be a crucial step. Removal of the energy of chemical excitation through collisional processes may not be sufficient to prevent unimolecular fragmentation at temperatures above 1500 K where PAH and soot are formed in pyrolysis systems [56, 57], diffusion flames [58] and the present $\phi = 1.8$ and $\phi = 2.0$ benzene flames. The unimolecular decomposition of the thermalized radical adducts can be very fast at 1500 K and above. For example, from transition state theory the half-life $t_{1/2}$ of a radical that decomposes unimolecularly through a process with an activation energy of 209 kJ mol^{-1} (50 kcal mol^{-1}) is only about 10^{-7}s at 1500 K if it is assumed that the entropy of activation is zero. This is of the same order as the time between collisions of a molecule having the collision diameter of toluene with other species of possible importance in the growth process (e.g., H, 2×10^{-6}s; H$_2$, 5×10^{-8}s; C$_2$H$_2$, 2×10^{-7}s; C$_4$H$_2$, 3×10^{-6}s; C$_4$H$_4$, 10^{-5}s, at 7.5 mm in the present $\phi = 1.8$ flame). Thus, unimolecular reactions with activation energies less than about 210 kJ mol^{-1} are much faster than bimolecular reactions. Therefore, the channel of decomposition of the radical adduct which generally has one or more bonds with dissociation energies less than 210 kJ mol^{-1} is important in determining whether growth into fused ring aromatic structures or fragmentation of the carbon framework occurs.

The exact role of the aromatic ring in stabilizing the radical adducts is uncertain. To gain insight into the rapid growth of aromatic structures, such as the form of the aromatic reactant (radical or stable molecule) and the nature of the "growth" species that are responsible for the major fraction of the mass that ends up as high molecular weight PAH and/or soot, several mechanisms have been considered on the basis of the enthalpies of reactions of the initial addition steps, the subsequent steps required to stabilize the adduct by formation of six-membered rings and other processes that may stabilize the adduct without forming six-membered rings. Mechanisms based on both the addition of non-aromatic radicals to stable aromatic structures (aromatic substitution reactions) and the addition of aromatic radicals to non-aromatic stable species have been considered [20].

For this analysis the thermochemical data for the stable structures considered were estimated from the group additivity method of Benson [59] and Stein $et\ al.$ [60]. The enthalpies of formation ΔH°_{f298} of radicals were estimated from ΔH°_{f298} of stable species and known bond dissociation energies [61] when available. The ΔH°_{f298} of C$_4$H$_3$ and C$_4$H$_5$ radicals (two isomers each) and those of similar structure were estimated from the methods of Cowperthwaite and Bauer [62], heats of hydrogenation from Skinner and colleagues [63-65] and updated thermochemical data [61] and are reported in reference 20. When dissociation energies were not available for bonds which break to give π-bonding resonance stabilized radicals, the resonance stabilization energies for alternant radicals were estimated using the methods of Stein and Golden [66].

To summarize the results of the preliminary analysis of the addition of the non-aromatic radicals C$_2$H$_3$, C$_3$H$_3$, C$_4$H$_3$ and C$_4$H$_5$ to benzene, fused ring compounds such as naphthalene and indene are not likely to be formed by rapid unimolecular rearrangement of the cyclohexadienyl-type radical adducts. In some cases, e.g., addition of the 1,3-butadien-1-yl (1-C_4H_5) radical to benzene (Rxn. 4), the favored channel is H-atom elimination from the cyclohexadienyl-type adduct (Rxn. 5) to form a stable structure with an unsaturated side chain (D) rather than formation of the fused ring compound (E) via Rxn. 6.

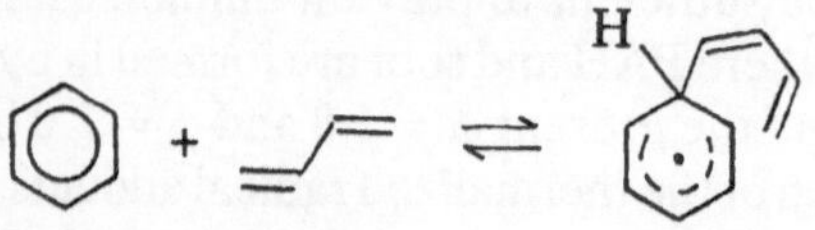

$$\Delta H^{\circ}_{Rxn.\ 298} \quad (Rxn.4)$$

$$-107 \text{ kJ mol}^{-1}$$

ΔH°_{f298} 83 kJ mol^{-1} 344   320

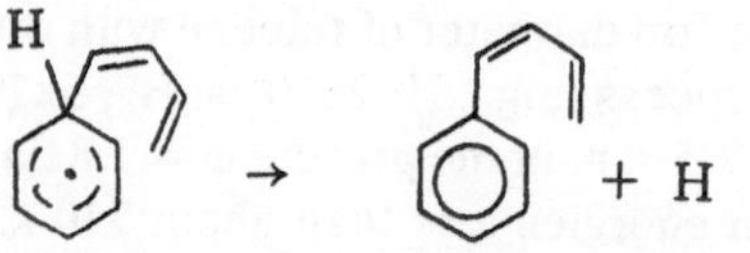

106   (Rxn. 5)

208   218
(D)

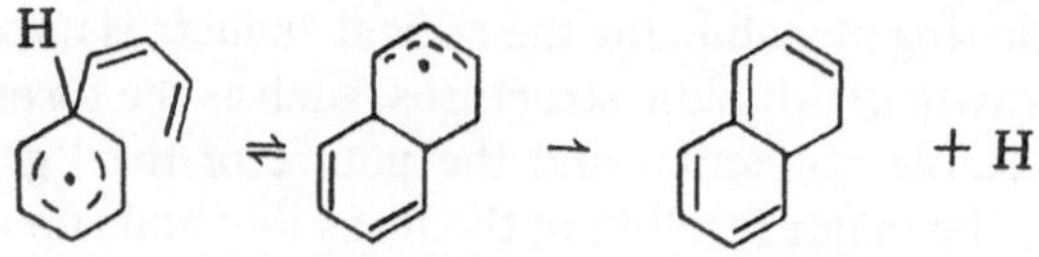

-19,   182   (Rxn. 6)

301   265
(E)

In other cases, e.g., C_3H_3 addition (Rxn. 7), decomposition of the adduct back to reactants is favored (Rxn. -7). Some of the substitution products

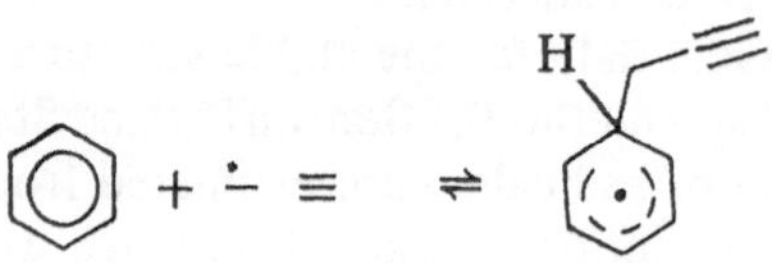

$$\Delta H^{\circ}_{Rxn.\ 298} \quad (Rxn.\ 7)$$

$$-39 \text{ kJ mol}^{-1}$$

ΔH°_{f298} 83 kJ mol^{-1} 362 ± 5   405 ± 21

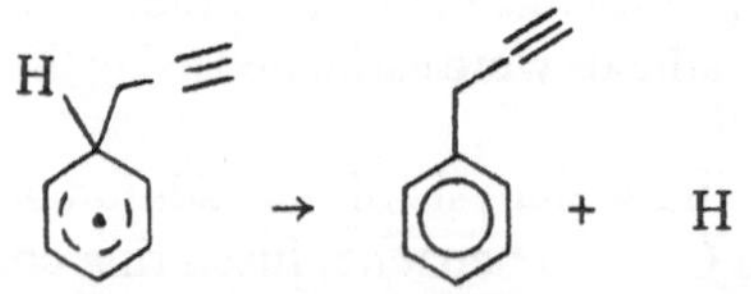

114   (Rxn. 8)

302   218

(formed by H-atom elimination from the cyclohexadienyl-type radical) with unsaturated side chains may participate in subsequent bimolecular reactions that yield radicals which may then unimolecularly form fused aromatic ring structures. For example, the 1-phenyl-cis 1,3-butadien-4-yl radical formed by H-atom abstraction from the 4-position of 1-phenyl-cis 1,3-butadiene (D) formed in Rxn. 5 may cyclize via an internal aromatic substitution reaction (Rxn. 9).

$$\Delta H^{\circ} \text{Rxn. 298}$$

$$\rightleftharpoons \quad \rightarrow \quad + \text{ H} \quad -112,42 \quad \text{(Rxn. 9)}$$

$$\Delta H^{\circ}_{f298} \quad 438 \text{ kJ mol}^{-1} \quad 326 \quad 150 \quad 218$$

The analysis of the addition of aromatic radicals to non-aromatic molecular species included addition of phenyl radical (C_6H_5) and benzyl radical (C_7H_7) to acetylene, butadiyne (C_4H_2), 1-buten-3-yne (C_4H_4), methyl acetylene (C_3H_4) and allene (also C_3H_4). Of those considered, there are only two mechanisms, C_6H_5 plus C_4H_4 and C_7H_7 plus C_3H_4 (allene), whereby unimolecular rearrangement of the initial adduct to form fused six-membered ring systems is favored over other decomposition channels. In general the initial adducts formed by C_7H_7 addition (e.g., to C_2H_2, Rxn. 10) are unlikely to stabilize by H-atom elimination (Rxn. 11) to form species with unsaturated side chains. The carbon framework is stabilized by H-atom elimination only after the initial adduct has undergone subsequent internal addition reactions such as cyclization to a five-member ring (Rxn. 12).

$$+ C_2H_2 \quad \rightleftharpoons \quad \quad \Delta H^{\circ} \text{Rxn. 298} \quad \text{(Rxn. 10)}$$

$$\Delta H^{\circ}_{f298} \quad 188 \text{ kJ mol}^{-1} \quad 227 \quad 370 \quad -45 \text{ kJ mol}^{-1}$$

$$\rightarrow \quad + \text{ H} \quad 150 \quad \text{(Rxn. 11)}$$

$$302 \quad 218$$

$$\rightleftharpoons \quad \rightarrow \quad + \text{ H} \quad -87, 108 \quad \text{(Rxn. 12)}$$

$$283 \quad 173$$

On the other hand, initial adducts formed by C_6H_5 addition (e.g., Rxn. 13) are likely to eliminate H-atoms directly and form species with unsaturated side chains (Rxn. 14) rather than form fused ring species.

$$\Delta H^{\circ}_{f298} \quad 328 \qquad 456 \qquad\qquad 558$$

$$\Delta H^{\circ} \text{ Rxn. } 298$$
$$-226 \text{ kJ mol}^{-1} \qquad \text{(Rxn. 13)}$$

$$(F) \qquad\qquad 558 \qquad\qquad +H \quad 218 \qquad \text{(Rxn. 14)}$$
$$\qquad\qquad\qquad\qquad\qquad\qquad 218$$

However, the vinyl-type radicals (e.g., F) formed from either phenyl radical addition to a triple bond (Rxn. 13) or H-atom addition to a triple bond of a side chain (Rxn.-14) may react with C_2H_2 (Rxn. 15). The resulting vinyl-type radical may then cyclize to form a fused ring structure through an internal aromatic substitution reaction (Rxn. 16).

$$\Delta H^{\circ}_{f298} \quad 558 \text{ kJ mol}^{-1} \qquad \begin{array}{c}+C_2H_2 \\ 227\end{array} \rightarrow \qquad 667 \qquad \text{(Rxn. 15)}$$
$$\Delta H^{\circ} \text{ Rxn. } 298$$
$$-118 \text{ kJ mol}^{-1}$$

$$\qquad\qquad \underset{535}{\overset{H}{}} \qquad\qquad 394 \qquad \begin{array}{c}+H \\ 218\end{array} \qquad -132, 77 \qquad \text{(Rxn. 16)}$$

This brings the discussion full circle — back to the competition between unimolecular and bimolecular reaction steps. The half-life of vinyl-type radicals formed either by C_6H_5 addition to triple bonds (e.g., Rxn. 13) or H-atom addition to the α-position of a triple bond conjugated with an aromatic ring (e.g., Rxn.-14) will be determined by the endothermicity of the H-elimination step (e.g., Rxn. 14). The four vinyl-type adducts formed by C_6H_5 addition to C_2H_2, C_3H_4 (methylacetylene), C_4H_2 and C_4H_4 have enthalpies of reaction for H-atom elimination ΔH_{H-e} of 164, 157, 218 and 184 kJ mol^{-1}, respectively. Only the vinyl-type radical formed from C_4H_2 (F) has a ΔH_{H-e} larger than 210 kJ mol^{-1} and a half-life greater than 10^{-7}s at 1500 K, which is of the same order as the time for a collision with C_2H_2 in the present $\phi = 1.8$ flame. It is the interaction of the unpaired electron with the triple bond of the vinyl-type radical (F) that is responsible for its added stability. Thus, aromatic structures with polyacetylenic side chains (C_4H and higher) may be more effective in forming fused ring structures in an environment of high C_2H_2, H-atom and C_4H_2 concentrations than species without the conjugated triple bonds.

The preceding analysis indicates that there are few mechanisms in which the favored channel for unimolecular stabilization of the radical formed by addition of aromatic and non-aromatic species might result in formation of fused aromatic ring systems. Many reactions may result in stable species with unsaturated side chains. The stability of radicals formed by addition of H-atoms to unsaturated side chains is important in determining whether bimolecular addition of more carbon, probably in the form of C_2H_2, can compete with unimolecular decomposition of the radical. Vinyl-type radicals formed from polyacetylenic side chains are more stable with respect to H-atom elimination than radicals from other types of side chains. Although phenyl radical addition to unsaturated non-aromatics is probably the dominant route for formation of aromatic species with side chains, the essential feature that the initial aromatic structure contributes to the PAH and soot formation process is the ability to undergo substitution reactions [67]. Internal aromatic substitution reactions (e.g., Rxn. 16) are important for stabilizing the growing carbon network. So the initial aromatic structure provides a foundation upon which a larger aromatic structure is built by addition of non-aromatics such as C_2H_2, C_4H_4 and perhaps C_4H_2. Other mechanisms for growth of fused ring aromatic structures that require methyl substitution and rely on internal aromatic substitution reactions for adduct stabilization have been discussed in a previous paper [21].

FORMATION OF PAH AND SOOT IN ACETYLENE FLAMES

The previous discussion of the role of aromatic species in PAH and soot formation was based on observations made in benzene flames. Recent measurements in our laboratory in a sooting $\phi = 3.0\, C_2H_2$ - O_2 - 3 mol % Ar flame that will be reported in detail elsewhere [68] suggest that similar processes may be occurring in flames of non-aromatic fuels. The structure of the near-sooting $\phi = 1.8$ benzene flame and the sooting $\phi = 3.0$ acetylene flame are compared in Fig. 11. Benzene is formed early in the C_2H_2 flame, in the region where the more hydrogenated species (e.g., C_4H_4 are in high concentration. In both flames, the mole fractions of the relatively small PAH (represented by $C_{14}H_{10}$ in the benzene flame and $C_{14}H_8$ in the acetylene flame) maximize in the presence of benzene, near the maximum C_4H_4 mole fraction and prior to the maxima of the high mass signals.

Although the high mass signals ($I_{M>700}$ and $I_{M>1000}$) are not directly comparable since different low mass cutoffs have been used in the two flames, there is more overlap between the profiles of the relatively small PAH and the high mass signals in the C_6H_6 flame than in the C_2H_2 flame. This may be due to the much higher C_6H_6 and PAH mole fractions early in the benzene flame and their sudden disappearance near 10 mm. Since the C_2H_2 flame is much richer, C_6H_6, C_4H_4, $C_{14}H_8$ and other PAH survive well into the burned gas region (Fig. 12). The slower increase in number of high mass species in the C_2H_2 flame persists longer and is perhaps limited

References pp. 135-137.

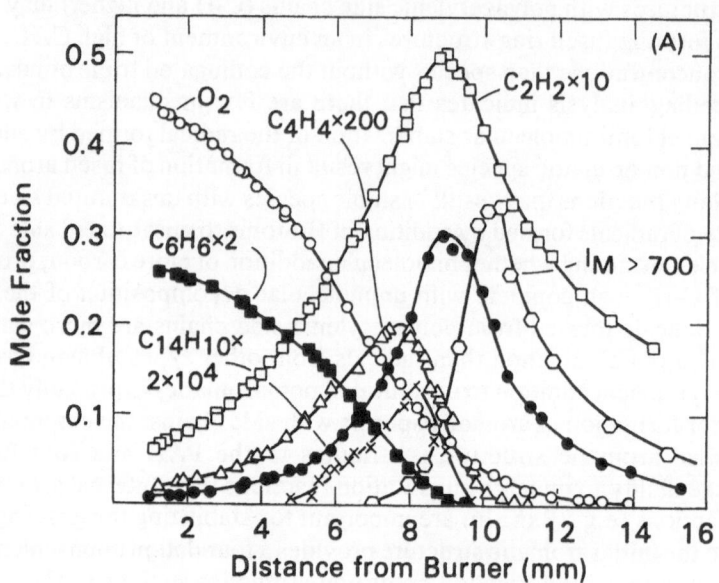

ϕ = 1.8 Benzene Flame

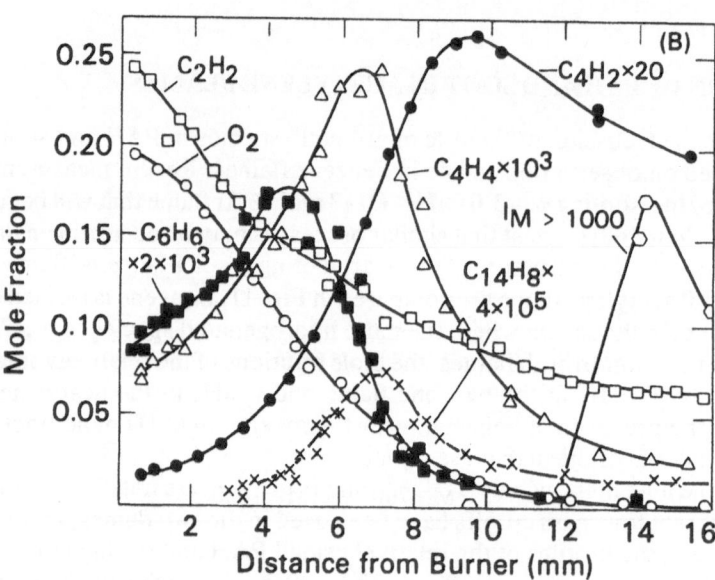

ϕ = 3.0 Acetylene Flame

Fig. 11. Mole fractions of several species and high mass signals ($I_{M \, > \, m}$, scale arbitrary) vs. distance from the burner in (A) a near-sooting (ϕ = 1.8) benzene (13.5 mol %) — oxygen (56.5 mol %) — argon (30.0 mol %) flame and (B) a sooting (ϕ = 3.0) acetylene (52.9 mol %) — oxygen (44.1 mol %) — argon (3.0 mol %) flame. Pressure = 2.67 kPa (20 Torr), cold gas velocity = 0.5 m/s for both flames.

by a decreasing radical mole fraction rather than the disappearance of aromatic species as may be the case for the benzene flame.

The mole fraction profiles of all the aromatic species measured in this sooting acetylene flame ($C_{10}H_8$, naphthalene; $C_{12}H_8$, acenaphthalene; $C_{14}H_8$, cyclopenta-cenaphthalene) exhibit similar behavior (Fig. 12). Rapid formation of relatively small aromatics occurs in the oxidation zone where their mole fractions maximize. Sharp declines occur prior to or as the $I_{M>1000}$ signal increases several orders of magnitude. Minima occur near the maximum of $I_{M>1000}$. The mole fractions of all aromatics then increase in the post flame zone much more slowly than in the oxidation zone.

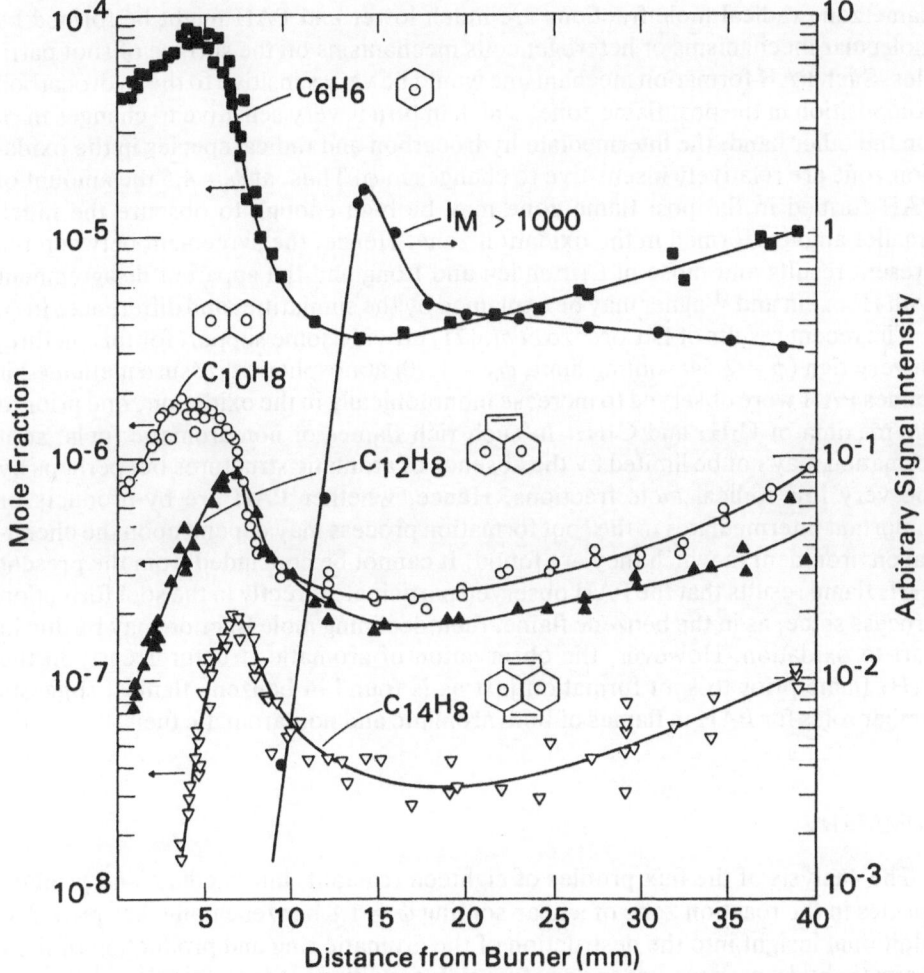

Fig. 12. Mole fractions of C_6H_6, $C_{10}H_8$, $C_{12}H_8$ and $C_{14}H_8$ vs. distance from burner in a sooting ($\phi = 3.0$) acetylene (52.9 mol %) — oxygen (44.1 mol %) — argon (3.0 mol %) flame. Pressure = 2.67 kPa (20 Torr). Cold gas velocity = 0.5 m/s.

References pp. 135–137.

These results are similar to those of Crittenden and Long [69] who studied a $\phi = 3.24$ C_2H_2 - O_2 - 7.3 mol % Ar flame at 5.3 kPa (40 torr) with a cold gas velocity of 0.25 m s^{-1} and observed PAH mole fractions that maximized in the oxidation zone. In apparent contrast, Homann and Wagner [70] have reported that PAH form later in the flame than polyacetylenes and that PAH mole fractions increase steadily behind the oxidation zone without going through any maxima. They show a mole fraction profile for $C_{14}H_8$ in a $\phi = 4.5$ C_2H_2 - O_2 flame at 2.67 kPa and a cold gas velocity of 0.57 m s^{-1} that increases monotonically between 20 and 40 mm from the burner. These seemingly conflicting results may be reconciled if it is assumed that two different mechanisms of production of PAH are operating in the oxidation zone and in the post flame zone. In the oxidation zone where radical concentrations are high the rapid production and decline of PAH is free radical in nature. In the post flame zone radical mole fractions are much lower and PAH might be formed by molecular mechanisms or heterogeneous mechanisms on the surface of soot particles. Such PAH formation mechanisms would be very sensitive to the hydrocarbon composition in the post flame zone, which in turn is very sensitive to changes in ϕ. On the other hand, the intermediate hydrocarbon and radical species in the oxidation zone are relatively insensitive to changes in ϕ. Thus, at $\phi = 4.5$ the amount of PAH formed in the post flame zone may be high enough to obscure the much smaller amount formed in the oxidation zone. Hence, the agreement between the present results and those of Crittenden and Long and the apparent disagreement with Homann and Wagner may be explained by the similarities and differences in ϕ.

The recent results of DiLorenzo et al. [71] provide some support for this picture. In very rich ($\phi = 2.54$; sooting limit, $\phi_c = 1.70$) atmospheric pressure methane-air flames PAH were observed to increase monotonically in the oxidation zone prior to the maxima of C_2H_2 and C_4H_2. In such rich flames of non-aromatic fuels, soot formation may not be limited by the absence of aromatic structures but perhaps by the very low radical mole fractions. Hence, whether PAH are by-products or important intermediates in the soot formation process may depend upon the chemical environment in which they are found. It cannot be concluded from the present C_2H_2 flame results that the PAH observed participate directly in the soot formation process since, as in the benzene flame, their declining mole fraction may be due in part to oxidation. However, the observation of aromatic structures early in the C_2H_2 flame prior to soot formation just as is found in benzene flames, suggests similar roles for PAH in flames of both aromatic and non-aromatic fuels.

SUMMARY

The analysis of the flux profiles of eighteen reactant, intermediate and product species in the reaction zone of a near-sooting $\phi = 1.8$ benzene flame has provided additional insight into the destruction of the aromatic ring and production of non-aromatic hydrocarbons under very fuel rich conditions. Since acetylene is not an early product in the ring fragmentation and it is preceded by CO production, mechanisms of ring destruction such as unimolecular fragmentation of phenyl and cyclohexadienyl radicals to produce C_2H_2 are not significant in this flame. The flux profiles of C_6H_6 and C_2H_2 are consistent with the rate coefficient of the OH + C_6H_6 reaction being much larger than that for OH + C_2H_2. Relationships between the fluxes of C_1, C_3 and C_4 species suggest that species of the same carbon number exchange hydrogen rapidly via reactions with H and H_2. The more hydrogenated species formed closer to the burner may be formed from less hydrogenated species diffusing upstream from near the end of the primary reaction zone.

Analysis of the structure of near-sooting and sooting benzene flames, and a preliminary analysis of reaction mechanisms suggests that the role of the aromatic hydrocarbon is to provide a structure capable of stabilizing by internal aromatic substitution reactions radicals formed from the addition of non-aromatic species. The presence of PAH in the oxidation zone of sooting C_2H_2 flames suggests that aromatics may play a similar role in flames of non-aromatic fuels. Formation of the initial aromatic rings may be an important part of the process. However, increasing mole fractions of PAH in the post flame zone of the sooting C_2H_2 flame and other data from even richer flames suggests that the major radicals H and OH and not aromatic species may become the limiting reactants in the soot formation process as the fuel equivalence ratio is increased.

ACKNOWLEDGEMENTS

We are grateful to the U.S. Enviornmental Protection Agency for support under Grant No. R803242 in the early part of this work. The acetylene flame data are from a study supported by Exxon Research and Engineering Co. We are grateful to Jerald A. Cole for performing the acetylene flame experiments.

REFERENCES

1. J.D. Bittner, and J. B. Howard, *Alternative Hydrocarbon Fuels: Combustion and Chemical Kinetics, (C. T. Bowman and J. Birkeland, ed.) Prog. in Astro. and Aero.*, Vol. 62 (1978), p. 335, AIAA.
2. H.F. Butze, and R. C. Ehlers, "*Effect of Fuel Properties on Performance of a Single Aircraft Turbojet Combustor*", NASA TM-71789, (1975).
3. W. S. Blazowski, *Sixteenth Symposium (International) on Combustion, The Combustion Institute*, Pittsburgh, Pa. (1977), p. 1631.
4. N. J. Friswell, *Comb. Sci. and Tech.*, Vol. 19 (1979), p. 119.
5. S. T. Minchin, *J. of Inst. Petrol. Tech.*, Vol. 17 (1931), p. 102.
6. A. E. Clarke, T. G. Hunter, and F. H. Garner, *J. of Petrol. Tech.*, Vol. 32 (1946), p. 627.
7. R. A Hunt, Jr., *Ind. and Eng. Chem.*, Vol. 45 (1953), p. 602.
8. R. L. Schalla, and G. E. McDonald, *Ind. and Eng. Chem.*, Vol. 45 (1953), p. 1497.
9. K. P. Schug, Y. Manheimer-Timnat, P. Yoccarino, and I. Glassman, *Comb. Sci. and Tech.*, Vol. 22 (1980), p. 235.
10. J. C. Street, and A. Thomas, *Fuel*, Vol. 34 (1955), p. 4.
11. J. J. Macfarlane, F. H. Holderness, and F. S. E. Witcher, *Comb. and Flame*, Vol. 8 (1964), p. 215.
12. F. J. Wright, *Twelfth Symposium (International) on Combustion, The Combustion Institute*, Pittsburgh, PA., (1969), p. 867.

13. W. S. Blazowski, R. B. Edelman, and P. T. Harsha, "Fundamental Characterization of Alternate Fuel Effects in Continuous Combustion Systems," Technical Progress Report prepared for DOE Contract EC-77-C-03-1543, (1978).

14. B. S. Haynes, H. Jander, and H. Gg. Wagner, Ber. Bunsenges. Phys. Chem., Vol. 84 (1980), p. 585.

15. G. Prado, J. Jagoda, K. Neoh, and J. Lahaye, Eighteenth Symposium (International) on Combustion, The Combustion Institute, Pittsburgh, PA, (1981), to be published.

16. S. C. Graham, J. B. Homer, and J. L. J. Rosenfeld, Proc. Roy. Soc. Lond., Vol. 344 (1975), p. 259.

17. D. B. Scully, and R. A. Davies, Comb. and Flame, Vol. 9 (1965), p. 185.

18. R. A. Davies, D. B. Scully, Comb. and Flame, Vol. 10 (1966), p. 165.

19. C. P. Fenimore, G. W. Jones, and G. E. Moore, Sixth Symposium (International) on Combustion, The Combustion Institute, Pittsburgh, PA, (1957), p. 242.

20. J. D. Bittner, Ph.D thesis, Department of Chemical Engineering, M.I.T., (1980).

21. J. D. Bittner, and J. B. Howard, Eighteenth Symposium (International) on Combustion, The Combustion Institute, Pittsburgh, PA, (1981), to be published.

22. R. M. Fristrom, and A. A. Westenberg, "Flame Structure", McGraw-Hill, N.Y. (1965), pp. 74-80.

23. D. F. Fairbanks and C. R. Wilke, Ind. and Eng. Chem., Vol. 42, (1950), p. 471.

24. R. M. Fristrom, and A. A. Westenberg, op. cit., p. 276.

25. R. A. Svehla, "Estimated Viscosities and Thermal Conductivities of Gases at High Temperatures," NASA Technical Report R-132 (1962).

26. J. O. Hirschfelder, C. F. Curtiss, and R. B. Bird, "Molecular Theory of Gases and Liquids", J. Wiley and Sons, N.Y., (1964), p. 1112.

27. L. S. Tee, S. Gotoh, and W. E. Stewart, Ind. and Eng. Chem. Fundam., Vol. 5 (1966), p. 356.

28. R. C. Reid, J. M. Prausnitz, and T. K. Sherwood, "The Properties of Gases and Liquids", 3rd ed., McGraw-Hill, N.Y., (1977), pp. 24 and 551.

29. R. S. Brokaw, Ind. Eng. Chem. Process Des. Dev., Vol. 8 (1969), p. 240.

30. A. Savitsky and M. J. E. Golay, Anal. Chem., Vol 36 (1964), p. 1627.

31. J. Steinier, and Y. Termonia, J. Deltour, Anal. Chem., Vol. 44 (1972), p. 1906.

32. R. A. Bonanno, P. Kim, J. H. Lee, and R. B. Timmons, J. Chem. Phys., Vol. 57 (1972), p. 1377.

33. T. M. Sloane, J. Chem. Phys., Vol. 67 (1977), p. 2267.

34. G. Boocock, and R. J. Cvetanovic, Can. J. Chem., Vol. 39 (1961), p. 2436.

35. P. Kim, J. H. Lee, R. J. Bonanno, and R. B. Timmons, J. Chem. Phys., Vol. 59 (1973), p. 4593.

36. M. C. Sauer, Jr., and B. Ward, J. Phys. Chem., Vol. 71 (1967), p. 3971.

37. R. A. Perry, R. Atkinson, and J. N. Pitts, Jr., J. Phys. Chem., Vol. 81 (1977), p. 296.

38. I. Mani, M. C. Sauer, Jr., Adv. Chem. Ser., Vol. 82 (1968), p. 142.

39. S. J. Sibener, R. J. Buss, P. Casavecchia, T. Hirooka, and Y. T. Lee, J. Chem. Phys., Vol. 72 (1980), p. 4341.

40. R. Cypres, and B. Bettens, Tetrahedron, Vol. 30 (1974), p. 1253.

41. F. Slemr, and P. Warneck, Ber. Bunsenge. Phys. Chem., Vol. 79 (1975), p. 152.

42. J. R. Kanofsky, D. Lucas, F. Pruss, and D. Gutman, J. Phys. Chem., Vol. 78 (1974), p. 311.

43. H. Gg. Wagner, and R. Zellner, Ber. Bunsenges. Phys. Chem., Vol. 76 (1972), p. 518.

44. K. H. Homann, W. Schwanebeck, and J. Warnatz, Ber. Bunsenges. Phys. Chem., Vol. 79 (1975), p. 536.

45. H. Niki, and B. Weinstock, J. Chem. Phys., Vol. 45 (1966), p. 3468.

46. K. H. Homann, J. Warnatz, and C. Wellman, Sixteenth Symposium (International) on Combustion, The Combustion Institute, Pittsburgh, PA, (1977), p. 853.

47. J. Peeters and G. Mahnen, "Combustion Institute European Symposium", Academic Press, N.Y., (1973), p. 53.

48. R. Atkinson and J. N. Pitts, Jr., J. Phys. Chem., Vol. 79 (1975), p. 295.

49. A. J. Colussi, J. L. Singleton, R. S. Irwin, and R. J. Cvetanovic, J. Phys. Chem., Vol. 79 (1975), p. 1900.
50. D. A. Hansen, R. Atkinson, and J. N. Pitts, Jr., J. Phys. Chem., Vol. 79 (1975), p. 1763.
51. F. H. Field, and J. L. Franklin, "Electron Impact Phenomena and the Properties of Gaseous Ions", Academic Press, N.Y., (1970), pp. 239-522.
52. S. M. Faist, "Analysis of Stable Species in a Benzene-Oxygen-Argon Laminar Premixed Flame by Chemical and Spectroscopic Techniques: Applications to Soot Formation and Combustion Chamber Deposits", M.S. thesis, Department of Chemical Engineering, M.I.T., (1979).
53. D. B. Olson, H. F. Calcote, Eighteenth Symposium (International) on Combustion, The Combustion Institute, Pittsburgh, PA, (1981), to be published.
54. J. D. Bittner, Comment following paper by P. Michaud, J. L. Delfau, and A. Barassin, Eighteenth Symposium (International) on Combustion, The Combustion Institute, Pitsburgh, PA, (1981), to be published.
55. P. Michaud, J. L. Delfau, A. Barassin, Eighteenth Symposium (International) on Combustion, The Combustion Institute, Pittsburgh, PA, (1981), to be published.
56. R. D. Smith, Comb. and Flame, Vol. 35 (1979), p. 179.
57. R. D. Smith, J. Phys. Chem., Vol. 83 (1979), p. 1553.
58. J. H. Kent, H. Jander, and H. Gg. Wagner, Eighteenth Symposium (International) on Combustion, The Combustion Institute, Pittsburgh, PA, (1981), to be published.
59. S. W. Benson, "Thermochemical Kinetics", J. Wiley and Sons, N.Y. (1976).
60. S. E. Stein, D. M. Golden, and S. W. Benson, J. Phys. Chem. Vol. 81 (1977), p. 314.
61. "Handbook of Chemistry of Physics", 60th ed., (R. C. Weast, ed.). CRC Press, Boca Raton, FL, (1979), p. F-231.
62. M. Cowperthwaite, and S. H. Bauer, J. Chem. Phys., Vol. 36 (1962), p. 1743.
63. T. Flitcroft, H. A. Skinner, M. C. Whiting, Trans. Faraday Soc., Vol. 53 (1957), p. 784.
64. T. L. Flitcroft, and H. A. Skinner, Trans. Faraday Soc., Vol. 54 (1958), p. 47.
65. H. A. Skinner, and A. Snelson, Trans. Faraday Soc., Vol. 55 (1959), p. 404.
66. S. E. Stein, and D. M. Golden, J. Org. Chem., Vol. 42 (1977), p. 839.
67. M. J. Perkins, in "Free Radicals", (J. Kochi, ed.), J. Wiley and Sons, N.Y., (1973), p. 231.
68. J. D. Bittner, J. A. Cole, J. P. Longwell, and J. B. Howard, manuscript in preparation.
69. B. D. Crittendon, and R. Long, Comb. and Flame, Vol. 20 (1973), p. 359.
70. K. H. Homann, and H. Gg. Wagner, Eleventh Symposium (International) on Combustion, The Combustion Institute, Pittsburgh, PA, (1967), p. 371.
71. A. Di Lorenzo, A. D'Alessio, V. Cincotti, S. Masi, P. Menna, and C. Venitozzi, Eighteenth Symposium (International) on Combustion, The Combustion Institute, Pittsburgh, PA, (1981), to be published.

DISCUSSION

C. Hunter (Ford Motor Company)

In view of the fact that you showed the polycyclics reaching a minimum in the case of the acetylene flame (Fig. 12) and in the case of the benzene flame you showed them decreasing and staying down (Fig. 8), it would suggest that either they're not precursors or that some are and some are not. How would you explain that?

Bittner

The behavior of the PAH that you describe may be due to the difference in fuels. The non-sooting benzene flame at $\phi = 1.8$ is just on the lean side of the sooting limit ($\phi_S = 1.9$). The heavily sooting $\phi = 3.0$ acetylene flame is much richer than the sooting limit ($\phi_S = 2.4$). As discussed in the paper, for the $\phi = 1.8$ and $\phi = 2.0$ benzene flames, the similar sensitivities to changes in ϕ of the benzene mole fraction and the high mass signal ($I_{M > 700}$) in the region of rapid increase in $I_{M > 700}$ (Figs. 9 and 10) suggest that the presence of aromatic structures is important. In the relatively lean benzene flames, the major radicals H and OH survive into the secondary reaction zone. In the much richer acetylene flames the radical mole fractions are expected to decline much more rapidly after reaching their maxima. In the rich acetylene flames, the conversion of lower molecular weight PAH to heavier species that may be soot precursors may be limited by the very low radical mole fractions, and not the absence of aromatic structures.

The gradual increase in the PAH mole fractions in the post flame zone may be due to formation through a mechanism that does not involve radicals but ions or heterogeneous reactions on the surface of the soot particles. In fact, Michaud et al. [56], have observed balanced ion-molecule reactions involving C_3H_3+, C_2H_2 and aromatic species in similar C_2H_2 flames. This suggests that the increase in PAH mole fractions may be caused by a shift in equilibrated reactions as the temperature drops or mole fractions of other species change in the post flame gases.

K. H. Homann *(Technische Hochschule Darmstadt)*

Since you mentioned our earlier work on these aromatic species and the first maximum, you must not forget this first maximum you found is at 5 millimeters above the burner. You're using a molecular beam nozzle that is quite big. We measured a lot of large species there, but we thought that it was not very significant because the probe disturbs the flame, especially if you're sampling from an oxidation zone and so near to the burner surface. I don't think you can rely on species that you find in the concentration of 10^{-7} mole fraction not to be artifacts there. I would be very careful interpreting this.

Bittner

Your objection is to the perturbation of the flame by the probe and not to the quality of the quench of the sample after it has been removed from the flame?

Homann

Yes, this can not easily be controlled. This is a small distance from the burner surface. We concentrated only on the profiles, from about 10 millimeters above the burner upwards and dismissed everything below that (except for very high concentration species like carbon monoxide or carbon dioxide). You have to be careful there.

Bittner

Crittenden and Long [70] used a much smaller probe with a much narrower angle near the orifice and also found maxima in PAH profiles early in the oxidation zone. I known that you have raised questions about the ability of their probe to quench the reactions of interest. It is difficult to envision that both probe perturbations on the flame and an inadequate quench would give similar results.

R. G. Gann *(National Bureau of Standards)*

Along the same line, the paper by Stepowski, *et al**. discusses how far away from the probe you have to be for OH to be wiped out. The reactivity of some of these species is probably comparable under these circumstances. But that's not why I got up to talk. In the last line of your abstract, you talked about a methyl substitution reaction. And in your presentation, you talked about OH reaction dominating the benzene breakdown. I have some problems with alkyl radicals or anything like that destroying an intact ring with no ligands already on it. OH seems a lot more reasonable.

* *D. Stepowski, D. Puechberty, and M. J. Cottereau, 18th Symposium (International) on Combustion, Waterloo, Canada (1980).*

Bittner

I wasn't proposing that methyl substitution is a primary route for benzene destruction. The reference to methyl substitution was its possible importance in aiding formation of closed ring compounds and not benzene destruction.

Gann

So your suggestion is that you start out with a benzene ring, the OH comes along and, I think you said, tends to add or substitute.

Bittner

It tends to add according to our phenyl radical profiles.

Gann

Do you have some hint about what comes next?

Bittner

Well in the reaction scheme presented, CO elimination produces a C_5 hydrocarbon. That is very uncertain. There is evidence from single reaction studies that both supports and refutes this step.

Gann

What fragment do you get down to that gives you that first two-ring compound?

Bittner

Naphthalene may be formed by phenyl radical addition to the terminal carbon of the triple bond in vinylacetylene as postulated in Mechanism II from reference [21].

Mechanism II: Phenyl Radical + Vinylacetylene

$\Delta H^{\circ}_{RXN}\ 298$

IIa   -194 kJ/mol

IIb   21

IIc   4, E_a = 180-188

IId   -209, 126

Gann

Does it seem reasonable that something like that could happen in less than 10 milliseconds?

Bittner

Yes. The rate limiting step in this scheme is the tran-cis isomerization (IIc). Assuming an activation energy of about 45 kcal/mol and a pre-exponential factor of about $10^{13} s^{-1}$, the characteristic time for this reaction is about 4×10^{-7} s at 1500 K.

A. S. Gordon *(University of California, San Diego)*

First of all, I think you did a very nice job here. I have two questions. First, there is a possibility, at least based upon some work that we did a number of years ago, that acetylene, in the presence of radicals, can readily form benzene, radicals being a catalyst for it. You have that kind of condition in your system. I would suggest that you either eliminate that possibility or address it.

Bittner

Certainly, the results from the acetylene flame (Fig. 12) show that benzene is formed from acetylene in rich flames. However, the small mole fractions of benzene (10^{-4} maximum) found in the $\phi = 3.0$ acetylene flame, indicate that this source of benzene production can be neglected in the analysis of the benzene destruction in the much leaner ($\phi = 1.8$) benzene flame.

Gordon

The other thing is: I would like to know why you feel that the benzene, or the aromaticity, in the presence of some radicals, adds somehow to make the PAH's. You mentioned this, but I don't see your rationale for it.

Bittner

The important characteristic of the aromaticity in the formation of PAH is the ease with which aromatic rings undergo substitution reactions. When radicals add to stable hydrocarbons, the chemically excited adducts will decompose unimolecularly back to reactants unless stabilized by other processes. Intramolecular aromatic substitution reactions (such as Rxn. 9 in the text) that form six-membered rings which are stabilized by H-atom elimination to extend the aromatic ring system, prevent the breakup of the carbon framework.

S. H. Bauer *(Cornell University)*

Where do all of the OH radicals that participate in the attack on benzene come from? What is their source? Do they diffuse upstream from a region downstream in which they are formed?

Bittner

Although not shown in the paper, the OH flux profile indicates a net diffusion of OH upstream from the region of net production (10 to 12.5 mm). There is a negative net flux of OH between the burner surface and about 11 mm. The negative slope of the OH flux profile with increasing distance from the burner between 6 and 10 mm indicates that there is a net consumption of OH in this region. The positive slope of the flux profile indicates a net production of OH between 10 and 12.5 mm. Benzene, other hydrocarbons, and CO all consume OH between 6 and 10 mm, but the major source of OH is not diffusion, since the maximum negative flux of OH (2.1×10^{-7} mol $cm^{-2} s^{-1}$) is much less than the initial benzene flux (7.3×10^{-6} mol $cm^{-2} s^{-1}$). The major source of OH between 6 and 10 mm is the sequence of chain branching reactions (Rxns. i and ii) which is initiated by H atoms diffusing upstream.

$$H + O_2 \rightarrow OH + O \qquad \text{(Rxn. i)}$$
$$O + H_2 \rightarrow OH + H \qquad \text{(Rxn. ii)}$$

H. Gg. Wagner *(Universität Göttingen)*

The larger the radicals are, the longer are their lifetimes for unimolecular decomposition. Very large aromatic radicals with extensive delocalization of the unpaired electron throughout the π-system may be as stable as molecules. The interesting question that remains is: how do these large, relatively stable radicals grow?

A. Snelson *(ITT Research Institute)*

Are you going to make any effort to look at the reaction of OH or O with benzene with the idea of identifying really what happens? At the present time it seems to me you don't know whether the OH is simply adding into the ring, whether it is abstracting a hydrogen, or what it's doing.

Bittner

Because of the complexity of flames it is difficult to identify specific reaction channels. However, the low mole fractions of phenyl radical (8×10^{-5} maximum) indicate that hydrogen abstraction cannot be a major route for benzene consumption. Reactions destroying phenyl radical are too slow to account for its low mole fractions if all benzene proceeded through phenyl radical. The OH plus benzene reaction has been studied by others under much cleaner conditions [e.g., 37]. We have no plans to investigate individual reactions of benzene with H, OH, or O.

PHYSICAL ASPECTS OF NUCLEATION AND GROWTH OF SOOT PARTICLES

G. PRADO*

Massachusetts Institute of Technology
Cambridge, Massachusetts

J. LAHAYE

Centre de Recherches sur la Physico-Chimie des Surface Solides, C.N.R.S.
Mulhouse, France

ABSTRACT

In order to shed light on the physical mechanisms of soot formation, two different soot producing systems were investigated: thermal decomposition of benzene and methane and laminar premixed flat flames. The experimental techniques used to measure the size and number density of soot particles were either electron microscopy or laser light scattering and absorption. In some experiments both techniques were used in a complementary fashion. Soot consists essentially of agglomerates of a number of small spherical carbon particles. Electron micrographs obtained from collected soot yield a reliable size distribution and number concentration of the individual spherical elements, providing information on particle nucleation, coalescent coagulation and surface growth. The *in situ* optical methods, on the other hand, see only the true agglomerate, providing complementary information on particle agglomeration.

A mechanism involving irreversible combination of colliding units appears to give satisfactory correlations of the experimental results. The chemical nature of soot nuclei is not well established. A possibility is that large polyaromatic molecules (PAH) undergo coagulation, forming the first spherical nuclei. In the early stages of particle growth, coalescent collisions between particles dominate the process resulting in a decreasing number density of particles of increasing diameter. During this stage, the growing particles maintain an approximately spherical shape. Dehydro-

Present address: Centre de Recherches sur la Pysico-Chimie des Surfaces Solides, C.N.R.S., Mulhouse, France.

References pp. 163-164.

genation occuring in the condensed phase has the effect of increasing the particle viscosity, and at a later stage, collisions are no longer coalescent. Chain-forming collisions become predominant. In both stages gas-phase hydrocarbon species attach to the surface of growing particles, providing the bulk of the soot mass fraction. Experimental results from the previously mentioned soot-producing systems are presented and discussed within the general frame of this mechanism.

INTRODUCTION

In combustion systems, soot formation occurs at a very fast rate. The initial hydrocarbons give rise in a few milliseconds to particles containing up to 10^7 carbon atoms. In thermal systems (inert conditions) spherical particles containing as many as 10^{10} carbon atoms have been observed. The build up of these particles involve chemical and physical processes [1-5].

The chemistry associated with the first steps of the soot formation process, and the role of the aromatic ring have been discussed in the previous paper (J.D. Bittner and J.B. Howard). Polyacetylenic and polyaromatic species are most often referred to as possible soot intermediates. Large polyaromatic molecules have been identified in soot forming systems, the largest being isomers of molecular weight equal to 448 (benzo[a]ovalene, pyreno[4, 3, 2-abc]-coronene) [6]. To our knowledge no compounds have been identified at larger masses, although significant amounts of neutral or ionized species have been reported in a mass range between the largest PAH molecules (MW 448) and the first spherical soot particles observed by electron microscopy or optical techniques, i.e. approximately 2 nm or MW 5000 [7-12].

The objective of this paper is to follow the fate of the soot particles from the first measurable spherical units to the final agglomerates. The soot producing systems investigated are, in order of growing complexity: thermal decomposition of benzene and methane, and laminar premixed flat flames. The experimental techniques used to follow the growing particles were *in situ* laser scattering and extinction, and electron microscopy of collected material. Both techniques are described in detail in other chapters of this book (A. D'Alessio for optical techniques, and J. Lahaye and G. Prado for electron microscopy).

Soot consists essentially of agglomerates of a number of small spherical particles. Electron micrographs yield a reliable size distribution and number concentration of the individual-spherical elements, providing information'on particle nucleation and surface growth. The *in situ* optical methods, on the other hand, see only the true agglomerate, providing information on particle agglomeration. In the text, the term "spherical units" refer to electron microscopy measurements, and "agglomerates" to *in situ* optical measurements.

For a better comprehension of the results, the concepts used to interpret the data are first presented.

THEORY OF PHYSICAL MECHANISMS OF SOOT FORMATION

Definition of Parameters — Three parameters can be used to characterize the soot phase:

- The soot volume fraction, f_V (cm³ soot/cm³ gas) which describes the total amount of carbonaceous material in the condensed phase.

- The particle number density, N, expressed as the number of particles per unit volume
- The particle size, D, which is the actual particle diameter for spherical units, or the diameter of a sphere of equal volume for aggregates. Each soot sample is in fact more accurately described by a size distribution which is usually either Gaussian or Log-Gaussian. In the first case (Gaussian distribution), the particle distribution is represented by its arithmetic mean diameter, D_A, and standard deviation, σ_A; in the second case (Log-Gaussian), the geometric mean diameter, D_G, and standard deviation, σ_G, are used.

For a given particle shape and distribution, the quantities f_V, N and D are mutually dependent (for monodispersed spherical particles, $f_V = \pi/6\ N\ D^3$), and any two are sufficient to characterize the system.

The global process of soot formation is usually divided in two stages: particle inception (formation), often referred to as nucleation, and particle growth. Two different processes may in turn contribute to particle growth: coagulation where particles collide and coalesce, and surface growth, which involves the attachment of gas-phase species to the surface of a particle and their incorporation into the particulate phase.

As will be seen later, particle nucleation and growth appear relatively well separated spatially in most soot forming systems. In other words, they are more or less consecutive and not parallel processes. The reason is that once the first particulate phase is formed, enough solid surface area is available for gaseous molecules deposition, and the probability of generating new particles is considerably reduced. An important consequence is that surface growth provides most of the soot mass loading. In the following section, the concepts used to describe the evolution of particle number density and size are reviewed.

Modeling of Particle Number Density — As discussed in detail in another chapter of this book (J. Lahaye and G. Prado), soot consists essentially of agglomerates of a number of small spherical particles. In some cases (thermal blacks for example), only spherical units are present. Two approaches have been used to predict the value of the particle number density: physical condensation [13, 14] and free-molecule coagulation [15, 16].

Physical Condensation — A relatively high concentration of gaseous molecules of several hundred atomic mass units are formed very early in soot forming systems. Physical condensation of these molecules into liquid droplets has been proposed to explain the appearance of the first condensed spherical particles observable by electron microscopy [13, 14]. The number density N_0 of these droplets, or "critical nuclei" can be derived from the equilibrium constant of the nucleation reaction:

$$aM \rightleftharpoons Ma$$

$$N_o = [M] \exp\left(-\Delta G^\circ/kT\right) \tag{1}$$

where

[M] = concentration of gaseous precursors
$\Delta G°$ = Gibbs free energy of formation of the critical nucleus
T   = temperature.

Before nucleation, the hydrocarbon precursors are in a supersaturation state (above equilibrium concentration). The formation of liquid nuclei drastically reduces the precursor concentration such that the formation of additional liquid nuclei becomes impossible. Intense surface growth through hydrocarbon deposition accounts for the rapid increase of particle size, the number density remaining essentially constant. The droplets are pyrolysed into a solid material, with some chain forming agglomeration occurring simultaneously.

As will be described later, this mechanism explains well the main features of soot formation in thermal systems at relatively low temperature. However, its validity has been strongly questioned for processes occurring at flame temperature [17, 18]. Equilibrated physical condensation at these high temperatures requires indeed molecules with a very high mass, typically 2000 to 3000 a.m.u. The mass of these molecules is so close to the mass of the first observed spherical particles that they correspond probably to the same entity. In other words, it appears more reasonable to postulate that the first particulate phase is formed through a long sequence of chemical reactions.

Free-molecule Coagulation – The rate of change of particle number density N with time t can be expressed as

$$\frac{dN}{dt} = \dot{N}_u - \dot{N}_c \tag{1}$$

where
$\dot{N}_u$ = the rate of nucleation, defined for practical purposes at the rate of appearance of the smallest observable particles (2 nm).
$\dot{N}_c$ = the coagulation rate.

$\dot{N}_u$ has an important contribution only very early in the process. Very rapidly, the number density is controlled by the coagulation rate, which can be expressed in the form of the Smoluchowski equation:

$$\frac{dN}{dt} = -k(D) N^2 \tag{2}$$

where the rate constant k depends on the particle diameter D, or equivalently, on the mean volume $\bar{v}$. For particles small compared with gas mean free path (Knudsen number > 10), Brownian motion and k are described in terms of free molecule theory; for soot particles in a typical flame environment this requires that $d \lesssim 60$ nm. For such a system in which uncharged spherical particles coalesce on every collision, this leads to a coagulation rate of:

$$\frac{dN}{dt} = -\frac{6}{5} k_{th} f_v^{1/6} N^{11/6} \tag{3}$$

where

$$k_{th} = \frac{5}{12} \left(\frac{3}{4\pi}\right)^{1/6} \left(\frac{6kT}{\rho}\right)^{1/2} G \cdot \alpha \tag{4}$$

and

f_v = particulate volume fraction
ρ = density of the particles
G = a factor which takes into account the inter-particle dispersion forces and can be expected to have a value of about 2 for spherical particles
α = a weak function of the particle size distribution, reflecting the variation in collision rate with different particle sizes.

Upon integration at constant volume fraction and temperature, equation (3) yields (S.I. units):

$$N = N_o [1 + 9.03 \times 10^{-13} N_o^{5/6} T^{1/2} f_v^{1/6} t]^{-6/5} \tag{5}$$

where
N_o = the initial particle number density
N = the particle number density at time t

For large enough values of N_0 (typically $N_0 > 10^{18}/m^3$), equation (5) can be reduced to:

$$N = 2.84 \times 10^{14} [T^{1/2} f_v^{1/6} t]^{-6/5} \tag{6}$$

Equation (6) predicts that the number of particles, N, is independent of the initial number N_0. The consequences of this with regard to the interpretation of experimental results are discussed later.

Coalescent coagulation is followed in many soot forming systems by chain-forming collisions. In this process, the colliding particles stick together but do not fuse and so give rise to an aggregate in which the individual particles retain their identity. As a first approximation, the above equation can be applied to describe this process, the aggregate being assimilated to a sphere of equal volume.

Modeling of Particle Growth — Two processes may account for the increase of soot volume fraction, f_v: generation of new particles (nucleation) and deposition of gas phase hydrocarbons onto the surface of particles (surface growth).

For a monodispersed population of spherical particles, the rate of increase of soot volume fraction can be expressed as:

$$\frac{df_v}{dt} = \frac{\pi}{6} D_o^3 \dot{N}_u + \pi N D^2 \dot{S} \tag{7}$$

where
- D_o = diameter of the initial nuclei
- $\dot{N}_u$ = rate of particle generation
- $\dot{S}$ = surface growth rate, including the contribution of surface reactions and the deposition of particles of diameter less than D_o.

As will be seen later, most of the increase in soot volume fraction is due to surface growth, and early in the process, equation (7) can be simplified as:

$$\frac{df_v}{dt} = \pi N D^2 \dot{S} \tag{8}$$

As mentioned previously, f_v, N and D are mutually dependent, and the rate of increase of particle diameter is readily derived:

$$\frac{d \ln D}{dt} = \frac{1}{3} \frac{d \ln f_v}{dt} - \frac{1}{3} \frac{d \ln N}{dt} \tag{9}$$

Equation (9) reflects the two independent processes responsible for particle size increase:

  i. surface growth, which results in increased volume fraction but constant number density, and
  ii. coagulation, where particles collide and coalesce or stick together, thereby decreasing N, f_v remaining constant.

A more complete treatment of equations (7)-(9), beyond the scope of the present paper, has been developed by Wersborg for polydispersed spherical particles [19, 20].

EXPERIMENTAL RESULTS

Pyrolytic Systems — Soot particles can be formed upon pyrolysis of hydrocarbons at relatively high temperature under inert atmosphere (absence of oxygen). This process is used in industry to produce a variety of carbon black (thermal black). It may also be relevant to reactions occurring in the inner core of a diffusion flame, where only very small amounts of oxygenated compounds are present.

The following results were obtained in a flow system, where hydrocarbons were pyrolysed diluted in nitrogen [14, 21-24]. The volume fraction, number density and particle size were measured as a function of the following parameters:

- Residence time (30 to 500 ms)
- Initial hydrocarbon volume fraction (0.25 to 100%)
- Temperature (1323 to 1723 K)
- Nature of initial hydrocarbon (benzene and methane)

In the next sections, selected results are presented to illustrate the major conclusions.

Influence of Residence Time – The material collected at the exhaust of the flow system was partitioned into extractible material (soluble in toluene) and carbonaceous residue. The yields of extractible material (tars) and carbonaceous residue as a function of reaction time are plotted in Fig. 1 for the thermal decomposition of 2% benzene in nitrogen at 1383 K. By measuring the amount of unreacted benzene and other reaction products (essentially pyrolytic carbon deposited on the inner walls of the reactor), we were able to close the mass balance to approximately 95%. Clearly,

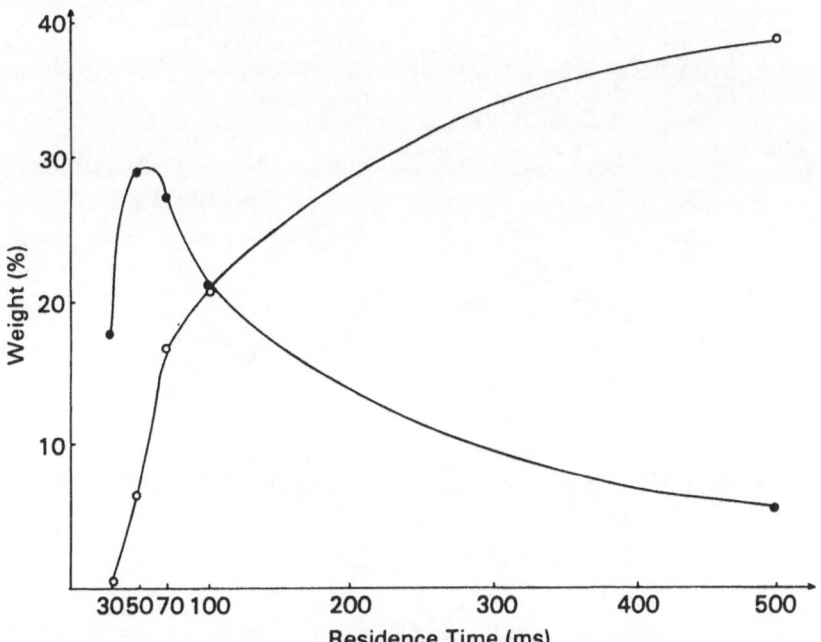

Fig. 1. Yields of carbonaceous residue (o) and tars (•) as a function of residence time. 2 % benzene in nitrogen. T = 1383 K.

References pp. 163-164.

in this system, the tars are reaction intermediates of soot formation. Electron microscopy observation of the carbonaceous residue reveals that the condensed phase is not a solid early in the process, but is rather viscous, as illustrated in Fig. 2 (a to c). Before 50 ms, the material coalesces, most probably mainly during collection and not in the reactor itself, and the usual chain-like structure is only observed at longer residence times.

An important conclusion is that at relatively low temperature (1383 K), the precursors of soot particles are viscous droplets of tars.

This coalescence during collection precludes the measurement of particle size and number density in the early part of the process with the technique used (electron microscopy measurement of collected carbonaceous residue). This information can be obtained only at the end of the process, where carbonization reactions in the condensed phase result in the formation of solid particles which resist coalescence. Unfortunately, at this point, most of the soot volume fraction is formed and only a limited particle growth is observed (Table 1).

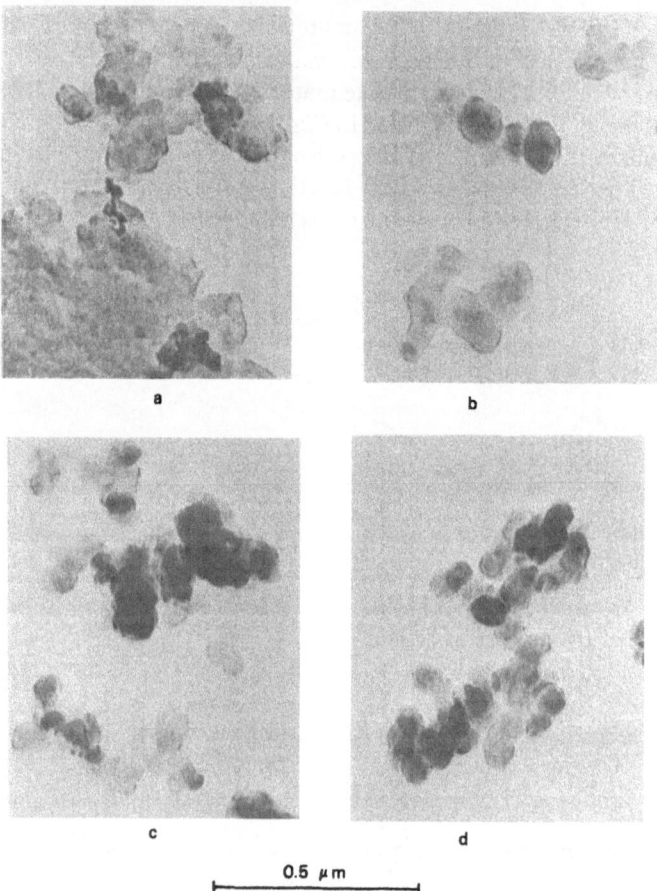

0.5 μm

Fig. 2. Carbonaceous residue collected during pyrolysis at 1383 K of a mixture of 2% benzene in nitrogen. Residence time: (a) 30 ms; (b) 50 ms; (c) 70 ms; (d) 100 ms.

TABLE 1
Benzene Pyrolysis. 6% Benzene in Nitrogen. T = 1383 K

Residence Time (s)	.1	.3	.5	1	2
10^6 x f_v	9.5	13.8	14.7	16.4	16.5
D(nm)	91.5	109.7	110.5	124.7	134.5
10^{-10}x N(cm^{-3})	0.87	0.91	0.87	0.75	0.55

Influence of Initial Hydrocarbon Volume Fraction – The effect of increasing the initial hydrocarbon volume fraction at constant temperature and residence time is summarized in Table 2. When increasing the proportion of benzene from 0.25 to 10%, the soot volume fraction increases by more than a factor of 200. This increase is totally accounted for by the increase in size of the spherical particles, their number remaining constant within the experimental uncertainty.

TABLE 2
Benzene Pyrolysis in Nitrogen. T = 1383 K. Residence Time = 0.5 s.

Benzene Volume Fraction	0.25	0.5	1	2	4	6	8	10
Soot Volume Fraction (10^7 x f_v)	0.94	3.02	12.2	31.4	70.0	147	172	222
D(nm)	25.8	37.8	52.4	71.7	91.4	100.5	132.3	138.5
10^{-10} x N (cm^{-3})	0.48	0.51	0.78	0.79	0.83	0.87	0.72	0.72
$\dfrac{D}{\sigma}$	5.5	5.5	5.8	6.2	5.4	5.7	5.6	5.5
χ^2LG								
χ^2G		2.7	1.5	1.5	1.3	0.4	0.7	

The evolution of particle size distributions (Fig. 3) with benzene volume fraction provides some insight into the mechanism of surface growth. These distributions can be characterized by the ratio of their mean diameter D and standard deviation σ. For a Gaussian distribution, this ratio is:

$$D/\sigma = \Sigma n_i d_i / \left\{ \Sigma [n_i (d_i - D)^2] \cdot \Sigma n_i \right\}^{1/2} \qquad (10)$$

where n_i is the number of particles in a diameter interval centered on d_i.

The size distribution widens considerably during surface growth (Fig. 3), with the ratio D/σ remaining constant (Table 2). A similar result was reported by Wersborg *et al.* [20] in a flat acetylene-oxygen flame. They found, as we did, that the value of D/σ is about 5 and independent of mean particle size as long as the particles are not aggregated.

This result is somewhat unexpected. Indeed, one would expect the amount of material deposited at the surface of a particle to be proportional to the particle

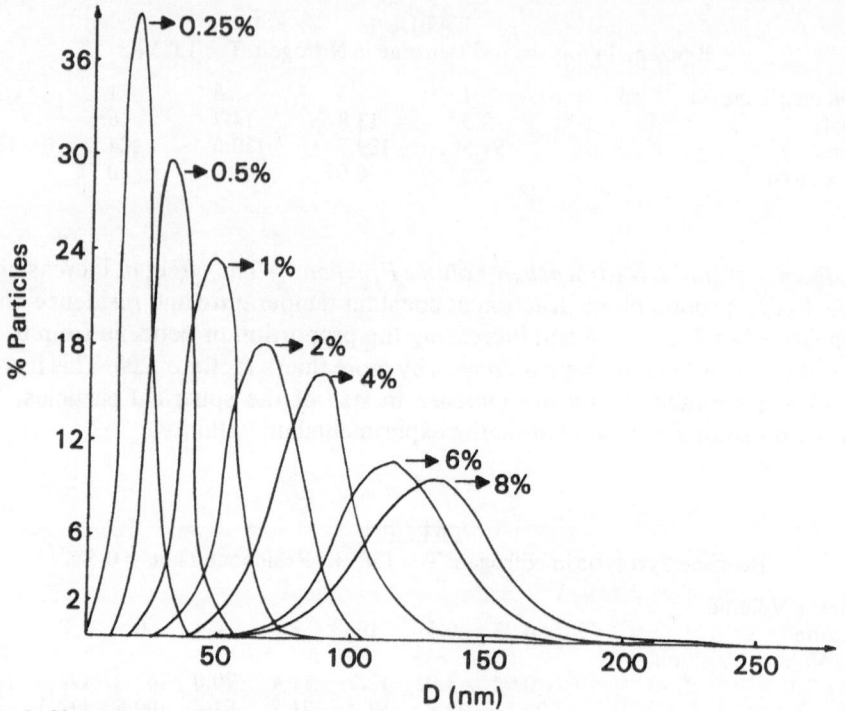

Fig. 3. Size distribution of soot particles. Benzene pyrolysis. T = 1383 K. Residence time = 500 ms. Initial benzene volume fraction : 0.25 to 8%.

external area. For a spherical particle of diameter d_i, the average surface growth rate $\dot{S}_i$ is:

$$\dot{S}_i = \frac{\Delta V_i}{S_i \Delta t} = \frac{1}{2} \frac{\Delta d_i}{\Delta t} \tag{11}$$

where

V_i = particle volume
S_i = particle external area

Consequently, if we assume $\dot{S}$ independent of d_i, during surface growth all the particle diameters of a given population should increase by the same quantity, Δd. In other words, the size distributions should not widen during growth but merely be translated from one another, with a constant standard deviation, and values of D/σ proportional to D. The constant value of D/σ when increasing the benzene volume fraction, and thereby the average surface growth rate at constant residence time (Table 2), indicates that such is not the case and that particle growth occurs according to:

$$d_i/d_i^{\,o} = \text{constant} \tag{12}$$

This result can be understood if we assume that, for a given set of conditions, the particle size distribution results from particle number distribution. In other words, the particle number density inside the reactor follows a Gaussian distribution, which is independent of the initial hydrocarbon volume fraction. In this case, it is possible to demonstrate that during surface growth, the particle size increases proportionally to its diameter [14, 23].

Another conclusion may be derived from the evolution of particle size distribution. For small particles (0. 25% benzene), the distribution is essentially Gaussian. For large particles (10%) a Log-Gaussian distribution is observed, the transition being continuous as indicated by the evolution of the ratio of the χ^2 test for Log-Gaussian ($\chi^2_{L.G.}$) and Gaussian (χ^2_G) distributions. By simulating particle collision and growth, we were able to demonstrate [23] that this distortion of the distribution curves can be explained by a limited number of partially coalescent collisions, involving only a few percent ($< 10\%$) of the total particle number. This was confirmed by phase contrast microscopy of the sample, which revealed partially coalesced particles, smoothed by surface growth [3, 21].

Influence of Temperature – The influence of temperature, at constant benzene volume fraction and residence time, on soot volume fraction, diameter and number density is summarized in Table 3. For the higher temperatures, a different oven which did not permit quantitative recovery of soot was used, and only the particle size could be measured. As apparent in Table 3, both the volume fraction and number density increase significantly with temperature, the particle size being drastically reduced. At the highest temperature, the particle diameter is close to particle size of soot formed in flames.

TABLE 3
Benzene Pyrolysis. 6% Benzene in Nitrogen. Residence Time: 0.5 s.

Temperature (K)	1311	1383	1419	1473	1523	1593	1700	1720
$10^6 \times f_v$	8.1	14.7	16.6	19.2	20.4			
D(nm)	97.3	100.5	93.4	79.7	76.9	47.3	34.4	28.6
10^{-10} N (cm^{-3})	0.76	0.87	1.42	2.37	2.75			

Influence of Initial Hydrocarbons – The results reported previously were obtained with benzene. In a second set of experiments, methane was used as representative of aliphatic hydrocarbons. Higher temperatures are necessary to form soot from methane as compared to benzene. Negligible amounts of soot were formed below 1450 K with methane. At higher temperatures, the same trends were observed with methane and benzene. The effect of initial hydrocarbon volume fraction on f_v, N and D are, for example, summarized in Table 4, for mixtures CH$_4$/N$_2$ at 1573 K for a residence time of 1 s. As for benzene, the increase in soot volume fraction is totally accounted for by increased surface growth, the particle number density remaining constant. At the same temperature, more spherical particles are present at the end of the process with benzene than with methane, the difference being larger at lower temperatures (Table 5, 1473 K). In Table 5, are also compared the particle number density measured when pyrolyzing methane, benzene and

TABLE 4
Methane Pyrolysis in Nitrogen. T = 1573 K. Residence Time = 1 s.

[CH₄] %	10	20	40	60	180	100
10^6 x f_V	1.90	4.89	9.75	12.4	16.7	18.5
D(nm)	47.3	62.6	81.4	83.4	93.0	98.6
10^{-10} x N (cm⁻³)	1.73	1.88	1.71	2.04	1.99	1.83

TABLE 5
Pyrolysis of Methane, Benzene and Methane/benzene Mixture in Nitrogen.
T = 1473 K. Residence Time = 1 s.

[CH₄] %	20	0	0	0	20	20	20
[C₆H₆] %	0	1	3	6	1	3	6
10^{-10} x N (cm⁻³) (experimental value)	0.56	2.16	2.37	2.37	0.68	0.87	1.37
10^{-10} x N (cm⁻³ (assuming cumulative effect)	--	--	--	--	2.72	2.93	2.93

benzene/methane mixture at 1473 K. The last line corresponds to the values which would be obtained if the hydrocarbons were reacting independently (cumulative effect). Although the soot volume fraction was strictly cumulative, the particle number density was not, the values obtained when pyrolysing a mixture lying within the range of values for pure methane (0.56 x 10^{10} cm⁻³) and pure benzene (2.37 x 10^{10} cm⁻³). In addition, the particle size distribution was unimodal (one maximum), indicating that the two hydrocarbons did not react independently with regard to physical mechanisms.

Particles Formed in Premixed Flames — An electron microscopy study of collected soot does not permit time-resolved measurements in the early part of the particle formation process. Optical techniques similar to those pioneered by D'Alessio *et al.* [25, 26] allow such measurements, given the availability of optical access to the system. We applied such techniques, together with electron microscopy, to elucidate physical processes occurring during soot formation in open, premixed flames.

Experimental techniques and detailed experimental results are published elsewhere [27, 28]. We will present here only the results which are pertinent to the discussion section below. These results were obtained on a propane/oxygen flame. Similar investigations, using only *in situ* optical techniques, have been carried out by other investigators, for different fuels, including methane [25, 26], ethylene and benzene [29, 30], with results essentially in good agreement with those reported here.

The quantities measured included the soot volume fraction, the size and number density of aggregates and spherical units, as defined in the introduction. Varying parameters were fuel equivalence ratio and temperature.

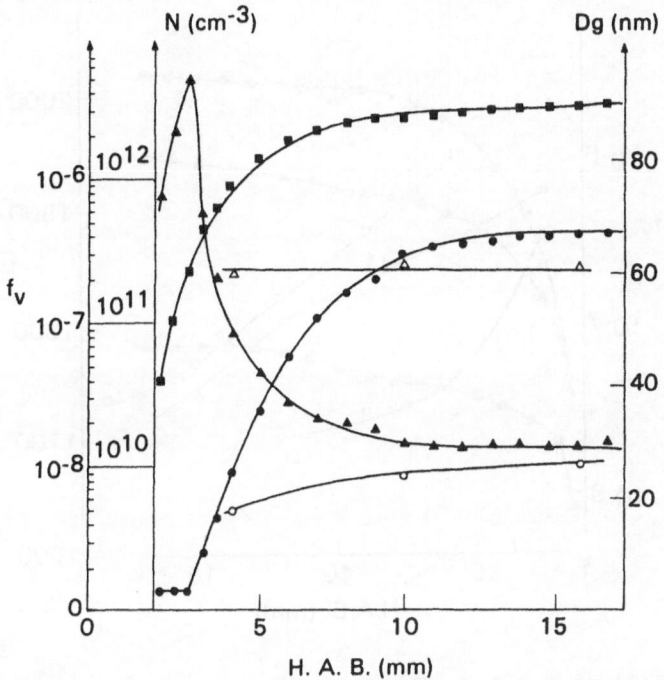

Fig. 4. Agglomerate volume fraction f_v (■). Number density N, cm^{-3} (▲). Diameter D_g, nm (●). Spherical units number density N, cm^{-3} (△). Diameter D, nm (O), as a function of height above burner (H.A.B., mm). Propane/oxygen flame. Fuel equivalence ratio = 2.9. Cold gas velocity = 5.5 cm/s.

All the information obtained on one flame is plotted in Fig. 4. The conditions are $\phi = 2.9$, v = 5.5 cm/s. The following trends, illustrated in Fig. 4, are representative of nearly all the flames studied. The agglomerate number density reaches a maximum very early in the process, whereas the number density of spherical units, when measurable, remains constant. Number densities are *in situ* values, measured at flame temperature. The agglomerate mean geometric diameter increases strongly and the sphere diameter increases slightly, depending on the increase of soot volume fraction. As with pyrolysis experiments, particles collected early in the process could not be measured by electron microscopy, due probably to intensive coalescence and/or surface deposition during sampling. In Fig. 4, 1 mm corresponds approximately to 1.2 ms.

Effect of Fuel-equivalence Ratio – The fuel equivalence ratio (ϕ) was varied from 2.1 to 3, by increasing the percentage of propane in the propane/oxygen mixture, at a constant cold gas velocity of 5.5 cm/s. Below $\phi = 2.1$, no soot was formed (blue flame); above $\phi = 3.0$, the flame was unstable, due probably to molecular diffusion. The flame temperature, measured by the Kurlbaum method, decreases when ϕ increases, and with increasing height above the burner. In the soot nucleation zone (3 mm), the flame temperature was only slightly affected by changes in ϕ (70 K difference from $\phi = 2.2$ to 2.9). The effect is stronger at larger distance above the

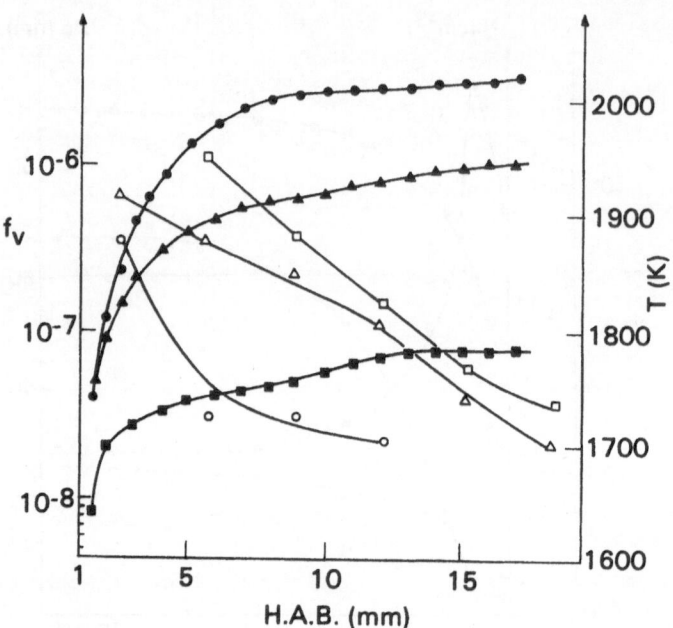

Fig. 5. Soot volume fraction (f_v, filled symbols) and temperature (T, open symbols) as a function of height above burner (H.A.B., mm). Propane/oxygen flame. Cold gas velocity = 5.5 cm/s. Fuel equivalence ratio = 2.2 (■), 2.5 (▲), 2.9 (●).

burner (160 K difference at 9 mm), and is attributed to increased radiative transfer with increased soot volume fraction.

For clarity, only three fuel equivalence ratios are plotted in Figs. 5 and 6. The soot volume fraction increases considerably with ϕ (Fig. 5), whereas the agglomerate number density changes only slightly (Fig. 6), with the exception of $\phi = 2.2$ which is at the very beginning of soot formation. For other values of ϕ (here 2.5 and 2.9), all the values of N fit on one unique curve. Furthermore, at the end of the reaction zone (18 mm), all the curves, including $\phi = 2.2$, converge towards a common limit, close to 10^{10} agglomerates/cm^3. The agglomerate mean geometric diameter increases considerably (Fig. 6) with increasing fuel-equivalence ratio and height above burner.

On examining the electron microscopy results, the picture is somewhat different (Table 6). The number of spherical units, when measurable, remains constant for $\phi = 2.2$ and 2.5, and increases by about 50% between 4 and 16 mm for $\phi = 2.9$. At 4 mm, $\phi = 2.2$ and 2.5, the material collected was poorly defined (no spherical units), and no measurements could be done. In Table 6, one can also compare results from electron microscopy and from the optical technique. Only at $\phi = 2.2$ chain-forming agglomeration is not dominant. At 10 mm, $\phi = 2.2$, the discrepancy in size is probably due to some particle surface growth during collection. At 16 mm, the diameters compare well.

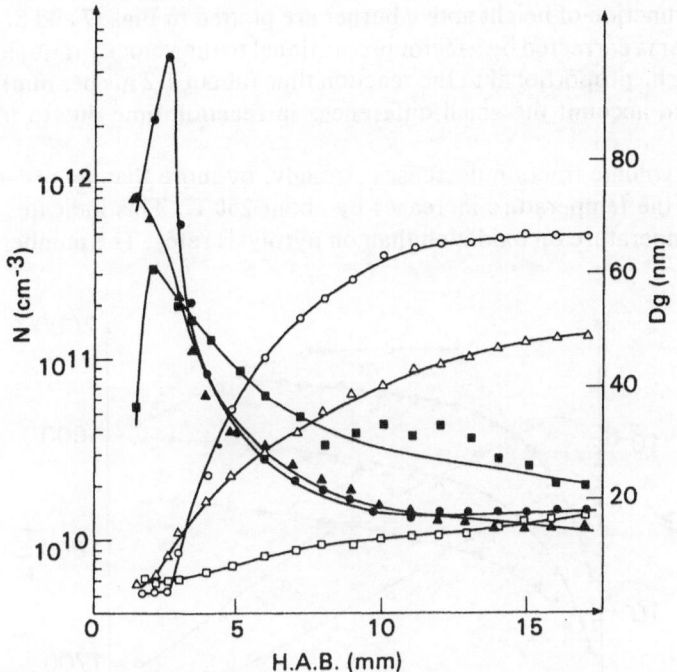

Fig. 6. Agglomerate number density (N, cm⁻³, filled symbols) and diameter (Dg, nm, open symbols) as a function of height above burner (H.A.B., mm). Propane/oxygen flame. Cold gas velocity = 5.5 cm/s. Fuel equivalence ratio = 2.2 (■), 2.5 (▲), 2.9 (●).

TABLE 6

Geometric Mean Diameter (D, nm) and Number Density (N, 10^{10} cm⁻³)
of Soot Particles Measured by Electron Microscopy and by In-situ
Optical Techniques. Propane/oxygen Flame. Cold Gas Velocity = 5.5 cm/sec.

Fuel equivalence ratio		2.2	2.2	2.2	2.5	2.5	2.5	2.9	2.9	2.9
Height above burner (mm)		4	10	16	4	10	16	4	10	16
Electron microscopy	D	—	16.2	15.4	—	17.3	18.5	18.2	24.4	26.4
	N	—	2.2	3.6	—	20.0	23.9	22.9	25.3	24.2
In-situ optical techniques	D	7.7	12.9	17.2	20.0	40.1	48.8	24.7	63.1	67.1
	N	10.2	4.7	2.2	6.5	1.6	1.3	8.6	1.5	1.5

Effect of temperature – The flame temperature was varied at a given fuel-equivalence ratio (2.5) by changing the cold gas velocity. Three velocities were used: 3.92, 5.50 and 6.76 cm/s, the maximum temperature differences in the nucleation zone (3 mm) was about 250 K. For these three velocities the values of T, f_V, N and D as a function of height above burner are plotted in Figs. 7 and 8. The height above burner is corrected by a factor proportional to the velocity ratio, in order that the abscissa be proportional to the reaction time (about 1.2 ms per mm). This does not take into account the small differences in reaction time due to temperature differences.

The soot volume fraction decreases strongly, by more than an order of magnitude, when the temperature increases by about 250 K. This indicates a stronger effect of temperature on oxidation than on pyrolysis rates. The number density of

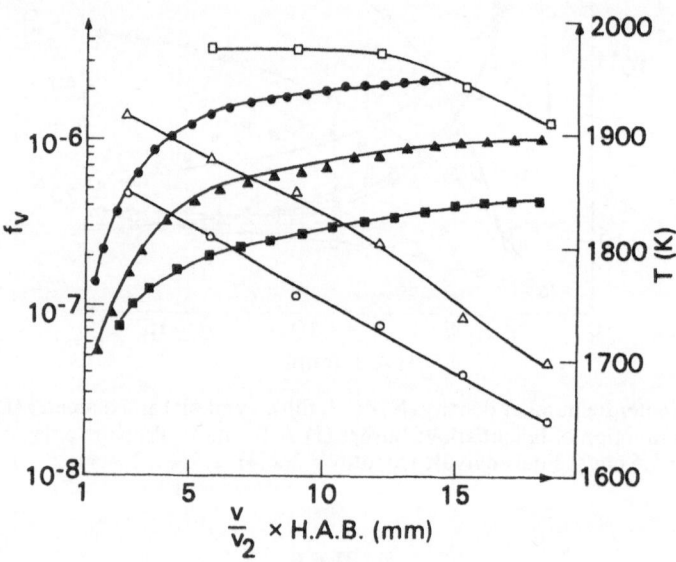

Fig. 7. Soot volume fraction (f_v, filled symbols) and temperature (T, open symbols) as a function of height above burner (H.A.B., mm). Propane/oxygen flame. Fuel equivalence ratio = 2.5. Cold gas velocity v_1 = 3.92 cm/s (●), v_2 = 5.5 cm/s (▲), v_3 = 6.76 cm/s (■).

TABLE 7
Geometric Mean diameter (D, nm) and Number Density (N, 10^{10} cm^{-3}) of Soot Particles Measured by Electron Microscopy and by In-situ Optical Techniques. Propane/oxygen Flame. Fuel Equivalence Ratio = 2.5. Height above Burner = 10 mm.

Cold gas velocity (cm/s)		3.92	5.5	6.26
Electron microscopy	D	21.0	14.3	16.4
	N	18.0	19.0	7.2
In-situ optical techniques	D	64.5	40.1	31.9
	N	0.9	1.6	1.5

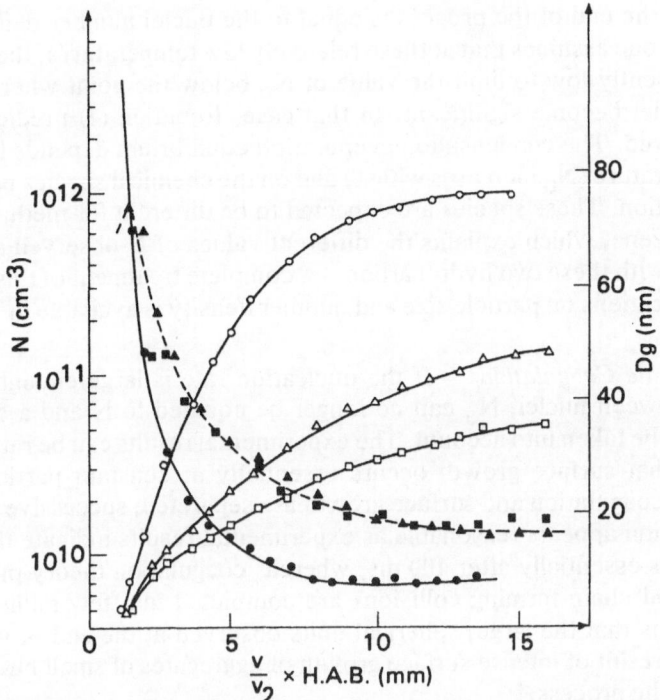

Fig. 8. Agglomerate number density (N, cm⁻³, filled symbols) and diameter (D_g, open symbols) as a function of height above burner (H.A.B., mm). Propane/oxygen flame. Fuel equivalence ratio = 2.5. Cold gas velocity v_1 = 3.92 cm/s ($\bullet$), v_2 = 5.5 cm/s ($\triangle$), v_3 = 6.76 cm/s ($\blacksquare$).

aggregates is not affected by temperature, with the exception of the lowest temperature (Fig. 8). The aggregate size (Fig. 8) as well as the sphere diameters (Table 7) decreases when the temperature increases.

DISCUSSION

In this section we interpret the experimental results described above in terms of the theoretical concepts outlined in the discussion of the physical mechanisms.

Pyrolysis Experiments — The major results, with regard to the spherical particle number density N measured in the pyrolysis experiments can be summarized as follows:

- At constant temperature, N is independent of residence time and initial hydrocarbon concentration.
- N increases with temperature.
- At a given temperature, N is smaller with methane than with benzene.

These results can be rationalized, in terms of both physical condensation and free-molecule coagulation theories.

Physical Condensation – To correlate the experimental results with physical condensation theory, it is necessary to assume that the particle number density N measured at the end of the process is equal to the nuclei number density N_0. In other words, one assumes that at these relatively low temperatures, the nucleation rate is sufficiently low to limit the value of N_0 below the point where collisions between nuclei become significant. In that case, Equation (1) predicts well the trends observed. The condensation/evaporation equilibrium depends indeed only on the temperature (N_0 increases with T) and on the chemical species participating in the nucleation. These species are expected to be different for methane as compared to benzene, which explains the different values of N observed at the same temperature with these two hydrocarbons. A complete treatment of this approach, with its implications on particle size and number density is available in References [3, 21-24].

Free Molecule Coagulation – If the nucleation rate is large enough to induce collisions between nuclei, N_0 can no longer be equated to N and a coagulation process must be taken into account. The experimental results can be rationalized if we assume that surface growth occurs essentially at constant particle number density, i.e., coagulation and surface growth are separated, successive processes. This assumption appears reasonable as experimental results indicate that surface growth occurs essentially after 100 ms, whereas coagulation theory predicts that coalescent and chain-forming collisions are completed in a few milliseconds. A consequence is that the large, spherical units observed at the end of the process would be the result of intense surface growth of aggregates of small nuclei formed very early in the process.

Due to relatively low temperature, the nucleation rate is probably lower than in flames, and the initial number of nuclei, N_0, is not expected to reach a large enough value to justify the assumption that the final particle number density is independent of N_0. Consequently, the particle number density, N, at the end of the process is dependent on nucleation rate and described by Equation (5). The constant value of N with regard to initial hydrocarbon concentration and its increase with temperature would indicate that the nucleation rate depends primarily on temperature. In the absence of oxygen, the nucleation rate is certainly expected to increase with temperature, as does the final number density of particles. Methane forms soot much less readily than benzene, and it is not surprising that particle nucleation rate, and consequently final number of particles are lower with methane than with benzene.

This interpretation is also supported by the final value of the particle number density, approximately $10^{10}/cm^3$, which is close to the number density at which collision rates, computed from the free-molecule collision theory, become negligible.

It is important to realize that the experimental results of the pyrolysis experiments are in agreement with the free-molecule coagulation theory only with the assumption that N represent the final number of aggregates at the end of the coagulation process. They cannot be rationalized with the assumption that the transition between coalescent and chain-forming collisions is the controlling factor for particle number density, as has been suggested in a recent publication [30]. As developed in this section, the nucleation rate appears more likely as the controlling factor at the relatively low temperatures encountered in pyrolytic systems.

In situ optical measurements of particle number density during the pyrolysis process should permit the assessment of the relative importance of physical condensation and free-molecule coagulation.

Premixed Flames — In the premixed flame experiments, there is no doubt that the particle number density decreases strongly during the soot-forming process, with values in good agreement with the predictions of the free molecule theory of coagulation (Equations (5) and (6)). In particular, Equation (6) predicts that the number of agglomerates, N, is the independent of the initial nuclei number density N_0. Due to the weak dependence on T and f_v, it follows that all the curves $N = f(t)$ should be very similar, in agreement with the experimental results. As an example, the theoretical and experimental agglomeration curves are plotted in Fig. 9 for $\phi = 2.5$. The agreement is certainly acceptable, with exception at the end of the process, where the number of particles measured optically might be inaccurate due to the nonsphericity of the agglomerates. Other reasons for the divergence occurring at larger residence time might also include:

i. a sticking coefficient for the collisions smaller than one and

ii. aggregates becoming too large for collisions to occur in the free-molecule regime [31].

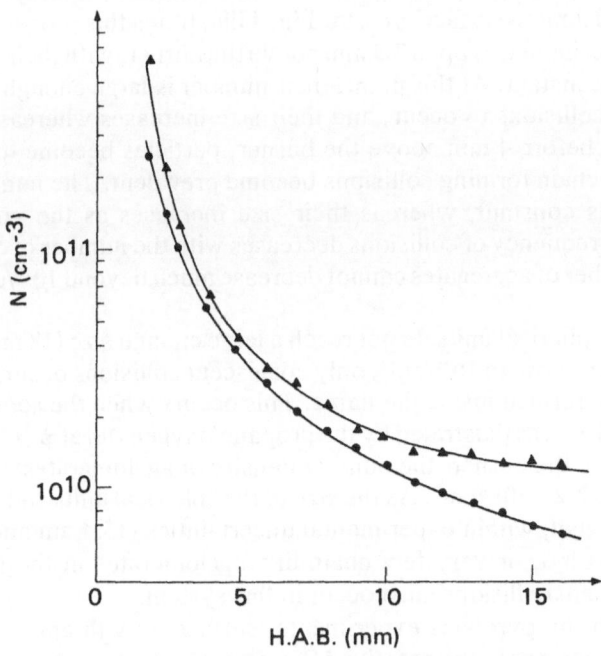

Fig. 9. Experimental (▲) and computed (●) number density of agglomerates (N, cm⁻³). Propane/oxygen flame. Fuel equivalence ratio = 2.5. Cold gas velocity = 5.5 cm/s.

References pp. 163-164.

Equation (6) is valid only for large enough values of N_O (typically $> 10^{12}/cm^3$) which implies a high nucleation rate ($\geq 2 \times 10^{21}$ m^{-3} s^{-1}). Due to the competition between oxidation and formation of soot intermediates, the nucleation rate probably decreases rapidly with increasing temperature. At the highest temperature studied, N_O might be sufficiently reduced so that Equation (5) cannot be simplified and N becomes dependent on N_O. This would explain the lower value of N measured at the highest temperature (Fig. 8).

The number of spheres constituting the agglomerates, as observed by electron microscopy, remains constant or increases slightly during a run, in sharp contrast with the behavior of the number of agglomerates. For most of the runs, this number of spheres is close to $2 \times 10^{11}/cm^3$, smaller than the initial number of "nuclei" ($\simeq 5 \times 10^{12}/cm^3$) and larger than the final number of agglomerates ($\simeq 10^{10}/cm^3$). These results support strongly the coalescent coagulation mechanism first advanced by Ulrich [15] and Graham [16, 17]. The first small "nuclei", detected optically, coalesce totally to form spherical carbon particles. The nature of these first "nuclei" remains unknown. However, the strong fluorescence observed in the first region of the flame, in this study as well as in other works [26, 29, 32], strongly suggests that they are large polyaromatic molecules. Only collisions involving at least one small particle are coalescent, whereas larger particles (D > 18 nm in this work) can no longer coalesce due to higher viscosity resulting from dehydrogenation [21] and/or longer coalescing time due to large amount of material involved. The number of these large spheres remains constant as they undergo chain forming aggregation and limited surface growth. Fig. 4 illustrates this process. The number of primary nuclei increases up to 2.5 mm above the burner, with their size remaining approximately constant. At this point, their number is large enough ($5 \times 10^{12}/cm^3$) for coalescent collisions to occur, and their size increases whereas their number decreases. Just before 4 mm above the burner, particles become too large to coalesce, and the chain forming collisions become prevalent. The number of carbon spheres remains constant, whereas their size increases as the volume fraction increases. The frequency of collisions decreases with the number of colliding units, so the final number of aggregates cannot decrease much beyond $10^{10}/cm^3$ in the time available.

If the carbon spherical units do not reach a large enough size (18 nm) before their number density is close to $10^{10}/cm^3$, only coalescent collisions occur, and there are no agglomerates formed inside the flame. This occurs when the soot volume fraction is small ($< 10^{-7}$), as illustrated by the propane/oxygen run at $\phi = 2.2$. Between 2 and 16 mm above the burner, the number density of agglomerates decreases from $3.2 \times 10^{11}/cm^3$ to $2.2 \times 10^{10}/cm^3$. As the size of the spherical units and agglomerates, at 16 mm, are equal, within experimental uncertainties (15.4 nm and 17.2 respectively), there are no, or very few chain-like agglomerates in the flame. Consequently, coalescent collisions must occur in this system.

In contrast to the pyrolysis experiments, surface growth and coagulation are simultaneous processes in the premixed flames studied, as indicated by the significant increase of volume fraction during the the decrease of particle number density. A consequence of surface growth is the building of continuous layers of carbon around contacting spheres, smoothing somewhat the spherical shape of individual particles. In these flames, the increase of f_v during aggregation however does not appear sufficient to conceal the spherical structure of the primary particles in the chain-forming collision regime. This is illustrated by the electron-microscopy results. The number of spherical units appear indeed constant (about $2 \times 10^{11}/cm^3$) for most conditions. Lower values observed at the highest temperature and lowest fuel-equivalence ratio are probably due to lower nucleation rates.

CONCLUSION

In premixed flames, soot particle number density and size are well correlated by the free-molecule coagulation theory. Coagulation and surface growth occur simultaneously. Nucleation rates as well as transition between coalescent and chain forming collisions are the controlling factors for spherical carbon unit number density and sizes.

In pyrolytic systems at lower temperatures, results can be explained both in terms of physical condensation or free-molecule coagulation. In the second case, coagulation and surface growth must be separate, consecutive processes. Irrespective of the theory, nucleation rates control the particle number density and sizes.

In flames as well as in pyrolytic systems, surface growth, and not formation of new particles, account for most of the soot volume fraction.

We propose that at relatively low temperature (1400 K), nucleation rates are low enough for physical condensation/reevaporation of soot intermediates to have a significant contribution. This contribution decreases with increasing temperature and at higher temperatures (possibly 1600-1700 K), free-molecule coagulation becomes largely predominant.

The paper has been limited to relatively simple soot forming processes. Application of the concepts described to more complex systems, closer to practical applications, is available in several recent publications [27, 33-35].

REFERENCES

1. H. B. Palmer and C. F. Cullis, Chemistry and Physics of Carbon, Vol. 1, (P. L. Walker, ed.), Marcel Dekker Inc., NY, (1965), p. 265.
2. J. B. Donnet and A. Voet, "Carbon Black. Physics, Chemistry and Elastomer Reinforcement", Marcel Dekker Inc., (1976).
3. J. Lahaye and G. Prado, Chemistry and Physics of Carbon, Vol. 14, (P. L. Walker and P. A. Thrower, eds.), Marcel Dekker Inc., NY, (1978), p. 168.
4. H. Gg. Wagner, Seventeenth Symp. (Int'l) on Comb., The Comb. Inst., Pittsburgh, (1979), p. 3.
5. B. S. Haynes and H. Gg. Wagner, Progr. in Energy and Comb. Sci., to appear.
6. P. A. Peaden, M. M. Lee, Y. Hirata, and M. Novotny, Analytical Chemistry, to appear.
7. B. L. Wersborg, L. K. Fox, and J. B. Howard, Comb. Flame, Vol. 24 (1975), p. 1.
8. G. Prado and J. B. Howard, Advances in Chemistry Series. No. 166, Evaporation – Combustion of Fuels, J. T. Zung, ed., American Chemical Society, (1978), p. 153.
9. K. H. Homann, Ber. Bunsenges. Phys. Chem., Vol. 83 (1979), p. 738.
10. J. L. Delfau, P. Michaud, and A. Barrassin, Comb. Sci. Tech., Vol. 20 (1979), p. 165.
11. D. B. Olson and H. F. Calcote, Eighteenth Symp. (Int'l) on Comb., The Comb. Inst., Pittsburgh, to appear.
12. J. D. Bittner and J. B. Howard, Eighteenth Symp. (Int'l) on Comb., The Comb. Inst., Pittsburgh, to appear.

13. C. W. Sweitzer and G. L. Heller, Rubber World, Vol. 134 (1956), p. 855.
14. J. Lahaye and G. Prado, Carbon, Vol. 12 (1974), p. 24.
15. G. D. Ulrich, Comb. Sci. Tech., Vol. 4 (1971), p. 47.
16. S. C. Graham, Sixteenth Symp. (Int'l) on Comb., The Comb. Inst., Pittsburgh, (1977), p. 663.
17. S. C. Graham, J. B. Homer, and J. L. J. Rosenfeld, Proc. Roy. Soc. Lond., A344 (1975), p. 259.
18. K. H. Homann, Comment, Sixteenth Symp. (Int'l) on Comb., The Comb. Inst., Pittsburgh, (1977), p. 717.
19. B. L. Wersborg, "Physical Mechanisms of Carbon Black Formation in Flames", Ph.D. thesis, M.I.T., (1972).
20. B. L. Wersborg, J. B. Howard, and G. C. Williams, Fourteenth Symp. (Int'l) on Comb., The Comb. Inst., Pittsburgh, (1973), p. 929.
21. G. Prado, "Mecanisme de formation des particules de noir de carbone lors de la decomposition thermique de benzene", Doctorate of Science Thesis, University center, Haut-Rhin and Louis-Pasteur University, Strasbourg, France, (1972).
22. G. Prado and J. Lahaye, J. Chim. Phys., Vol. 72 (1973), p. 1678.
23. G. Prado and J. Lahaye, J. Chim. Phys., Vol. 75 (1975), p. 483.
24. J. Lahaye and G. Prado, C. R. Acad. Sci., Paris, Ser. C., Vol. 283 (1976), p. 425.
25. A. D'Alessio, A. DiLorenzo, A. F. Sarofim, F. Beretta, S. Masi, and C. Venitozzi. Fifteenth Symp. (Int'l) on Comb., The Comb. Inst., Pittsburgh, (1975), p. 1427.
26. A. D'Alessio, A. DiLorenzo, A. Borghese, F. Beretta, and S. Masi. Sixteenth Symp. (Int'l) on Comb., The Comb. Inst., Pittsburgh, (1977), p. 695.
27. I. J. Jagoda, G. Prado, and J. Lahaye. Comb. Flame, Vol. 37 (1980), p. 261.
28. G. Prado, J. Jagoda, K. Neoh, and J. Lahaye, Eighteenth Symp. (Int'l) on Comb., The Comb. Inst., Pittsburgh, to appear.
29. B. S. Haynes, H. Jander, and H. Gg. Wagner, Seventeenth Symp. (Int'l) on Comb., The Comb. Inst., Pittsburgh, (1979), p. 1365.
30. B. S. Haynes, H. Jander, and H. Gg. Wagner, Ber. Bunsenges. Phys. Chem., Vol. 84 (1980), p. 585.
31. G. D. Ulrich. Comment, Eighteenth Symp. (Int'l) on Comb., The Comb. Inst., Pittsburgh, to appear.
32. K. Muller-Dethlefs, "Optical Studies of Soot Formation and The Addition of Peroxides to Flames", Ph. D. Thesis. Imperial College, London, April (1979).
33. G. P. Prado, M. L. Lee, R. A. Hites, D. P. Hoult, and J. B. Howard, Sixteenth Symp. (Int'l) on Comb., The Comb. Inst., Pittsburgh, (1977), p. 649.
34. B. S. Haynes and H. Gg. Wagner, Ber. Bunsenges. Phys. Chem., Vol. 84 (1980), p. 489.
35. J. H. Kent, H. Jander, and H. Gg. Wagner, Eighteenth Symp. (Int'l) on Comb., The Comb. Inst., Pittsburgh, to appear.

DISCUSSION

P. J. Pagni (*University of California, Berkeley*)

Gilles, you distinguish between surface growth and coagulation, and we have a free molecular coagulation theory that appears to work. Physically exactly what is surface growth, and how would you model it?

Prado

Surface growth is a process involving the deposition and attachment of gas-phase molecules to the surface of a particle and their incorporation into the particulate phase. Surface growth results in increased soot volume fraction but does not affect particle number density which is controlled by coagulation. The modeling of surface growth, and consequently of soot volume fraction, would require the knowledge of the controlling rates of the chemical reactions involved. As these reactions are largely unknown, more work is needed before establishing a predictive model. On the contrary, the physical processes controlling particle number density appear to obey the free molecular theory and can be modeled with a good accuracy. In other words, soot particle number density can be predicted in many soot forming systems, but not particle size.

These two processes (coagulation and surface growth) may be simultaneous or consecutive processes. They are simultaneous in premixed flames and possibly in thermal systems at high temperature. In that case, continuous layers of turbostratic carbon are deposited on the surface of the coagulating spherules, as suggested by phase contrast electron microscopy. In pyrolytic systems at lower temperature, coagulation (if any — more work is needed to clarify this point) should occur first, within about 50 ms as predicted by the free coagulation theory and be followed by surface growth which occurs up to 1 s. in the experiments reported in this paper. The chain like structure of the particles formed at the beginning of the process would disappear during surface growth, resulting in large (up to 500 nm) non-agglomerated spherical particles.

Pagni

What species are these that are formed?

Prado

The species responsible for surface growth are not yet clearly identified. Possible candidates are polyacetylenic species and polyaromatics, especially alkylated ones. More work along the lines reported in the previous paper (Bittner and Howard) is needed to understand the chemical nature of soot precursors.

R. D. Kern (University of New Orleans)

I do not quite understand Fig. 1 concerned with yields of reaction products. What is that which is peaking and coming down?

Prado

Extractable matter. We collect all the reaction products at the end of the flow tube, carbonaceous residue and tars which are deposited on the filters. The tars are extracted with toluene, and their amount weighted. For the conditions corresponding to Fig. 1, 30% of the initial amount of benzene was converted to extractable tars.

Kern

What are the conditions of this experiment as far as pressure and temperature are concerned?

Prado

This experiment was performed at atmospheric pressure and 1110°C.

Kern

Do you do this type of experiment as you raise the temperature?

Prado

Yes.

Kern

How did the residence time for the onset of soot shift? Here it takes 30 milliseconds before you see any soot at all.

Prado

Due to experimental limitations, we could not obtain time-resolved measurements below 30 ms. However, we measured a significant increase in the soot yield at 30 ms when increasing the temperature. This suggests that for the temperature range under investigation (1300 to 1700 K), the residence time for the onset of soot decreases with increasing temperature.

K. Otto *(Ford Motor Company)*

If you would add artificial nuclei to your system, how would your particles change? I'm thinking here in terms of the formation of diesel soot, where you have all kinds of inorganic particles — such as phosphorus, zinc, or iron, typically about 1μm in diameter. What difference would such nucleation tend to cause?

Prado

One might expect pyrocarbon deposition on inorganic particles such as the ones you mentioned. However, 1μm is much larger than a typical soot spherule diameter, and the material obtained will be very different from what is usually referred as soot.

There is no direct experimental evidence that metallic additives affect soot nucleation in laboratory flames investigated so far. Experimental results have indicated that metallic additives introduced in a molecular form affect the coagulation and burnout of soot, due to ionic effects and catalytic formation of OH radicals.

In an experiment some years ago, we introduced freshly formed soot particles as artificial nuclei, in an oven where benzene was being pyrolyzed. We observed a surface deposition of carbonaceous material on these particles, with simultaneous formation of new particles in the gas phase. In other words, soot intermediates can deposit on artificial nuclei added to the system without preventing formation of new particles. This was a carbon-carbon deposition, and mechanisms might be different for carbon-metal deposition.

H. Gg. Wagner (Universität Göttingen)

Could you please give me the numbers about the rates of chain formation. And would you comment on your statement that your concept also holds in flames when you go towards the limits of the soot formation?

Prado

The rate of chain formation is equal to the one predicted from the free molecule coagulation theory as expressed by equations (3) to (6) in the paper. In Fig. 9, which shows the agreement between experimental results and theoretical computations, no fitting parameter is used. The sticking coefficient is assumed to be one, and a factor of two is used to account for interparticle dispersion forces.

C. Hunter (Ford Motor Company)

Were you suggesting that the growth of a spherule was simply because of gas phase condensation and that you couldn't have spherules come together, coalesce, and those form the spherules? Not just a role of spherules to form the agglomerates, but a dual role — to coalesce and to agglomerate.

Prado

I am suggesting that both processes, coalescence and surface growth occur. At the beginning of the process, the spherules are small and fluid enough to coalesce upon collision, forming larger spherules. This coalescent process becomes less important as the particles age, due to their solidification through dehydrogenation. It is well established that the C/H ratio of soot particles increases as they age in flames or in thermal systems. At some point, the spherules are too large and too viscous to coalesce, and the size of the spherules increases only through surface growth. Subsequent collisions between spherules result in the formation of the usual chain-like structure.

G. D. Ulrich (University of New Hampshire)

I'd like to endorse your balanced treatment of the growth process where you consider nucleation, surface growth, and coagulation as well as coalescence. I think that traditionally in carbon black and in soot formation literature nucleation has been emphasized. In many cases we call that which we don't understand "nucleation". From the oxide formation perspective, which is very alive and active, we have essentially instantaneous production of nuclei and then a stable system which consists of liquid droplets, and so we don't need to be timid about coagulation and coalescence. I think the two fields — particularly in the pre-mixed flame system you dealt with — the systems are very similar and the particles are very similar. What I'd like to emphasize is that in many regimes in soot formation you can disregard nucleation because the coagulation process wipes it out. It's not very often that you can disregard coagulation. That would be done in rare systems where it is very dilute or where the particles are very large. In conclusion, I'd like to register a vote against "spherules" since there're not really spherical anyway and "primary particles" has been established in the literature already.

A. S. Gordon *(University of California, San Diego)*

I believe it has been established that when you take carbon particles out of the flame they have a lot of ESR activity. This means that there are free radicals there. Professor Swarc some years ago talked about "living polymers" which were quite identical to this in that the ESR signals would be the growth centers for material from the gas phase to continue the polymerization process. I wonder why we don't consider that in carbon formation.

Wagner

It has been considered in detail. Important information has been taken from the particles. You can see when particles grow older; then the ESR signals decay.

Gordon

When you have oxygen available, for instance, you know that perhaps the oxygen complex then will determine the signal. But inside the flame where the particles are growing wouldn't you expect that these free radical sites would be attached to unsaturated molecules?

Wagner

Yes, they are. You may recall that I mentioned there is one mechanism whereby you can build these particles: You add radicals to species with multiple bonds, keeping the radical character. That was my point. When the particles are larger and start to coagulate, a fraction of these radical sites are brought into the interior and then, the total ESR signal goes down.

J. Lahaye *(Centre National de la Recherche Scientifique)*

Carbon blacks are able to react with free radicals (molecules and macromolecules); on the other hand, they are paramagnetic. One could expect free radicals to react with unpaired electrons present on carbon surfaces. Actually, the E.S.R. signal of carbon blacks remains unaffected after reaction with radical species; it means that bonds between carbon surfaces and radicals do not result from radical-radical reactions. To sum up, carbon blacks are paramagnetic and are able to react with radicals, but it does not mean that free radicals are present on a carbon black surface (I am not talking about young soot immediately after formation but about industrial carbon blacks).

S. H. Bauer *(Cornell University)*

Opinions have been expressed during the symposium that, while critical nuclei must first be generated to initiate particle growth, the detailed mechanism of this primary process and the number densities of such nuclei are of little or no consequence in determining the density, morphology and characteristics of the clusters comprising the soot samples which are collected or ultimately burned. To establish orders of magnitude we estimate that critical size nuclei appear within a few microseconds after extended pyrolysis of the fuel had occurred, at densities 10^{13}-10^{14} cm^{-3}; that these grow by accretion very rapidly over a time scale which is determined by the diffusion of the monomeric units over mean distances in the range of 10^3 A. At that stage the primary particles are approximately spherical in size. The subsequent clustering occurs over a millisecond time scale such that the final particle density is decreased to 10^{10}-10^{12}cm^{-3}. The clusters mostly consist of compacted, short strings of beads. Granted that the experimentally observed properties of soot refer to the latter type of clusters, one may question whether for all types of soot all remembrance of their origins are lost. Is there one or are there several types of (primary) critical nuclei, and do these imprint some of their structural features on the subsequent clusters?

I propose that nucleation and development of the carbonaceous spherules is analogous in some respect (although markedly different in others) to the nucleation and condensation of fine metal particles from their supersaturated vapors. Indeed, the "self-consistent" kinetic model which we developed* can be extended to the C/H particles whereas classical nucleation theory obviously cannot. [What meaning could one assign to the surface tension parameter for the C/H system?] That the appearance of a condensed phase in this system introduces a very striking change in the equilibrium composition of the vapor, thus marking the catastrophic event, is evident from a comparison of calculated equilibrium compositions for two systems shown in the accompanying figure [C/H = 3/1; p = 0.1 atm; no C_{solid} present] and [H$_2$; p = 0.1 atm; C_{solid} present] as a function of the temperature. In the absence of a solid phase between 1500-2500 K the dominant species are the polyacetylenes: C_2H_2, C_4H_2, C_6H_2 and the radicals, C_4H, C_3H, C_6H. In contrast, these species are practically non-existent when the solid is present; the gas phase consists almost entirely of H$_2$ and H atoms. I propose that the mole fractions indicated in the left half of the figure gives us clues as to what species are most likely to be involved in

---

* S. H. Bauer and D. J. Frurip, J. Phys. Chem. Vol. 81, p. 1015 (1977).

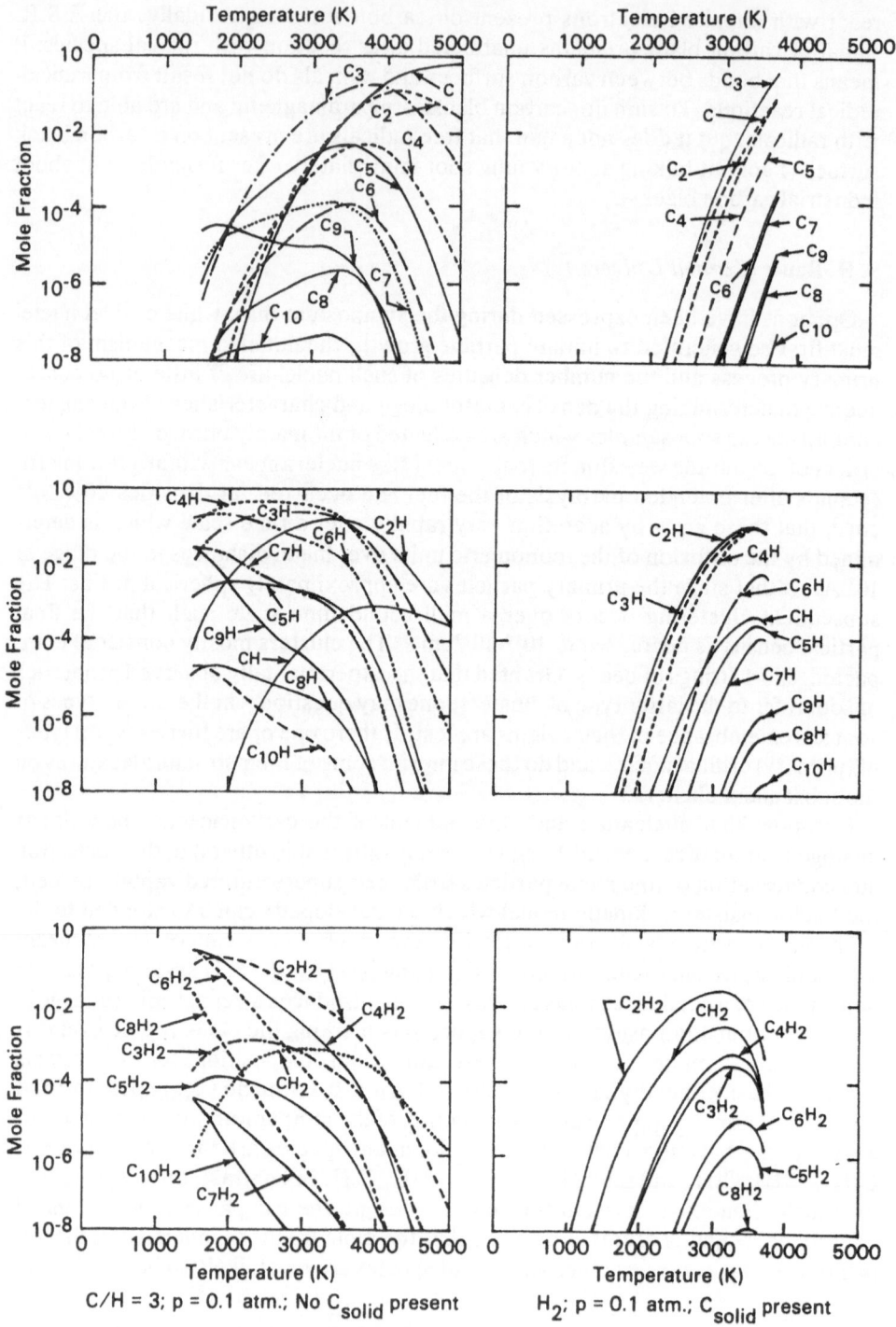

Calculated Equilibrium Compositions for Two Systems

the primary condensation process. We assume that the higher molecular weight fuels are rapidly, but not necessarily completely, decomposed to near equilibrium mixtures of compositions indicated in the figure.

Let us postulate that: (i) in the gas phase the dominant "monomeric" species are those present at equilibrium in the C/H = 3 system (these are essentially unchanged for C/H = 2), and (ii) the condensed phase arises from successive accretion of these gaseous species, with an occasional addition of a larger incompletely pyrolyzed fragment. This is schematically indicated by

$$
\begin{array}{l}
C_2H_2 + C_4H \xrightleftharpoons{k_1(M)} C_6H_3 \\
\qquad\qquad\qquad\;\;\; \uparrow k_2(M) \\
\qquad\qquad\qquad\qquad\;\; \longrightarrow C_{10}H_5 \\
\qquad\qquad C_4H_2 \\
\qquad\qquad\qquad\qquad\quad\; \uparrow k_3(M) \\
\qquad\qquad\qquad\quad C_3H \longrightarrow C_{13}H_6 \\
\qquad\qquad\qquad\qquad\qquad\qquad \longrightarrow \ldots C_{nx}H_{ny} \ldots
\end{array}
$$

the index n denotes the step number in the accretion sequence. These are representative of more complex steps, which could involve hydrogen elimination or partial oxidation. One need not regard the higher species as being acetylenic; indeed, the associations are probably followed by rearrangements to ring structures[†]. It is essential also to postulate that (iii) these association *reactions are reversible*. Thus critical size nuclei are developed and attain a significant concentration only when the system is highly supersaturated. The kinetics of avalanche condensation is best illustrated with a monoatomic model

$$
A + A \overset{(M)}{\rightleftharpoons} A_2 \qquad\qquad\qquad A_{n-1} + A \rightleftharpoons A_n
$$

$$
A_2 + A \overset{(M)}{\rightleftharpoons} A_3 \qquad\qquad\qquad A_n + A \rightleftharpoons A_{n+1}
$$

After a short time let the population distribution of n-mers be designated by N_n (prior to condensation). The time evolution of this distribution function is controlled by a set of simultaneous first order differential equations. Computer solution of these equations shows that the system will attain a special type of *steady state*. There will be a critical size n^*, such that for $n \leqslant n^*$, N_n remains essentially unchanged because —

† *However, refer to R. Hayastu, R. G. Scott, M. H. Studier, R. S. Lewis and E. Anders, Science Vol. 209, p. 1515, 1980 - on the identification of carbynes in meteorites.*

$$\left.\begin{array}{l} k_{n-1,n}(N_{n-1})\,(N_1) \approx k_{n,n-1}\,(N_n) \\[2mm] k_{n,n+1}(N_n)\,(N_1) \ll k_{n-1,n}(N_{n-1})\,(N_1) \end{array}\right\} \text{for } n < n^*,$$

whereas for $n > n^*$, N_n remains essentially unchanged because

$$k_{n-1,n}(N_{n-1})\,(N_1) \approx k_{n,n+1}\,(N_n)\,(N_1) \gg k_{n+1,n}\,(N_{n+1})$$

Graphically, the condition is illustrated by

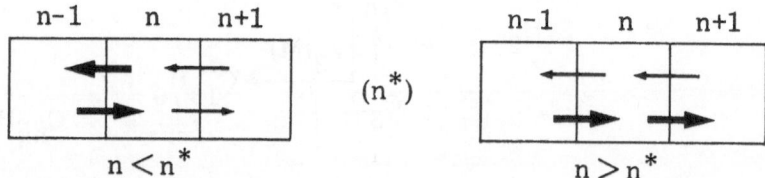

N_n^* is lower than any N_n ($n < n^*$), so that n^* represents a "bottle neck". Since the steady state condition for $n < n^*$ departs little from the equilibrium, one can estimate both n^* and N_{n*} from *thermodynamic* parameters ($\Delta H^\circ_{n-1,\,n}$; $\Delta S^\circ_{n-1,\,n}$) if these are known (or estimated, as Dr. D. B. Olson does in his paper.) The criterion for n^* is

$$\left\{dN_n/dn = 0\right\} \quad \begin{array}{c} \text{eq} \\ n=n^* \end{array}$$

For $n \geq n^*$, in constrast, the system is *kinetically controlled*. Condensation occurs because the corresponding rate constants increase more rapidly than N_n's decrease. The number density of spherules is determined by the steady state flux of clusters which get past the bottle-neck

$$J_{n*}^{ss} \approx k_{n*,\,n*+1}[M]\;(N_{n*})\,(N_1)$$

This is a measure of the critical size nuclei which are generated per $cm^{-3}s^{-1}$. The mean size attained by the spherules is determined by the level of supersaturation, i.e., by the magnitudes of N_{n*} and N_1, being larger for small levels, since the number of nuclei upon which the available material can condense is then smaller.

To pursue the application of this model to the generation of soot spherules one must somehow obtain values for enthalpies, entropies and association rate constants as functions of n. We hope that some of these can be measured; most of them will have to be estimated by applying additivity rules. For monatomic condensations, the following empirical relations were fitted to computed functions. For $nA = A_n$, per mole of A_n---

$$\Delta H_n^\circ = - nQ (1-n^{- .25})$$

Q is the heat of vaporization (sublimation) of bulk material

$$\Delta S_n^\circ = + n^3 \Omega (n^2 + 32 - \Omega)^{-1}$$

Ω is Trouton's constant

To estimate $k_{n, n+1}$ we used the standard termolecular expression for a binary association rate constant. This requires an estimate of the potential energy function between a monomer and an n-mer $[U_{n, 1}(r)]$.

The application of some form of kinetic analysis of C/H condensation remains a challenging problem.

R. G. Gann *(National Bureau of Standards)*

Dr. Bauer's model [it is less a model than an approach for coming to grips with the problem] is an intriguing way to start. However, two aspects need to be considered. The first is that in proceeding from an n-mer to the next higher one in the agglomeration sequence, the rate constants for the various steps may be grossly different. Some of the association steps will be followed by intramolecular rearrangements and this may lead to a branched sequence rather than to linear growth. Second, the kinetic analysis starts with the assumption that the concentrations of the unit fragments can be calculated assuming the system is in thermodynamic equilibrium; that needs to be tested.

I have one minor question. In looking at the profiles of mole fraction *vs* temperature, it appears as though two mole fractions added to more than unity.

Bauer

I would like to respond. With regard to the last question. Dr. Gann may not have noticed that the vertical scale was logarithmic; the sum of mole fractions is always unity. I did not show the curves for H_2 and H, which are major species at the higher temperatures.

I agree that one cannot directly extend the model we developed for nucleation-condensation of a single atomic species to the growth of soot particles. At this time we do not know whether the same concepts apply to condensation processes wherein the simplest units are radicals or small reactive molecules, but this should be tested. The possibility that branching may occur, leading to a wide range of structures for the condensate is a serious difficulty. However, the occurrence of intramolecular rearrangement is not troublesome provided their rates are faster than the rates of accretion. I emphasize that what matters in this model are *products* of rate constants and the concentrations of the particular species involved — not merely the rate constants. As you well know, the onset of rapid condensation is determined by the slowest step in the growth sequence — the subsequent rates being much faster exert no control. Since the "bottle-neck" generally appears at low n-mers, there is a reduced chance that complex structural problems will introduce difficulties.

Lahaye

I wish to point out that the nucleation of a solid material, soot or a metal, can occur under either of two extreme conditions. In one case a kind of equilibrium is reached, followed by supersaturation; then condensation occurs according to Toschev, Palmer, Zeldovich, and so on. In the other extreme the condensed phase is generated by a sequence of classical chemical reactions. In the first situation thermodynamics of reversible systems applies whereas the second depends on the chemistry of the system. In his talk Dr. Prado mentioned these two possibilities. If soot is formed at low temperatures, say under thermal-pyrolyzing conditions at $\approx$ 1000°C, there is some evidence that this type of equilibrium exists. However, at higher temperatures, as in flames, we believe that equilibrium is not reached. Then nucleation follows the second type to which I referred.

Bauer

I respectfully differ with you relative to the dicotomy which you proposed. In all homogeneous nucleation, once overall steady state is reached, there are low n-mer condensation steps which are nearly (but not quite) in thermodynamic equilibrium, and high n-mer condensation steps which are in steady state but the concentrations of the transient species are very far from values calculated *via* statistical thermodynamics. The transition from one type to another occurs at some generally small n^*, called the critical size nucleus. The "irreversible" feature appears in all systems for $n > n^*$ because then steady state exists because the rate of production of an n-mer from (n-1) mer is essentially equal to the rate of removal of the n-mer by generation of an (n+1)mer. I see no difference in principle between a process such as:

$$A_n + A_1 \underset{k_{n,\,n+1}}{\overset{k_{n+1,n}}{\rightleftarrows}} A_{n+1} \quad \text{(A denotes atoms)}$$

and

$$(C_2H)_n + C_4H \rightleftharpoons C_{2n+4}H_{n+1}.$$

If one uses the index n (concentration N_n) to characterize a transient species on its way from monomers to a large aggregate, the first question to be answered, is there an n^*, for which N_{n^*} is a minimum (i.e., is there a bottle-neck in the rate)? If the values n^* and N_{n^*} are known, one can estimate the rate of production of the primary spherules.

Wagner

For the description of vapor condensation the nucleation concept, developed since Becker and Doering to a rather high degree of perfection, has proved to be a very useful one. Prof. Bauer mentions it in connection with soot formation where it has been applied in the past several times without knowing what "monomer" really means. Since then, many experimental details have become available which indicate that it may be an interesting problem in itself. For soot formation in combustion, however, one should not worry too much about the very first part of the process because it does not determine the final result and is not rate determining. Roughly 90% of the mass of soot growth is by surface reaction on particles when the particle number has decreased via coagulation to approximately 10 times the final particle number.

IONIC MECHANISMS OF SOOT NUCLEATION
IN PREMIXED FLAMES

D. B. OLSON and H. F. CALCOTE

AeroChem Research Laboratories, Inc.
Princeton, New Jersey

ABSTRACT

Although numerous chemical mechanisms have been postulated for the formation of soot in hydrocarbon combustion, none have received quantitative support. Understanding of this process is complicated by the large number of species involved and by the phase change associated with nucleation.

An experimental program is underway at AeroChem to determine the role of ionic processes in soot formation in premixed hydrocarbon flames. The hypothesis being tested is that ions produced via chemi-ionization serve as nuclei for initial formation of soot particles. Rapid ion-molecular kinetics, rapid structural rearrangements, and thermochemistry favorable toward growth to larger species are characteristics of the ion chemistry, in agreement with observations about the soot formation process.

Mass spectrometric measurements of flame ion concentration profiles have been made in low pressure rich and sooting acetylene/oxygen and benzene/oxygen flames. As both flames are made increasingly fuel rich and approach sooting the predominant ion $C_3H_3^+$ is replaced by large positive aromatic ions with mass greater than 300. It is argued that these ions are the soot precursors.

Similarities and differences betweeen the ion spectra of rich and sooting acetylene and benzene flames are discussed along with possible reasons for dual maxima in ion concentration profiles observed in sooting flames. A detailed chemical mechanism considering both neutral and ionic flame chemistry is being developed for which it is necessary to estimate the thermochemical parameters for large ionic species. Preliminary results of computer simulations of the chemical kinetics using this scheme are presented. The results are consistent with an ionic mechanism of soot formation.

References pp. 200-201.

INTRODUCTION

The mechanism of soot formation has been recognized for many years as an extremely complex problem and the nucleation step — the transition from molecular species where chemical reactions dominate to incipient soot particles where physical processes dominate — has been recognized as the least understood step in the process. Many diverse theories have been put forward to explain the nucleation and all of them seem to have been refuted [1, 2]. One of the troublesome problems is that there may be many mechanisms operating simultaneously or that different mechanisms may dominate under different conditions. We are exploring the hypothesis [2, 3] that the basic nucleation mechanism in most systems involves growth through rapid ion-molecule reactions producing larger and larger ions. The initial ion is assumed to be the chemi-ion $C_3H_3^+$ which is observed in large concentrations in rich flames. Thus the energetics which drive the process are originally derived from the nonequilibrium chemi-ionization process. In the ionic hypothesis, the large ions are subsequently neutralized by ion-electron and ion-ion recombination to become incipient particles. As the neutral particles grow via coagulation and/or surface growth to a critical size they are thermally ionized. Thus there are two stages of ionization, a situation which has led to some of the confusion in interpreting previous experiments where it has not been clear whether the chemiions or thermally ionized particles are being observed. In two recent publications [2, 3] we have compared the chemi-ionization soot nucleation mechanism to the various neutral species mechanisms and shown the ionic mechanism to be more consistent with observations. In this paper we present further evidence for the proposed ionic mechanism and examine some of its problems. These problems include the lack of a satisfactory pathway for the production of $C_3H_3^+$ in rich flames, the value of the absolute concentration of chemi-ions in sooting flames, and the inadequately explained observation that, at equivalence ratios near sooting, individual large ions show double-peaked profiles as a function of distance through the flame.

Two basic non-ionic mechanisms are currently postulated. One involves the growth of large polyacetylenes (or other unsaturated molecules) and subsequent ring formation to produce species with aromatic structures. These then grow by addition of unsaturated groups to form larger and larger polycyclic aromatic hydrocarbons, eventually to become soot particles. The other mechanism, consistent with the observation that aromatics soot more readily than do aliphatics , assumes the continual growth of larger and larger aromatic structures by condensation of intact polycyclic aromatic hydrocarbons. There are severe problems with either of these mechanisms. Carbon blacks and soot are not large linear polyacetylene molecules but are multi-ring structures, so the polyacetylenes must in some way produce a ring structure. Any reaction scheme one can envision for this process would be expected to be slow because of the complicated rearrangement geometry involved. Furthermore, the concentration of polyacetylenes remains large even downstream in the flame beyond the point where soot nucleation has ceased. If no other ingredient were required, the number of particles would be expected to continue to increase as long as significant concentrations of polyacetylenes were present. Polycyclic hydrocarbon nucleating (condensation) mechanisms are similarly inconsistent because the concentration of these molecules continues to increase in the flame front beyond the point where soot nucleation ceases. Furthermore, thermodynamic considerations appear to limit the growth of larger and larger polycyclic aromatic hydrocarbons in an intermediate size range.

Much of the evidence for an ionic nucleation mechanism is circumstantial, and much of what has been interpreted as verification of ionic nucleation is related to large charged particles. As we noted above, these particles can arise from thermal ionization and have nothing to do with the nucleation process. The essential remaining evidence supporting an ionic nucleation mechanism can be summarized [2, 3] as:

1. A strong positive correlation is observed in premixed flames between the growth of large ions and the appearance of soot.
2. Carbon tetrachloride (and other electrophilic molecules) increase the concentration of large ions in diffusion flames and the subsequent onset of soot formation.
3. Fuels which produce ions also produce soot, while fuels which do not produce ions do not produce soot.
4. Reactions of O atoms with excess acetylene demonstrate very rapid ionic polymerization to form $C_{2n}H_{2n+1}^+$ and $C_{2n}H_{2n-1}^+$ ions.
5. Large concentrations of ions are also produced in pyrolysis experiments.
6. The location of the molecular ion peak concentration in the flame front corresponds to the maximum concentration of a group of poorly characterized molecules previously identified as soot precursors.
7. The concentration of chemi-ions has been estimated[2] to be large enough and at an appropriate location in the flame to account for the rate and concentration of particles formed.

Additional thermodynamic and chemical kinetics arguments will be presented in this paper.

FORMATION OF CHEMI-IONS

The ion concentrations observed in hydrocarbon flames are far in excess of what can be accounted for by thermal ionization. There is overwhelming evidence that in lean and close-to-stoichiometric compositions this is due to the chemi-ionization reactions first proposed [4] by Calcote

$$CH + O \rightarrow CHO^+ + e^-$$ (1)

The rate coefficient [5] for this reaction in the temperature range 2000 to 2400 K is about 2.3×10^{-13} cm^3 s^{-1}. The observed concentrations of CHO$^+$ are, however, almost three orders of magnitude smaller than the maximum total ion concentration because of very rapid ion-molecule reactions, especially

$$CHO^+ + H_2O \rightarrow H_3O^+ + CO \tag{2}$$

The ions are removed mostly by dissociative recombination:

$$H_3O^+ + e^- \rightarrow H_2O + H \tag{3}$$

The rate constant [6, 7] for this reaction is 2×10^{-7} cm^3 s^{-1} at 2000 K with a temperature dependence of $T^{-\frac{1}{2}}$.

Many other superequilibrium concentrations of ions are observed [8, 9] due to very rapid ion-molecule reactions with the neutral species present which are in local equilibrium with each other. The various ion concentrations all decay at nearly the same rate, probably by reaction with O, OH, O_2, or H_2O to form H_3O^+ with a bimolecular rate coefficient [10] of 4×10^{-11} cm^3 s^{-1}. Some of these species are displayed [11] in Fig. 1. In lean flames the dominant ion is $C_2H_3O^+$ which is formed by

$$CHO^+ + CH_2CO \rightarrow C_2H_3O^+ + CO \tag{4}$$

However, in even slightly rich flames (where the equivalence ratios, ϕ, are small compared to those at which soot is formed), the dominant ion is $C_3H_3^+$. Note that the maximum equivalence ratio shown in Fig. 1 is 1.3 and soot formation begins at $\phi \approx 2.6$ (see Fig. 2 for example). At concentrations approaching those where soot is formed $C_3H_3^+$ is still the dominant ion, and H_3O^+ and $C_2H_3O^+$ are no longer important. They have been replaced by non-oxygen-containing ions like $C_5H_3^+$ and larger ions containing only carbon and hydrogen. In an ethylene/oxygen flame at 0.5 kPa Miller [10] measured the $C_3H_3^+$ concentration at $\phi = 1.9, 1.3$, and 1.5 to be about 1×10^{10} ions cm^{-3}; he observed a slightly higher value at $\phi = 1.3$ but within experimental error, his observations are all equal.

In lean and stoichiometric flames the appearance of $C_3H_3^+$ is generally explained [8, 9] by a series of reactions deriving from CHO^+

$$CHO^+ + CH_2O \rightarrow CO + CH_3O^+ \xrightarrow{C_2H_2} C_3H_3^+ + H_2O \tag{5}$$

$$CHO^+ + CH_4O \rightarrow CO + CH_5O^+ \xrightarrow{C_2H_2} C_3H_3^+ + H_2O + H_2 \tag{6}$$

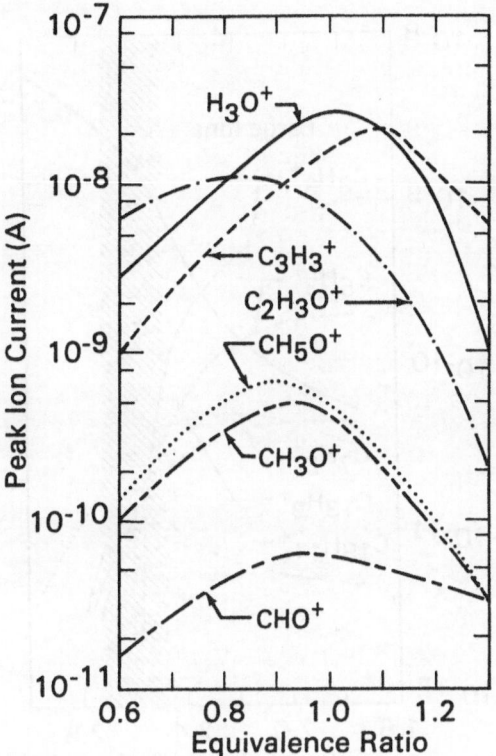

Fig. 1. Effect of flame composition on peak ion currents for 0.5 kPa acetylene/oxygen flames. (From Reference 11).

$$CHO^+ + C_2H_2O \rightarrow CO + C_2H_3O^+ \xrightarrow{C_2H_2} C_3H_3^+ + CO + H_2 \quad (7)$$

Reaction (7) probably goes through two steps:

$$C_2H_3O^+ + O \rightarrow H_2O + C_2HO^+ \xrightarrow{C_2H_2} C_3H_3^+ + CO \quad (8)$$

However, all of the above reactions require oxygen-containing species and it seems unlikely that in very rich flames the concentration of oxygenated species is sufficient to account for the large observed concentrations of $C_3H_3^+$. One possible reaction route which does not require oxygenated species is

$$CHO^+ + CH_2 \rightarrow CO + CH_3^+ \xrightarrow{C_2H_2} C_3H_3^+ + H_2 \quad (9)$$

References pp. 200-201.

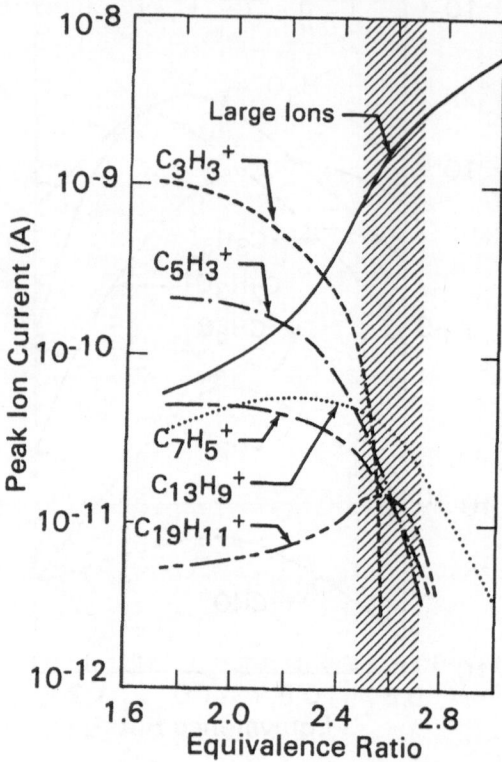

Fig. 2. Effect of equivalence ratio on peak ion currents for 2.0 kPa acetylene/oxygen flames. The shaded area indicates the minimum equivalence ratio for soot formation.

Besides the CH/O reaction (1), other pathways to account for the large concentration of $C_3H_3^+$ observed in flames have been proposed. The most commonly quoted is [12, 13]

$$CH^* + C_2H_2 \rightarrow C_3H_3^+ + e^-$$ (10)

Efforts by several groups to find supporting evidence for the occurrence of this reaction by correlating the concentrations of CH in the $A^2\Delta$ or $B^2\Sigma$ state with the rate of ion formation have been unsuccessful. Miller[10] thoroughly considered the mechanism of $C_3H_3^+$ formation in C_2H_4/O_2 and C_2H_2/O_2 flames and concluded that it is complex. He measured the dependence of the $C_3H_3^+$ on concentrations of C_2H_4 or C_2H_2 added to H_2/O_2 flames and found a $[C_2H_4]^{3.5}$ and a $[C_2H_2]^2$ dependence. Yet in both cases the CH* dependence on the hydrocarbon concentration was 2. If Reaction (10) were the dominant mechanism the order should be 3 in both cases (C_2H_2 and C_2H_4). It was suggested [10, 14] that proton transfer from some unidentified species to C_3H_2 might account for the formation of $C_3H_3^+$ but insufficient data were available to examine this hypothesis.

Evidence for other proposed independent sources [15] of $C_3H_3^+$, e.g.

$$C_2^* + CH_3 \rightarrow C_3H_3^+ + e^- \tag{11}$$

is nonexistent.

On the basis of thermodynamics there are other conceivable chemi-ionization reactions which can explain some of the large ions observed in the flame front of fuels approaching sooting, e.g.,

$$C_{16}H_{10} + C_6H \rightarrow C_{22}H_{11}^+ + e^- \qquad \Delta H \approx -30 \text{ kcal mol}^{-1} \tag{12}$$

While reactions of this type cannot be eliminated with available information it seems unlikely that in such a large molecule all of the energy of reaction should be concentrated in one electronic state.

For the purposes of developing an ionic mechanism of soot formation in flames we accept the presence of large concentrations of $C_3H_3^+$ as the initial ion but recognize that its source has yet to be explained.

ABSOLUTE ION CONCENTRATIONS

To quantitatively evaluate our proposed mechanism of soot nucleation it would indeed be desirable to know not only the absolute ion concentrations but also the rates of formation of the chemi-ions which we propose initiate the growth of soot nuclei. It is difficult to obtain absolute ion concentrations with a mass spectrometer, especially over a mass range for which a good calibration is impossible, as in the case of large positive ions in sooting flames. Absolute measurements with electrostatic probes are also difficult in the presence of a large range of ion masses.

Previously we have relied on the results of Howard and coworkers [16, 17] for establishing that the ion concentrations were sufficiently large to be important in soot nucleation. Recent measurements, however, by Delfau *et al.* [18] and Homann [19] disagree with the earlier results. It thus seems reasonable in view of the importance of these results to review those measurements. We chose four sets of data obtained in rich and sooting low pressure acetylene/oxygen flames as close to the same experimental conditions as possible; they are summarized in Table 1.

Howard and associates used both an electrostatic probe [16] and a molecular beam sampling system [17] with a Faraday cage detector. Delfau *et al.* [18] used an

TABLE 1
Comparison of Total Ion Concentrations in Acetylene Flames

	Electrostatic Probe		Molecular Beam Sampling	
	Prado et al. [16]	Delfau et al. [18]	Wersborg et al. [17]	Homann [19]
Oxidizer	O_2	$N_2/O_2 = 0.6$	O_2	O_2
Pressure, Torr	20	45	20	20
Flow velocity, cm s^{-1}	50	27	38	44
Equivalence ratio	3.5	3.0	3.0	2.9
Equivalence ratio at soot point	2.4	2.2	?	2.5
Height in flame for peak ion concentration, cm	2.0	1.5	2.0	1.3
Description of what is measured	≈ 4 nm particles	large ions > 700 amu	ions > 300 amu	large ions about 40-200 amu
Ion concentration, ions cm^{-3}	4 x 10^{10}	1.3 x 10^{10}	8 x 10^{10}	3 x 10^8

electrostatic probe while Homann [19] used a molecular beam sampling system with an ion collector immediately behind the flame sampling orifice with electrostatic grids to control the sampling process. First, the data have been obtained over such a range of conditions that it is surprising that they agree as well as they do; the highest ion concentration was measured by Howard *et al*, and the lowest by Homann, both with molecular beam sampling. The difference is about a factor of 40.

The Prado and Howard electrostatic probe data were taken under the conditions we have previously chosen [2] for comparisons because of the availability of data from various investigators at those conditions. It thus is instructive to compare the conditions of this work and other data sets in the table to see if the variations are reasonable, i.e., expected. The Delfau *et al*. ion concentration result would be expected to be lower because of the diluent N_2 and the lower equivalence ratio, while the higher pressure and flow velocity would increase the value. Wersborg *et al.'s* molecular beam data would be expected to be higher because of the higher flow velocity but lower because of the lower equivalence ratio. Homann's data would be expected to be lower because of the lower flow velocity and lower equivalence ratio. Finally, it is interesting that from Wersborg *et al*. the effect of equivalence ratio is considerably greater than in the data of Homann or Delfau *et al*.!

It is significant that the authors describe the quantity they are measuring as quite different. For example, Prado and Howard describe their data as "total positive ions" but assume in the calculation that the measured value represents 4 nm diameter charged particles. They present another set of data in the same paper for charged particles which peaks at about 3 cm with a maximum value of about 4×10^9 cm^{-3}, rather close to the concentration of what Homann calls "positively charged soot particles," 2.4×10^9 cm^{-3} peaking at 2.0 cm.

The very large difference between Homann's concentration value, Table 1, and the others is difficult to explain. Indeed, the greatest difference in ion concentrations is between the two sets of molecular beam sampling data [17, 19]. Homann's use of the equation for isentropic flow through the sampling nozzle as the basis for absolute calculation of ion flux and neutral gas flux may not, as he recognizes, be valid since ions are probably lost on every collision with the wall. Wersborg *et al*. similarly assumed isentropic nozzle flow and made absolute concentration measurements in the flame from measured molecular beam flux. Comparing Homann's data directly with that of Wersborg *et al*. is somewhat uncertain because it is not clear that the same mass range was being measured. Furthermore, the different measurement conditions (equivalence ratio and flow velocity) would emphasize any other differences because of the idiosyncrasies of individual burner systems; for example, increasing the flow velocity caused the flame to move closer to the burner rather than further away. Thus, Homann found that increasing the unburned gas flow velocity by about 10 cm s^{-1} decreased the observed peak concentration of charged soot particles by a factor of 2 to 3. Wersborg *et al*. on the other hand observed a decrease of a factor of about 100 for a similar change in flow velocity. Apparently these workers are not measuring the same quantities and in any case, given the high sensitivity of their systems to various parameters, it is amazing that the values in Table 1 are as close as they are. Incidentally a value of about 10^{10} ion cm^{-3} is all that seems to be required for ionic soot nucleation.

Probably none of the measurements in Table 1 are relevant to the question of ionic nucleation. The proper question is whether there are sufficient small chemi-ions such as $C_3H_3^+$ to quantitatively account for the proposed mechanism. To answer this question one must look at data such as presented in Fig. 2. The instrument used in these experiments was not calibrated but the relative concentrations are approximately correct. A rough estimate of the calibration factor using an assumed ion-electron recombination rate constant (that for H_3O^+) and the observed ion disappearance rate leads to peak ion concentrations of about 10^{10} cm^{-3}.

However, there is another apparent difference between our work [3] and the work of Delfau *et al.* [18], Wersborg *et al.* [17], and Homann [19] which demands resolution. This is the change in total ion concentration with equivalence ratio near the critical equivalence ratio for soot formation. If the data reported in Fig. 2 are taken by operating the mass spectrometer in a mode to observe ions in two ranges, 13 to 300 amu and greater than 300 amu, the ion concentration in the range 13 to 300 falls sharply at the soot point (see Fig. 3) while the concentration above 300 amu increases such that the total ion concentration is essentially constant from $\phi = 1.7$ to

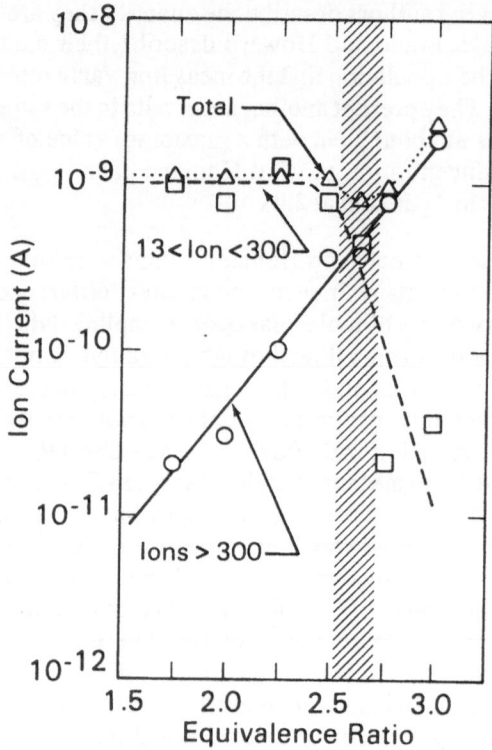

Fig. 3. Effect of equivalence ratio on total ion currents divided into two broad mass classes for 2.0 kPa acetylene/oxygen flames. The shaded area indicates the minimum equivalence ratio for soot formation.

3.0 with only a small increase between the soot point and $\phi = 3.0$. In contrast Delfau *et al,* observed more than a factor of 10 decrease in maximum ion concentration from $\phi = 1.7$ to the soot point and then an increase of more than a factor of 10 from the soot point to $\phi = 3.0$. Homann observed an increase from his soot point to $\phi = 2.9$ of a factor of about 25 and Wersborg *et al,* observed an increase of about 25 from their soot point to $\phi = 3.0$. We suspect that in all of the measurements, including our own, the sensitivity of the measuring technique to different ion masses has not been accurately accounted for.

Clearly there is need for some very careful measurements of ion concentrations, especially individual species or groups of species, in rich and sooting hydrocarbon flames. In the meantime the data in the literature are supportive, although shaky, of an ionic mechanism of soot nucleation.

ION PROFILES

The low pressure flame apparatus and ion sampling quadrupole mass spectrometer, shown in Fig. 4, have been described previously [3, 20]. Briefly, the flames

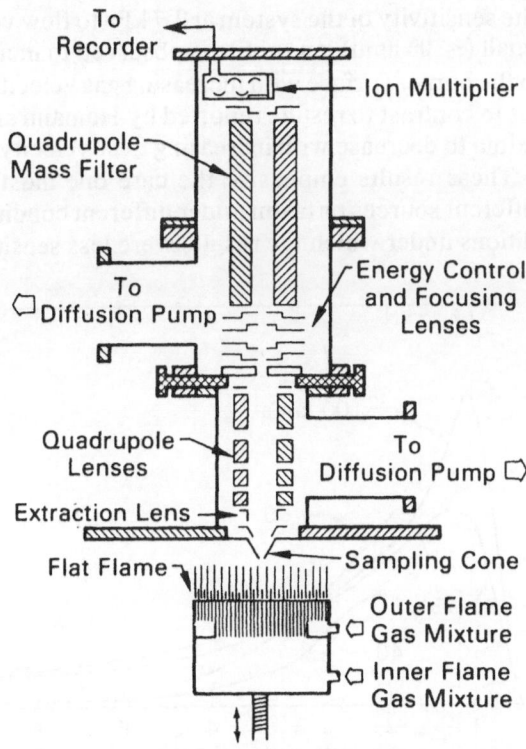

Fig. 4. Flame ion sampling mass spectrometer.

were stabilized on a 12 cm diameter flat flame burner and were surrounded by a fuel-rich annular shield flame 16 cm in diameter to prevent air or combustion product entrainment. The mass spectrometer sampling cone was constructed from 316 stainless steel with a 90° outer angle and a 0.25 mm orifice. The ion beam is focused in the first stage (P $\approx$ 30 mPa) using quadrupole lenses to minimize mass discrimination onto a 1 mm diameter aperture into the second pumping stage. The custom-made quadrupole mass filter uses 2.5 cm diameter rods and is characterized as a relatively low resolution, high transmission instrument, operating at 0.5 MHz and covering the mass range 13 to 300 amu. A Bendix Model 306 magnetic electron multiplier is used off-axis as detector; its output is amplified by a Keithley 602 electrometer and displayed on an X-Y recorder. Ion profiles through the flame were obtained by moving the burner with respect to the mass spectrometer sampling cone. Gases were controlled using critical flow orifices while benzene was metered by a precision syringe pump. The liquid benzene was sprayed into a low pressure flash evaporator and the vapor fed to the burner-mixing chamber through heated lines.

Initial experiments were performed at 2.7 kPa with an unburned gas velocity of 50 cm s⁻¹ at the burner. Rather serious interactions of the flame with the burner were reduced by lowering the total pressure. Unless otherwise noted the work reported here corresponds to a pressure of 2.0 kPa and unburned gas velocity of 50 cm s⁻¹.

An example of the sensitivity of the system at 2.7 kPa to flow velocity is shown in Fig. 5. The total small (<300 amu) ion profile is observed to increase in magnitude and move closer to the burner surface with increasing gas velocity. This increase in maximum current is in contrast to results reported by Homann and Wersborg *et al.* who found the maxima to decrease with increasing flow velocity at 4.0 kPa and 2.7 kPa, respectively. These results emphasize the care one must exercise in comparing data from different sources or taken under different conditions. We selected experimental conditions under which the results were less sensitive to flow velocity.

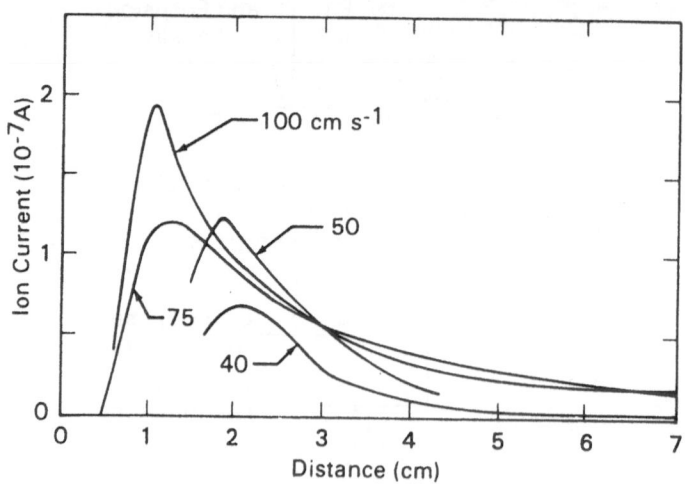

Fig. 5. Effect of unburned gas velocity, cm s⁻¹, on total small ion profile for 2.7 kPa acetylene/oxygen flames.

Detailed individual ion species profiles were measured for masses up to 300 amu at equivalence ratios above and below the critical equivalence ratio ϕ_c, for soot formation as determined by the appearance of the yellow-orange continuum soot emission. In addition, the spectrometer was operated as a high-pass mass filter and profiles were recorded of ions with mass larger than 300 up to cutoff of the filter, estimated to be about mass 1000. Ion current data from high-pass operation of the filter are not directly comparable to individual mass profiles because the throughput of the instrument is greater when acting as a high-pass mass filter. Relative ion currents are reported instead of ion concentrations because of the difficulties in quantitatively and reproducibly calibrating the instrument, although quantitative evaluation of soot nucleation mechanisms will require knowledge about absolute levels of ion concentration in future experiments, and these are planned.

Data were taken by scanning the mass range at fixed distances from the burner to establish which ions were present at high concentrations; individual concentration profiles of selected species were then recorded by setting the spectrometer to that mass and moving the burner. Fig. 6 shows spectra taken in nonsooting and sooting acetylene/oxygen flames at 1 cm distance in the blue-green oxidation zone. The essential features of the spectra are the dominance of mass 39, $C_3H_3^+$, followed by a series of peaks spaced every 13 or 14 amu. Only odd numbered ion masses are observed in contrast to neutral species sampled from similar flames [21] whose masses (number of hydrogens) are even. For most species the identity may be assigned with good confidence although the assigned structures [3] are somewhat speculative. At lower masses more than one isomer is observed, e.g., 63/65 amu, 89/91 amu. This latter pair is interesting because mass 89 must be an open-chained

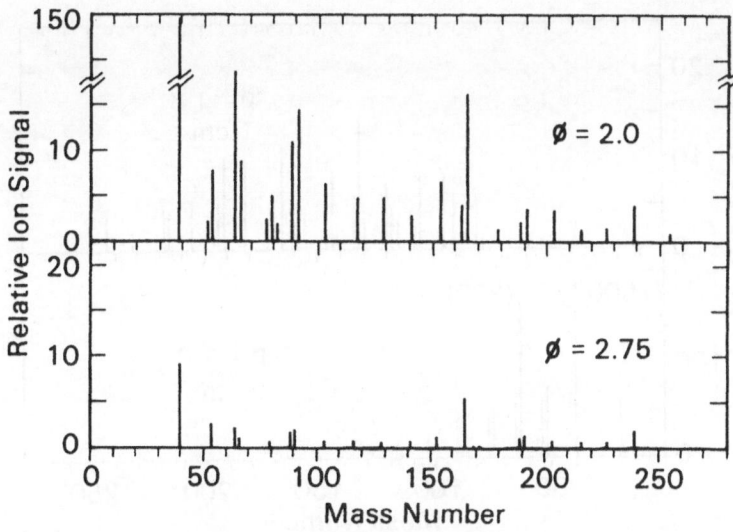

Fig. 6. Ion mass spectra taken 1.0 cm from the burner surface for a nonsooting and a sooting 2.0 kPa acetylene/oxygen flame. $\phi_c = 2.6$.

References pp. 200-201.

isomer but 91 can be either open-chained, a seven-membered ring, or a toluene-derived species. The most prevalent heavier species observed are masses 165 and 239 corresponding respectively to relatively stable three and five ringed compounds. In these rich flames, H_3O^+ is observed only at low concentrations. At $\phi =$ 2.75 the concentrations of the masses shown in Fig. 6 have all dramatically fallen and have been replaced by ions of larger mass.

Fig. 7 shows mass spectra taken at two different positions in a nearly sooting benzene flame. In the oxidation zone at 1 cm from the burner unique oxygenated ionic species are observed at masses 95, 109, 131, etc., whereas mass 39, $C_3H_3^+$, is essentially absent. At 2 cm the $C_3H_3^+$ concentration has grown several orders of magnitude and the ions unique to benzene have disappeared. These ions are identified as containing oxygen, and appear to be derived from the neutral C_6H_6O species previously observed in similar flames [21]. Notable at 1 cm is the lack of species smaller than C_6 indicating little ring opening has occurred in this region. The ion spectra in benzene flames beyond the flame front are quite similar to those of acetylene flames with little or no information about the parent fuel persisting. One speculates that it may be a general rule that rich flames differ in the relative concentrations of ionic species (as opposed to identities of ions present) downstream of the oxidation zone. That is not to say that if ions nucleate soot that all flames would soot at the same equivalence ratio since the chemistry of the oxidation zone may dominate the process or the relative concentration of some ions may be more important than others.

Figs. 8 and 9 show profiles of selected ions in nonsooting, $\phi = 2.0$, and sooting, $\phi = 2.7$, acetylene flames ($\phi_c \approx 2.6$). In Fig. 8 each ion profile is observed to rise rapidly at the beginning of the flame front, at 0.6 cm. The concentrations peak and slowly decay due mainly to ion-electron recombination beyond ≈ 2 cm where oxidation is no longer occurring.

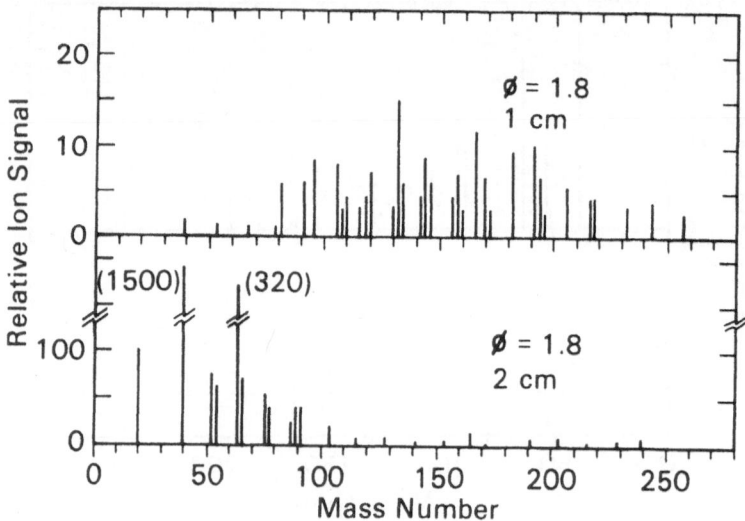

Fig. 7. Ion mass spectra taken at 1.0 cm and 2.0 cm from the burner surface for a nearly sooting 2.0 kPa benzene/oxygen flame. $\phi_c = 1.85$.

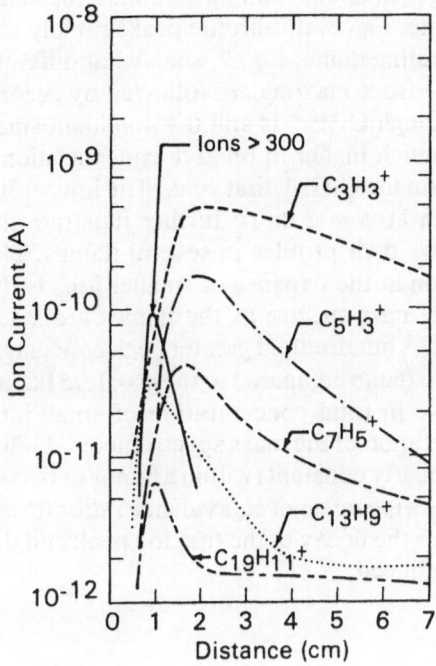

Fig. 8. Ion profiles for a 2.0 kPa nonsooting, $\phi = 2.0$, acetylene/oxygen flame. Calculated adiabatic flame temperature is 2300 K.

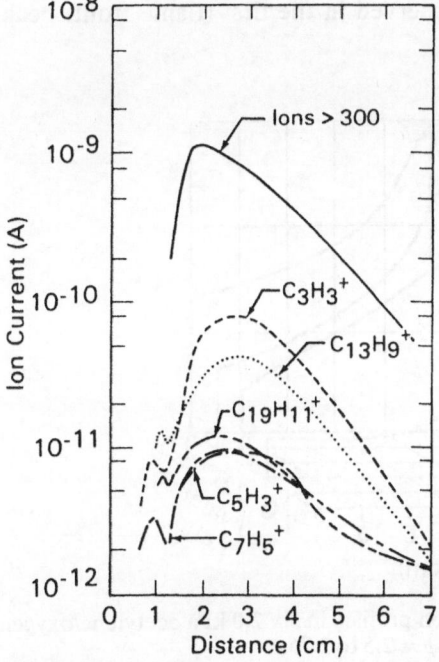

Fig. 9. Ion profiles for a 2.0 kPa sooting, $\phi = 2.7$, acetylene/oxygen flame. Calculated adiabatic flame temperature is 2170 K.

Also shown in these figures is the signal obtained using high-pass operation of the mass spectrometer, attributed to ions larger than 300 amu up to about 1000 amu (labeled ions > 300). In the nonsooting flame, Fig. 8, this profile peaks rapidly and disappears by 2 cm downstream. In the sooting flame, Fig. 9, somewhat different ion profiles are observed. The initial flame front maxima are followed by second larger, broad peaks, at about 2-4 cm. Although $C_3H_3^+$ is still the dominant small ion, the heavier species (> 300 amu) are much higher in relative concentrations. Note that the $C_{13}H_9^+$ first peak is larger than the $C_3H_3^+$ first peak. The ions > 300 are ten times higher in concentration than at $\phi = 2.0$. To further illustrate the behavior of these larger ions, Fig. 10 shows their profiles in several flames. The dramatic increase in large ion concentration at the expense of smaller ions is the most significant change that occurs in the mass spectra as the flames are made increasingly richer and finally produce soot. A hundredfold greater peak concentration for the large ions is obtained in a $\phi = 3.0$ flame compared to the $\phi = 1.75$ flame, accompanied by a corresponding decrease in total concentration of small ions (determined separately using bandpass operation of the mass spectrometer, 13-300 amu). The total ion concentration remains nearly constant (within a factor of two — the precision of the measurements) over the wide range of equivalence ratios investigated, see Fig. 3. Fig. 10 also demonstrates the decay of the first ion peak and the growth of the second ion peak as soot is produced.

SECOND ION PEAKS

Comparing the ion profiles from a nonsooting acetylene flame, Fig. 8, with a sooting flame, Fig. 9, the appearance of the second broad maxima is puzzling. Benzene flames have the same behavior, as shown in Fig. 11. The ions observed in the second peak are the same as those observed in the first (flame front) peak,

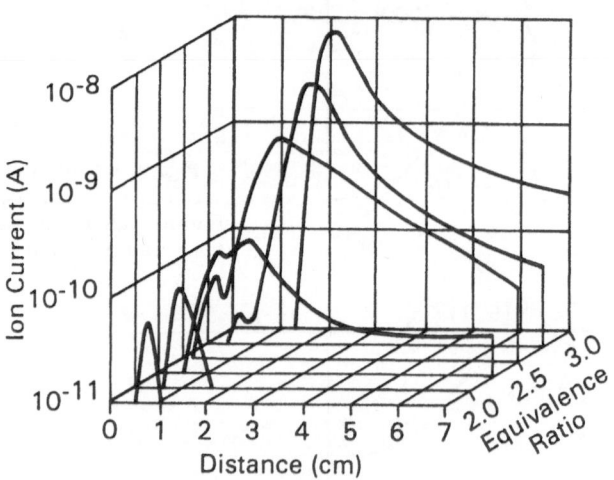

Fig. 10. Effect of equivalence ratio on large ion profiles in six 2.0 kPa acetylene/oxygen flames. Transition to sooting occurs in the range $\phi = 2.5$ to 2.75.

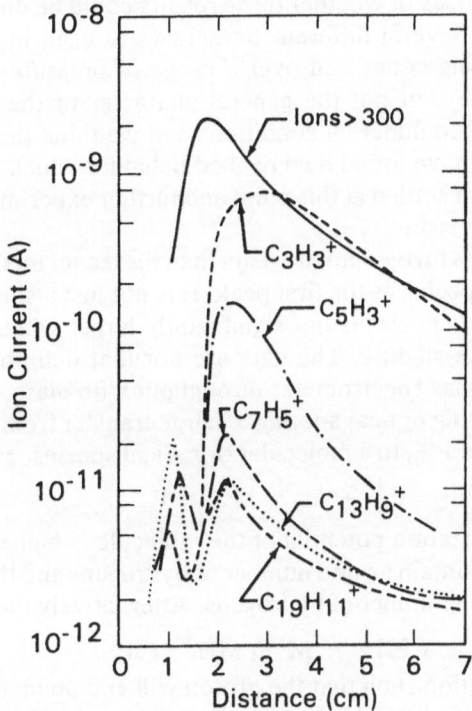

Fig. 11. Ion profiles for a 2.0 kPa nearly sooting, $\phi = 1.8$, benzene/oxygen flame. Calculated adiabatic flame temperature is 1900 K.

although relative levels of different species change. Double peaks are observed in profiles of a single species. Delfau *et al.* [18] have observed dual maxima in electrostatic probe measurements in low pressure flames, but only present data where the identity of the ions in the two peaks may be different. Neutral beam sampling from benzene flames does not show dual peaked behavior [21]. On the other hand, Homann [19] performed ion beam sampling from C_2H_2/O_2 flames in a simple mass discriminator and found a strong increase in ionization at the beginning of the sooting zone (a second peak).

The source of this second peak is not clear. It cannot be due to the primary chemi-ionization reaction of $CH + O$ since the O-atom and CH concentration are very small beyond the flame front. Other proposed chemi-ionization steps such as $CH^* + C_2H_2$ would also have peaked in the primary reaction zone. Thermal ionization of soot particles probably occurs when the particles become large enough to be characterized by ionization potentials near that of graphite. Calcote [2] used experimentally observed particle size distributions, temperature profiles, and estimated ionization potential vs. particle size variations to calculate charged particle concentrations in sooting low pressure acetylene flames. The agreement with observed values was excellent. This means that the molecular-sized ions which we are observing are not necessary to explain charged soot particles. However, it does not explain the source of the second peak for small ions.

References pp. 200-201.

A difficult question arises of whether these results could be due to experimental problems in sampling. Several different burners were used in conjunction with several different sampling cones and over a range of pressures. Specific details could be changed some, but not the general character of the results. The dual peaked behavior persisted under all conditions and we think the results are valid. Other laboratories [22] have found dual peaked behavior which they believe to be false. The question is not settled at this point and further experimental work will be necessary to resolve the issue.

The second peak arises further downstream than the leaner mixture first peak and includes the same ion species as the first peak. It is not just a shift in location. The total number of charged species is not significantly larger when the second peak appears; it may increase slightly. The data are not that definitive because of an unknown variation in mass spectrometer throughput with mass. Flames with dual maxima are always sooting or near sooting. Charge transfer from a small thermally ionized charged particle, P^+, to a molecular or radical species, M

$$P^+ + M = M^+ + P$$

is unlikely since the ionization potential of the molecule is higher than that of the particle. Also, the ions contain an odd number of hydrogens and the neutral species observed contain an even number of hydrogens. Alternatively the process

$$PH^+ + M = MH^+ + P$$

will proceed in the direction such that the proton will end up on the species which has the highest proton affinity (PA). For hydrocarbon molecules the PA generally increases with size. Also the PA of molecules in a homologous series is inversely related to the ionization potentials of the members [23] so this would also indicate that large molecules (or particles) would have high PAs and therefore scavenge charge from molecular-sized ions rather than create them.

Condensation reactions of the type

$$M_1 + M_2 = M_3$$

produce M_3 molecules with large excess internal energies. The M_3 species may dissociate if the collision rate is insufficient for it to be stabilized. We here point out that such condensation processes involving large molecules (particles?) release sufficient energy in some cases for ionization to occur. The reaction

$$C_{16}H_{10} + C_6H \rightarrow C_{22}H_{11}^+ + e^- \qquad \Delta H \approx -30 \text{ kcal mol}^{-1} \qquad (12)$$

is a (merely illustrative) example.

The sudden appearance and disappearance of "ions > 300" in the flame front, e.g., Fig. 8, followed by ions of lower mass, which must be intermediates in forming "ions > 300," appears to be incongruous. The formation of these large ions so early in the flame front implies extremely rapid ion-molecule reactions — a basic tenet of our argument. The initially produced small ions rapidly produce larger ions before they can be detected. This also occurs in lean flames where the chemi-ion CHO^+ is not the first or most dominant ion observed [7]. The rapid decay of these large ions, presumably due to ion-electron recombination or oxidation — leaving a smaller ion — and the persistence of smaller ions, can be interpreted as a decrease in the availability of building blocks necessary for sustaining the > 300 ions downstream. The smaller ions can continue to be formed from $C_3H_3^+$. More quantitative data will be required to test this interpretation.

THERMODYNAMIC ARGUMENTS

Tanzawa and Gardiner [24, 25] studied the pyrolysis of acetylene and were able to develop a kinetic mechanism capable of explaining their data and all previous shock tube data from 600 to 3400 K. As a minor part of this work they examined which species were likely to play an important role in soot formation. They concluded thermodynamically that large C_nH_2 and C_nH species would never reach high enough concentrations to form soot at the observed rate. A kinetic system of faster reactions and involving species of greater stability at high temperature was required.

Stein [26] investigated the high temperature equilibrium of selected polycyclic aromatic hydrocarbons (PCAH) with acetylene and hydrogen using the group additivity techniques developed by Benson and coworkers [27] to obtain thermochemical parameters. He found that above 1700 K, for a series of PCAH the equilibrium concentrations pass through a minimum dependent upon the temperature and concentration of C_2H_2 and H_2. Essentially a benzene condensation mechanism forming large cyclic molecules was studied and some characteristics of a nucleation process found. The molecules for which lowest concentrations were predicted could act as critical species and be involved in the rate limiting reaction. The identity of these critical species is dependent upon specific conditions, but they were found to be approximately in the 20 to 50 carbon range.

Thermodynamic arguments are very useful, but it is likely that the dynamics of intact benzene ring condensation are too slow for this simple picture to be very realistic. The crucial question is how to obtain molecular species, almost certainly PCAHs, larger than some critical size on a very short time scale. We think this is the essence of the nucleation problem of soot formation and that these molecules may grow as ions and be neutralized after they have passed the critical size. Beyond that point they would continue to grow into particles.

It is informative to look at the thermodynamics of the ionic species observed in rich and sooting flames. These data are also required for kinetic modeling of the process. We find, however, that experimental data are not available for many of these low H/C ratio species. Most ionic hydrocarbon thermodynamic information has been obtained by electron impact at room temperature so that flame ion species are relatively uncharacterized. For example, a recent comprehensive source of thermodynamic data [28] on gaseous ion lists 22 odd and 61 even mass species between C_{14} and C_{30} which have been studied. Flame ions are all odd mass species. Since tabulated thermodynamic data are not available for many ionic species observed in our rich flames, we have used the group additivity technique to calculate these data for some of the species experimentally observed and proposed to be important in soot formation. Properties of the relevant neutral molecules were calculated as functions of temperature and the ionization potential (IP) used to calculate the ΔH_f of the ionic species. To estimate the IP of molecules for which experimental values are not available we examined the available data [28] for trends and used a smooth curve through the experimental results to estimate ($\pm$ 50 kJ mol^{-1}) needed values. There are probably much better methods to estimate these data. Even so, the major uncertainty in estimating heats of formation is not the technique but the choice of isomers at a particular mass. Even though ions are known to rearrange very rapidly to more stable structures [23, 29, 30] (an advantage of the ionic mechanism over neutral mechanisms), we do not know where kinetic constraints will be present so that the observed ion is a less stable isomer.

Fig. 12 shows estimated heats of formation (298 K) for hydrocarbon ions between C_1 and C_{25} along with selected neutral species. The data are plotted vs. carbon number where linearly increasing trends in the ΔH_f of $C_x H_n$ vs. X indicate that group additivity is followed. Data for $C_n H_2$, $C_n H$, and PCAH neutrals are also shown. The curve for the largest ions. $C_n H_{12}{}^+$, is seen to be approximately parallel to that for PCAH neutrals and separated by 550-650 kJ mol^{-1} (6 to 7 eV), a reasonable ionization potential for such molecules. One sees that thermodynamically one cannot create larger and larger molecules (or ions) with the same number of hydrogens (such as increasingly larger polyacetylenes); the lines are simply too steep. On the other hand, for the proposed ionic mechanism the major species would move across this figure at about the same level of heat of formation, $\pm$ about 100 kJ mol^{-1}. Clearly more careful thermodynamic considerations are warranted using better data (estimated or measured) and free energy rather than heat of formation and including other reactants and products.

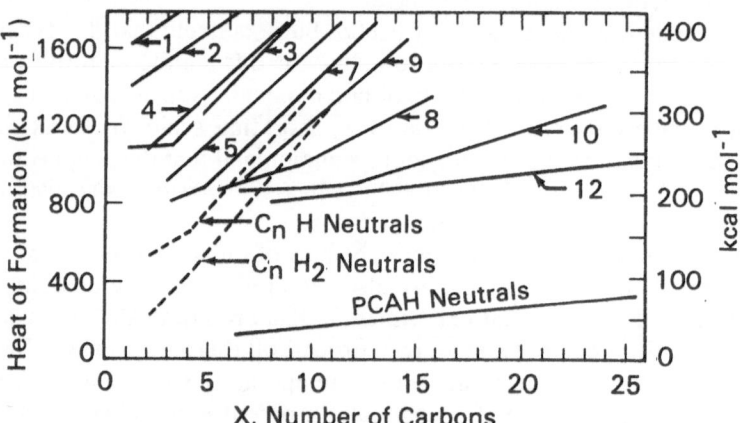

Fig. 12. Heat of formation trends for $C_x H_y{}^+$ families where the number of hydrogens, y, is indicated for each line. Also shown are $C_n H$, $C_n H_2$, and PCAH neutral species trends. Uniquely stable ions differ somewhat from these lines.

KINETIC ARGUMENTS

A chemical kinetic mechanism for soot formation is necessarily very complicated since many species are involved and the physical change from gaseous to condensed matter requires many steps. Nucleation phenomena in general are only poorly understood. Our basic hypothesis is that, in parallel to a neutral-radical chain reaction mechanism of C_2H_2 pyrolysis and combustion, there exists an ion-molecule mechanism which very rapidly produces large aromatic ions and by ion recombination large aromatic molecules (incipient soot particles) which then grow further by neutral and/or ion-molecule surface growth and coagulation. The objective is to develop a mechanism of "getting over the hump" to these large molecules; this is the nucleation process.

A series of ion molecule reactions is shown in Table 2 which comprises a subset of an overall mechanism considering only gaseous species. The primary ions are assumed to be produced by the usual flame chemi-ionization reactions although (cf. above) these are not well established under fuel rich conditions. Whatever the

TABLE 2
Proposed Ion-Molecule Mechanism

$$CH + O = CHO^+ + e^-$$
$$CH^* + C_2H_2 = C_3H_3{}^+ + e^-$$
$$C_2 + CH_3 = C_3H_3{}^+ + e^-$$
$$C_3H_3{}^+ + e^- = \text{neutral products}$$

$$C_3H_3{}^+ + \begin{cases} C_2H_2 = C_5H_5{}^+ \\ C_4H_2 = C_5H_3{}^+ + C_2H_2 \\ C_4H_2 = C_7H_5{}^+ \end{cases}$$

$$C_5H_3{}^+ + \begin{cases} C_2H_2 = C_7H_5{}^+ \\ C_4H_2 = C_7H_3{}^+ + C_2H_2 \\ C_4H_2 = C_9H_5{}^+ \end{cases}$$

$$C_5H_5{}^+ + \begin{cases} C_2H_2 = C_7H_7{}^+ \\ C_4H_2 = C_7H_5{}^+ + C_2H_2 \\ C_4H_2 = C_9H_7{}^+ \end{cases}$$

$$C_7H_5{}^+ + \begin{cases} C_2H_2 = C_9H_7{}^+ \\ C_4H_2 = C_9H_5{}^+ + C_2H_2 \\ C_4H_2 = C_{11}H_7{}^+ \end{cases}$$

$$C_7H_7{}^+ + \begin{cases} C_2H_2 = C_9H_9{}^+ \\ C_4H_2 = C_9H_7{}^+ + C_2H_2 \\ C_4H_2 = C_{11}H_9{}^+ \end{cases}$$

$$C_9H_7{}^+ + \begin{cases} C_2H_2 = C_{11}H_9{}^+ \\ C_4H_2 = C_{11}H_7{}^+ + C_2H_2 \\ C_4H_2 = C_{13}H_9{}^+ \end{cases}$$

etc.

References pp. 200-201.

actual chemi-ionization reaction, we assume $C_3H_3^+$ to be produced and propose the ion-molecule sequence for growth to large PCAH ions. Most of the growth flux is probably due to reaction of ionic species with C_2H_2 and C_4H_2, although C_2H, C_4H, and C_6H_2 may be present in large enough concentrations to contribute under some conditions.

In order to test such a mechanism we have performed preliminary computer simulations of an acetylene/oxygen system. The objective was not realistic modeling of a flame but simply to see if this kinetic scheme would show rapid growth of large ions, up to about C_{30}. Particles have not been considered up to now.

Briefly, the computations were done using a computer program simulating an adiabatic one-dimensional flow tube. Thermodynamic parameters for neutral species were obtained from the JANAF tables [31] or other published data [32], whereas ionic species data were estimated as discussed above (using experimental data where available). The calculation was initiated at 1100 K with a $C_2H_2/O_2/N_2$ mixture at 2.7 kPa containing small amounts of CO, H_2, and H atoms. About 65 reactions of 35 species were considered. This mixture was calculated to ignite after 20 ms (1 cm distance at 50 cm s^{-1})* reaching a maximum temperature of 2200 K. The equivalence ratio of the initial mixture was 3.5.

The neutral portion of the reaction mechanism consists basically of an acetylene pyrolysis scheme [24] combined with an acetylene oxidation scheme [33]. Thirty ion molecule reactions were added to test whether species up to $C_{37}H_{11}^+$ could be produced. Only the CH/C_2H_2 chemi-ionization reaction was used. All ion-molecule reactions were arbitrarily given the same rate constant, $k = 2 \times 10^{-11}$ cm^3 s^{-1}.

A discussion of the results of these calculations should be prefaced by a clear statement that (1) we do not intend that this is a computationally accurate description of a flame, (2) the combustion mechanism is not complete and was not checked against any data, and (3) the ionic mechanism is only a part of what would be required to describe ion chemistry under these conditions. What we did want to test was whether, given a mechanism such as in Table 2 with reasonable rate constants, large ions would grow to high enough concentrations and rapidly enough to support our hypothesis. If so, calculations similar to these may help to analyze the complex processes going on in a sooting flame.

Results of a typical simulation run are shown in Fig. 13 for selected neutral and ionic species concentrations and the temperature. The total ion concentration peaks at about 1.8 cm at a value of 4×10^9 cm^{-3} with most species rapidly disappearing after the flame zone and $C_3H_3^+$ persisting and only slowly decaying. The neutral species profiles are reasonable for a nonsooting flame with large concentrations of polyacetylenes calculated. Except for the $C_7H_7^+$ profile, the ionic profiles are very similar to those observed experimentally in rich nonsooting acetylene flames with heavy species rapidly peaking and $C_3H_3^+$ persisting. Although we have only limited data on absolute concentrations of these species, the calculated values

---

*Diffusion of radical species from the flame zone toward the burner into the unburned gases is, of course, important in ignition processes. Realistic simulations of flame structure must consider mass and thermal diffusion. Our simplified calculation does, however, serve to investigate chemical processes on a semiquantitative basis.

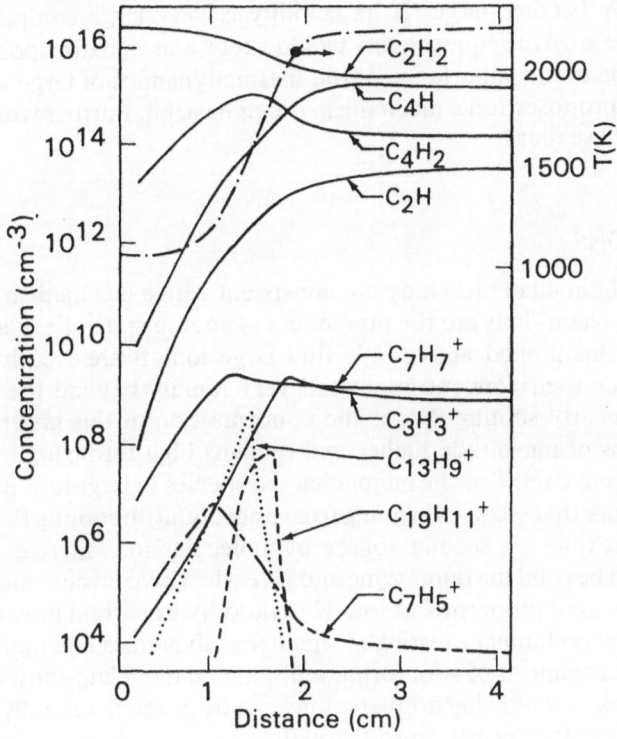

Fig. 13. Calculated concentration profiles for a 2.7 kPa acetylene/oxygen/nitrogen flame. The simulation is initialized at 1100 K and considers 65 reactions of 35 species. All ion-molecule reactions are given a k = 2 x 10^{-11} cm^3 s^{-1}. No condensed phase is considered.

are close to experimental results. All of the ions are produced via the CH*/C$_2$H$_2$ reaction (with CH* from C$_2$H + O = CH* + CO) with other possible sources ignored. The ion concentrations may therefore be artificially low. These results clearly point out a mistake that is often made deducing mechanistic information from relative positions of peak concentrations or even relative concentrations. This mechanism produces C$_3$H$_3$$^+$ as the only primary ion, yet the concentration of this ion is not always the highest even early in the reaction and in particular its profile does not peak prior to other species. In a complex reacting system the ranking of species profile peaks may bear little or no relationship to the sequence in which they are produced.

The C$_7$H$_7$$^+$ concentration shown in Fig. 13 is much larger than observed in experiments mainly due to a low heat of formation attributed to this ion (assumed to be benzyl, ΔH_f = 910 kJ mol^{-1}). This unusually stable ion therefore forms a sink in the ion-molecule sequence and its concentration reaches a large value. Since high concentrations of this ion are not observed we conclude that the actual ion is not benzyl but some less stable isomer. The calculated results were found to be very sensitive to assumed thermodynamic properties of the more stable species leading us to the prediction that if there is a large ion of very high stability its concentration could quickly become large. If its stability is very large compared to the next species in the growth sequence this would act as a terminating point in the growth sequence. Again the importance of the thermodynamics of large ions to the arguments of the proposed ionic mechanism is emphasized. Further work will be necessary to test these ideas.

CONCLUSIONS

The data obtained in this study are consistent with a mechanism for soot formation in which chemi-ions are the precursors of soot in spite of several unexplained observations mentioned above. We find large ions formed early in nonsooting flames and see their concentrations fall very rapidly beyond the flame zone. In sooting and nearly sooting flames the concentration of this group of ions builds several orders of magnitude higher and remains high throughout the flame. The abrupt shift from $C_3H_3^+$ as the major charged species to large ions predominating is the main feature that changes in going from nonsooting to sooting flames. In sooting flames an unexplained second source of molecular ions increases the total ion concentration beyond the flame zone and gives dual peaked concentration profiles.

Thermodynamic properties of low H/C ratio hydrocarbon ions have been estimated to allow preliminary testing of the thermodynamics and kinetics of the proposed ionic mechanism of soot formation. Kinetic modeling shows that sufficient concentrations of multi-ring aromatic ions can be produced rapidly. Further work will consider particle formation and growth.

We conclude that the ionic mechanism of soot formation warrants further study.

ACKNOWLEDGMENTS

This research was sponsored by the Air Force Office of Scientific Research (AFSC), United States Air Force, under Contract F49620-77-C-0029. The United States Government is authorized to reproduce and distribute reprints for governmental purposes notwithstanding any copyright notation herein. The authors also gratefully acknowledge the participation of Dr. W. J. Miller in the initial experiments and discussions.

REFERENCES

1. H. B. Palmer and C. F. Cullis, in "Chemistry and Physics of Carbon," P. L. Walker, Ed., Marcel Dekker, New York, Vol. 1 (1965), p. 265.
2. H. F. Calcote, March, 1980; Accepted for publication in Combust. and Flame.
3. D. B. Olson and H. F. Calcote, "Eighteenth Symposium (International) on Combustion," The Combustion Institute, Pittsburgh, (in press).
4. H. F. Calcote, "Eighth Symposium (International) on Combustion," Williams and Wilkins Co., Baltimore, (1962), p. 184.
5. J. Peeters and C. Vinckier, "Fifteenth Symposium (International) on Combustion," The Combusition Institute, Pittsburgh, (1975), p. 969.

6. J. W. McGowan, P. M. Mul, V. S. D'Angelo, J. B. A. Mitchell, P. Defrance and H. R. Froelich, Phys. Rev. Lett., Vol. 42 (1979), p. 373.
7. H. F. Calcote and W. J. Miller, in "Reactions Under Plasma Conditions," M. Venugopalan, Ed., John Wiley & Sons, New York, Vol. 2 (1971), p. 327.
8. J. M. Goodings, D. K. Bohme and T. M. Sugden, "Sixteenth Symposium (International) on Combustion," The Combustion Institute, Pittsburgh, (1977), p. 891.
9. H. F. Calcote, in "Ion-Molecule Reactions," J. L. Franklin, Ed., Plenum Press, New York, Vol. 2 (1972), p. 673.
10. W. J. Miller, "Eleventh Symposium (International) on Combustion," The Combustion Institute, Pittsburgh, (1967), p. 311.
11. H. F. Calcote, S. C. Kurzius and W. J. Miller, "Tenth Symposium (International) on Combustion," The Combustion Institute, Pittsburgh, (1965), p. 605.
12. G. P. Glass, G. B. Kistiakowsky, J. V. Michael and H. Niki, J. Chem. Phys., Vol. 42 (1965) p. 608.
13. P. F. Knewstubb and T.M., Sugden, "Seventh Symposium (International) on Combustion," Butterworths, London, (1959), p. 247.
14. W. J. Miller, Oxid. Combust. Rev., Vol. 3 (1968), p. 97.
15. R. J. Bowser, and F. J. Weinberg, Combust. Flame, Vol. 27 (1976), p. 21.
16. G. Prado and J. B. Howard, in "Evaporation-Combustion of Fuels," J. T. Zung, Ed., American Chemical Society, Washington, DC, (1978) p. 153.
17. B. L. Wersborg, A. C. Yeung and J. B. Howard, "Fifteenth Symposium (International) on Combustion," The Combustion Institute, Pittsburgh, (1975), p. 1439.
18. J. L. Delfau, P. Michaud and A. Barassin, Combust. Sci. Technol., Vol. 20 (1979), p. 165.
19. K. H. Homann, Ber. Bunsenges. Phys. Chem., Vol. 83 (1979), p. 738.
20. D. E. Jensen and W. J. Miller, J. Chem. Phys., Vol. 53 (1970), p. 3287.
21. J. D. Bittner, and J. B. Howard, "Eighteenth Symposium (International) on Combustion," The Combustion Institute, Pittsburgh, (in press).
22. P. Michaud, personal communication, (1980).
23. S. G. Lias and P. Ausloos, "Ion-Molecule Reactions. Their Role in Radiation Chemistry," The American Chemical Society, Washington, DC, (1975), p. 94.
24. T. Tanzawa and W. C. Gardiner, Jr., "Seventeenth Symposium (International) on Combustion," The Combustion Institute, Pittsburgh, (1979), p. 563.
25. T. Tanzawa and W. C. Gardiner, Jr., J. Phys. Chem., Vol. 84 (1980), p. 236.
26. S. E. Stein, J. Phys. Chem., Vol. 82 (1978), p. 566.
27. S. E. Stein, D. M. Golden and S. W. Benson, J. Phys. Chem., Vol. 81 (1977), p. 314.
28. H. M. Rosenstock, K. Draxl, B. W. Steimer and J. T. Herron, J. Phys. Chem. Ref. Data, Vol. 6 (1977), Suppl. No. 1.
29. R. G. McLoughlin, J. D. Morrison and J. C. Troeger, Organic Mass Spectr. Vol. 14 (1979), p. 104.
30. R. D. Bowen and D. H. Williams, J. Amer. Chem. Soc., Vol. 100 (1978), p. 7454.
31. D. R. Stull and H. Prophet, "JANAF Thermochemical Tables," Second Ed., NSRDS-NBS-37, National Bureau of Standards, Washington, DC, (1971).
32. R. S. Duff and S. H. Bauer, J. Chem. Phys., 36 (1962), p. 1754.
33. C. J. Jachimowski, Combust. Flame, Vol. 29 (1977), p. 55.

DISCUSSION

G. D. Ulrich (*University of New Hampshire*)

I have two questions. 1) The ion concentrations seem lower than typical particle concentrations in flames. Is it compatible to assume that ionization contributes to nucleation when the ion concentration is several orders of magnitude less than the particle population? 2) Is thermal ionization from soot particles a possible explanation for your second ion peak?

Olson

In answer to the first question: The data which I presented were ion current profiles and not absolute ion concentrations. It is not clear that the ion concentrations are less by several orders of magnitude than the particle concentration. Available data indicate that in a 2.7 kPa acetylene oxygen flame at $\phi = 3.5$ the soot particle concentration peaks* at about 1.4×10^{10} particles cm^{-3} and the ion concentrations** in flames at close to the same conditions are (dependent upon the experiments) 4×10^{10}, 1.3×10^{10}, 8×10^{10}, and 3×10^8. On close examination of all these experiments, however, there are reasons for questioning the actual numbers reported. Further experiments are needed to clarify the situation and we are currently pursuing them.

The second question dealt with thermal ionization. Thermal ionization is certainly acting in the flames to produce charged *particles*. The question reduces to how could the presence of charged particles of mass $> 10^4$ amu produce the charged molecular species which we observe in the dual peaks? Neither charge transfer nor proton transfer would drive the reaction in the direction of producing molecular ions from charged particles. Such reactions always move in the direction of creating larger charged particles rather than smaller charged molecules. We don't have a good explanation of the dual peaks at this time.

*H. F. Calcote, to be published in Combust. and Flame.
**D. B. Olson and H. F. Calcote, Eighteenth Symposium (International) on Combustion (The Combustion Institute, Pittsburgh, in press).

D. B. Kittelson (University of Minnesota)

Could your double peak suggest that you're not looking at one single ion or one single family of ions, but that you have a transition from one family to another? In the case of a single molecular weight (you have a fairly low resolution unit) couldn't you have a transition from one predominant ion at that molecular weight to another one to give you a double peak?

Olson

Not for the individual ion profiles. The resolution is one mass unit so that we are certainly looking at specific ions.

Kittelson

A second question. Did you include the dissociative recombination of ions in your kinetic scheme?

Olson

The dissociative recombination of $C_3H_3{}^+$ was included but not of larger ions. The ion-molecule sequence we studied was greatly condensed and certainly there were a lot of steps omitted.

D. C. Sing (General Motors Research Laboratories)

I have a general question. In your flame, or a flame like it, is the predominant negative species an electron or is it a negative ion?

Olson

Very interesting question. In stoichiometric flames the dominant negative species are electrons. In particle-containing, i.e., soot-containing flames, I think there are negative species other than electrons present. We plan to look for negatively charged ions in the near future.

K. H. Homann (Technische Hochschule Darmstadt)

Results which I recently published* relate directly to this point. We found that behind the maximum in the ionization curves most of the negative charges in the flame are ions — not electrons. The negative charges are originally formed as free electrons; then they are attached to some larger molecules or particles and are present as large negative ions.

*K. H. Homann, Ber. Bunsenges. Phys. Chem., Vol. 83 (1979), p. 738.

J. Lahaye (Centre National de la Recherche Scientifique)

You have ion current and not ion concentration profiles. Don't you believe it would be reasonable to delay your conclusion about the role of ions in soot formation until you calibrate?

Olson

No. However, we would not use the word "conclusion." We are hypothesizing that ions play a key role in soot nucleation based on a lot of evidence already in the literature.* Three of the reported measurements of ion concentration in flames give ion concentrations sufficient to support the hypothesis; a fourth value is too low (see above). As already stated all of the four experiments are subject to criticism. Clearly, a reliable measurement is highly desirable to further support our hypothesis, as is a quantitative demonstration that the reaction rates and concentration of the species are sufficient to account for the observed rate of soot formation.

*H. F. Calcote, to be published in Combust. and Flame.

K. C. Smyth *(National Bureau of Standards)*

I'm confused as to how we interpret the profile data that we've been seeing. Jim Bittner showed a lot of species profiles and made an argument based on the sequence in which these various profiles occur. You showed both experimental and model profiles in which $C_3H_3{}^+$, the proposed initial ion, peaks after the larger ions. Why is it not then proper to conclude that $C_3H_3{}^+$ is not the source of these larger species because it does not appear to be consumed by the time they peak?

Olson

Arguments based on the relative position of peaks in concentration profiles are very weak. This is demonstrated by the model profiles, Fig. 13, in which the only source of ions was $C_3H_3{}^+$, yet other larger ions peaked ahead of it. In flame ion chemistry it is very generally accepted that the source of ions is the chemi-ionization reaction $CH + O \rightarrow CHO^+ + e^-$, yet the CHO^+ profile peaks downstream of larger ions and is always very small. The initial species in a series of reactions will peak downstream of other species in the reaction chain when reactions consuming the initial species are very fast and the initial species continues to be produced beyond the position at which the reactant removing the initial species has peaked.

F. E. Jamerson *(General Motors Research Laboratories)*

Have you given any consideration to the influence of ionic mechanisms on the agglomeration of the individual spherules?

Olson

No.

A. S. Gordon *(University of California, San Diego)*

Earlier we heard about pure diffusion flames with essentially no oxygen inside the flames. There is no problem in generating large amounts of carbon from many of these flames. Surely I don't think you would try to tell me there is a chemi-ionization mechanism associated with the formation of carbon under these circumstances. Then why do we have to look at a different mechanism when we have oxygen in a premixed flame?

Olson

There may in fact be ions produced during pyrolysis; both Abrahamson and Kennedy* and Bowser and Weinberg** have reported large ion concentrations in pyrolysis experiments. In the paper we present two reactions, 10 and 11, which do not require oxygen. On the other hand there is evidence[***] that small concentrations of oxygen diffusing into diffusion flames could create ions and promote soot formation.

[*] *J. Abrahamson and E. R. Kennedy, in Twelfth Biennial Conference on Carbon, Extended Abstracts, American Carbon Society and the School of Engineering, University of Pittsburgh, Pennsylvania State University, (1975), p. 167.*
[**] *R. J. Bowser and F. J. Weinberg, Combust. and Flame, Vol. 27 (1976), p. 21.*
[***] *E. T. McHale and E. G. Skolnik, Atlantic Research Corp., Final Technical Report, ARC No. 47-5748, October 1979.*

U. Bonne *(Honeywell)*

You showed a three-dimensional graph where ionization was plotted versus distance above the burner for various fuel-air ratios. Was that a premixed flame, and was it total ionization or specific ionization?

Olson

All were premixed flames, acetylene/oxygen, at 2.0 kPa. The profiles in Fig. 10 were the current signals from heavy ions of mass greater than 300 amu as a group.

LASER LIGHT SCATTERING AND FLUORESCENCE DIAGNOSTICS OF RICH FLAMES PRODUCED BY GASEOUS AND LIQUID FUELS

A. D' ALESSIO

University of Naples
Naples, Italy

ABSTRACT

Rich combustion systems present a much more complex chemistry than stoichiometric or lean ones: compounds from unsaturated C_2 hydrocarbons to polynuclear aromatic hydrocarbons (PAH) are formed and destroyed, soot particles nucleate, grow and agglomerate and eventually burn out at different stages of the combustion. Furthermore, the fuel is introduced in the form of liquid droplets in many practical systems.

The present paper presents initially a short summary of the scattering and extinction properties of spherical particles, from the Rayleigh regime to the ray optics limit (geometric regime), referring particularly to the simultaneous presence of soot particles and fuel droplets.

Problems connected with polydispersion of spheres and the presence of particles of different shapes (spheroids, clusters, chains, etc.) are also briefly discussed.

The paper reviews subsequently the u.v./visible absorption and fluorescence spectra of some PAH compounds and discusses their interferences with the extinction due to soot particles and/or droplets and the spontaneous Raman effects due to low molecular mass compounds. The experimental section presents initially some laser light scattering (LLS) and extinction measurements carried out on large diesel oil spray flames. A procedure for distinguishing between fuel droplets and soot particles is illustrated, and average size and number concentration of soot particles are evaluated. The effects of swirl on the spatial distribution of soot particles and fuel droplets for these types of flames are also shown. Optical results obtained on premixed and diffusion CH_4O_2 flames are compared in the next section: LLS measurements provide evidence for zones of early formation of soot particles while u.v. absorption and fluorescence spectra are related to the build-up of PAH compounds.

The final part of the paper compares the experimental results with the theoretical models and discusses agreement and limits; suggestions are

advanced both for more refined theoretical approaches and for more sophisticated and systematic experiments.

LIST OF SYMBOLS

C	cross section [cm^2 sr^{-1}]
C_λ	monochromatic cross section [cm sr^{-1}]
D	particle diameter [Å], [nm], [m]
E	efficiency: $4C/\pi D^2$
$f(\alpha)$	normalized size density distribution of soot
$f(D)$	size density distribution of droplets
$i(\vartheta, \alpha, m)$	scattering function: $4\pi Q/\lambda^2$
I_0	incident light intensity [W cm^{-2}]
k	imaginary part of refractive index
m	complex refractive index
n	real part of refractive index
n_{ext}	dispersion coefficient: $d \ln E_{ext}/d\ln\lambda$
$p(\vartheta, m)$	phase function: $4\pi C(\vartheta, m)/C_{scatt}$
Q	scattering or fluorescence coefficient [cm^{-2} sr^{-1}]
Q_λ	monochromatic scattering or fluorescence coefficient [cm^{-2} sr^{-1}]
α	size parameter: $\pi D/\lambda$
α_m	mean size parameter
γ	polarization ratio: Q_{HH}/Q_{VV}
Δ	asymmetry ratio: $Q_{VV}(\vartheta_1)/Q_{VV}(\vartheta_2)$
ΔV	scattering volume [cm^3]
$\Delta\Omega$	detection solid angle [sr]
λ	light wavelength [nm]
Λ	wavelength ratio: $Q_{VV}(\lambda_1)/Q_{VV}(\lambda_2)$
σ	standard deviation

SUBSCRIPTS

abs	absorption
ext	extinction
f	fluorescence
H	horizontally polarized light
scatt	scattering
V	vertically polarized light

SUPERSCRIPTS

el	elastic
fl	fluorescence
d	droplets
g	gas
s	soot

INTRODUCTION

The optical diagnostics of practical combustion devices should play, in the next years, a decisive role in improving the efficiency and safety of combustion-based technologies.

Almost all practical systems possess, in space and/or time, rich combustion regions or conditions where pyrolytic as well as oxidative paths in the chemical evolution of the fuel must be considered. Thus hydrocarbons fragments, high molecular mass intermediates and products and, under certain conditions, soot particles are present in non-negligible amounts.

Specific optical methods are necessary in order to study those combustion regimes which are chemically much more complex than the lean and stoichiometric ones. In this paper the emphasis will be put on those light scattering and absorption methods which should contribute to the study of two classes of problems:

- a. characterization of soot particles in terms of shape, size distribution, and concentration during nucleation, growth and agglomeration.
- b. identification and concentration measurement of the intermediate hydrocarbon compounds produced in the pyrolytic combustion paths.

In a practical combustion system the fuel itself is often in a disperse form (liquid droplets, coal particles, etc.); its vapors or volatile phase are a complex mixture of hydrocarbons, and in the combustion processes other classes of particles, such as cenospheres, fly ash, metal particles, etc., are produced. Furthermore, some optical techniques are not able to distinguish between particles and gas phase compounds so that interferences arise between the aerosol characterization and the chemical analysis.

Therefore any approach to these general problems is somewhat limited by our present knowledge. The systematic work in this area is quite recent, and consequently many of the problems and difficulties are only now being appreciated.

The initial part of this paper presents a theoretical background for understanding the scattering and extinction properties of small particles, with particular reference to soot particles and hydrocarbon fuel droplets. Classical electromagnetic theory is the basis of "elastic" scattering effects which do not produce any change of frequency relative to that of the incident light beam. Emphasis is given to spherical particles in the Rayleigh, Mie and ray-optics regime, but different shapes, such as spheroids or chains, are also briefly considered. Finally, the role of polydispersivity and the interferences among different classes of elastic scatterers such as gas molecules, soot particles and fuel droplets, is discussed. The theoretical discussion is then extended to "inelastic" scattering effects which produce a frequency change between the incident and scattered photons. The fluorescence and u.v./ visible absorption properties of polycyclic aromatic hydrocarbons (PAH) are briefly addressed in relation to their interference with the scattering and extinction properties of the soot particles.

The experimental section contains results obtained mainly by the group working in the Naples Combustion Laboratory (with which the author is affiliated) and is intended to indicate the level of correspondence between theoretical schemes and experiments rather than to be a systematic exposition. Laminar premixed and diffusion CH_4/O_2 flames and larger diesel oil turbulent flames are considered and the peculiarities of situations discussed in relation both to the different combustion regions and to the different optical techniques (laser light scattering, absorption, fluorescence, Raman).

The paper concludes with considerations of the present state-of-the-art of these optical techniques and some personal observations of the author concerning the future prospects for the techniques.

THEORETICAL BACKGROUND

Laser Light Scattering (LLS) effects may be considered at different levels of theoretical sophistication in which the distinction between elastic and inelastic effects is not so clear cut. However, in the present paper we are concerned with the intensity properties of the scattered light and with frequency shifts detectable by conventional spectroscopic apparatus. The basic theory of the elastic effects is therefore that contained in the classic books by Van de Hulst [1] and Kerker [2] and applied to combustion problems by Hottel and Sarofim [3]. The subject has been by reviewed by Jones [4], with references to the recent literature.

Quasi-elastic LLS effects, detectable with interferometric or optical-mixing techniques [5, 6] are not considered here. The application of these effects to combustion problems has been considered recently by Penner et al. [7], Driscoll et al. [8] and Gouesbet et al. [9]. Laser Doppler Anemometry and particle sizing based on heterodyne detection systems are also outside the field covered by this paper [10-12].

Our discussion of inelastic LLS effects concerns large molecules (mostly aromatic) more than small radicals. Electronic-vibronic spectra of polyatomic molecules are discussed by Herzberg [13] and by Gaydon [14] whose book is specifically related to flame conditions. More specific descriptions of the fluorescence and absorption properties of aromatic molecules are found in the treatises by Clar [15] and Berlman [16] and review articles by Birks [17] and Stockburger [18]. Spontaneous Raman Scattering is briefly mentioned here; its application to combustion has been described by Lapp and Penney [19] and Ledermann [20], among others. Although non-linear Raman effects, such as Coherent AntiStokes Raman Spectroscopy (CARS), have a major relevance to the diagnostics of practical combustion systems, they are not addressed here; Eckbreth [21] has very recently reviewed the CARS techniques as applied to flame conditions.

Definitions — A typical apparatus for light scattering and extinction measurements is shown schematically in Fig. 1. The incident beam should be defined through its Stokes parameters [22], but, more simply, we assume it to be monochromatic with a wavelength λ_0 and intensity I_0, and to be linearly polarized. A beam of intensity I_{0V} is taken to be vertically polarized when its electric field oscillates in a plane perpendicular to the scattering plane, identified by the directions of the incident and scattered beams.

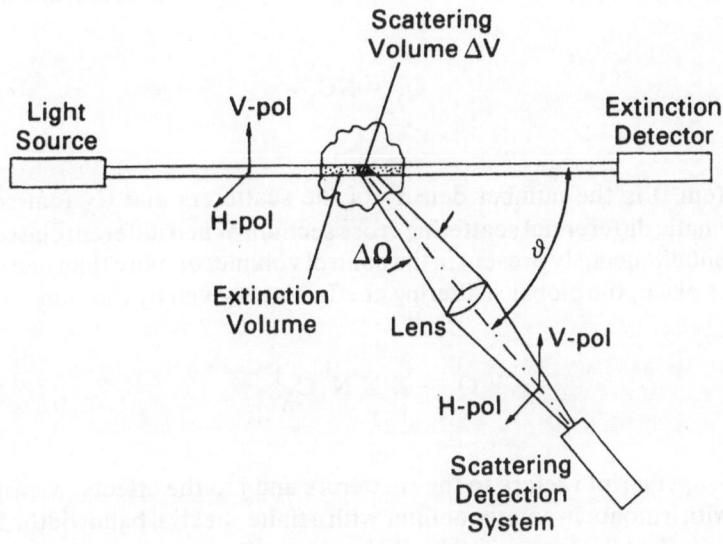

Fig. 1. Schematic of apparatus for light scattering and extinction measurements.

Similarly, a horizontally polarized beam of intensity I_{OH} is obtained when its polarization is parallel to the scattering plane. When a detector is located in the scattering plane at a scattering angle ϑ with respect to the incident beam direction, it measures the monochromatic energy flux of the scattered light $F(\vartheta)\lambda$, given by the expression:

$$F(\vartheta)_\lambda = Q_\lambda \Delta V \Delta \Omega I_{o\lambda_o} \tag{1}$$

where ΔV is the scattering volume and $\Delta \Omega$ the solid angle aperture of the detecting optics.

The quantity Q_λ, the monochromatic scattering coefficient, (of dimensions cm^{-2} sr^{-1}) is the monochromatic energy flux scattered at ϑ per unit solid angle, from a unit volume, for a unit incident energy flux density; it is obtained from the experimental data through a calibration procedure.

The polarization state of the scattered beam is measured by means of an analyzer. In particular Q_{VV} and Q_{HV} are the scattering coefficients when the incident light is vertically polarized and the vertical or horizontal components of scattered beam respectively are measured. Similarly, Q_{HH} and Q_{VH} are obtained when the incident light is horizontally polarized and the scattered light is analyzed in the horizontal and the vertical polarization plane respectively.

We consider, in addition, that the scattering volume contains a relatively large number of scatterers but that the cloud is sufficiently dilute so that, on average, the

light scattered by a single particle will leave the system without further scattering.
Then

$$Q_\lambda = NC_\lambda \tag{2}$$

where N (cm^{-3}) is the number density of the scatterers and C_λ (cm sr^{-1}) is the
monochromatic differential scattering cross section. When different classes of scatterers are simultaneously present in the control volume, or more than one scattering
effect takes place, the global scattering coefficient is given by the sum

$$Q_\lambda = \sum_i \sum_j N^i C_\lambda^{i\,j} \tag{3}$$

where the superscript i refers to the scatterers and j to the effects. Measurements
are made with a monochromator or filter with a finite spectral bandwidth, $\Delta\lambda$, which
is much larger than the true halfwidth of the elastically scattered line when a laser is
used as source. Therefore spectrally integrated scattering coefficients, Q, and cross
sections, C, are used for these effects:

$$Q = \int_{\Delta\lambda} Q_{\lambda_o}^{el} \, d\lambda \qquad C = \int_{\Delta\lambda} C_{\lambda_o}^{el} d\lambda \cdot \tag{4}$$

Q and C have the dimensions of cm^{-1} sr^{-1} and cm^2 sr^{-1} respectively.

The tranmissivity $I_\lambda (L)/I_\lambda (0)$ of a collimated beam is obtained when the
detector is located at $\vartheta, = 0$, and its value is given by the Lambert-Beer law when
the medium of thickness L is homogeneous:

$$\frac{I_\lambda(L)}{I_\lambda(0)} = \exp\left[- L \sum_i \sum_j N^i C_{ext}^{i,\,j} \right] = \exp\left[-LK_{ext} \right] \tag{5}$$

Here K_{ext} is the extinction coefficient (cm^{-1}), and C_{ext} is the extinction cross
section (cm^2), which is the sum of the absorption cross section C_{abs} and the total
scattering cross section C_{scatt}, obtained by integration of the previously defined
differential cross section over 4π solid angle, when elastic effects are considered.

Explicit expressions for the scattering and extinction cross sections and coefficients are given in the following sections for a variety of elastic and/or inelastic
scatterers.

Rayleigh Scattering by a Sphere — When the scatterers have characteristic
dimensions much smaller than the wavelength of the incident radiation, the elastic
scattered field is equivalent to that emitted by a oscillating electric dipole induced
by the incident field (Rayleigh effect) [1-3].

In the Rayleigh approximation the scattering angular cross sections C_{VV} and C_{HV} are independent of the scattering angle ϑ, while the C_{HH} cross section varies with ϑ according to the expression:

$$C_{HH} = C_{VV} \cos^2 \vartheta + C_{HV} \sin^2 \vartheta \,. \tag{6}$$

Moreover from the reciprocity law C_{HV} is equal to C_{VH} [2].

The ratio $\rho_v = C_{HV}/C_{VV}$ is the depolarization ratio. When the scatterer does not present any optical or geometrical anisotropy, its polarizability is a scalar quantity and the depolarized cross sections C_{HV} and C_{VH} are equal to zero. Consequently the horizontal component C_{HH} at $\vartheta = 90°$ is absent.

For a spherical absorbing particle the polarizability is a function of the diameter D and its complex refractive index m=n-ik. The angular scattering cross section for vertical polarization of the light is given by the expression:

$$C_{VV} = \frac{\lambda_o^2}{4\pi^2} \left| \frac{m^2-1}{m^2+2} \right| \alpha^6 \tag{7}$$

and the absorption cross section is equal to:

$$C_{abs} = \frac{\lambda_o^2}{4\pi} \, \mathrm{Im} \left\{ \frac{m^2-1}{m^2+2} \right\} \alpha^3 \tag{8}$$

where $\alpha = \pi D/\lambda$ is the ratio of the perimeter of the particle to the wavelength of the incident light.

In the theory of the scattering and absorbing properties of spherical particles, it is customary to refer to dimensionless parameters like E_{scatt} and E_{abs}, efficiencies for scattering and absorption, which are defined as the ratio of the optical cross section to the geometrical cross section of the particle:

$$E_{scatt,abs} = \frac{C_{scatt,\,abs}}{\frac{\pi D^2}{4}} \tag{9}$$

Moreover the angular cross sections are expressed through dimensionless factors $i(\vartheta)$ by the expressions:

$$C_{VV} = \frac{\lambda_o^2}{4\pi} \, i_{VV}$$
$$\tag{10}$$
$$O_{HH} = \frac{\lambda_o^2}{4\pi} \, i_{HH}$$

References pp. 254-256.

Therefore the Rayleigh theory of an absorbing spherical particle can be summarized by the following equations:

$$E_{abs} = A\alpha$$

(11)

$$E_{scatt} = B\alpha^4$$

$$i_{VV}(\vartheta) = \frac{3}{8} \cdot B\alpha^6 = i_{HH}(\vartheta) / \cos^2(\vartheta)$$

where A and B are algebraic functions of the complex refractive index of the material of the particles.

The complex refractive index of soot is not known with accuracy. This complicated subject has been reviewed in a previous paper [23]. For the purposes of our calculations we use the index determined from the optical properties of soot in the visible by Dalzell and Sarofim [24] which give $m = 1.56 - i0.56$. However, there is still considerable ignorance about the variation of the optical properties with the chemical composition of soot and with the flame temperature.

Lee and Tien [25] have very recently given a different estimate of the complex refractive index $[m = (1.9 \pm 0.1) - i(0.55 \pm 0.1)]$ from in-flame spectral extinction measurements. Since the extinction coefficient in the flames is due also to gas phase absorbers, their analysis does not appear completely convincing for the values in the visible, as will be seen below.

Using $m = 1.56 - i0.56$ in eq. (11), we obtain $A = 1.04$ and $B = 0.52$. Therefore in the Rayleigh regime $(\alpha < 1)$ the absorption cross section is much larger than that for scattering. The size of the sphere is obtained by taking the ratio of the measured scattering and extinction coefficients since this quantity is proportional to the volume of the particle (see eqs. 7-8), and the number concentration of soot particles is given by each of the coefficients if the system is monodisperse. No other quantitative information is obtainable from the angular or polarization properties of the scattering.

Rayleigh Scattering by Molecules — The elastic scattering property of gas phase atoms and molecules is just a particular case of the Rayleigh theory. It is easy to see, by the use of the Lorenz-Lorentz formula, that the polarizability can be expressed in terms of the overall refractive index of the gaseous medium [26], and that the C_{VV} cross section takes the simple form:

$$C_{VV} = \frac{4\pi^2 (n-1)^2}{\lambda_o^4 N_o^2} \frac{3}{3 - 4\rho_V}$$

(12)

where n is the gas refractive index and $N_o = 2.687 \times 10^{19}$ cm^{-3} is the number density of the scatterers, both at NTP.

Rudder and Bach [27], in careful ruby-laser scattering experiments, have measured the C_{VV} cross section and the depolarization ratio of different gaseous compounds and verified the above formula to within 2%.

Table 1 reports the vertical C_{VV} cross sections and the vertical depolarization ratios for a number of gaseous compounds involved in the high temperature oxidation of methane at the wavelength of 514.5 nm. The values were computed from eq. 14 interpolating the refractive index values given by the Landolt-Bornstein tables [28]. The values are a critical average of the determinations by Rudder and Bach, by Bridge and Buckingham [29] and the older values quoted by Bhagavantam [30].

TABLE 1
Rayleigh Scattering Cross Sections; λ_o = 515.5 nm

Compound	C_{VV} (cm² sr⁻¹)	ρ_V
CH_4	1.57×10^{-27}	0
C_2H_6	4.6×10^{-27}	2×10^{-3}
C_2H_4	4.14×10^{-27}	1.2×10^{-2}
CO_2	1.39×10^{-27}	4×10^{-2}
CO	8.93×10^{-28}	5×10^{-3}
O_2	6.03×10^{-28}	3×10^{-2}
H_2	1.54×10^{-28}	3.3×10^{-3}
H_2O	5.08×10^{-28}	1×10^{-2}
10 Å Soot sphere	6.76×10^{-25}	0
100 Å Soot sphere	6.75×10^{-19}	0

The values of the cross sections for gaseous compounds are all in the range 10^{-28} to 10^{-27} cm² sr⁻¹; for comparison also reported are the cross sections of soot spheres of 10 Å and 100 Å diameter, computed from eq. 7, with m=1.56-i0.56.

Rayleigh cross sections of hydrocarbons of high molecular mass (600 to 1000 a.m.u.) can be evaluated by additivity rules for the bond polarizabilities [26]; C_{VV} values are on the order of 10^{-25} to 10^{-24} cm² sr⁻¹, which are comparable to those predicted for an absorbing sphere with the same molecular mass.

The scattering coefficients for vertical and depolarized conditions for a mixture of i gas components are given by the expressions:

$$Q_{VV}^g = N^g \sum_i x^i \, C_{VV}^i \qquad (13)$$

$$Q_{HV}^g = N^g \sum_i x^i \, \rho_V^i \, C_{VV}^i \qquad (14)$$

where N^g is the local number gas density and i the mole fraction. Eq. 13 allows the determination of the total density, and hence of the temperature, where no particulate scatterers are present, in the case of a single component or when the composition is known. When only two gaseous components are present, the additional measurement of the depolarized coefficient permits, in principle, the simultaneous determination of N^g and the molar ratio $x^{(1)}/x^{(2)}$.

Rayleigh Scattering by an Absorbing Spheroid — If a particle is geometrically and/ or optically anisotropic, its polarizability is a tensor expressed by the corresponding ellipsoid.

A spheroid (A, B, C,) is characterized by the values of the two semiaxes. The oblate spheroid has $A < B = C$, the semiaxis ratio is defined as $p = B/A$, and its limit is the infinite disc when $B = C = 0$. The prolate spheroid has $A > B = C$, the semiaxis ratio is $p = A/B$, and its limit is the infinite cylinder. The spheroid degenerates into a sphere for $p = 1$.

The scattering and absorption cross sections are functions of $\alpha = 2\pi S/\lambda$, where S is the major semiaxis. The dipolar model for a small spheroidal particle has been developed by Rayleigh and Gans (see ref. [2]). It is convenient to consider the ratio of the cross sections for spheroids to those of spheres of equal volumes. In the case of a spatially random orientation of the particles we have [2]:

$$\left(\frac{C_p}{C_1}\right)_{scatt} = \frac{16}{27}\,\pi^2 \left(\frac{|\beta'|^2 + 2\,|\beta''|^2}{\left|\dfrac{m^2 - 1}{m^2 + 2}\right|^2}\right) \tag{15}$$

$$\left(\frac{C_p}{C_1}\right)_{abs} = \frac{4}{9}\,\pi\,\frac{\mathrm{Im}\,\{-\beta' - 2\beta''\}}{\mathrm{Im}\left\{\dfrac{m^2 - 1}{m^2 + 2}\right\}} \tag{16}$$

β' and β'' are the diagonal components of the polarizability unit tensor and are given by the equations:

$$\beta' = \frac{m^2 - 1}{4\pi + (m^2 - 1)\,p'} \qquad \beta'' = \frac{m^2 - 1}{4\pi + (m^2 - 1)\,p''} \tag{17}$$

The shape factor p' for a prolate spheroid is

$$p' = \frac{4\pi}{p^2 - 1}\left\{\frac{p}{(p^2 - 1)^{1/2}}\,\ln\left\{[p + (p^2 - 1)^{1/2}] - 1\right\}\right\} \tag{18}$$

for an oblate spheroid

$$p' = \frac{4\pi}{p^2 - 1} \left\{ p^2 - \left(\frac{p^4}{p^2 - 1} \right)^{1/2} \arcsin \left[\frac{(p^2 - 1)^{1/2}}{p} \right] \right\} \quad (19)$$

and $p'' = (4\pi - p')/2$.

Eqs. 15 and 16 are only formally independent since it may be shown that (C_p/C_1)scatt = (C_p/C_1)abs [31]. This identity can be put in the form

$$\left(\frac{C_{scatt}}{C_{abs}} \right)_{spheroid} = \left(\frac{C_{scatt}}{C_{abs}} \right)_{sphere} \quad (20)$$

Therefore the ratio Q_{VV}/K_{ext} is proportional to the volume of the Rayleigh scatterers, and it is independent of its shape.

The scattering and absorption cross sections of a spheroid are always slightly higher than the corresponding properties of the equal volume sphere; their ratio

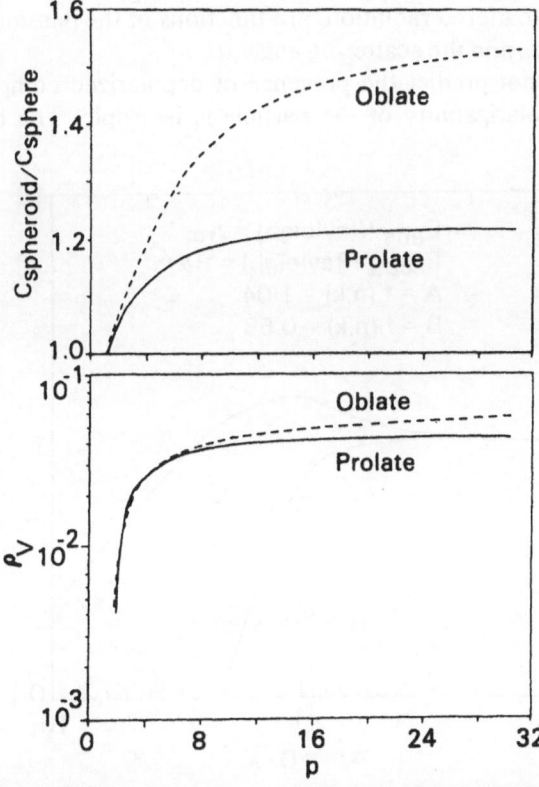

Fig. 2. Computed ratio of spheroid-sphere cross sections and depolarization ratio as a function of the semiaxis ratio p, for m = 1.56 - i 0.56. Adapted from Ref. [31].

reaches asymptotic values of 1.2 for a prolate spheroid and 1.6 for an oblate spheroid when the semiaxis ratio p is larger than 30, (28 is reported in Fig. 2). This effect may have some importance in determining an increase of emissivity, for a fixed amount of soot, when the particles are elongated, as is pointed out by Jones [32].

A more interesting effect, from a diagnostic point of view, is the presence of depolarized components due to the tensor nature of the polarizability. The computed depolarization ratio $\rho_V = Q_{HV}/Q_{VV}$ is also reported in Fig. 2 as a function of the semiaxis ratio p when m = 1.56-i0.56: ρ_V is equal to zero for p=1 (sphere), but its value grows rapidly up to 3×10^{-2} for p~6, for both oblate and prolate spheroids. Thereafter it is not sensitive to shape effects for greater elongation of the particle. A complete discussion of the effect of the complex refractive index on the ρ_V-p relationship is presented by Ferrara [31].

Lorenz-Mie Scattering — The exact solution of the Maxwell equations for an absorbing particle of arbitrary size is possible only for the sphere and is known as the Lorenz-Mie theory. Only the more significant physical aspects of the numerical results computed for soot particles, in the diameter range 10-3000 Å, are reported here since the formal structure of the equations is reported in the literature [1-3].

The extinction efficiency Q_{ext}, the scattering efficiency Q_{scatt} and the dimensionless parameters i_{VV} and i_{HH}, which define the angular distribution and polarization state of the scattered radiation, are functions of the parameter α, the complex refractive index, and the scattering angle ϑ.

The theory does not predict the presence of depolarized components since it assumes that the polarizability of the particle is isotropic. The behavior of the

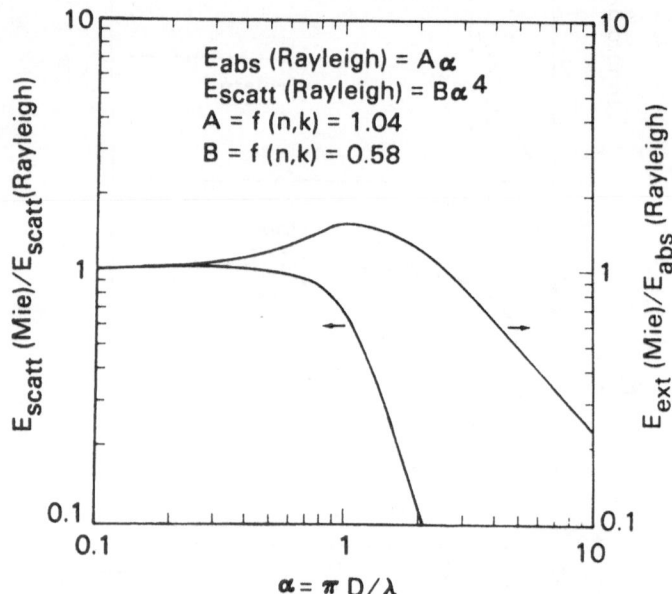

Fig. 3. Computed ratios between Lorenz-Mie and Rayleigh efficiencies of scattering and extinction as a function of size parameter α for m = 1.56 - i 0.56.

Lorenz-Mie parameters is better presented in comparison with the corresponding Rayleigh parameters, as reported in eq. 11.

The ratios $E_{ext}/A\alpha$ and $Q_{scatt}/B\alpha^4$ are plotted in Fig. 3 versus the parameter α for m=1.56-i 0.56. The extinction efficiency ratio increases with α in the range 0.3-3.0, but it declines for larger sizes. The fourth power dependence of the scattering efficiency with respect to α is maintained up to α =0.5, but when the particles become larger, it increases at a much slower rate. The ratio $i_{VV}^{Mie}/i_{VV}^{Ray}$ is shown in Fig. 4 for different scattering angles. The curves are very similar up to α=0.3 but they diverge noticeably for larger sizes thus evidencing an asymmetry with i_{VV} (ϑ<90°) greater than i_{VV} (ϑ>90°).

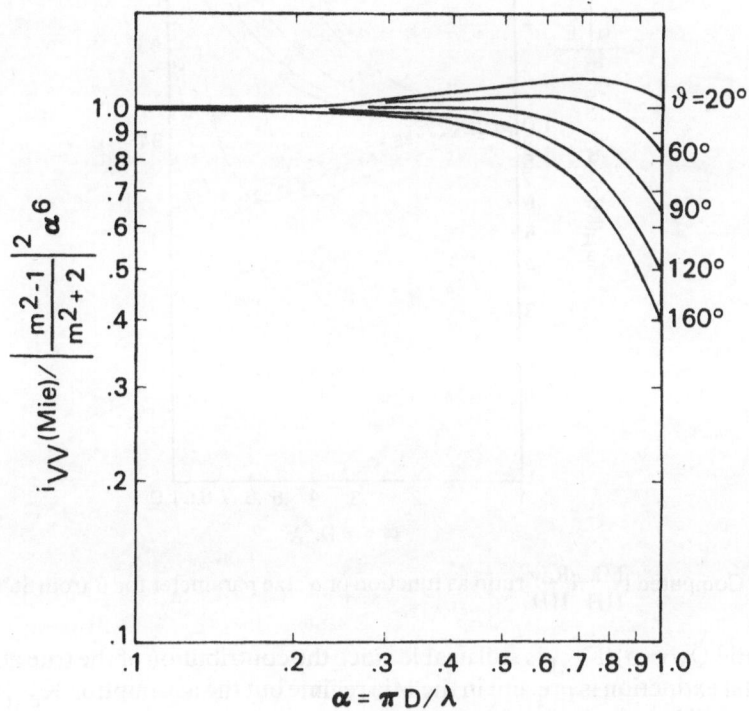

Fig. 4. Computed i_{VV} (ϑ) angular scattering function ratio between the Lorenz-Mie and Rayleigh conditions, as a function of size parameter α for m = 1.56 - i 0.56.

Another interesting feature of the Lorenz-Mie analysis is the strong variation of the horizontal components of scattering around 90° with respect to the Rayleigh case, as is evident in Fig. 5 where the ratio $i_{HH}^{Mie}/i_{HH}^{Ray}$ is plotted for ϑ from 86° to 94° in the range $0.1 \leqslant \alpha \leqslant 1$. The i_{HH} coefficients increase very strongly with respect to the Rayleigh ones; a non-zero component is present also at 90°. Furthermore the minimum in the i_{HH}-ϑ curves shifts from 90° to higher angles, and the effect, shown in Fig. 12, is quite significant, e.g. α=1, ϑ_m=94°.

The ratios $\Delta=Q_{VV}(\vartheta)/Q_{VV}(\pi-\vartheta)$ and $\gamma=Q_{HH}$ (ϑ)/$Q_{VV}(\vartheta)$ are both suitable for determining α and hence the size of the particle. The sensitivity of the α-ratio is higher for low scattering angles whereas the γ ratio has its maximum sensitivity near 90°.

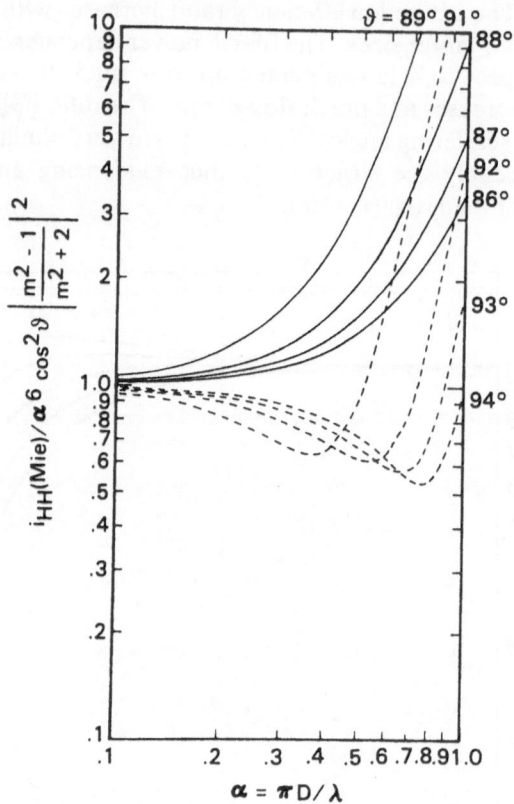

Fig. 5. Computed $i_{HH}^{Mie}/i_{HH}^{Ray}$ ratio as function of α size parameter for ϑ from 86° to 94°.

The ratio $Q_{VV}(\vartheta)/K_{ext}$ is still usable since the contribution of the true absorption to the total extinction is present in the Mie regime but the assumption $K_{ext}=K_{abs}$ is no longer valid.

All the Mie parameters are influenced by the choice of the complex refractive index and for the extinction efficiency, E_{ext}, this is particularly relevant. For example Fig. 6 shows the dependence of the dispersion parameter $n_{ext}=d\ln Q_{ext}/d\ln\alpha$ upon α for several values of the complex refractive index; 1.95-i 0.66 is the value determined in the visible on electrode carbon by Senftleben and Benedict [33] and 1.74-i 0.74 has been found by Taft [34] for glassy carbon. The parameter n_{ext} is equal to 1 for very small particles; it has a maximum, which is more pronounced the higher the real part of the refractive index, for intermediate size particles, and tends to zero or assumes negative values for larger sizes. The n_{ext} parameter and the corresponding parameter $n_{scatt}=d\ln E_{scatt}/d\ln\alpha$ have particular relevance when the wavelength dependence of the particle extinction or scattering are measured, as is discussed in the experimental section.

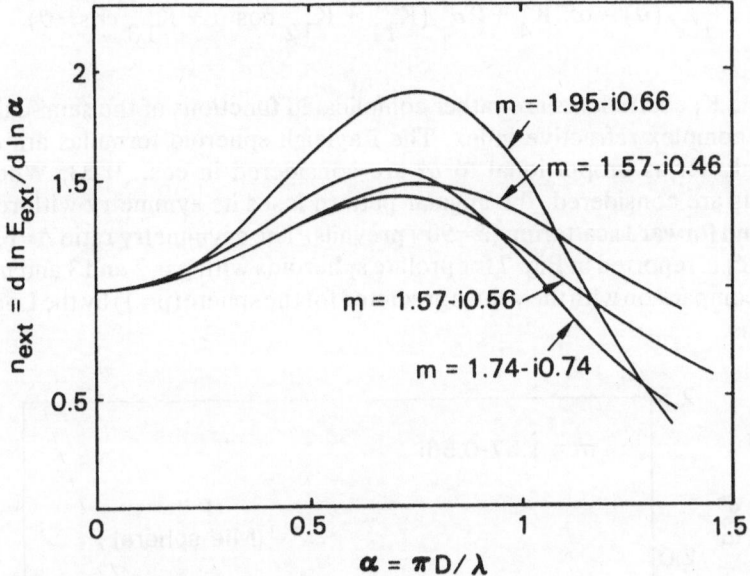

Fig. 6. Dependence of dispersion parameter upon α, for different values of refractive index. Adapted from Ref. [55].

Scattering by Large Spheroids (Stevenson Approximation) — The exact electromagnetic solution for an absorbing spheroid with size comparable to the wavelength presents great mathematical difficulties, and limited results have been obtained for a fixed spatial orientation of the particle [35]. Stevenson [36] proposed an approximate solution for ellipsoids in the form of a power series in the ratio of size to wavelength, truncated after the first three terms. This expansion can be applied to spheroids with the larger axis up to 1/3 of the wavelength in the medium ($\alpha \leqslant 0.95$) without having errors in excess of 5%.

Heller and Nakagaki [37] have applied the Stevenson theory to the case of internally isotropic, nonabsorbing, oriented spheroids with m up to 1.4, by integrating over all possible positions of the spheroid with respect to an incident polarized beam.

The computations have been extended to absorbing spheroids with refractive indices n=1.4 to 2 and k=0.4 to 1 by Ferrara [31].

In the Stevenson expansion the i parameters are given by the equations:

$$i_{VV}(\vartheta) = \alpha^6 \, K_1 + 2\,\alpha^8\,(K_2 + K_3 \cos\vartheta) \tag{21}$$

$$i_{HH}(\vartheta) = \alpha^6 \, (K_4 + K_5 \cos^2\vartheta) + 2\,\alpha^8\,(K_6 + K_7 \cos\vartheta + $$
$$K_8 \cos^2\vartheta + K_9 \cos^3\vartheta + K_{10} \cos^2\vartheta) \tag{22}$$

$$i_{HV}(\vartheta) = \alpha^6 K_4 + 2\alpha^8 (K_{11} + K_{12} \cos\vartheta + K_{13} \cos^2\vartheta) \qquad (23)$$

where the K_i coefficients are rather complicated functions of the semi-axis ratio p and the complex refractive index. The Rayleigh spheroid formulas are obtained when only terms proportional to α^6 are considered in eqs. 21-23. When larger spheroids are considered, the angular pattern loses its symmetry with respect to $\vartheta < 90°$ and forward scattering ($\vartheta = 90°$) prevails. The asymmetry ratio $\Delta = i_{VV}(20°)/i_{VV}(160°)$ is reported in Fig. 7 for prolate spheroids with p = 2 and 3 and m = 1.57-i0.56 in comparison with the values predicted for the sphere (p=1) by the Lorenz-Mie equations.

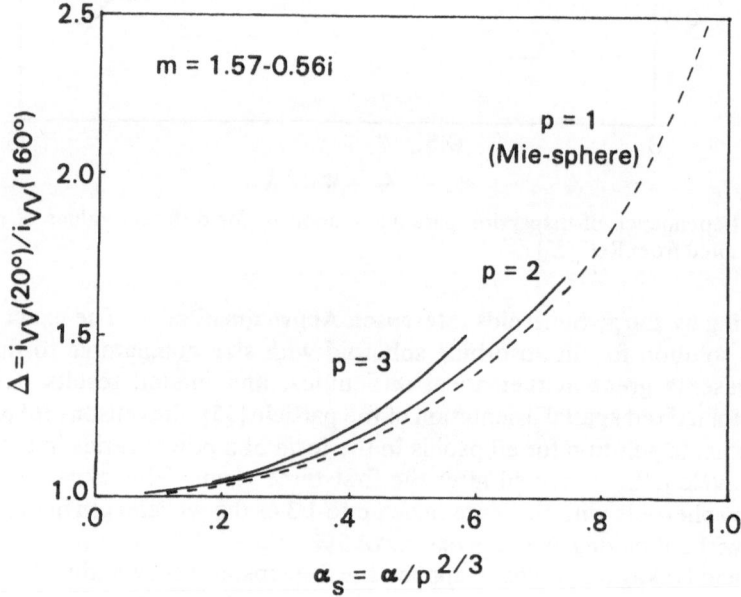

Fig. 7. Computed asymmetry ratio for prolate spheroids with p = 2 and 3 in comparison with the values predicted for a sphere (p = 1) by the Lorenz-Mie equations. Adapted from Ref. [31].

The comparison has been made for particles with equal volume so that the ratio is reported as a function of $\alpha_s = \alpha/p^{2/3}$. It is worthwhile to note that the asymmetry of spheroids with moderate elongation does not appear to be very different from that of the sphere.

The influence of size on the depolarization ratio $\rho_v = i_{HV}/i_{VV}$ has also been investigated at $\vartheta = 90°$, as shown in Fig. 8 for prolate spheroids; ρ_v does not depend significantly upon size, and its dependence on the particle shape factor p is equal to that predicted for the Rayleigh spheroid.

Another interesting feature of the Stevenson spheroids is the differences between i_{HV} and i_{HH} at $\vartheta = 90°$, which are equal in the Rayleigh limit (see eq. 6). For a prolate spheroid i_{HV} is always greater than i_{HH}, and the reverse is true for an

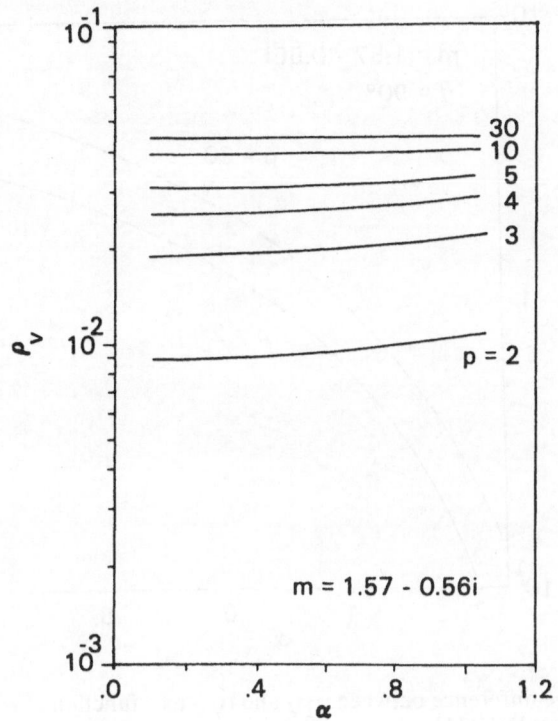

Fig. 8. Computed depolarization ratio at $\vartheta = 90°$ for prolate spheroids of different semiaxis ratios p as a function of α. Adapted from Ref. [31].

oblate spheroid so that this type of measurement should allow the unambiguous determination of the shape of the spheroid.

Fig. 9 shows the percentage difference between i_{HV} and i_{HH} as a function of α for two extreme p values. The effect is relatively small, but it becomes larger when the volume of the particle increases whereas it is not greatly influenced by the particle shape.

Finally, it should be noted that the Stevenson analysis predicts that the minimum of the i_{HH} (ϑ) angular pattern occurs at $\vartheta=90°$ and does not allow backward shifts as in the Mie-Lorenz theory.

Scattering by Clusters and Chains of Spherical Particles — Loose aggregates of soot particles might be treated as chains or clusters of smaller spherical elementary particles with diameters around 200 to 300 Å. However, only in the case of scattering by two spheres randomly positioned relative to the direction of the incident polarized beam has there been a rigorous solution for spheres of arbitrary size [38]. Approximate solutions for linear chains and clusters formed by N absorbing Rayleigh spheres have been presented by Ravey [39] and Jones [40, 41]. The latter author calculates the angular patterns of i_{VV}, i_{HV}, i_{VH} and i_{HH} and some of their

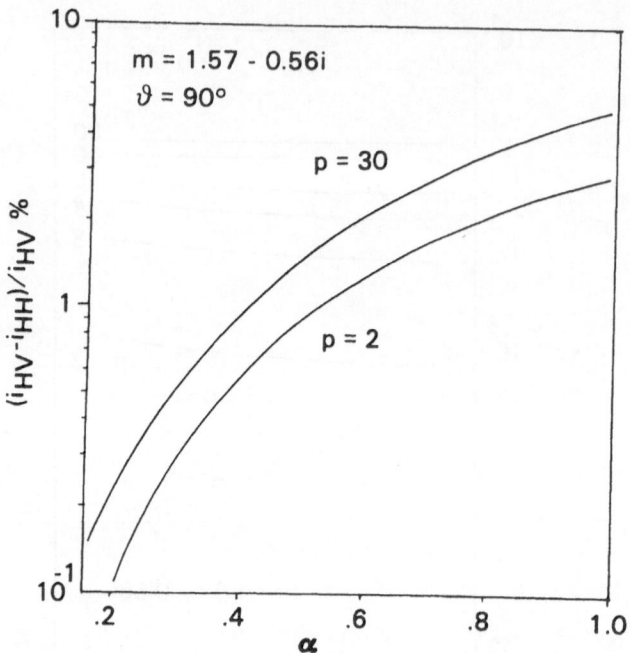

Fig. 9. Percentage difference between i_{HV} and i_{VV} as a function of α for two extreme p values. Adapted from Ref. [31].

ratios as functions of the diameter of the elementary Rayleigh particle for up to ten particles with complex refractive index values near to those typical of soot particles. His model predicts that the symmetry of a two-particle system can be likened to a sphere of diameter $D_S = \sqrt{2}D$, a long chain to $D_S = N^2D$ and a cluster to $D_S = ND$. The depolarization ratio ρ_V at $\vartheta = 0$ due to a long chain is equivalent to that exhibited by a prolate Rayleigh spheroid with semiaxis ratio slightly greater than two. Furthermore, Jones states that the extinction efficiency of long chains does not change with N while the scattering efficiency increases linearly with it.

A final result of his analysis is that i_{HV} is a strong function of ϑ and N and has a deep minimum at $\vartheta = 120°$. It does not obey the Krishnan relation [42]

$$i_{HV} = i_{VH} \cdot \tag{24}$$

The behavior of the polarization components is qualitatively and quantitatively different from that computed by Ravey for the same model and no explanation is furnished for this discrepancy.

Scattering by Large Particles (Ray Optics Approximation) — Although the Lorenz-Mie theory also gives an exact description of the scattering properties for large spheres ($\alpha > 50$), its use is particularly cumbersome, since the Mie series converges with the number of terms on the order of α. Geometrical optics predicts satisfactorily all the features of the angular pattern, besides the regions of diffraction, glory and rainbows, even the position and polarization of the rainbows [1, 43, 44]. The forward diffracted light is described by the well-known Fraunhofer theory while glory and rainbows have been treated by Nussenzveig [45] by the complex angular momentum method.

The scattering angular cross sections $C_{VV}(\vartheta)$ or $C_{HH}(\vartheta)$ for large spheres is proportional to α^2 [46] and is usually expressed in terms of phase functions $p(\vartheta, m)$ defined as:

$$p(\vartheta,m) = \frac{C(\vartheta, m)}{C_{ext}/4\pi} \tag{25}$$

In this limit

$$C_{ext} = 2 \cdot \frac{\pi D^2}{4} \tag{26}$$

so that

$$C_{VV}(\vartheta, m) = \frac{1}{8} D^2 \, p_{VV}(\vartheta, m) \tag{27}$$

In Fig. 10 the phase function $p_{VV}(\vartheta)$ is plotted as a function of the scattering angle for $m = 1.5$ and $m = 1.6$. The forward scattered radiation is higher than the backward contribution and is mainly due to reflected and refracted contributions; the contributions due to light which is internally reflected two or more times is detected in the rainbow region.

The angular position of the rainbows and the light cut-off of the forward lobe are determined by the refractive index m and are shifted backward as m increases, except for the two-internal-reflections rainbow. The polarization ratio p_{HH}/p_{VV}, plotted in Fig. 11 shows that, besides the rainbow regions where γ has values lower than 0.4, p_{HH} has the same behavior as p_{VV} and it is higher than p_{VV} in the forward lobe for a scattering angle $\vartheta > 30°$. The scattering and extinction coefficients integrated over the size distribution are proportional to the total surface area of the large spheres.

$$Q(\vartheta, m) = \frac{1}{8} p(\vartheta, m) \, N \int_0^\infty D^2 f(D) dD \tag{28}$$

$$K_{scatt}(m) = \frac{\pi}{2} N \int_0^\infty D^2 f(D) dD$$

Since the phase function is not dependent on the particle size, it is not possible to obtain the dimension of the particles by the polarization and asymmetry ratio.

This information can be deduced from the Fraunhofer pattern [47], because the position of the interference rings is linearly proportional to α^{-1}.

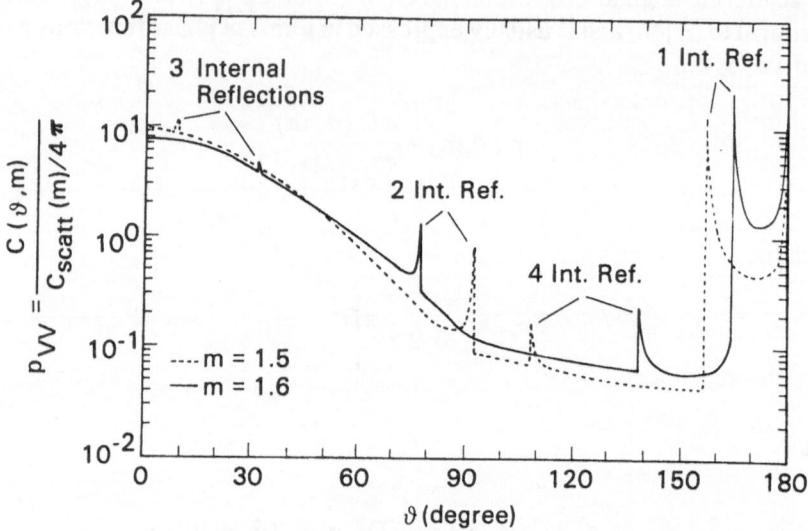

Fig. 10. Phase function p_{VV} computed from geometric optics as a function of ϑ scattering angle for refractive indices m = 1.5 and 1.6. Adapted from Ref. [83].

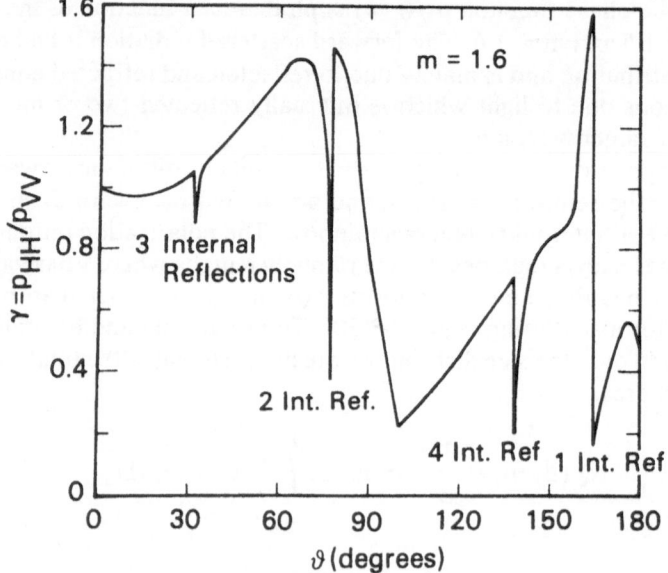

Fig. 11. Polarization ratio $\gamma = p_{HH}/p_{VV}$ computed from geometric optics as a function of scattering angle ϑ. Adapted from Ref. [83].

Polydispersion of Lorenz-Mie Spheres — When spherical particles of different sizes are present in the measurement volume it is necessary to integrate the Lorenz-Mie parameters over the size distribution. So we define the averaged quantities:

$$\overline{Q}\,(\alpha_m,\sigma) = \int_o^\infty Q\,(\alpha)\,f\,(\alpha)\,d\alpha$$

$$\overline{i}\,(\alpha_m,\sigma) = \int_o^\infty i\,(\alpha)\,f\,(\alpha)\,d\alpha$$

$$\overline{C}\,(\vartheta) = \frac{\lambda_o^2}{4\pi}\,\overline{i}\,(\vartheta) \tag{29}$$

$$\overline{C}_{ext} = \frac{\lambda_o^2}{4\pi}\,\overline{\alpha^2\,E_{ext}}$$

where $f\,(\alpha)$ is the normalized size distribution function, which assumes, for the present calculation, a log-normal function with median mean size α_m and geometrical mean standard deviation σ.

The values of the ratio between the averaged cross sections and those computed for $\alpha = \alpha_m$ for $\alpha \rightarrow 0$ may be computed analytically from the Rayleigh expression:

$$\frac{\overline{i_{VV}}}{i_{VV}\,(\alpha_m)} = \exp\left[18\sigma^2\right] \qquad \frac{\overline{\alpha^2\,E_{ext}}}{\alpha_m^2\,E_{ext}\,(\alpha_m)} = \exp\left[\frac{9}{2}\,\sigma^2\right] \tag{30}$$

The ratios of the scattering cross sections are very high in the Rayleigh regime, when the monodisperse values increase proportionally to α^6, and then decline for higher α_m due to their slower increase with α in the Mie regime [23].

Many characteristics of the Mie scattering, outlined above, are enhanced by polydispersion; for example Fig. 12 illustrates how ϑ_m, the scattering angle where i_{HH} has a minimum, moves as α_m is increased from 0 to 1.3 for σ which changes from 0 (monodisperse case) to 0.7. The ratio $\gamma\,(90°) = i_{HH}\,(90°)/i_{VV}\,(90°)$ is a very sensitive function of α_m and σ also for systems with a narrow distribution.

Figs. 13 and 14 illustrate respectively how polydispersion influences the asymmetry ratio $\Delta = i_{VV}\,(20°)/i_{VV}(160°)$ and the scattering/extinction ratio $i_{VV}\,(20°)/\overline{\alpha^2 E_{ext}}$. The former is not very sensitive for small α_m while it changes rapidly for $\alpha > 0.5$ and $\sigma > 0.3$, and the reverse is true for the scattering/extinction ratio.

References pp. 254-256.

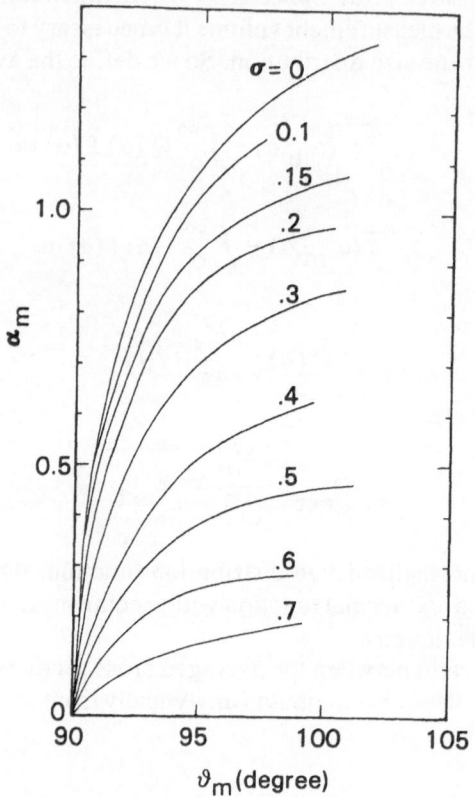

Fig. 12. Scattering angle ϑ_m, where i_{HH} has a minimum, as a function of the mean size parameter α_m and the geometrical mean standard deviation σ.

There are more than two independent ratios of measurable parameters so that, in principle, it is possible to obtain simultaneously the two parameters of the distribution α_m and σ from different experimental measurements.

Interferences among Different Scatterers — As was stated in the Introduction, soot particles coexist in the scattering volume with other types of scatterers such as gaseous compounds and condensed-phase particles of dispersed fuel. In this section we consider only liquid condensed phases. Analogous considerations could be put forth for solid phases.

The scattering coefficient when multiple scattering can be neglected is the sum of the scattering coefficients of the single species

$$Q = N^g C^g + N^s C^s + N^d C^d \tag{31}$$

where g, s and d stand for gas, soot and droplet respectively.

The largest cross section for low molecular weight gases (e.g., H_2, CO_2, H_2O, CH_4 etc.) is about 10^{-27} cm^2 sr^{-1} whereas the largest one for heavy molecular weight gases (e.g. $C_{16} H_{32}$) is 10^{-24} cm^2 sr^{-1}; these values yield maximum scattering coefficients of 10^{-8} cm^{-1} sr^{-1} and 10^{-6} cm^{-1} sr^{-1} and, depending on the experimental conditions, they have to be compared to the values for soot and liquid droplets. The cross sections for soot range between 6.76×10^{-25} and 6.75×10^{-19} cm^2 sr^{-1} for spheres with diameters between 10 and 100 Å while they range between 10^{-8} and 10^{-6} cm^2 sr^{-1} for liquid droplets of 10 and 100μm diameter. The number of scatterers expected in the scattering volume on the basis of physical considerations yields an order of magnitude estimate of the scattering coefficients, as is reported in the experimental section.

While this analysis indicates that the scattering coefficient cannot be attributed to one category of scatterers, some discrimination between different species might be possible by independent scattering measurements which do not introduce new unknowns.

If the first measurement is of the VV components at $\vartheta = 90°$ for incident beam wavelength λ_1, and only soot and droplets are present in the scattering volume, we

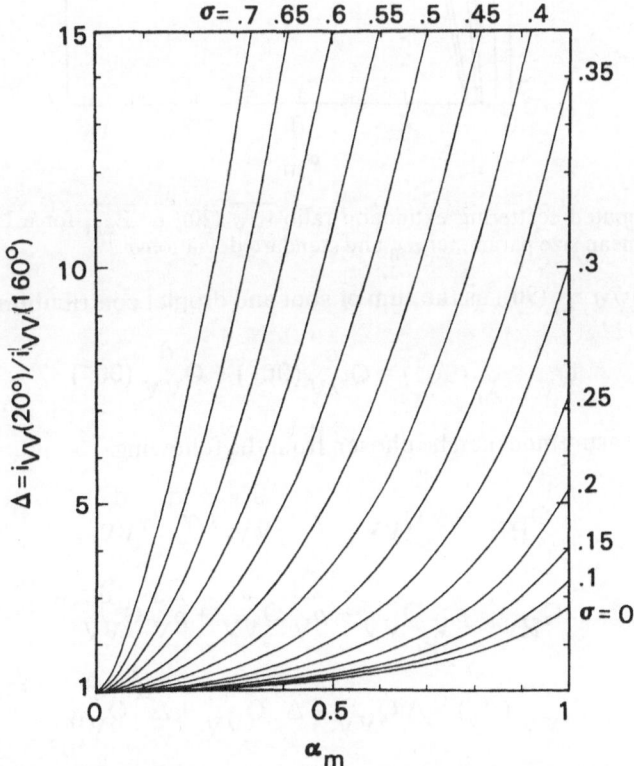

Fig. 13. Computed asymmetry ratio for a log-normal size distribution of mean size parameter α_m and standard deviation σ.

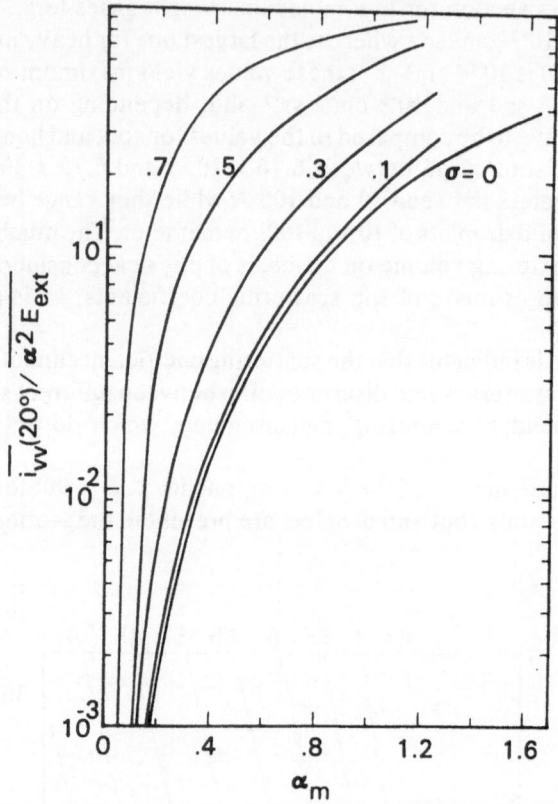

Fig. 14. Computed scattering extinction ratio $\overline{i_{VV}}\,(20)/\ \overline{\alpha^2\,E_{ext}}$ for a log-normal 'size distribution of mean size parameter α_m and standard deviation σ.

can express $Q_{VV,\,\lambda,}\,(90°)$ as the sum of soot and droplet contribution

$$Q_{VV,\,\lambda,}\,(90°) = Q_{VV}^{s}\,(90°) + Q_{VV}^{d}\,(90°) \tag{32}$$

The second measurement can be chosen from the following:

$$Q_{HH} = \gamma\,Q_{VV} = \gamma^{s}\,Q_{VV}^{s} + \gamma^{d}\,Q_{VV}^{d}$$

$$Q_{HV} = \rho_{V}\,Q_{VV} = \rho_{V}^{s}\,Q_{VV}^{s} + \rho_{V}^{d}\,Q_{VV}^{d} \tag{33}$$

$$Q_{VV}\,(\vartheta_{2}) = \Delta\,Q_{VV} = \Delta^{s}\,Q_{VV}^{s} + \Delta^{d}\,Q_{VV}^{d}$$

$$Q_{VV,\,\lambda_{2}} = \Lambda\,Q_{VV} = \Lambda^{s}\,Q_{VV}^{s} + \Lambda^{d}\,Q_{VV}^{d}$$

where γ, ρ_V, Δ, Λ are the polarization, depolarization, asymmetry and wavelength ratios defined for the total scattering coefficient and for the partial contributions due to soot and droplets when the superscripts s and d are present.

The γ^d, ρ^d, Δ^d, Λ^d ratios are constant in geometrical optics as shown in a previous paragraph and their values are ~ 0.8, 0, ~ 40, 1 respectively, so that we have to choose a second measurement whose γ^s, ρ^s, Δ^s, Λ^s values have to be considered constant compared to the corresponding droplet value.

The γ^s polarization ratio is predicted to be zero from the Rayleigh and Lorenz-Mie theory, whereas the Rayleigh and the Stevenson spheroid approximations predict a function of the particle shape which has an asymptotic value dependent on the refractive index and attains a maximum value of 4×10^{-2}. Thus, compared to $\gamma^d = 0.8$, γ^s can be considered constant and the two orders of magnitude of difference entail enough sensitivity to discriminate between the two components.

Analogous considerations for ρ^s, Δ^s and Λ^s can show that the sensitivity is reduced compared to the choice of γ^s, whenever they can be considered constant.

Electronic States of Large Molecules — A brief description of the electronic states of large molecules is necessary in order to describe their absorption and fluorescence spectra in the near ultraviolet and visible region (200 nm-800 nm). Only valence electrons and orbitals are to be considered because the electrons occupying the filled inner shell of atoms contribute very weakly to molecules electronic structure.

A great deal of energy is necessary to promote electrons engaged in σ bonding to vacant molecular orbitals so that their electronic transitions take place well into the vacuum u.v. region; for molecules containing isolated π bonds, the electronic transitions have a lower energy, but they are still in the vacuum u.v. Partially delocalized electron systems exhibit spectra in the near u.v. regions whereas completely delocalized π electrons, as in aromatic molecules, have electronic spectra ranging from the near u.v. to the near i.r., depending on their size and structure.

Therefore interest is mainly focused on polynuclear aromatic hydrocarbons (PAH) and some conjugated-double-bond linear molecules which are well-known stable components of many rich combustion systems [48, 49, 50].

The π electronic structure of aromatic molecules is usually described in terms of free-electron molecular orbitals [17] and has singlet electronic states (S_0, S_1, S_2, S_3 . . .) and triplet states (T_1, T_2, T_3 . . .) where S_0 designates the ground state. According to Platt's notation [51] $S_0 = {}^1A$ and the subsequent higher singlet states are ${}^1L_b$, ${}^1L_a$ and ${}^1B_b$; ${}^1L_a$ and ${}^1L_b$ states have lower energy than ${}^1B_b$, whereas the energy differences between ${}^1L_a$ and ${}^1L_b$ depends on the structure of the ring systems and the type of substituents [17]. A manifold of vibrational and rotational states are associated with each electronic state so that transitions between levels do not give pure electronic spectra but vibronic ones, whose shape as determined both by the population of the vibronic states and their transition probability, through their Franck-Condon factors.

Absorption Spectra of Aromatic Molecules — The absorption of visible or u.v. radiation by a molecule is accompanied by the promotion of an electron from the ground state to an upper electron configuration; in most aromatic molecules three transitions are normally observed:

a. ^1A - ^1B_b transition in the u.v. (called also α — band in the Clar notation [15]) which has very high intensity with the maximum monochromatic absorption cross section being about 10^{-16} to 10^{-15} cm²;

b. ^1A - ^1L_b transition which is observed in the near u.v. or visible with cross sections C_{abs} around 10^{-17} cm² and designated by Clar as the β-band.

c. ^1A - ^1L_a transition, observed always in the long wavelength limit of the spectrum, with absorption cross sections between 5×10^{-19} and 5×10^{-18} cm², which is the weakest transition and is designated by Clar as the p-band.

Fig. 15 (adapted from Clar) shows how these bands appear for a series of PAH compounds like benzene, tetracene, pentacene at room temperature in the liquid phase.

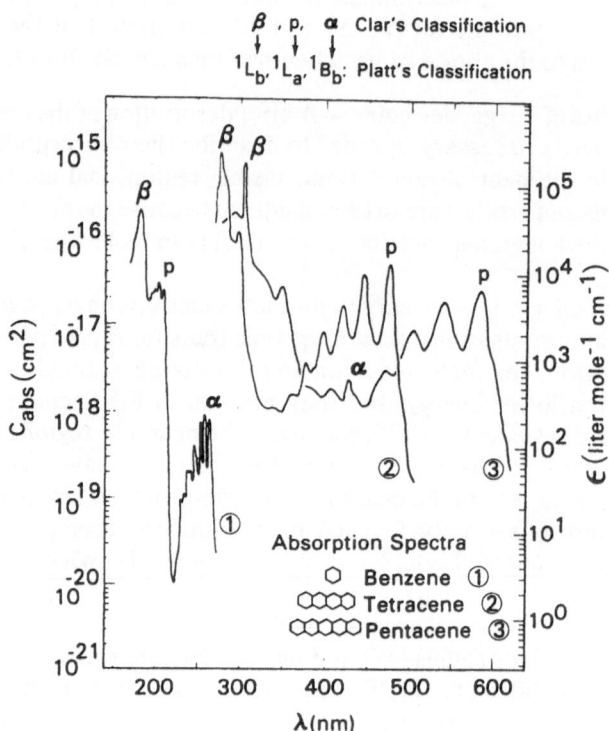

Fig. 15. Absorption cross sections of visible and u.v. radiation for three aromatic molecules. Adapted from Ref. [15].

Unfortunately measurements of the absorption spectra in the vapor phase at high temperature are very scarce. It should be noted that at high temperature the upper vibronic levels of the ground states are appreciably populated with respect to the v = 0 level so that radiation of longer wavelength than that of the normally prominent (0, 0) band will also be absorbed and vibronic coupling with the upper excited state will sometimes increase the radiative transition probabilities.

Fluorescence by Aromatic Molecules — When a molecule is electronically excited by absorption of a quantum of light, on a 10^{-15} sec time scale, it decays to its equilibrium state in a much longer time through a complex sequence of radiative and non-radiative relaxation processes. Radiative transitions from singlet excited states to the ground singlet state produce fluorescence. A non-radiative process which brings the molecule from its initial vibrational state to the equilibrium one in the same electronic level is called vibrational relaxation and is completed in some tenths of picoseconds. Radiationless electronic transitions from higher to lower states are called internal conversion when they take place between singlet states and intersystem crossing when a singlet-triplet transition is involved. The rate of the non-radiative processes is directly related to the coupling between the lower vibrational levels of the upper electronic states and the higher vibrational levels of the lower electronic state.

The fluorescence intensity and its spectral position and distribution therefore depend not only on the absorption properties of the molecule but also on the competition between the radiative and non-radiative processes outlined above. Aromatic molecules exhibit fluorescence much more commonly than other delocalized electrons molecules because their rigid ring structures hamper vibrational dissipation of excitation energy, thus slowing down the internal conversion processes. When a fluorescence experiment is carried out at room temperature and in presence of a solvent, internal conversion and vibrational relaxation are very efficient so that fluorescence transitions always start from the lowest vibrational state of the first excited level, even if the incident photon has a much higher energy (Kasha rule). The final levels are the different vibrational levels of the ground state and the spectral distribution of fluorescence reflects the relative Franck-Condon factors. Consequently the fluorescence band has a mirror symmetry of the absorption band and is shifted toward the red (Stokes shift), under the further hypothesis that the first excited state S_1 and the ground state S_0 have similar equilibrium nuclear configuration and spacing between vibronic levels.

Experiments carried out in the gas phase and at high temperature show a different behavior however. Higher vibronic levels of the ground and excited states are appreciably populated and the initial vibronic distribution tends to relax to the equilibrium one. This broad vibronic distribution here favors coupling between electronic states so that internal conversion processes are no longer irreversible.

Inverse electronic relaxation and reverse intersystem crossing take place so that it is also possible to observe fluorescence from S_2 to S_0 (dual fluorescence effect) [52]. Furthermore the normal S_1 -S_0 emission may have as an upper level higher vibronic states giving rise to spectral features (hot bands) shifted towards higher energy than the incident one (antiStokes shift).

In a very recent paper Coe and Steinfeld [53] reported for the first time fluorescence excitation and emission spectra of different PAH in an atmospheric flame at temperatures between 1200 K and 1500 K. They noted the presence of dual fluorescence for pyrene and that, furthermore, the spectra did not differ much from those obtained at 400 K. Acenaphthylene and fluoranthene in a flame were easily excited at 488 nm even though they do not exhibit absorption spectra at room temperature in this λ- region.

References pp. 254-256.

The most significant parameter of fluorescence is the quantum yield ϕ_f which is defined as the fraction of excited molecules that returns to the ground state by fluorescent emission; it is given by the ratio between the mean observed fluorescence lifetime τ_f (which depends on the competiton of radiative and non-radiative processes) and the longer purely radiative lifetime τ_0 (which is the inverse of the Einstein transition probability). Thus:

$$\phi_f = \tau_f / \tau_0. \tag{34}$$

Berlman [16] gives numerical values for ϕ_f and τ_0 for a large variety of aromatic compounds from measurements in solution of the fluorescence and absorption spectra; τ_f may also be evaluated directly by measuring the fluorescence decay after a pulse excitation of the medium. Generally ϕ_f is in the range $5 \cdot 10^{-2}$ to 1 while τ has a value between 1 and 100 nsec.

Fluorescence quantum yield is also defined as the ratio between the number of photons emitted to the number of photons absorbed so that where the system is dilute it is given by the expression:

$$\phi_f = \frac{4\pi \int_0^\infty (Q_{f\lambda} \, \Delta\lambda) \, d\lambda/\Delta\lambda}{K_{abs}} \tag{35}$$

where $Q_{f\lambda} \Delta\lambda$ is the measured fluorescence coefficient under the instrumental band width $\Delta\lambda$ and the integral is extended to the band and multiplied by 4π because of the isotropic nature of the emission.

EXPERIMENTAL ASPECTS

Laser light scattering and spectral extinction measurements were carried out with a laminar burner for gaseous fuels and on relatively large scale for liquid fuels, where commercial nozzles were employed. The reason for this last choice is that it is very hard to design basic experiments for liquid fuel combustion which still retain the features peculiar to practical systems. Furthermore many of the properties of the PAH-soot systems are studied more easily for a gaseous fuel combustion and the same results may be translated to the more complex situation in a coordinated way.

Burners — Methane and oxygen were burned on the usual sintered porous plate developed by Kaskan and used in previous work by this group. Fig. 16 shows schematically how it is used for generating both premixed and diffusion flames: in the first case methane, oxygen, and in some cases diluting nitrogen, were introduced in the 50 mm diameter central region, while shielding nitrogen flowed in the external annulus (Fig. 16a). Diffusion flames were obtained introducing only meth-

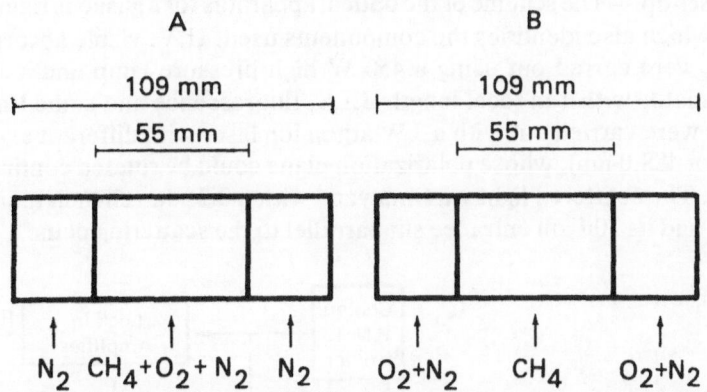

Fig. 16. Scheme of the flat flame burner; A) Premixed flames; B) Diffusion flames.

ane in the centre while a mixture of oxygen and nitrogen was fed to the external zone (Fig. 16b).

The burner system for liquid fuel is described in Fig. 17: number 2 oil was fed at a rate of 7 kg/hr through a 60° semisolid commercial nozzle at 14 atm. Inlet air was introduced, at room temperature, into two separate tubes around the nozzle: the inner duct confined a pure axial flow, whose rate was 10% of the total one, while the outer one was connected to a variable swirl generator, consisting of an array of blades, whose inclination with respect to the radial direction could be varied continuously.

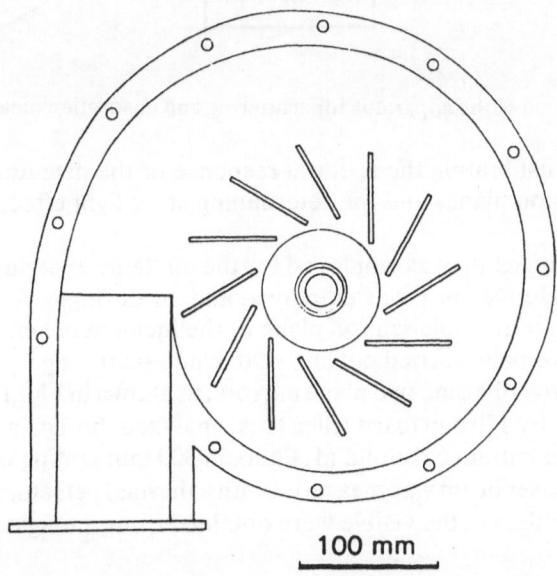

Fig. 17. Horizontal section of the oil spray burner.

Optical Set-up — The scheme of the optical apparatus for a gaseous flame is shown in Fig. 18 which also identifies the components used. U.V. visble absorption measurements were carred out using a 450 W high pressure lamp and a Jarrell-Ash monochromator with 1 m focal length; LLS, fluorescence and some Raman measurements were carried out with a 3 W argon ion laser with different exciting lines (514.5 nm or 488.0 nm), whose polarization plane could be rotated continuously via a λ/2 plate. The scattered light was analyzed with a 0.25 Jarrell-Ash monochromator, which had its 100 μm entrance slit parrallel to the scattering plane.

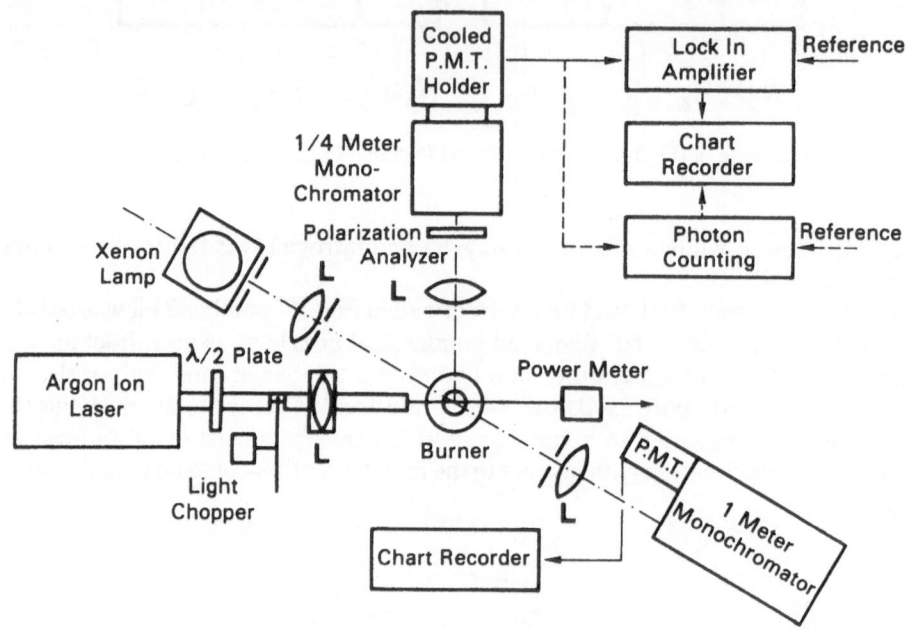

Fig. 18. Scheme of the apparatus for scattering and absorption measurements.

Procedures for calibrating the spectral response of the detection system in the different polarization planes and for determining stray light effects were reported elsewhere [54].

A slightly simpler set up was employed for the oil flame experiment: a vertically polarized 15 mw He-Ne laser was employed and an Ehringhaus compensator allowed the rotation of the polarization plane of the incident beam. Scattering measurements were normally carried out at $\vartheta = 90°$ while scattering in the forward lobe was performed at $\vartheta = 10°$ using two plane mirrors, as shown in Fig. 19. The scattered light was collected by 150 mm fused silica lens, analyzed through a HN 22 polarizer and focused on the entrance slit of a McPherson 300 mm grating monochromator. Extinction by the laser beam was measured with a thermal detector while additional spectral extinction data in the visible were obtained using a tungsten strip filament lamp.

Lock-in detection systems were employed in both experiments, the incident beams being chopped.

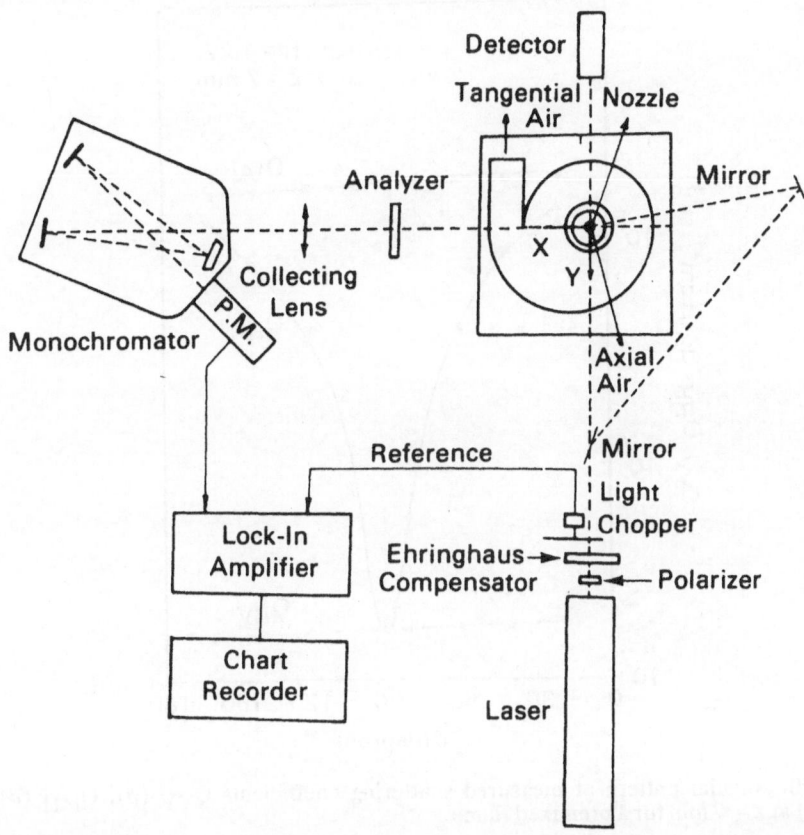

Fig. 19. Scheme of the apparatus for $\vartheta = 90°$, $\vartheta = 10°$ scattering and extinction measurements.

RESULTS

A variety of experimental results have been obtained in recent years using the optical methods outlined above [55-58] so that only a few specific topics will be examined here.

LLS Measurements in Premixed Flames — Fig. 20 illustrates the angular distribution of the scattering coefficients $Q_{VV}(\vartheta)$, $Q_{HH}(\vartheta)$ and $Q_{HV}(\vartheta)$ typical of the soot particle growth region ($z = 7$ mm), while the pattern of scattered radiation in a region where coagulation is significant ($z = 12$ mm) is shown in Fig. 21.

The constancy of $Q_{VV}(\vartheta)$ in the first diagram implies that the particles are still in the Rayleigh regime and the presence of a non-zero and almost equal Q_{HH} and Q_{HV} components at $\vartheta = 90°$ is a clear indication of anisotropy. The depolarization ratio ρ_V separated from fluorescence effects, is equal to 1×10^{-2} and, apart from an initial increase near the first nucleation zone, it remains almost constant.

References pp. 254-256.

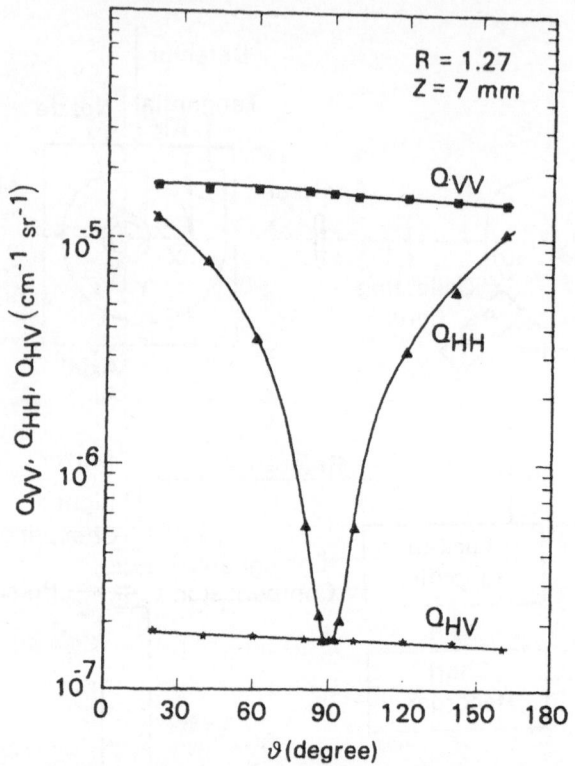

Fig. 20. Angular pattern of measured scattering coefficients $Q_{VV}(\vartheta)$, $Q_{HH}(\vartheta)$ and $Q_{HV}(\vartheta)$ at $z = 7$ mm for a premixed flame.

When these data are interpreted in terms of Rayleigh prolate spheroids in the limit of very elongated particles ($\rho > 10$), very low and unrealistic values of the complex refractive index are required, whereas the literature values indicate that the semiaxis ratio should be nearly two. Haynes *et al.* [59] found recently that while ρ_V is also on the order of $(2 - 3 \times 10^{-2})$ for ethylene flames, it has a much lower value $(2 - 3 \times 10^{-3})$ for benzene flames so that they attribute this effect to intrinsic anisotropy of spherical particles.

Muller-Dethlefs [60] also attributes his ρ_V measurements on ethylene premixed flames to intrinsic anisotropy, but since he found higher values in counter-diffusion ethylene flames increasing radially, he concluded that in this case the anisotropy was due to the shape of the particles.

Evaluation of the particle volume by the scattering/extinction method is still valid, but the soot volume fraction is slightly over-evaluated when the system is simulated by a cloud of Rayleigh spheres (see Fig. 2).

The morphological analysis of aggregated particles presents much more difficult problems; Fig. 21 shows that an asymmetry arises in both $Q_{VV}(\vartheta)$ and $Q_{HH}(\vartheta)$ components and becomes more and more pronounced as the residence time increases [56]. Other evidences of non-Rayleigh behavior come from the wavelength variation of the ratio Q_{HH}/Q_{VV} at $\vartheta = 70°$ and $\vartheta = 80°$ and the departure from the

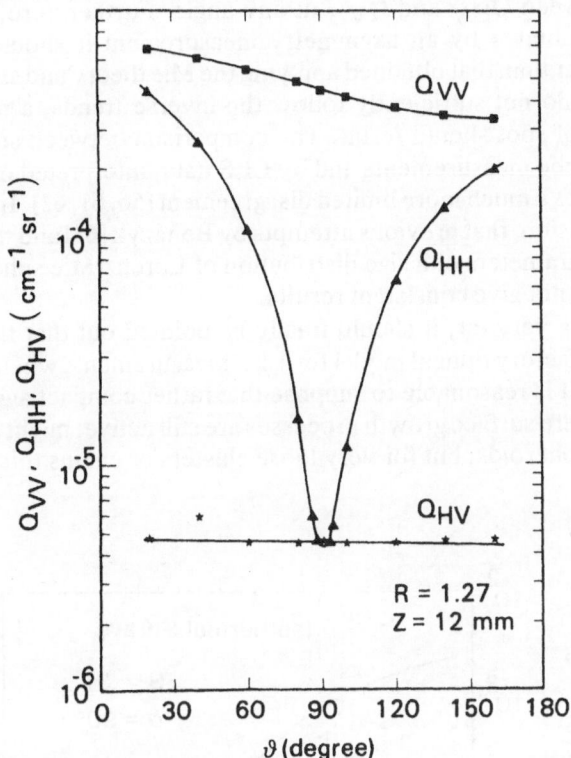

Fig. 21. Angular pattern of measured scattering coefficients Q_{VV} (ϑ), Q_{HH} (ϑ) and Q_{HV} (ϑ) at $z = 12$ mm for a premixed flame.

λ^{-4} law of Q_{VV} in the range 270-510 nm [55]. The first approximation taken for these features is monodisperse Mie scattering which is able to predict those effects in a qualitatively consistent way. However the asymmetry ratio yields particle sizes about 60 per cent higher than those determined by the scattering extinction technique [56]. Fig. 21 shows also two other characteristics which are not predicted at all by the Lorenz-Mie theory: there is still a Q_{HV} component at all the scattering angles, and the minimum of Q_{HH} is invariably at $\vartheta = 90°$, and therefore the shift of the minimum expected for larger spheres (see Fig. 12) does not take place. The Stevenson spheroid model predicts an asymmetry comparable with that of the equivolume sphere, has a minimum of Q_{HH} at $\vartheta = 90°$ and a depolarization ratio ρ_V which is not really influenced by the particle size (see Fig. 8). It was impossible to measure with sufficient accuracy the difference Q_{HV} (90°) — Q_{HH} (90°) in order to check another characteristic of this theory (see Fig. 9).

The evidence of electron microscopy does not strongly support this interpretation, and models like clusters or chains are physically more appealing.

In this connection Jones [40] noted that the depolarization of a chain tends to that of a spheroid with p = 2; but, apart from this encouraging result, many aspects of his theory are in contradiction with the experiments; Q_{HV} does not exhibit the marked minimum at 120° but is rather constant or slightly decreasing with ϑ, and there is no difference between Q_{HV} and Q_{VH} at any angle. Furthermore, if the volume of scatterer is evaluated by an asymmetry measurement it should be an order of magnitude lower than that obtained applying the Mie theory and since the scattering cross sections do not sufficiently follow the inverse trends, a much lower mass concentration of soot should result. The comparison between soot concentration obtained by probe measurements and by LLS data, interpreted on the basis of the Mie-theory gives a much more limited disagreement [56, 61, 62]. It is not surprising, after this discussion, that previous attempts by Bonczyk [63] and the author's group to obtain the parameters of a size distribution of Lorenz-Mie spheres from experimental data did not give consistent results.

Without going very far, it should finally be pointed out that there is not at the moment a satisfactory optical model for LLS measurements when the particles are agglomerated. It is reasonable to suppose that rather compact aggregates, generated in zones where surface growth processes are still active, might be approximated by spheres or spheroids, but for very loose clusters or chains this interpretation is misleading.

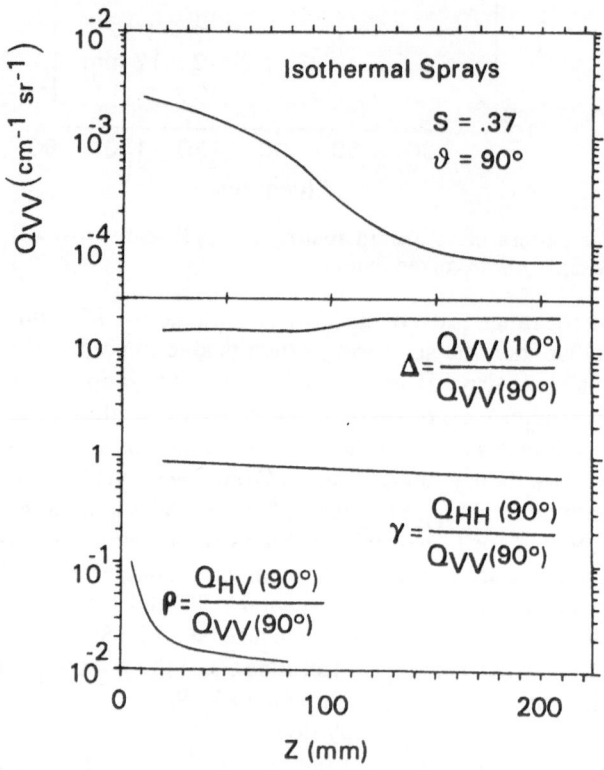

Fig. 22. Measurements along the axis on isothermal spray with swirl number $S = 0.37$. The upper part shows the Q_{VV} ($90°$) scattering coefficient at $\lambda = 632.8$ nm. The lower part shows the asymmetry, polarization and depolarization ratio (Δ, γ, ρ_V) as a function of height. Adapted from Ref. [64].

LLS Measurements in Oil Spray Flames — Different LLS characteristics of the droplets were obtained from measurements carried out on the burner, previously described, for isothermal and burning conditions. Fig. 22 illustrates the scattering coefficient Q_{VV} (90°), the asymmetry ratio Δ (10°, 90°), the polarization ratio γ and the depolarization ratio ρ_V along the axis of a moderately swirled jet.

Q_{VV} has an initial value of 10^{-2} to 10^{-3} cm^{-1} sr^{-1} near the nozzle, where droplet formation takes place, and then it decreases to 10^{-4} to 10^{-5} cm^{-1} sr^{-1} for the expansion of the fuel jet and its entrainment by the air flow. The asymmetry ratio remains remarkably constant along the axis with values around 20; the polarization ratio is nearly 1 initially and later declines gradually, reaching a final value of 0.7.

The constancy of the ratios Δ and γ over a large range of the scattered intensities and different fluid dynamic conditions suggests that these properties are very insensitive to droplet size, as is predicted by the ray optics model (see above). The high values of Δ are in qualitative agreement with the ratio p_{VV} (10°)/p_{VV} (90°) and the measured γ seems to agree also on a quantitative basis with the predicted one. However, the experiment is not sufficiently accurate at this stage to confirm the theory, and the refractive index of the droplets has not been measured. The ρ_V depolarization ratio is quite high near the nozzle (10^{-1}), but it stabilizes at a lower

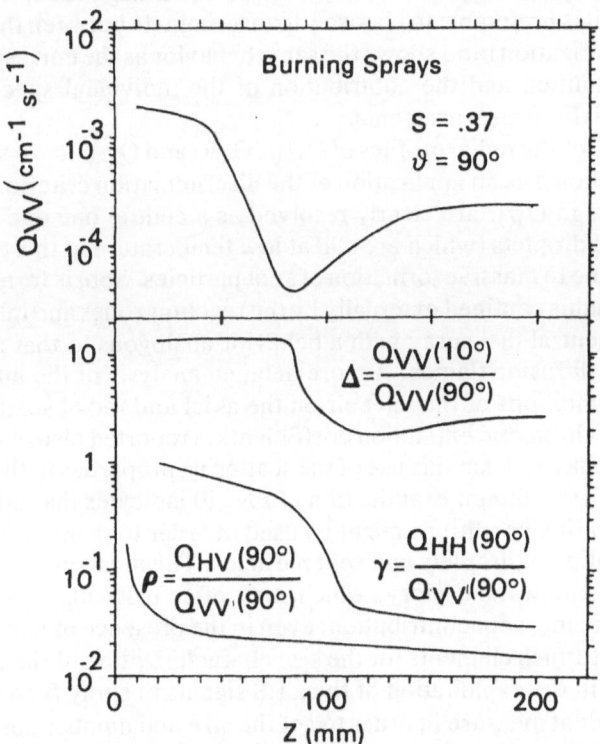

Fig. 23. Measurements along the axis on a flame with swirl number S = 0.37. The upper part shows the Q_{VV} (90°) scattering coefficient at λ = 632.8 nm. The lower part shows the asymmetry, polarization and depolarization (Δ, γ, ρ_V) as a function of height. Adapted from Ref. [64].

value of 2×10^{-2} when the intensity of the scattered light decreases. The higher initial values of ρ_V and γ are probably due to multiple scattering effects which become negligible as for the droplets are dispersed by the air flow; it is not clear at the moment if the residual depolarization in the thin spray region has to be attributed to shape effects of droplets or to intrinsic properties of the fuel.

The corresponding measurements of the scattering coefficient, the polarization and depolarization ratios are shown in Fig. 23 for a burning spray condition with the same inlet fluid dynamics.

The scattering coefficient Q_{VV} (90°) has the same behavior as for the cold spray only in the first part of the flame. However, it falls to a minimum at $z = 80$ mm and subsequently increases in contrast with the previous profile, where the decrease of the scattered intensity was continuous. The Δ and γ ratios range very clearly between the two limits corresponding to the exclusive presence of droplets in the first part of the flame ($\Delta \approx 20$, $\gamma \approx 1$) and exclusive presence of soot ($\Delta \approx 2$, $\gamma \approx 2 \times 10^{-2}$) in the final part.

The simultaneous presence of fuel droplets and soot in the central volume yields the intermediate values of Δ and γ in the central part, and a discrimination between the scattering coefficients due to droplets and soot can be obtained by the technique outlined previously. Since γ varies over two orders of magnitude, whereas Δ only over one order, it is preferable to choose γ to discriminate between the two species.

The ρ_V depolarization ratio shows the same behavior as the corresponding curve for the cold condition and the contribution of the individual species cannot be distinguished by these measurements.

Fig. 24 illustrates the radial profiles of Q_{VV}, Q_{VV}^s and Q_{VV}^d at $\vartheta = 90°$ obtained at $z = 40$ mm, and shown as an application of the discrimination criterion.

The two peaks in Q_{VV} are clearly resolved as a central one due almost exclusively to the fuel droplets (which are still at low temperature in this condition) and an off-axis one due to massive formation of soot particles. Soot is formed in the high temperature annulus confined externally by the reaction zones and internally by the relatively cold central fuel spray with a behavior analogous to that of moderately swirled gaseous diffusion flames. A more detailed analysis of the influence of the fluid-dynamic conditions of the inlet air on the axial and radial scattering profiles and on the monochromatic extinction coefficients is reported elsewhere [64].

It must be emphasized that this use of the scattering properties in the droplet-soot system is just a first attempt; examination of Fig. 10 indicates that other scattering angles or polarization conditions might be used in order to depress or enhance the relative contribution of droplets and soot particles. The most promising angles are obviously the rainbows and glory, which may offer particularly suitable "windows" for evidencing, soot contribution, even in the presence of larger amounts of droplets, and/or furnish elements for the size characterization of the droplets [45].

The next step in the exploitation of the LLS signals in spray flames is to obtain another independent measure in order to get the size and number concentration of the soot particles from the evaluated Q^S. For example the measured asymmetry ratio Δ in the final part of the flame of Fig. 23 indicates that the particles are outside the Rayleigh range and are therefore aggregated. Other suitable LLS measurements are the horizontal component Q_{HH}^s at angles between 70° and 80° or the wavelength dependence of Q_{VV}^s.

In the present case the extinction coefficient was used as additional measurement so that the soot volume fraction and size were evaluated with the scattering extinction method [65]. Unagglomerated particles with sizes between 20 and 40 nm were detected in the initial high temperature combustion zone just downstream from the nozzle, whereas larger particles around 100nm were found in the final part of the flame, where their properties are controlled more by mixing and recirculation inside the flame than by a local production process.

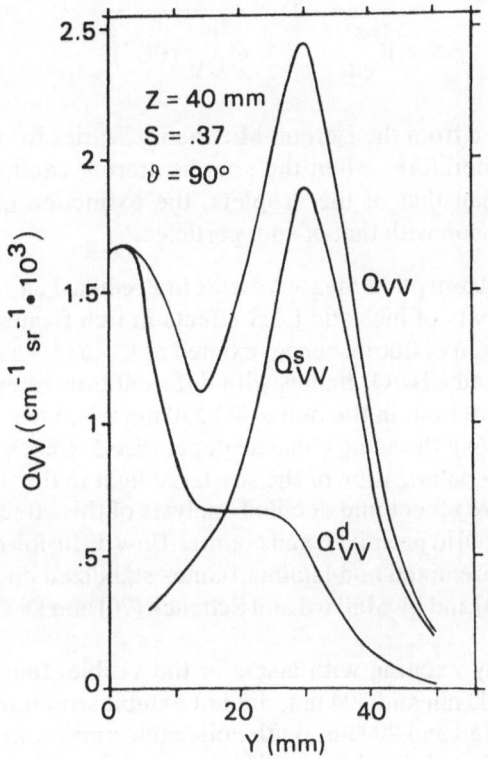

Fig. 24. Measurements of the scattering coefficients Q_{VV}, Q_{VV}^s, Q_{VV}^d, along the radius of a flame with swirl number S = 0.37 where s and d refer to soot and droplets. Adapted from Ref. [64].

The local K_{ext} values were obtained by an Abel inversion procedure although they might be evaluated by a differential scattering method [66] in flame conditions which only approximate axial symmetry. Additional problems are encountered in distinguishing the extinction due to droplets from that due to soot particles and the true absorption of high molecular mass compounds. This last aspect will be fully discussed in the next paragraph so that it is sufficient here to anticipate that at λ = 632.8 nm the gas phase contribution should be negligible.

It is easy to demonstrate that the ratio K_{ext}^d/K_{ext}^s is given by the expression [64]:

References pp. 254-256.

$$\frac{K_{ext}^{d}}{K_{ext}^{s}} = 11.76\alpha^3 \frac{Q_{VV}^{d}(90°)}{Q_{VV}^{s}(90°)} \tag{36}$$

when the soot particles are in the Rayleigh region. For larger particles

$$\frac{K_{ext}^{d}}{K_{ext}^{s}} = f(\alpha)\frac{Q_{VV}^{d}(90°)}{Q_{VV}^{s}(90°)} \tag{37}$$

where $f(\alpha)$, calculated from the Lorenz-Mie theory, varies from 0.16 for $\alpha = 0.4$ to 1.33 for $\alpha = 1.3$. Therefore, when the soot scattering coefficient is an order of magnitude higher than that of the droplets, the extinction of the latter may be neglected in comparison with that of soot particles.

Fluorescence and Absorption Measurements in Premixed and Diffusion Flames — The first measurements of inelastic LLS effects in rich flames were done in 1975 [67]; the axial evolution of fluorescence, excited at $\lambda = 514.5$ nm, was followed at $\lambda = 490$ nm for premixed CH_4/O_2 flames with different flow rates. Its spectral distribution was studied later on, in the range 400-550 nm when the laser excitation was done at $\lambda = 488$ nm [68]; the same inelastic depolarized effect was invoked in order to explain the high depolarization of the scattered light in the first reaction zone of rich flames [57]. More recent and detailed analysis of this effect has been reported by Müller-Dethlefs [60] in premixed and counter-flow diffusion ethylene-air flames, by Haynes et al. in premixed and laminar flames stabilized on a double Wolfhard-Parker burner [59, 69] and by Mallard and Schenck [70] and Di Lorenzo et al. [54] in premixed flames.

All the authors, by exciting with lasers in the visible, found that the spectra, extended between 400 nm and 700 nm, did not exhibit structure and had a spectral half-width between 150 and 200 nm, with noticeable anti-Stokes components. The peak emission wavelength is slightly shifted toward the red as the exciting wavelength λ_0 is increased [69, 70] and the influence of the parent fuel on the spectrum shape is equally low [59]. Excitation spectra in the visible, carried out by Cincotti and Menna [71], also show that the monochromatic emission is higher as λ_0 moves toward the blue.

Fig. 25 is given as an example of the fluorescence spectra in premixed flames, taken with the relatively narrow bandwidth of 0.32 nm. The values are normalized with respect to the maximum at $\lambda = 555$ nm and show that the shape is very slightly influenced by the residence time inside the flame.

It is also interesting to compare the fluorescence, spontaneous Raman, and elastic scattering in the early part of premixed flames. The results reported in Fig. 26 refer to CH_4/O_2 flames diluted with nitrogen (in order to decrease the burning velocity and expand the reaction zone). The measurements at $\lambda = 605.3$ nm were

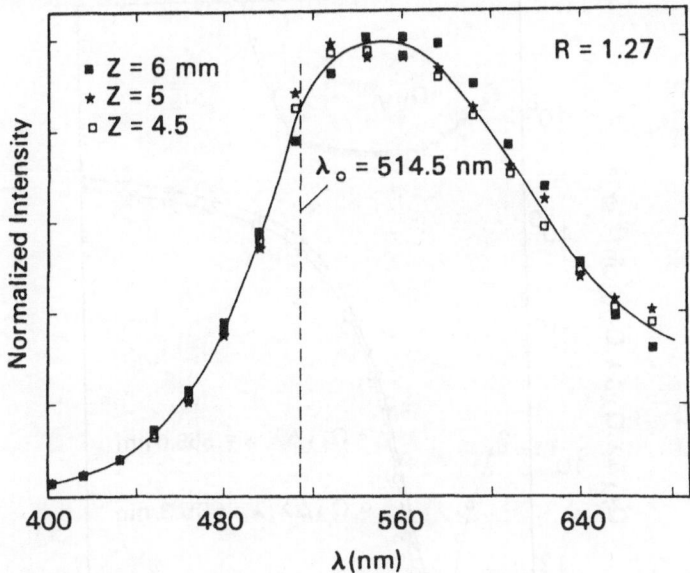

Fig. 25. Spectral profiles of fluorescence excited at $\lambda = 514.5$ nm measured at different heights inside a premixed $CH_4/O_2 = 1.27$ flame, normalized at $\lambda = 555$ nm. Adapted from Ref. [73].

taken with a 32 nm bandwidth in order to include both the Stokes vibrational Raman emission of methane and broadband fluorescence, while the measurements at $\lambda = 555$ nm, carried out with $\Delta\lambda = 0.32$ nm are in a region where there is only a fluorescence contribution. The light scattered at $\lambda = 605.3$ nm starts with a value of 10^{-11} cm^{-1} sr^{-1} near the burner (which is just the value expected for Raman of methane at room temperature[19])decreases for both CH_4 disappearance and temperature increases up to $z = 2$ mm and subsequently strongly increases where fluorescence prevails, as is also indicated by the light scattered at $\lambda = 555$ nm. It appears from the above example how limited is the possibility of detecting Raman signals from major components in rich flames since the background fluorescence may be higher by orders of magnitude.

Fig. 26 also illustrates the interference between the elastic scattering due to gas phase compounds and soot particles in the nucleation zone. The decreasing part and the plateau of the Q_{VV} curve is controlled exclusively by the Rayleigh scattering of the reagent and/or products whose total density decreases as the temperature increases in the preheating and reaction zones (see earlier discussion of Rayleigh Scattering by Molecules.) Since this level is always about $2 - 4 \times 10^{-9}$ cm^{-1} sr^{-1} it limits heavily the detectability of soot particles in the early nucleation zone. It is easy to see that if the early soot nuclei have diameters of about 1.0 to 1.5 nm a number concentration higher than 10^{13} cm^{-3} is required in order to detect them [68], so that it is impossible to follow the very early stages of nucleation of soot particles with LLS measurements. Extinction measurements have a greater potential in principle. With an extinction coefficient of 2×10^{-4} cm^{-1} the number concentration is 4×10^{12} cm^{-3} for 1.0 nm particles, which is more than an order of magnitude lower

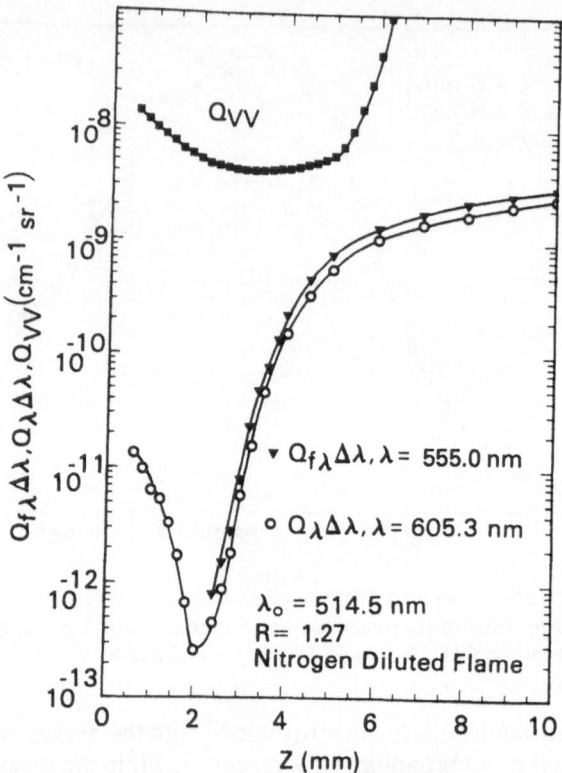

Fig. 26. Axial profiles of elastic scatter, fluorescence and inelastic emission at the Stokes Raman wavelength of methane measured in a premixed $CH_4/O_2 = 1.27$ flame diluted with 165 Nlt/hr of nitrogen. Adapted from Ref. [73].

than the minimum value detectable with the LLS technique. However, in the same region, there is interference between absorption by the particles and that due to the molecular species, as will be discussed later.

Diffusion flames give very interesting conditions for a more controlled study of inelastic scattering effects since the formation of soot particles and high molecular weight hydrocarbons are not so strongly coupled as in premixed flames.

Fig. 27 shows the radial profiles, at two heights above the burner, of the elastic Q_{VV} (90°) and fluorescence emission at $\lambda = 555$ nm for a diffusion CH_4/O_2 flame realized by the scheme of Fig. 16 [72, 73].

At $z = 6$ mm the scattering is practically constant in the central region, on the fuel side, and it drops to lower values in the outer annulus, for the temperature increase in the reaction zone. At $z = 12$ mm the relevant peak, in the same zone, indicates the localized presence of soot particles. Fluorescence is almost constant in the central region, peaks slightly nearer the center than the soot maxima and disappears outside the reaction zone.

The present type of diffusion flame is similar in some features to that studied in detail by Haynes and Wagner [69] and Kent, Jander and Wagner [74] but presents a much more extended region on the fuel side at moderate temperatures.

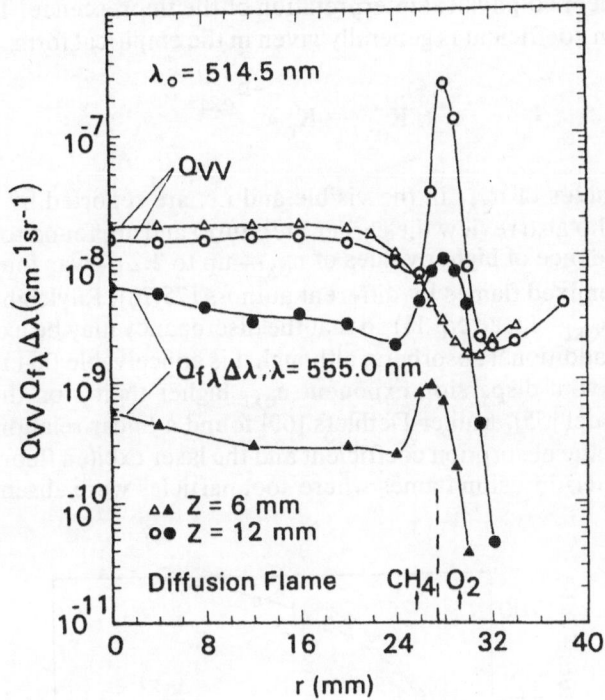

Fig. 27. Radial profiles of the elastic scatter and fluorescence measured at z = 6 mm and z = 12 mm above the burner for a diffusion methane-oxygen flame. Adapted from Ref. [73].

Fig. 28, where the axial profiles of the elastic and inelastic properties at the center of the burner are shown, is a clear indication of this statement. Q_{VV} (90°) is always higher than 10^{-8} cm^{-1} sr^{-1} and, up to z= 12 mm, is due almost exclusively to methane, whose temperature is increasing. In fact, when T is evaluated along the lines discussed in the theory section, it agrees quite well with those determined from thermocouple measurements, so that a temperature range from 300K to 800K is spanned. Fluorescence onset takes place for temperatures slightly higher than 500K, as is evident from the measurements at λ=555.0 nm, and, at the same height, it prevails over the Raman CH$_4$ signals, which are much higher than in the premixed case, as testified to by the measurements at λ=605.3 nm.

Fluorescence spectra determined at moderate temperature exhibit noticeable differences with respect to the previously discussed high temperature measurements. Fig. 29 illustrates the evolution of the fluorescence spectra, normalized at λ=520 nm, with the temperature at center of the burner. At z=6 mm the half width is relatively narrow, the antiStokes components are low, and a pronounced banded structure at λ=520 nm is evident. An increase in temperature produces a slight shift, both in the blue and in the red regions of the emission: the banded structures tend to merge together and the spectral shape tends to be similar those found in premixed or diffusion flames at high temperature.

The spectral distribution of the extinction spectra should be discussed before attempting a chemico-physical interpretation of the fluorescence. The monochromatic extinction coefficient is generally given in the empirical form:

$$K_{ext} = K_o \lambda^{-n_{ext}}$$

Numerical values of n_{ext} in the visible and i.r. are reported by Wolfhard and Gaydon [75] (who also review the earlier literature) and are found to vary between 1.5 and 1.0. Evidence of higher values of n_{ext}—up to 2.2 — was found in the early zone of rich premixed flames by different authors [75-78]. Rayleigh soot particles should exhibit $n_{ext} = 1$ (see eq. 11) so that the discrepancy may be explained by the contribution of additional absorbers, although it is conceivable that the agglomerated particles have a dispersion exponent n_{ext} higher than 1 on the basis of the Lorenz-Mie model [55]. Müller-Dethlefs [60] found a linear relationship between the monochromatic absorption coefficient and the laser excited fluorescence emission in his counter-diffusion flames where soot particles were absent. This impor-

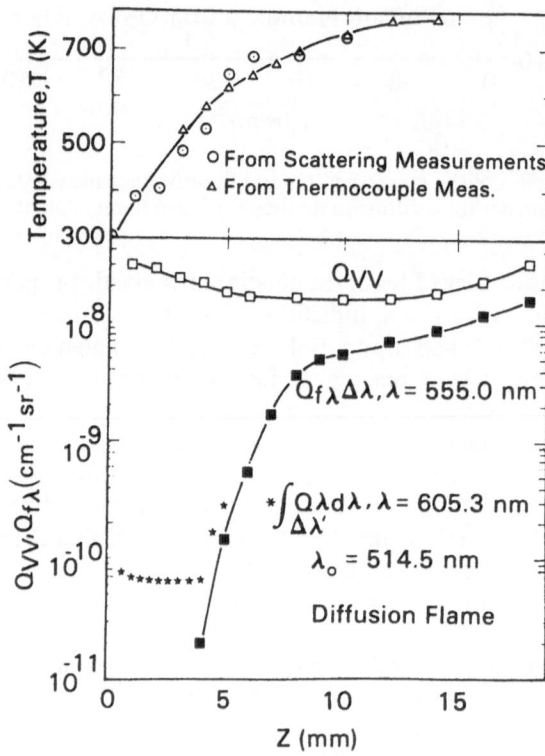

Fig. 28. Lower part: Measurements of the elastic scatter, fluorescence and methane Raman Stokes emission taken at the center of the burner as function of height for a diffusion flame (CH₄ = 187 Nlt/hr, O₂ = 1476 Nlt/hr). Upper part: Axial temperature profiles determined from the CH₄ Rayleigh scatter and from thermocouple measurements. Adapted from Ref. [73].

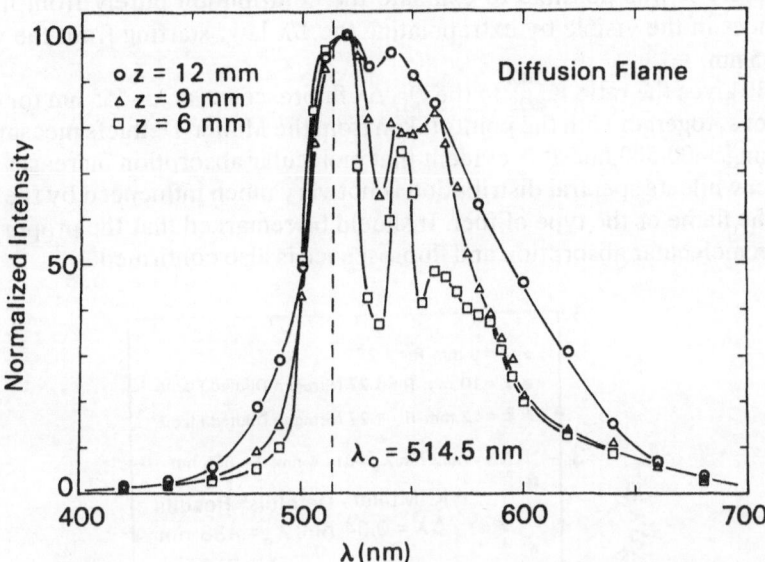

Fig. 29. Spectral profiles of fluorescence excited at $\lambda = 514.5$ nm measured at different heights inside the diffusion flame, normalized at $\lambda = 520$ nm. Adapted from Ref. [73].

tant result establishes the molecular nature of this absorption and furthermore enabled it to be calibrated through fluorescence measurements, even in premixed flames where soot particles were also present. He was able to demonstrate that the soot particle absorption strictly obeys Rayleigh's $1/\lambda$ law, at least before the massive agglomeration region. The remainder is consequently due to compounds whose monochromatic absorption decays toward the red at a faster rate.

On this basis a more systematic comparison between absorption and fluorescence was carried out by trying to measure the total quantum efficiency of the process in premixed and diffusion flames; this quantity is given by:

$$\phi^{tot} = \frac{4\pi \int_{\Delta\lambda} (Q_{f\lambda} \Delta\lambda) \dfrac{d\lambda}{\Delta\lambda}}{\left(K_{ext}^{soot} + K_{abs}^{mol} \right)}$$

where the numerator is readily evaluated by a graphical integration of the measured spectra of fluorescence and the denominator is the measured extinction at $\lambda = 514.5$ nm. ϕ^{tot} in premixed flames ranged between 10^{-2} and 10^{-3}, apart from higher values in the early combustion zone ($z \leqslant 2$ mm) [54]. The percentage extinction in the diffusion flame was below 1%, at the limit of detectability of the present set-up, so that the lower estimate of the quantum efficiency was around 0.1. Müller-Dethlefs estimated a quantum efficiency of 0.11 in his ethylene-oxygen measurements [60] so that it is reasonable to conclude that in the investigated premixed flames the extinction due to soot particles is at least one order of magnitude higher than that due to molecular compounds at $\lambda = 514.5$ nm.

References pp. 254-256.

Therefore it was possible to estimate the contribution purely from molecular substances in the visible by extrapolating the $1/\lambda$ law, starting from the values at $\lambda = 514.5$ nm.

Fig. 30 gives the ratio K_{abs}^{mol} to the $Q_{f\lambda}\Delta\lambda$ fluorescence at $\lambda = 555$ nm for different conditions, together with the points taken from the Müller-Dethlefs measurements, in the range 400-500 nm. It is evident that molecular absorption increases towards the blue, while its spectral distribution is not very much influenced by the position inside the flame or the type of fuel. It should be remarked that the proportionality between molecular absorption and fluorescence is also confirmed.

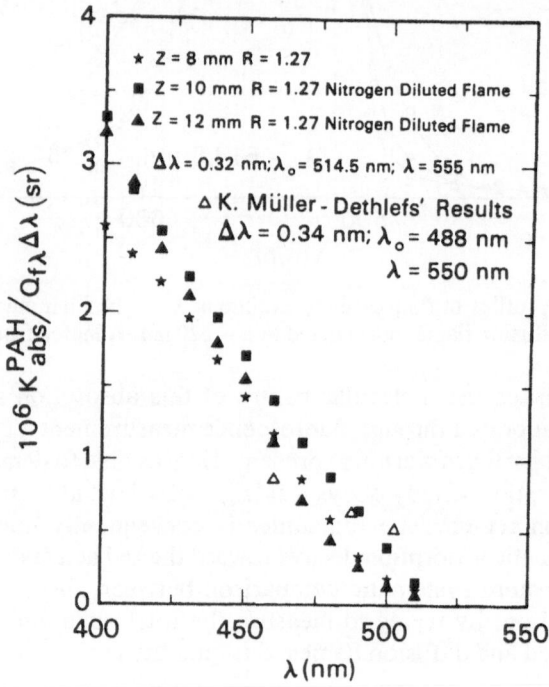

Fig. 30. Measured ratio between PAH absorption coefficient and monochromatic fluorescence emission as function of wavelength for different flame conditions; the data obtained by Müller-Dethlefs [60] are also shown.

A more complete picture of the absorption is obtained when the measurements are extended in the near u.v., down to 200 nm. In fact the presence of u.v. absorption features which are not connected exclusively with soot particles has been reported, for instance, by Laud and Gaydon [79] and this topic is also discussed in the last edition of Gaydon's book [14]. More recent and systematic work has been done by Menna in his thesis [80] and it is partly reported in ref. [54].

Fig. 31 shows an example of a typical spectrum, taken in the soot formation region: the soot contribution has been computed, with the hypothesis outlined above, for a size of 30 nm and the variation of the refractive index with the wavelength taken into account according to ref. [24]. The large peak in the region 240-250 nm is surely due to molecular absorbers; it shifts slightly to the blue in the early nucleation zone and for leaner flames, as is discussed in detail in ref. [54] and [80].

The measured molecular absorption shows u.v. peaks and also has appreciable components in the visible so that it agrees qualitatively with the PAH spectra at low temperature as appears from the example reported in Fig. 15; unsaturated compounds exhibit only an u.v. absorption which does not extend into the visible. Fluorescence is easily excited by absorption of visible radiation as happens for single PAH compounds, as reported by Coe and Steinfeld [53], and is proportional to it. The relatively high quantum efficiency is also a characteristic of aromatic molecules so that the present evaluation is in agreement with the low temperature literature data.

Therefore it is reasonable to deduce from a spectroscopic analysis that PAH compounds are responsible for fluorescence and the anomaly of the extinction coefficient in the visible and the u.v. peaks, downstream from the reaction zone. Further, and perhaps stronger, support for this attribution is given by direct sampling and chemical analysis in the flame; PAH's are the only compounds in CH_4/O_2 rich flames which also grow continuously beyond the reaction zones, and it was found that their evolution was almost exactly paralleled by fluorescence emission [54]. Analogous conclusions were drawn by Haynes *et al.* in the comparative study of premixed flames with aliphatic and aromatic fuels[59].

An estimate of the absorption coefficient in u.v. was done on the basis of the measured concentration of individual PAH compounds up to four rings and the absorption cross sections reported in the literature. The results were compared with the values reported in Fig. 31; the agreement was quite good. It is noteworthy

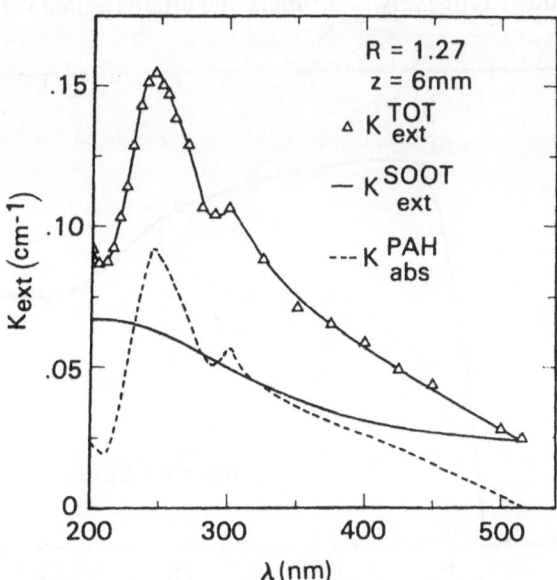

Fig. 31. Measured total extinction coefficient, K_{ext}^{TOT}, in the spectral range 200 nm - 520 nm; soot contribution is evaluated for 30 nm diam. particles (full line). The PAH contribution is obtained by difference (dashed line).

that compounds like diacetylene, which has a relatively high concentration in the same region of the flame, present an u.v. absorption cross section which is much lower than those of PAH [81]. An unambiguous attribution of the fluorescence spectra is not possible at the moment. However, progress in measurements of spectroscopic properties of individual PAH compounds at high temperature can be expected in the near future. The spectra taken at low temperature in fact show structure, thus suggesting that their merging at other conditions is due more to temperature effects than to the overlap of the emission due to many different compounds.

Another interesting feature is the slight influence of temperature on the spectral width and consequently on the AntiStokes emission. The most reasonable explanation for this effect is that although the population of the upper vibronic level is much higher, the Franck-Condon factor still favors $\Delta V \approx 0$ transitions, so that the spectra are not enlarged.

Evidence of optical inelastic effects are relatively scarce; Fujiwara *et al.* [82] have reported measurements of laser-induced fluorescence in kerosene/air and gasoline/air flames. They found fluorescence spectra in the region 320-370 nm, and the highest intensities were near the nozzle. Both results suggest that in this case fluorescence arose from vapor or partially decomposed vapor of liquid fuel rather than from polycyclic aromatic hydrocarbons.

It should be noted that the spectral distribution of the extinction in the visible along the axis of the oil flame, previously examined, might also be interpreted on the basis of both elastic and inelastic effects. Fig. 32 [83] shows how the dispersion exponent n_{ext} varies along the flame: it is almost zero in the initial part, thus showing that it is due exclusively to droplets and attains a final value of 1.4, typical

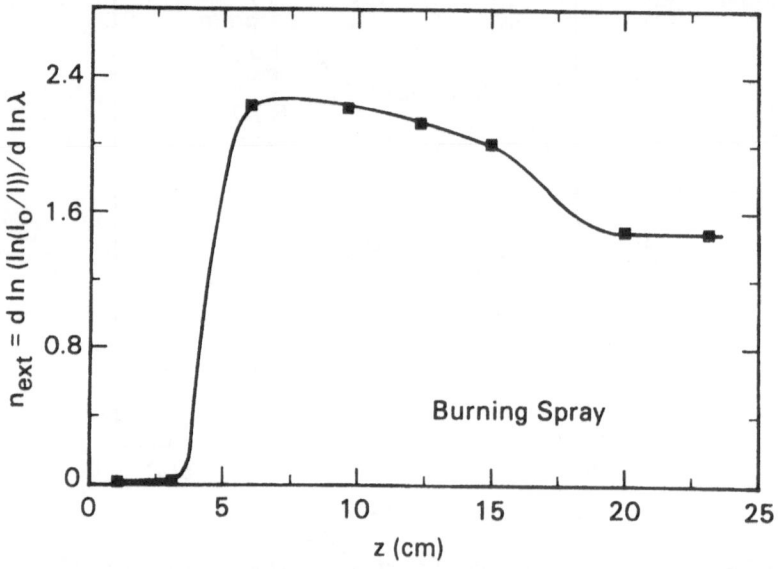

Fig. 32. Measured dispersion coefficient for an oil-spray flame as a function of the height above the burner.

of agglomerated soot particles. However it reaches up to 2.2 in the intermediate region where the vaporization of fuel and its low temperature pyrolysis prevails over soot nucleation, so that is is reasonable to deduce that this inelastic effect is also connected to PAH formation in liquid fuel flames. Direct laser-excited fluorescence experiments, presently underway, should confirm and detail this point of view.

This section has shown the complex and not yet fully understood interference between effects due to soot particles and PAH. The results of Lee and Tien [25], Chippet and Gray [84] and others, who obtained size distributions and complex refractive index values exclusively from the λ-dependence of the extinction coefficient, are thus to be considered somewhat cautiously unless there is good evidence that PAH were negligible under their conditions and until their results can be confirmed by other optical measurements.

CONCLUSIONS

Just a few points will be emphasized as concluding remarks:

i. LLS is a really powerful technique for following not only soot formation but also the general flame structure. Its potential is particularly evident in diffusion flames and for combustion with complex aerodynamics and transient features.

ii. Morphological analysis of soot aggregates with optical techniques are still in a developmental stage. Progress is expected particularly in the development of more advanced theoretical models for light scattering and extinction by absorbing particles of many different shapes.

iii. It also seems possible to study with LLS and other spectroscopic techniques practical combustion systems, which contain fuel droplets, coal particles and ash. More detailed experimental and theoretical work on large particles is therefore required.

iv. Fluorescence and u.v. absorption measurements are useful for studying the formation and spatial distribution of high molecular weight products of pyrolysis.

The presence of PAH seems to explain reasonably most of the features observed in rich gaseous flames. The optical measurements are more useful for conditions where no direct sampling is possible.

Use of powerful pulsed dye-lasers should both extend the wavelength regions to the near u.v. and furthermore allow measurements of fluorescence decay times. High temperature spectroscopic studies on pure aromatic compounds are also necessary before this effect can be fully confirmed.

ACKNOWLEDGMENTS

The Author wishes to thank all his coworkers in Naples for material help in the preparation of the manuscript. He is particularly grateful to A. Cavaliere, V. Cincotti and P. Menna, who kindly furnished yet unpublished results.

He is also greatly indebted to Prof. A. Sarofim, of M.I.T., for a long series of stimulating discussions on both sides of the ocean.

REFERENCES

1. H. C. Van de Hulst, "Light Scattering by Small Particles," Wiley; New York, 1957.
2. M. Kerker, "The Scattering of Light," Academic Press, New York, 1969.
3. H. C. Hottel and A. F. Sarofim, "Radiative Heat Transfer," McGraw-Hill, New York, 1967.
4. A. R. Jones, Prog. Energy Combust. Sci., Vol. 5 (1969), p. 73.
5. B. Chu, "Laser Light Scattering," Academic Press, New York, 1974.
6. B. J. Berne and R. Pecora, "Dynamic Light Scattering," Wiley, New York, 1976.
7. S. S. Penner, J. M. Bernard and T. Jerskey, Acta Astronautica (1976), p. 93.
8. J. F. Driscoll, D. M. Mann and W. K. McGregor, Combust. and Techn., Vol. 20 (1979), p. 41.
9. G. Gouesbet, P. Flament, M. Weill, Internal. Report TTI/GFW/79/11/15 Laboratoire de Thermodynamique Unversité de Rouen, 1979.
10. F. Durst, A. Melling, J. H. Whitelaw, "Principles and Practice of Laser-Doppler Anemometry", Academic Press, New York, 1976.
11. N. A. Chigier, K. Dvorak, Fifteenth Symp. on Comb., The Comb. Institute, (1974), p. 573.
12. W. M. Farmer, Appl. Optic, Vol. 11 (1972), p. 2603.
13. G. Herzberg, "Electronic Spectra and Electronic Structure of Polyatomic Molecules" Van Nostrand, Princeton, New Jersey, 1966.
14. A. G. Gaydon "The Spectroscopy of Flames" Chapman and Hall, London, 1974.
15. E. Clar, "Polycyclic Hydrocarbons" Vol. 1 and 2, Academic Press, London, 1974.
16. I. B. Berlman, "Fluorescence Spectra of Aromatic Molecules," Academic Press, New York, 1971.
17. J. B. Birks, "The spectroscopy of the π-electronic states of aromatic hydrocarbons" in "Organic Molecular Photophysics" J. B. Birks (ed.) Wiley, New York, Vol. 1 (1973), p. 1.
18. M. Stockburger, "Fluorescence of aromatic molecular vapours" in "Organic Molecular Photophysics" J. B. Birks (ed.) Wiley, New York, Vol. 1 (1973), p. 57.
19. M. Lapp and C. M. Penney, (eds.) "Laser Raman Gas Diagnostics," Plenum Press, New York, 1974.
20. S. Lederman, Progr. Energy and Comb., Vol. 3 (1977), p. 1.
21. C. Eckbreth, P. A. Bonczyk and J. F. Verdieck, Progr. Energy and Comb., Vol. 5 (1979), p. 253.
22. M. Born and E. Wolf, "Principles of Optics," Academic Press, New York, 1975.
23. A. D'Alessio and M. Ambrosio, La Termotecnica, Vol. 30 (1976), p. 460.
24. W. H. Dalzell and A. F. Sarofim, Trans. ASME, Vol. 91 (1969), p. 100.
25. S. C. Lee and C. L. Tien, to be published in 18th Symposium (Int.) on Combustion.
26. E. U. Condon, "Molecular Optics" in "Handbook of Physics" E. U. Condon and H. Odishaw, (eds.), McGraw-Hill, New York, 1967.
27. R. R. Rudder and D. R. Bach, J. Opt. Soc. Am., Vol. 58 (1968), p. 260.
28. Landolt-Börnstein "Zahlenwerthe und Funktionen aus Physik, Chemie, Astronomie, Geophysik und Technik" II Band, 8 Teil, Optische Konstanten, Springer-Verlag, Berlin, 1968.
29. H. J. Bridge and A. D. Buckingham, Proc. Roy. Soc., A Vol. 295 (1966), p. 334.
30. S. Bhagavantam, "Light Scattering and the Raman effect," Chemical Publishing, New York, 1942.
31. P. Ferrara, "Tesi di Laurea in Ingegneria Chimica," Napoli, 1977.
32. A. R. Jones, "Combustion Institute European Symposium," F. J. Weinberg (ed.) Academic Press, London, (1973), p. 376.

33. H. Senftleben and E. Benedict, Ann. Phys. (Leipzig) IV, Vol. 54 (1917), p. 65.
34. E. A. Taft, personal communication 1969, see also ref. [23].
35. S. Asano and G. Yamaroto, Appl. Optics, Vol. 14 (1975), p. 29.
36. A. F. Stevenson, J. Appl. Phys., Vol. 24 (1953), p. 1134 and p. 1143.
37. W. Heller and M. Nakagaki, J. Chem. Phys. Vol. 60 (1974), p. 3297; Vol. 61 (1974), p. 3619.
38. S. Levine and J. Olaofe, J. Coll. Interf. Sci., Vol. 27 (1968), p. 442.
39. J. C. Ravey, J. Coll. Interf. Sci., Vol. 46 (1974), p. 139.
40. A. R. Jones, Proc. Roy. Soc. Lond A., Vol. 366 (1979), p. 111.
41. A. R. Jones, to be published in J. Phys. D.
42. R. S. Kirshnan, Proc. Ind. Acad. Sci., Vol. A7 (1938), p. 91.
43. J. V. Dave, Appl. Opt. Vol. 8 (1969), p. 155.
44. S. T. Shipley and J. A. Weinman, J. Opt. Soc. Am., Vol. 68 (1978), p. 130.
45. H. M. Nussenzveig, J. Opt. Soc. Am., Vol. 69 (1979), p. 1068.
46. I. Shimizu, 1st International Conference on Liquid Atomization and Spray Systems, Tokyo, Japan, 27-31 August, 1978.
47. J. Swithenbank, et al. A.I.A.A. 14th Aerospace Sciences Meeting Paper 76-79, Washington D.C., Jan. 26-28, 1976.
48. B. D. Crittenden and R. Long, Comb. and Flame, Vol. 20 (1973), p. 359.
49. A. Di Lorenzo and S. Masi, Second European Symposium on Comb., Orleans, (1975), p. 368.
50. J. D. Bittner and J. B. Howard, Eighteenth Symposium (International) on Combustion, Abstracts of papers, (1980), p. 41.
51. J. R. Platt, "Systematics of the Electronic Spectra of Conjugated Molecules," Wiley, New York, 1964.
52. T. Deinum, C. J. Werkhoven, R. P. H. Longelaar, J. Rettschick and J. D. W. Van Voorst, Chem. Phys. Letts., Vol. 27 (1974), p. 552.
53. D. S. Coe and J. I. Steinfeld, Chem. Phys. Letts. to be published.
54. A. Di Lorenzo, A. D'Alessio, V. Cincotti, S. Maşi, P. Menna, C. Venitozzi, Eighteenth Symposium (Int.) on Combustion, Abstracts of papers (1980), p. 17.
55. A. D'Alessio, A. Di Lorenzo, F. Beretta and C. Venitozzi, Fourteenth Symposium (Int.) on Combustion, The Combustion Institute, (1973), p. 941.
56. A. D'Alessio, A. Di Lorenzo, A. F. Sarofim, F. Beretta, S. Masi and C. Venitozzi, Fifteenth Symposium (Int.) on Combustion, The Combustion Institute, (1975), p. 1427.
57. A. D'Alessio, A. Di Lorenzo, A. Borghese, F. Beretta and S. Masi, Sixteenth Symposium (Int.) on Combustion, The Combustion Institute, (1977), p. 695.
58. F. Beretta, A. Cavaliere, A. D'Alessio and C. Noviello, Comb. Sci. Techn., Vol. 22 (1980), p. 1.
59. B. S. Haynes, H. Jander and H. Gg. Wagner, Ber. Bunsenges. Phys. Chem. Vol. 84 (1980), p. 585.
60. K. Muller-Dethlefs, Ph.D. Thesis, Imperial College, London, 1979.
61. G. Prado, J. Jagoda, K. Neoh and J. Lahaye, Eighteenth Symposium (Int.) on Combustion, Abstracts of papers, (1980), p. 41.
62. H. Bockhorn, F. Fetting, U. Meyer, R. Reck and G. Wannemacher, Eighteenth Symposium (Int.) on Combustion, Abstracts of papers, (1980), p. 42.
63. P. A. Bonczyk, Comb. and Flame, Vol. 35 (1979), p. 191.
64. F. Beretta, A. Cavaliere, A. D'Alessio, C. Noviello and C. Scodellaro, paper presented at Italian Flame Days, Cervia (Italy), June 1980, in press in "Rivista dei Combustibili".
65. F. Beretta, A. Cavaliere, A. Ciajolo, A. D'Alessio, A. Di Lorenzo, C. Langella and C. Noviello, Eighteenth Symposium (Int.) on Combustion, Abstracts, of papers, 1980.

66. A. Cavaliere and A. D'Alessio, Comb. Science and Techn., Vol. 21 (1980), p. 153.
67. A. D'Alessio, A. Di Lorenzo, G. Micera and F. Beretta, 2° Simposio Internazionale di Dinamica delle Reazioni Chimiche su "Le Fiamme quali Reazioni in Flusso – The Flames as Reaction in Flow," Padova – ITALY – 15 – 17 dicembre (1975), p. 147, C.N.R. Padova 1976.
68. F. Beretta, A. Borghese, A. D'Alessio and C. Venitozzi, Workshop on "Laser Measurements in Combustion," Urbino – ITALY – 7 – 9 September, 1977.
69. B. S. Haynes and H. Gg. Wagner, Ber. Bunsenges. Phys. Chem., Vol. 84 (1980), p. 499.
70. W. G. Mallard and P. K. Schenck, personal communication, 1979.
71. V. Cincotti and P. Menna, unpublished results, 1980.
72. V. Cincotti, Tesi di Laurea in Ingegneria Chimica, Napoli – ITALY – 1980.
73. V. Cincotti, A. D'Alessio, P. Menna and C. Venitozzi, paper presented at Italian Flame Days, Cervia (Italy), June 1980, to be published in "La Rivista dei Combustibili".
74. J. H. Kent, H. Jander and H. Gg. Wagner, Eighteenth Symposium (Int.) on Combustion, Abstract of Papers, (1980), p. 41.
75. A. G. Gaydon and H. G. Wolfhard, "Flames, Their Structure, Radiation and Temperature," Chapman and Hall, 1970.
76. R. C. Millikan, "Temperature, Its Measurements and Control in Science and Industry," Vol. III, Part 2, Reinhold, New York (1962), p. 497.
77. U. Bonne and H. Gg. Wagner, Ber. Bunseges. Phys. Chem., Vol. 69 (1965), p. 35.
78. A. D'Alessio, F. Beretta and C. Venitozzi, Comb. Sci. Techn., Vol. 5 (1972), p. 263.
79. B. B. Land and A. G. Gaydon, Comb. and Flame, Vol. 16 (1971), p. 55.
80. P. Menna, Tesi di Laurea in Ingegneria Chimica, Napoli, 1979.
81. H. J. Haink and M. Jungen, Chem. Phys. Letts., Vol. 61 (1979), p. 319.
82. K. Fujiwara, N. Omenetto, J. B. Bradshaw, J. N. Bower and J. D. Winefordner, Applied Spect., Vol. 34 (1980), p. 85.
83. A. Cavaliere, unpublished results.
84. S. Chippet and W. A. Gray, Comb. and Flame, Vol. 31 (1978), p. 149.

DISCUSSION

B. S. Haynes (Massachusetts Institute of Technology)

I want to direct some comments to the nature of the ubiquitous green fluorescence seen by many people now on irradiation of fuel-rich combustion systems with visible(∿500 nm) laser light. Qualitatively, these observations seem to require that a large species (to give a substantial absorption coefficient in the visible) be present at rather high concentrations (to account for the intensity of the fluorescence). Unfortunately, however, probe measurements of sooting flames would suggest these requirements to be mutually contradictory.

At M.I.T., there has been a concerted effort to understand the spectroscopy of the polycyclics at high temperatures*,** and this has led to some initially surprising conclusions as to the nature of the fluorescing species.*** Firstly, Coe and Steinfeld* have shown that certain species, namely those which possess a "hidden transition'' (i.e., exhibit ''anomalous fluorescence'') exhibit a red shift in their excitation spectra as the temperature is raised. The fluorescence spectrum remains basically unchanged, with a slight blue shift in some cases. These results can be understood, with the aid of Fig. 1, as follows: in a molecule possessing a hidden transition, absorption takes place from the ground electronic state (S_0) to the second excited singlet (S_2). Absorption to the first excited singlet (S_1) is very weak because of unfavorable Franck-Condon factors. Upon excitation to S_2, the molecule relaxes rapidly to S_1 and fluoresces to hot vibrational bands in S_0 which have favorable Franck-Condon factors.

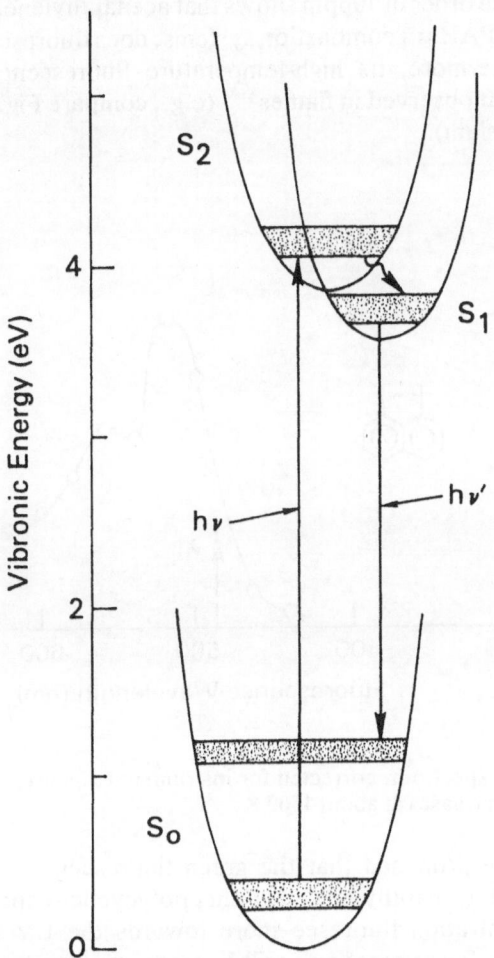

Fig. 1. Schematic illustration of absorption and fluorescence in a polycyclic aromatic hydrocarbon with a ''hidden transition''.

At typical flame temperatures ($T\sim1500\,\text{K}$), the vibrational energy distribution extends some tens of thousands of wavenumbers (cm^{-1}) so that the hot vibrational levels of S_0 will be populated. By the reverse of the anomalous fluorescence argument, these bands have favorable Franck-Condon factors for excitation directly to S_1. Therefore, in the flame, new features appear in the excitation spectrum to the red of the room temperature absorption, as demonstrated for pyrene where an excitation band at 340 nm appears in the flame-temperature spectrum where absorption is negligible at room temperature*. Generalizing this result, we can expect a species to absorb in the visible at flame temperatures if it fluoresces in that region at room temperature (and possesses a hidden transition).

An examination of polycyclic species expected to be present in sooting flames at concentrations of the order of 10ppm shows that acenapthylene, one of the simplest and most prevalent PAHs in combustion systems, does fluoresce at 510 nm at room temperature. Furthermore, its high-temperature fluorescence spectrum corresponds well with that observed in flames*** (e.g., compare Fig. 2 with D'Alessio's Fig. 29, this Symposium).

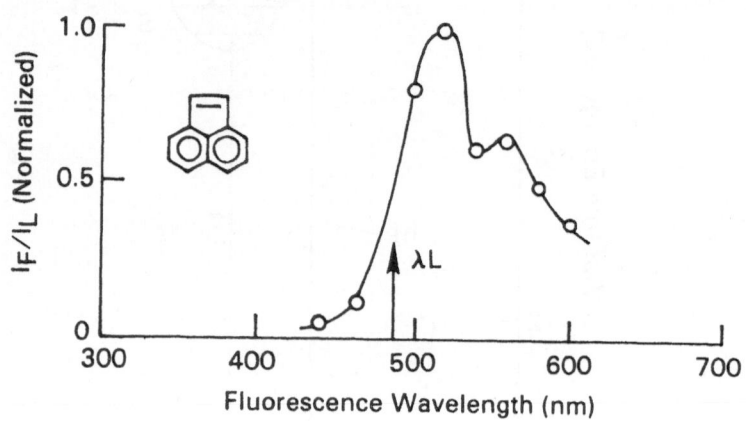

Fig. 2. Fluorescence spectrum, corrected for instrument response, of acenapthylene injected into fuel-rich flame gases at about 1500 K.

Therefore we have proposed that the green fluorescence always has a large contribution from hot acenapthylene***. Other polycyclics expected to be present at significant concentration fluoresce more towards the UV (e.g., 460 nm for fluoranthene; 390 nm for pyrene) and will become important contributors to the total fluorescence signal only as the exciting wavelength is reduced.

*D. S. Coe and J. Steinfeld in "Laser Probes for Combustion Chemistry", A.C.S. Symposium Series, Vol. 134, (D. J. Crosley, ed.), (1980), p. 247.
**D. S. Coe and J. Steinfeld, "Fluorescence Excitation and Emission Spectra of Polycyclic Aromatic Hydrocarbons in an Atmospheric Flame," to appear in Chem. Phys. Letts.
***D. S. Coe, B. S. Haynes and J. Steinfeld, "Identification of a Source of Argon-Ion-Laser Excited Fluorescence in Sooting Flames," to appear in Combustion and Flame.

D'Alessio

I agree basically with the interpretation given by Dr. Haynes; we have also found that acenaphthylene is a major component among PAHs [54]. It is possible to estimate the order of magnitude of the absorption cross section of the fluorescing compound at $\lambda = 514.5$ nm, starting from the measured concentration, the total fluorescent emission and the evaluated quantum efficiency [60], assuming that it is only acenaphthylene. It appears that we need a cross section between 10^{-19} and 10^{-18} cm^2 in order to justify the observed fluorescence. At room temperature acenaphthylene presents cross sections of this magnitude in the range 400 - 450 nm while its absorption disappears beyond 500 nm*. Therefore a high temperature absorption mechanism between upper vibronic levels of S_0 to S_1 may easily produce this relatively small shift to the red. However, we should note that knowledge of the high temperature spectral properties of aromatic molecules is just in its initial development stage from a fundamental chemical-physical point of view. New classes of effects such as inverse electronic relaxation and reverse intercrossing, have been found quite recently. Therefore I am expecting surprises both from further experimental measurements of laser excited fluorescence and from their interpretation.

*E. Heilbronner, J. P. Weber, J. Michl and R. Zahradnik, Theor. Chim. Acta, Vol. 6 (1966), p. 141.

SOOT OXIDATION IN FLAMES

K. G. NEOH, J. B. HOWARD * and A. F. SAROFIM

Massachusetts Institute of Technology
Cambridge, Massachusetts

ABSTRACT

The kinetics and mechanisms of soot oxidation were studied at temperatures of 1575 to 1865 K and O_2 mole fractions of 10^{-5} to 0.05 in a two-stage atmospheric-pressure combustion system in which soot-laden gas from a primary flame was partially cooled, mixed with oxygen-containing gas, and burned in a downstream premixed flame. The rate of oxidation of soot was measured using light scattering and absorption techniques to determine the particle size and concentration of soot as a function of distance, and hence of time, in the secondary flame. Complementary measurements of soot particle size were performed using probe sampling, electron microscopy, and electrical mobility analysis, and the specific surface area of the soot at different stages of burnout was measured by the BET technique. Temperature and gas composition profiles in the secondary flame were determined using coated thermocouples with electrical compensation for radiation loss, chromatographic analysis of sampled stable gases, and a spectroscopic technique combined with partial equilibrium calculations for radical species. It was found that OH radical is the principal oxidant of soot under these conditions, with O_2 being of secondary importance. The average value of the collision efficiency of OH with soot is found to be 0.28 if the optical (equivalent sphere) diameter of the soot aggregates is used in the calculations, or about 0.13 if the diameter of the individual spherical units within the aggregates is used. The actual value of the collision efficiency is expected to be bounded by these two values. The results indicate that soot burnout rates predicted from the Nagle and Strickland-Constable formula, which assumes O_2 to be the oxidant, are drastically underestimated under fuel-rich flame conditions owing to the neglect of the OH contribution. Breakup of soot aggregates during burnout was observed under fuel lean conditions.

To whom correspondence should be addressed.

INTRODUCTION

Soot formation is difficult to suppress in the early stages of combustion, particularly with the increasing trend toward the use of high C/H fuels and staged combustion for NO_x control. There is therefore increasing interest in soot oxidation under fuel rich or near stoichiometric conditions. Although this problem has been addressed by a number of investigators there is a residual uncertainty about the relative importance of O_2, O, OH, CO_2 and H_2O in soot gasification and also about the changes in particle size and number density during soot consumption.

Previous investigations of soot oxidation were carried out in flames and in non-flame environments. A summary of the major investigations [1-4] of pyrolytic graphite and soot oxidation in non-flame environments is given in Table 1. Other investigations that are not listed in Table 1 include those of Otterbein and Bonnetain [5], Walls and Strickland-Constable [6] and Olander et al. [7]. The major studies of soot oxidation in flames are those of Lee, Thring and Beer [8], Tesner and Tsibulevsky [9, 10] and Fenimore and Jones [11]. A summary of the experimental conditions and the conclusions reached is given in Table 2.

Appleton et al. [12, 13] compared the rates of soot oxidation in flames [8, 9, 11], pyrolytic graphite oxidation using O_2 [2, 6], and carbon black oxidation in a shock tube [4], with the rates predicted by the Nagle and Strickland-Constable formula [1]. They concluded that the magnitude of most of the soot and pyrolytic graphite oxidation rates can be adequately correlated by the Nagle and Strickland-Constable formula with the notable exceptions of three slightly fuel rich runs by Fenimore and Jones. Radcliffe and Appleton [13] suggested that the data from the slightly fuel rich flames could be explained by oxygen concentrations greater than those estimated by Fenimore and Jones based on equilibrium concentrations. On the basis of our calculations, in order for the Nagle and Strickland-Constable formula to predict the observed rates in the rich flames, the O_2 concentration would have to be 400 to 4000 times higher than the equilibrium values. However, the weak dependence on oxygen concentration which Fenimore and Jones observed under fuel-lean conditions was accounted for by the Nagle and Strickland-Constable formula.

Although the Nagle and Strickland-Constable formula seems adequate for predicting oxidation rates of soot and pyrolytic graphite by O_2, there is a lack of evidence that this formula will correctly predict soot oxidation rates in flames where there are more than one potential oxidant. In fuel rich flames the O_2 concentration may be comparable to the O concentration and may be much less than the OH concentration. The reaction probability of soot with O and OH can be expected to be higher than with O_2. In addition to the uncertainty of the roles of the various chemical species in soot oxidation in flames, the effect of the oxidation on the physical structure of the soot is also unclear. Although there is evidence of internal burning of soot particles during oxidation in flames [14], and small particles of graphitized carbon black at 873 K [15], this effect has not been adequately studied. Soot burnout rates were often calculated from mass concentrations assuming a shrinking sphere model, and there is a lack of information on how the soot particle size and number concentration change during the oxidation process.

TABLE 1

Previous Studies of Pyrolytic Graphite and Soot Oxidation in Non-Flame Environments

Investigator	Carbon Type	Oxidant Used	Temperature Range	Comments
(1) Nagle and Strickland-Constable [1]	Pyrolytic Graphite	O_2	1273-2273 K	Formulated semi-empirical formula based on 2 types of reaction sites on C surface
(2) Rosner and Allendorf [2]	Pyrolytic Graphite	(i) O_2 (ii) O	1200-2100 K	O atoms shown to be more reactive than O_2. Maximum reaction probability of O atom is $\sim$0.5
(3) Wright [3]	(i) Soot (ii) Pyrolytic Graphite	O	300- 850 K	Reaction probability of O atoms between 5×10^{-4} to $> 10^{-2}$
(4) Park and Appleton [4]	Carbon Black	O_2	1700-4000 K	Rates correlated by Nagle and Strickland-Constable formula

TABLE 2

Previous Studies of Soot Oxidation in Flames

Investigator	Combustion System	Temperature Range	O_2 Mole Fraction	Comments
(1) Lee, Thring and Beer [8]	Product gases of diffusion flames of hydrocarbon mixture	1300-1700 K	0.04 - 0.12	First order dependence of rates on O_2. E = 165 kJ/mol (39.3 kcal/mol)
(2) Tesner and Tsibulevsky [9]	Laminar diffusion flames of C_2H_2 - H_2 and C_2H_2 - H_2O	1400-2000 K	not specified	First order dependence of rates on O_2. E = 167 kJ/mol (40 kcal/mol)
(3) Tesner and Tsibulevsky [10]	Upper part of laminar diffusion flames of C_2H_2 - N_2 and C_2H_2 - CO_2	1800-1940 K	not specified	First order dependence of rates on CO_2. E = 314 kJ/mol (75 kcal/mol)
(4) Fenimore and Jones [11]	Premixed flames with separation of soot generation and oxidation stages	1530-1890 K	$0.3 - < 10^{-4}$	Weak dependence of rates on O_2. Proposed OH as principal oxidant with a collision efficiency of $\sim$0.1

In the present study, both fuel-rich and fuel-lean flames were studied in order to determine which of the species H_2O, CO_2, O_2, O and OH are most important in soot oxidation. The changes in soot size and number concentration during soot oxidation were followed by optical techniques. With the ability to measure the soot number concentration, it was not necessary to use the assumption made by previous investigators that the number concentration remains constant throughout the burnout process. The possibility of internal burning was also explored in this work.

EXPERIMENTAL

The apparatus used in this study is depicted schematically in Fig. 1. Soot for the burnout studies was generated by burning premixed CH_4 and O_2 with a fuel equivalence ratio of 2.10 in a primary burner consisting of a water-cooled sintered bronze disk. The soot-laden primary combustion gases were partially cooled and passed into a secondary mixing chamber into which O_2 and N_2 were injected at high velocity through four nozzles. Occasionally additional CO_2 and CH_4 were introduced in the secondary mixing chamber in order to achieve the desired ranges of experimental temperatures and CO_2 concentrations. The mixture of soot, primary combustion gases and secondary gases was then reignited above a secondary burner* composed of a stainless steel tube bundle. The soot burnout in the residing laminar secondary flame was measured to heights limited to 10 mm above the

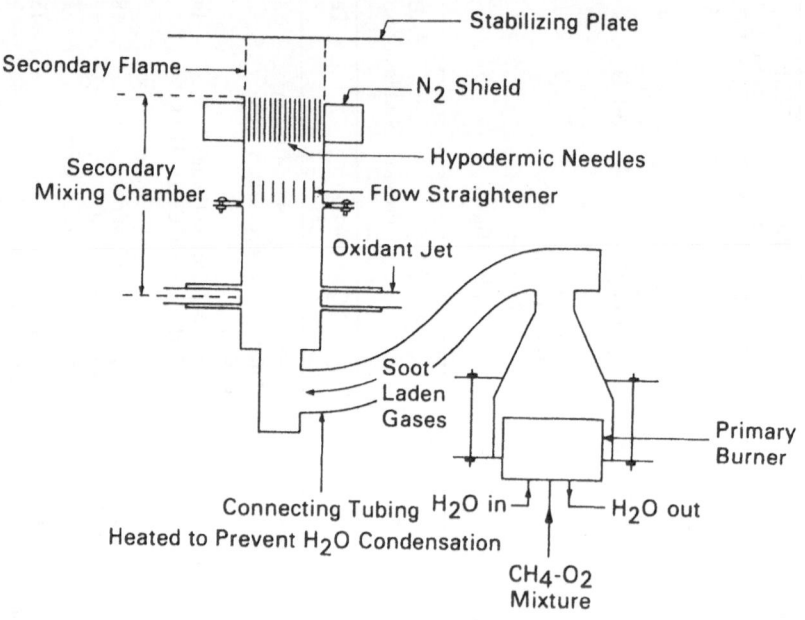

Fig. 1. Design of combustion system for soot burnout studies.

*The equivalence ratios in the primary and secondary flames are designated $1°\phi$ and $2°\phi$ (see also Table 3.)

burner, a region in which the system could be treated as being one-dimensional. Nitrogen flow through an annular ring shielded the combustion products from the ambient air. A total of seven secondary flame conditions was studied with the primary flame maintained at the conditions mentioned above. A summary of the operating conditions is given in Table 3.

The experiments carried out to characterize the soot and the secondary flames are summarized in Table 4. The soot burnout rate was followed by measuring the soot size and concentration using optical methods similar to those developed by D'Alessio et al. [16]. Independent measurements of particle size were obtained for selected runs using an electrical mobility analyzer or by electron microscopy of samples collected on a microscope grid placed in a quartz microprobe sampling train. BET surface area and pore size distributions were obtained for the soot collected from the primary burner and from the richest secondary flame. The gas

TABLE 3

Summary of Operating Conditions for Soot Burnout Studies

Primary Flame: CH_4 - O_2 Premixed Flame
Primary Fuel Equivalence Ratio ($1°\phi$) = 2.10

Secondary Flames:

Flame Notation	Fuel Equivalence Ratio of Secondary Flame ($2°\phi$)	Gases Added to Primary Combustion Products
0.85	0.85	O_2, N_2
0.95	0.95	O_2, N_2
1.05	1.05	O_2, N_2
1.10	1.10	O_2, N_2
1.15	1.15	O_2, N_2
1.15($+2°$ CH_4)	1.15	O_2, N_2, CH_4
1.15 ($+2°$ $CH_4 + 2°$ CO_2)	1.15	O_2, N_2, CH_4, CO_2

TABLE 4
Summary of Experiments

Experiments	Parameters Measured
1. Light scattering and extinction	Soot size and concentration
2. Coated thermocouples with electrical compensation for radiation	Flame temperature
3. Probe sampling and gas chromatography	Stable gas phase species (except H_2O)
4. Atomic absorption with Li/LiOH method	H atom concentration
5. Study of collected soot: electron microscopy, electrical aerosol analysis, surface (N_2 - B.E.T.) and pore analysis, and elemental analysis	Soot size, surface area, pore size and composition

was sampled using a quartz microprobe and the concentrations of N_2, CO, CO_2, H_2, O_2 and hydrocarbons, when present, were analyzed by gas chromatography. The H atom concentration was measured using LiCl addition and absorption to measure Li concentration following the procedures developed by Bulewicz *et al.* [17] and McEwan and Phillips [18]. Temperatures were measured using an electrically compensated coated thermocouple. The H_2O concentration was obtained from hydrogen and oxygen elemental balances. Using partial equilibrium considerations, OH and O concentrations were estimated from H, H_2O and H_2 concentrations in rich flames and from H, H_2O and O_2 concentrations in lean flames. In rich flames the O_2 concentrations were too low for accurate measurement by gas chromatography and were also calculated from partial equilibrium considerations. A detailed description of the experiments and the data obtained is given by Neoh [19].

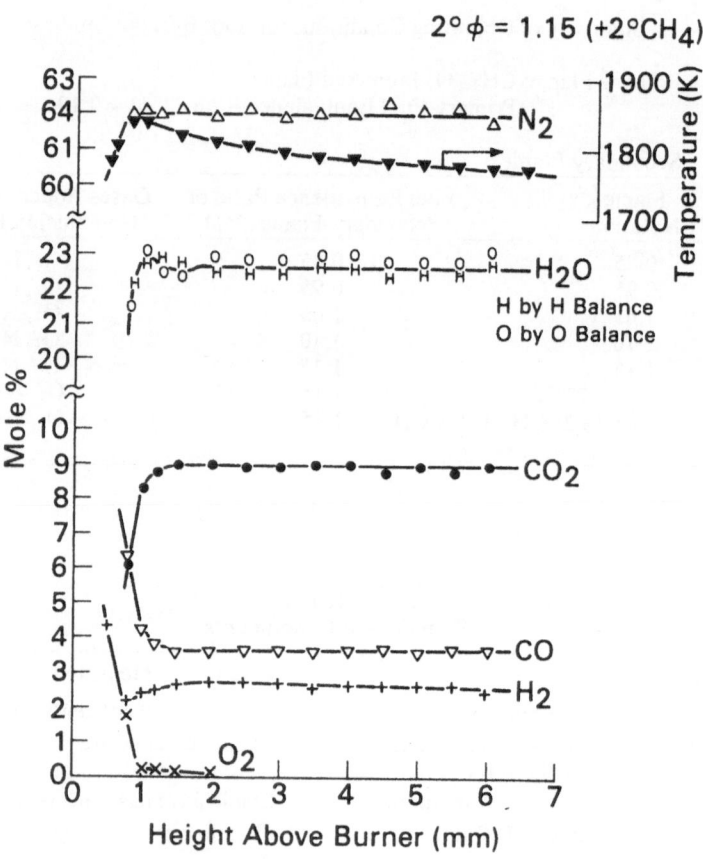

Fig. 2. Temperature profile and composition profies of stable gas phase species for $2°\phi = 1.15 (+2° CH_4)$ flame.

RESULTS

The temperature profile and the composition profile of the stable gas phase species for one flame [$2°\phi = 1.15 (+ 2° CH_4$)] are shown in Fig. 2. The H_2O concentrations estimated by hydrogen and oxygen balances (denoted by the symbols H and O, respectively, in Fig. 2) for all the flames agreed to within 3%. For all seven flames the H_2O concentration in the burnt gas region was fairly constant with a value between 20% and 26%. For six of the seven flames, the CO_2 concentration was also fairly constant, with a value between 9% and 11%. For the seventh flame additional CO_2 was added to the primary combustion gases to bring the concentration in the secondary flame to approximately 20%. The temperature range covered by the seven flames is 1575 — 1865 K and the range of oxygen mole fraction is 10^{-5} — 0.05. Fig. 3 shows the mole fraction of OH, O_2 and O estimated from partial equilibrium for the $2°\phi = 1.15 (+2°CH_4)$ flame. In the rich flames the O_2 concentration was of the same order or less than that of OH and the O concentration was lower than that of either OH or O_2 in the temperature range investigated.

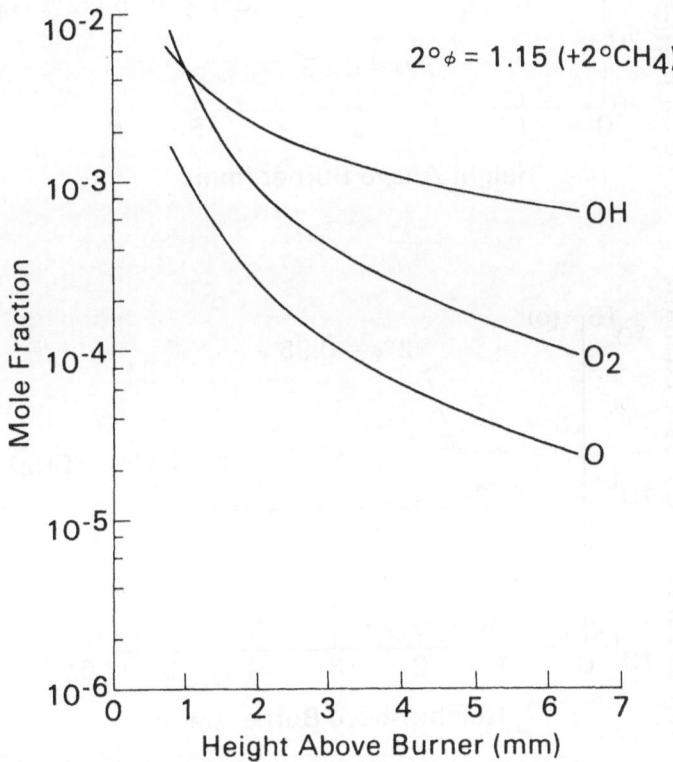

Fig. 3. Profiles of OH, O_2 and O mole fraction calculated assuming partial equilibrium for $2°\phi = 1.15 (+2° CH_4)$ flame.

The variation of soot size and number concentration with height above the burner for two of the flames, $2°\phi = 0.95$ and $2°\phi = 1.15 (+ 2°CH_4)$, is shown in Fig. 4. The calculations were carried out assuming a lognormal distribution of particle sizes with a standard deviation of 0.3. In a lean flame, the reduction in soot sizes was very rapid and the soot particles were totally consumed within a few milliseconds. In lean flames, the number density of the soot was observed to increase sharply after approximately 80% of the original mass of the soot had been oxidized, but for rich flames no such increase was observed even after 80% burnout. This effect is postulated to be due to the breakup of soot particles or agglomerates under lean conditions but not under rich conditions, a phenomenon discussed in more detail elsewhere [19].

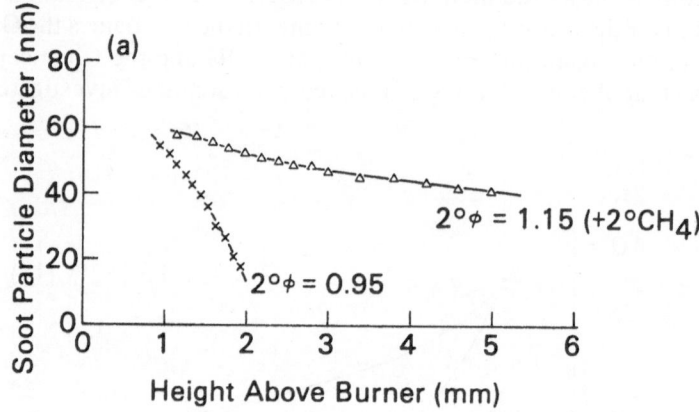

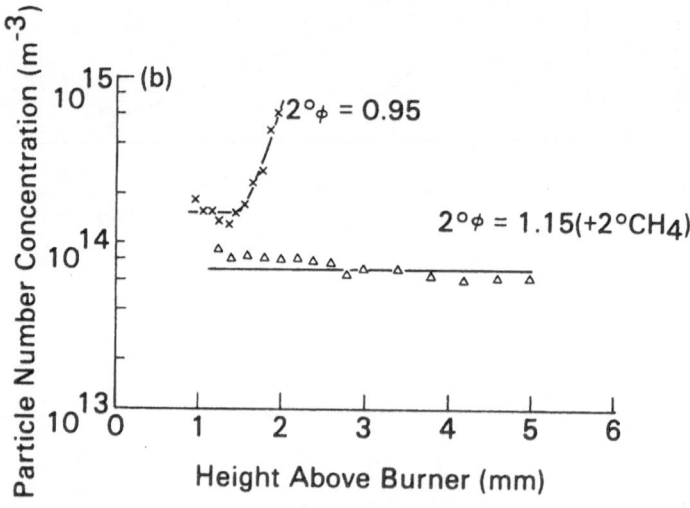

Fig. 4. Profiles of (a) soot particle diameter and (b) number concentration for $2°\phi = 0.95$, 1.15 ($+2°$ CH$_4$) flames. (For $2°\phi = 0.95$ flame, 1 mm, ~ 0.66 ms; for $2°\phi = 1.15$ ($+2°$ CH$_4$) flame, 1 mm ~ 0.61 ms).

DISCUSSION

The specific soot burnout rate, ω (kg/m²s) is defined as

$$\omega = \frac{1}{A} \frac{dm}{dt} \tag{1}$$

where A is the available surface area (m²/m³), m is the mass concentration of soot (kg/m³) and t denotes the time (s). For a lognormal distribution of sizes with standard deviation σ, Eq. (1) becomes

$$\omega = \frac{1}{N\pi \, d_m^2 \, \exp[2\sigma^2]} \frac{d}{dt} (N \, \frac{\pi\rho}{6} d_m^3 \, \exp[\frac{9}{2}\sigma^2]) \tag{2}$$

where d_m is the geometric mean diameter (m), N is the number concentration (m⁻³), ρ is the density (kg/m³) of the soot particles. Since N may not be constant with time, Eq. (2) can be expressed in terms of d_m and m,

$$\omega = \frac{\rho}{6} d_m \exp(\frac{5}{2}\sigma^2) \frac{d \ln m}{dt} \tag{3}$$

The gas phase species which are likely to be potential oxidants of soot are H_2O, CO_2, O, O_2 and OH. In order to determine which of these species is the principal oxidant, the collision efficiency (or reaction probability) required for each species acting alone to account for the observed burnout rate was calculated as the ratio of the experimentally observed reaction rate to the collision rate from kinetic theory. The collision efficiency, γ_i, is defined as the probability that when an oxidant molecule or radical collides with the soot, reaction will occur so as to remove a C atom. Fig. 5 shows the values of γ_i calculated for i = H_2O, O, O_2 and OH. The values of γ_{CO_2} exhibit the same trend as that of γ_{H_2O} [19] and so are not shown here.

From Fig. 5, H_2O can be dismissed as the principal oxidant of soot in this investigation because of the following reasons:

1. the rates from the leanest to the richest flames differ by a factor of 10 to 100 while the H_2O concentrations show little variation;
2. it is unlikely that the rapid decrease in γ_{H_2O} with height above the burner could be due to the decrease in temperature.

For example, using the case of the 2°ϕ = 1.15 (+2°CH₄) flame, in order to explain the decrease in γ_{H_2O} in terms of a temperature effect would require an unlikely activation energy of 1260 kJ/mol (300 kcal/mole). The values of γ_{CO_2} show the same variation with height above the burner as does γ_{H_2O}, as mentioned above, and for six of the seven flames the CO_2 concentrations also remained fairly constant. Therefore the two arguments against H_2O can also be used to dismiss CO_2 as an important oxidant. In the case of O atoms, the values of collision efficiency range

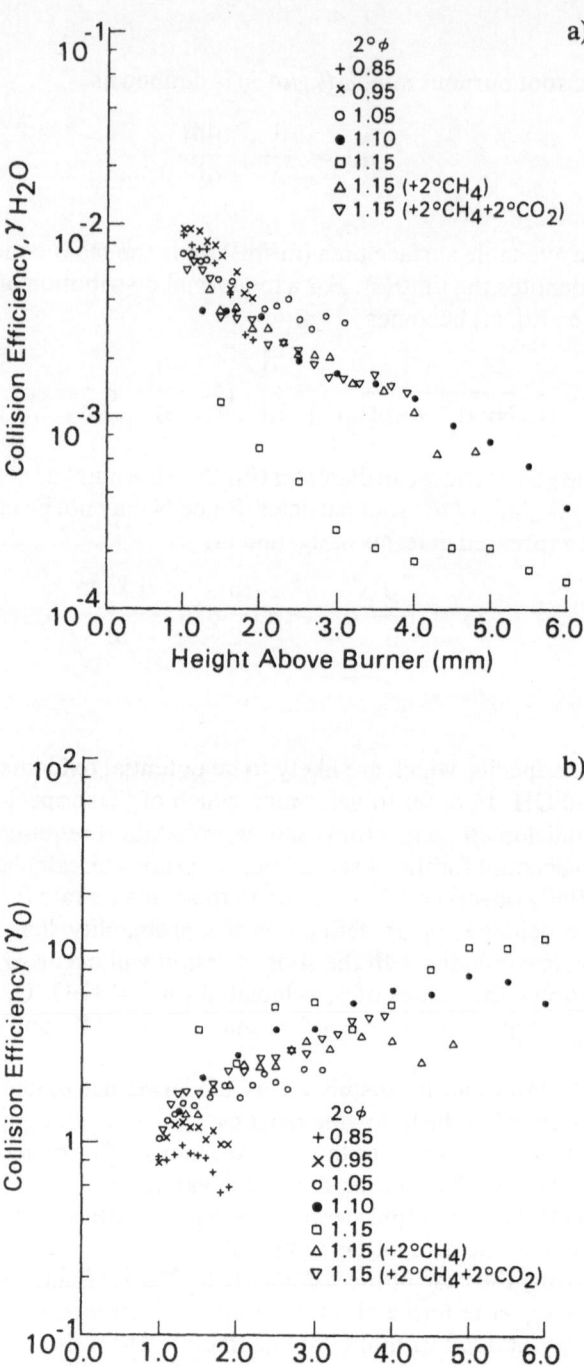

Fig. 5. Collision efficiencies, assuming soot burnout is due to attack by (a) H_2O, (b) O, (c) O_2, and (d) OH as a function of height above burner.

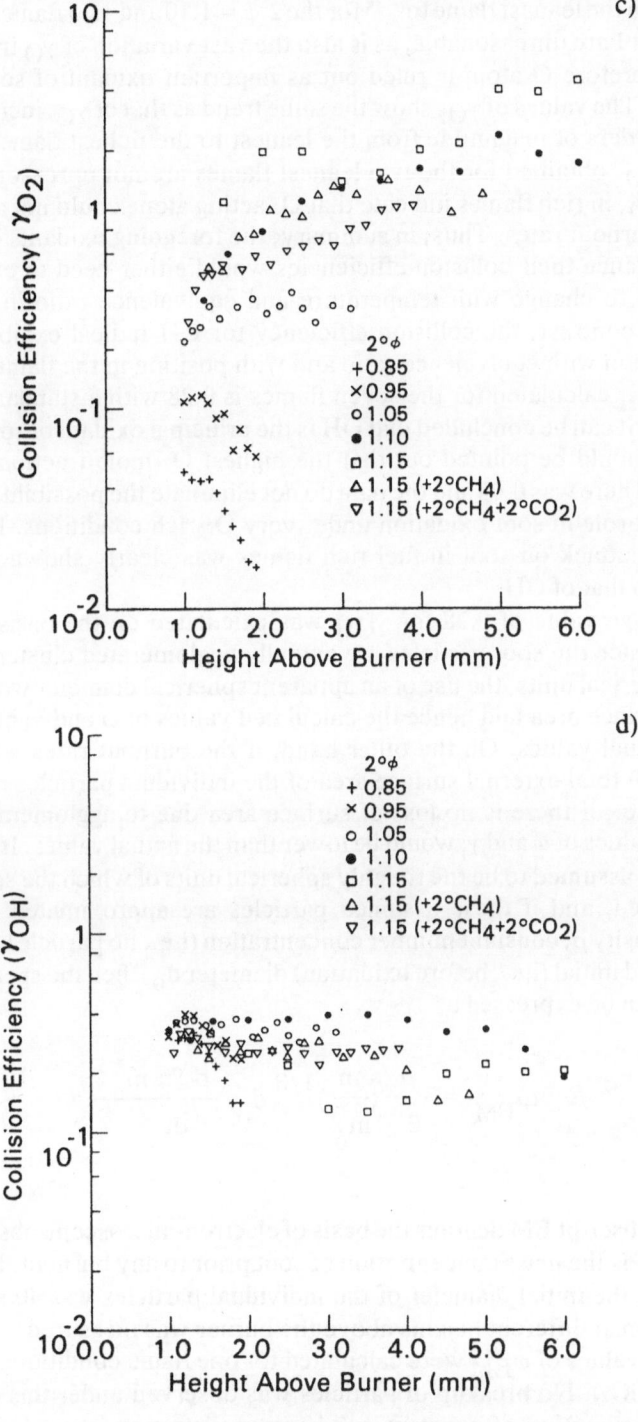

Fig. 5 (Cont'd.)

References p. 277

from ~ 0.8 in the leanest flame to > 5 for the $2°\phi = 1.10$ and 1.15 flames. Values of γ_O greater than 1 are unreasonable, as is also the vast variation of γ_O from lean to rich flames. Therefore O atom is ruled out as important oxidant of soot under these conditions. The values of γ_{O_2} show the same trend as that of γ_O, increasing steadily over two orders of magnitude from the leanest to the richest flame. Although the values of γ_{O_2} obtained for the two leanest flames are not unreasonable, the high values of γ_{O_2} in rich flames indicate that O_2 acting alone could not account for the observed burnout rates. Thus, in summary, the foregoing oxidants cannot explain the results since their collision efficiencies would either need to exceed unity or would need to change with temperature and equivalence ratio in an unrealistic manner. In contrast, the collision efficiency for OH radical exhibits reasonably small variation with equivalence ratio and with position in the flame. The average value of γ_{OH} calculated for the seven flames is 0.28 with a standard deviation of 0.07. Hence it can be concluded that OH is the principal oxidant of soot for all seven flames. It should be pointed out that the highest O_2 mole fraction in the flames investigated here was 0.05 and the data do not eliminate the possibility of O_2 playing a significant role in soot oxidation under very O_2-rich conditions. But the importance of O_2 attack on soot in fuel rich flames was clearly shown to be minimal compared to that of OH.

The average value of 0.28 for γ_{OH} was calculated on the basis of an optical diameter. Since the soot particles are actually agglomerated clusters or chains of roughly spherical units, the use of an apparent spherical diameter would underestimate the surface area and hence the calculated values of ω and γ_i could be higher than the actual values. On the other hand, if the burnout rates were calculated assuming the total external surface area of the individual particles is available for oxidation, i.e., if there is no loss of surface area due to agglomeration, then the calculated values of ω and γ_i would be lower than the actual values. If the individual particles are assumed to be the roughly spherical units of which the soot aggregates are composed, and if these assumed particles are approximated as spheres of constant density ρ, constant number concentration (i.e., no particle breakup during burnout), and initial (i.e., before oxidation) diameter d_0, then the specific oxidation rate ω_{EM} can be expressed as

$$\omega_{EM} = \frac{\rho}{6} \left(\frac{m}{m_o}\right)^{1/3} d_o \frac{d \ln m}{dt} \tag{4}$$

where the subscript EM denotes the basis of electron-microscope observed particle sizes and m_o is the mass concentration of soot prior to any burnout. From electron microscopy, the initial diameter of the individual particles was 40 nm. The mass concentration at different heights above the burner was measured by optical techniques. The values of ω_{EM} were calculated for one flame condition, $2°\phi = 1.15$ (+ $2°CH_4 + 2°CO_2$). No breakup of particles was observed under this condition and hence the assumption of constant number concentration is valid. The values of ω_{EM} were found to be a factor of approximately 0.49 smaller than the values calculated using the optical diameters. Hence the collision efficiencies based on the electron microscopic diameter will be 0.49 of the values calculated using the optical diameters. Therefore the value of γ_{OH} for agglomerated soot particles (i.e., the actual soot aggregates) can be expected to be between 0.28 and 0.13. It is interesting to note that in the $2°\phi = 0.85$ flame where particle breakup was seen to occur after 1.5 mm above the burner, the value of γ_{OH} decreased from about 0.3 at heights less than 1.5 mm to 0.14 at the tail of the soot zone at 1.9 mm. This is to be expected since before the breakup occurred the surface area calculated from the optical diameter would be lower than the actual area, but when the soot aggregates broke up into smaller, more rounded fragments, the optical diameter would be close to the actual diameter of each fragment and the value of γ_{OH} would approach the actual value.

It is of interest to compare the contribution of OH with that of O_2, as calculated from the Nagle and Strickland-Constable formula, to soot burnout rates in flames. In Fig. 6 the ratio of the observed rate, ω, to the rate predicted by the Nagle and Strickland-Constable formula, ω_{NSC}, at the observed temperature and O_2 concentration is plotted as a function of O_2 mole fraction in the flame. The points denoted by FJ are from Fenimore and Jones [11]. This plot shows that in flames with high O_2 concentration, predictions from the Nagle and Strickland-Constable formula are close to the measured rates but the discrepancy between predicted and observed rates becomes larger as the flames are made more fuel rich. From this plot it is not surprising, as described below, that Appleton *et al.* found good agreement

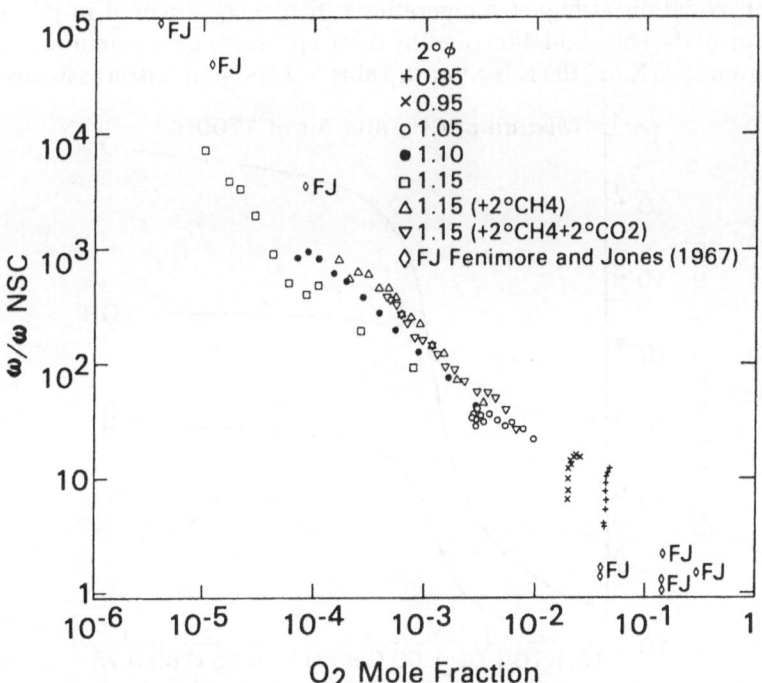

Fig. 6. Comparison of experimental rate, ω, to rate predicted by Nagle and Strickland-Constable formula, ω_{NSC}.

between the Nagle and Strickland-Constable formula and the soot oxidation rates of Lee *et al.* [8] and Tesner and Tsibulevsky [9], carbon black oxidation rates in a shock tube [4], and the fuel lean runs of Fenimore and Jones [11], but not for the three fuel rich runs of the latter. For the lean flames of Fenimore and Jones and the diffusion flames of Lee *et al.* the O_2 mole fraction was of the order 0.04 and higher. For Tesner and Tsibulevsky's flames, Radcliffe *et al.* [13] estimated the O_2 mole fraction to be 0.01. In the shock tube oxidation of carbon black [4], Park and Appleton used a O_2-Ar mixture and hence the soot oxidation occurred in the absence of OH. In the fuel rich flames of Fenimore and Jones, although the O_2 concentrations were probably higher than the equilibrium values, the principal oxidant was likely to be OH instead of O_2.

PRACTICAL IMPLICATIONS

The accurate prediction of soot burnout rates in flames requires knowledge of the relative importance of OH, O_2 and possibly O atoms under the prevailing conditions. Fig. 7 shows the variation of the calculated equilibrium concentrations of OH, O_2 and O with fuel equivalence ratio for a CH_4 — air mixture at 1700 K. A change of fuel equivalence ratio from 1.15 to 0.75 produces an increase of a factor of 10^6 in equilibrium O_2 concentration compared to an increase of a factor of $\sim$14 for OH and 10^3 for O atoms. To obtain an idea of the relative importance of OH, O_2 and O in soot oxidation at higher temperatures than those covered in this work, a comparison of the soot oxidation rates by these species at their equilibrium concentrations from 1500 K to 2100 K is given in Table 5. This comparison assumes that the

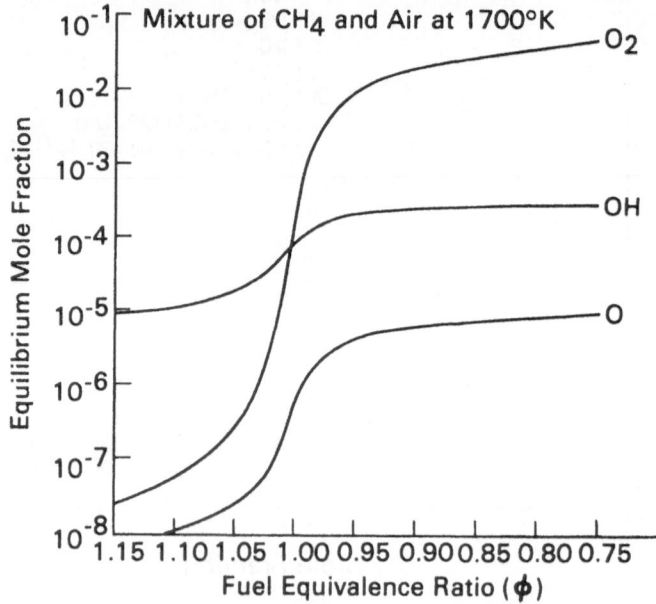

Fig. 7. Variation of equilibrium concentrations of OH, O_2 and O with fuel equivalence ratio for a CH_4-air mixture at 1700K.

TABLE 5

Comparison of Soot Oxidation Rates by OH, O_2 and O at Their Equilibrium Concentrations

T\φ	1.15		1.05		0.95		0.75	
	$\omega O_2/\omega OH$	$\omega O/\omega OH$	$\omega O_2/\omega OH$	$\omega O/\omega OH$	$\omega O_2/\omega OH$	$\omega O/\omega OH$	$\omega O_2/\omega OH$	$\omega O/\omega OH$
1500K	1.53×10^{-7}	3.84×10^{-4}	8.59×10^{-7}	4.36×10^{-4}	1.70	2.14×10^{-2}	2.30	3.59×10^{-2}
1700K	2.70×10^{-6}	3.16×10^{-3}	1.58×10^{-5}	3.63×10^{-3}	1.22	5.01×10^{-2}	1.91	8.43×10^{-2}
1900K	2.65×10^{-5}	1.68×10^{-2}	1.69×10^{-4}	1.93×10^{-2}	0.47	9.90×10^{-2}	1.36	0.17
2100K	1.72×10^{-4}	6.49×10^{-2}	1.26×10^{-3}	7.30×10^{-2}	9.82×10^{-2}	0.18	0.51	0.29

rate of soot oxidation by O_2 is given by the Nagle and Strickland-Constable formula and the collision efficiencies of OH and O are 0.2 and 0.5, respectively, throughout the temperature range considered. Since the concentrations of these species in many systems may not be the equilibrium values, Table 5 is intended only as an estimate of the relative importance of O_2, OH, and O in soot oxidation. From the equilibrium concentrations it is concluded that OH is the dominant oxidant under fuel rich conditions, that O atom is at most only of secondary importance at temperatures below 2000 K, and that O_2 can become a significant factor under fuel lean conditions, particularly at lower temperatures. If allowance were made for superequilibrium concentrations of radicals, the conditions at which O_2 becomes of equal importance to OH would be shifted to leaner conditions. Accurate estimates or measurements of temperature and of concentration of the principle oxidant are of course required for accurate prediction of soot burnout rates.

In practical systems the prediction of soot burnout rates or times can be complicated by the effect of turbulence. Previous investigators of turbulent diffusion flames [20, 21, 22, 23] have proposed a model in which eddies of soot-containing gases mix with oxygen rich eddies and the soot particles then burn away very rapidly. OH radicals can be expected to play an important role in the oxidation of the soot contained in the fuel rich eddies. The use of the Nagle and Strickland-Constable formula to predict soot burnout rates under these conditions would result in overestimation of the burnout times. Using the $2°\phi = 1.15 (+ 2°CH_4)$ flame as an example, at a temperature of 1790 K, the mole fractions of OH and O_2 were 1.2 x 10^{-3} and 3×10^{-4}, respectively. The observed burnout time for a 40nm particle under these conditions is $\sim$ 4ms, while the predicted value from the Nagle and Strickland-Constable formula is 2.5s.

CONCLUSION

The present experiments have shown OH to be more important than O_2 in the oxidation of soot in atmospheric pressure flames between 1575 K and 1865 K and an O_2 mole fraction between 0.05 and 10^{-5}. Using the optical (equivalent sphere) diameter of the soot aggregates, the average value of collision efficiency of OH with soot is found to be 0.28. If the diameter of the individual spherical units within the aggregates is used, the corresponding value is about 0.13. The actual value of the collision efficiency is expected to be bounded by these two values. The Nagle and Strickland-Constable formula which has been shown to predict adequately soot burnout rates in flames with a high O_2 content or in O_2 — containing mixtures in the absence of OH will drastically underestimate the rate of soot oxidation in fuel rich flames due to the neglect of the OH contribution.

ACKNOWLEDGEMENTS
We are grateful to the National Science Foundation, Division of Engineering, Particulate and Multiphase Processes Program, for financial support under Grant

No. NSF ENG 75-22679 AO1, and to Dr. A. D'Alessio of the Universita di Napoli for many helpful discussions.

REFERENCES

1. J. Nagle, and R. F. Strickland-Constable, Proc. of the Fifth Carbon Conf. I (1962), p. 154.
2. D. E. Rosner, and H. D. Allendorf, AIAA Journal, Vol. 6 (1968), p. 650.
3. F. J. Wright, "Fifteenth Symposium (International) on Combustion," The Combustion Institute, Pittsburgh, (1975), p. 1449.
4. C. Park, and J. P. Appleton, Combustion and Flame, Vol. 20 (1973), p. 369.
5. M. Otterbein, and L. Bonnetain, Carbon, Vol. 7 (1969), P. 539.
6. J. R. Walls, and R. F. Strickland-Constable, Carbon, Vol. 1 (1964), p. 33.
7. D. R. Olander, W. Siekhaus, R. Jones and J. A. Schwarz, J. Chem. Phys., Vol. 57 (1972).
8. K. B. Lee, M. W. Thring, J. M. Beer, Combustion and Flame, Vol. 6 (1962), p. 137.
9. P. A. Tesner and A. M. Tsibulevsky, Combustion, Explosion, and Shock Waves, Vol. 3 (1967), p. 163.
10. P. A. Tesner, and A. M. Tsibulevsky, Combustion and Flame, Vol. 11 (1967), p. 227.
11. C. P. Fenimore and G. W. Jones, J. Phys. Chem., Vol. 71 (1967), p. 593.
12. J. P. Appleton, "Soot Oxidation Kinetics at Combustion Temperatures", Paper presented at AGARD Propulsion and Energetics Panel, 41st Meeting, London, England, April (1973).
13. S. W. Radcliffe, and J. P. Appleton, Combust. Sci. and Tech., Vol. 4 (1971), p. 171.
14. G. Reinhold, P. Szargan, P. Nau, and K. Richter, Chem. Techn., Vol. 23 (7) (1971), p. 416.
15. W. J. Thomas, Carbon Vol. 3 (1966), p. 435.
16. A. D'Alessio, A. DiLorenzo, A. F. Sarofim, F. Beretta, S. Masi, and C. Venitozzi, "Fifteenth Symposium (International) on Combustion," The Combustion Institute, Pittsburgh, (1975), p. 1427.
17. E. M. Bulewicz, C. G. James, and T. M. Sugden, Proc. Roy. Soc. A235 (1956), p. 89.
18. M. J. McEwan, and L. F. Phillips, Combustion and Flame, Vol. 9 (1965), p. 420.
19. K. G. Neoh, "Soot Burnout in Flames", Sc.D. Thesis, Department of Chemical Engineering, Massachusetts Institute of Technology, in preparation.
20. W. H. Dalzell, G. C. Williams, and H. C. Hottel, Combustion and Flame, Vol. 14 (1970), p. 161.
21. H. A. Becker, and S. Yamazaki, "Sixteenth Symposium (International) on Combustion," The Combustion Institute, Pittsburgh, (1977), p. 681.
22. B. F. Magnussen, "Fifteenth Symposium (International) on Combustion," The Combustion Institute, Pittsburgh, (1975), p. 1415.
23. B. F. Magnussen, and B. H. Hjertager, "Sixteenth Symposium (International) on Combustion," The Combustion Institute, Pittsburgh, (1977), p. 719.

DISCUSSION

P. J. Pagni (University of California, Berkeley)

I have a question about applying your results to a diffusion flame. If I have a Wolfhard-Parker over-ventilated flame, soot forms on the fuel-rich side and then moves into the reaction zone where it probably meets with oxygen. Before hearing your results I would have tried to use Appleton's correlation in the local oxygen concentration to calculate the oxidation of the soot. But I suspect you're telling me that first I have to calculate the intermediate OH concentration and then calculate its rate in oxidizing the soot.

Howard

Yes, that is correct. In the laminar diffusion flame you described or in a turbulent diffusion flame where there are fuel-rich pockets into which oxidizing species diffuse, the soot burning rate may be much faster in the very fuel rich zones than would be predicted based on oxidation by O_2, and the difference can be as much as 3 to 4 orders of magnitude.

A. D'Alessio (University of Naples)

Like Professor Pagni, I have a question regarding the very interesting effect you notice. On going to relatively lean conditions, you observed particle breakup where the number increases and the size decreases. That conclusion comes from the optical measurements. So I think that the oxygen side of a diffusion flame should be an ideal position to check for the particle breakup effect that you found in another situation. We had some suspicions that we had found these effects when we analyzed our optical data obtained in large oil flames. We used the scattering/extinction method and found that at increasing distances from the burner the particle size was going down and the number was going up. We were so surprised that we thought these trends were in some way connected with very low accuracy of the data, which was reasonable since we had very high turbulence. I would like to see the experiment performed in the oxygen side of a diffusion flame in order to confirm this mechanism for oxygen rich flames.

Howard

I agree with that. I think Dr. Prado has a point relative to this discussion.

G. Prado (Massachusetts Institute of Technology)

In France we have conducted such measurements in a diffusion flame by the photo-optical technique. We reported the data a few months ago in Combustion and Flame. We indeed find that at the tip of the flame the number of particles increases by nearly a factor of 10 and the size decreases. Our data would clearly support the view that at the tip of the flame the aggregates are breaking down. Therefore your conclusion from a premixed flame is consistent with our data from a diffusion flame.

R. Stevenson (General Motors Research Laboratories)

I have a question relating to morphology. You talked about the breakup of aggregates or agglomerates into the primary particles. Do you also see the hollow structure described by Professor Lahaye in the discussion following his paper? It looked as though the center was cut out of a ball of ribbon so all you could see were a couple of circles, of hollow spheres. You do not see the carbon black type picture of which we have seen so many examples.

Howard

We have seen some evidence of internal burning in spherules, but not routinely or consistently. Therefore I must admit that the extent to which these spherules react internally is not quantitatively known. It would be interesting to know whether there is some internal area available within the spherules before the burnout starts. Another kind of burning is reaction on the outer surface of the spherules located deep within large aggregates of spherules. An OH radical having three to ten collisions within the aggregate before reaction occurs can penetrate to some depth into the aggregate. Much less reactive O_2 would be expected to penetrate to the remote regions of even the largest aggregate under many flame conditions of practical interest. Such behavior presumably leads to a hollowing out, so to speak, of the aggregate.

Stevenson

Let me rephrase the question. Is there a characteristic morphological change that you can associate with oxidation? In other words, if I were to see in one of my particles some characteristic morphology can I immediately say that the particle has undergone oxidation?

Howard

Yes. A decrease in the size of the spherules sampled at increasing times of oxidation or any hollowing out of spherules would be indicative of oxidation. Also, the occurrence of relatively small (say 10 nm diameter) spherules separated from accompanying aggregates would, according to our experience, imply not only oxidation but a substantial extent (perhaps 80%) of burnout.

H. Gg. Wagner *(Universität Göttingen)*

You mentioned that you did surface determination and chemical analysis during this process. Would you mind telling us a little bit about the results of the chemical analysis? How much hydrogen do you lose in the particles?

Howard

Unfortunately, too few chemical analyses have been performed to provide the basis for a statement here. With regard to the surface analysis (which was performed with gas adsorption) we do see evidence either of surface area being available within the spherules or of capillary condensation in regions between adjacent spherules. The apparent surface area is in some cases as much as a factor of 2 or so larger than that corresponding to the superficial area of the spherules, but because of the unknown extent of capillary condensation we cannot conclude from these data alone that the spherules possess accessible pores.

L. C. Yang *(Jet Propulsion Laboratory)*

I have a very simple question. Did you happen to record the flow speed in your system: in other words, what was the residence time of the particles in your second flame?

Howard

The velocity was calculated. One millimeter of distance above the burner is equivalent to about 0.6 milliseconds, depending on flame conditions. So our probing distance of about 10 millimeters corresponds to about 6 milliseconds. Observed burnout times for the more fuel lean conditions were under one millisecond.

A. I. Medalia *(Cabot Corporation)*

I was very interested in the morphological question. If you study oxidation at comparatively low temperatures where you can get reactions in minutes or longer — say 400 to 500°C, no higher — then you get this hollowing out or sponginess, depending on the size. So the indication is there that the rate of oxidation is controlled by the imperfection of the layers. Maybe at the higher temperature then the rate-determining step is simply diffusion of the OH to the surface. Wherever it hits, it reacts and doesn't care how reactive the site itself is. Do you really then see oxidized "aciniform" carbon which is not spongy, is not hollowed out?

Howard

As you go from the fuel-rich to more lean conditions, that is, as more and more oxygen is present, then the contribution of O_2 becomes increasingly important. Since oxidation by O_2 is much less efficient than oxidation by OH, this trend leads to an increasing tendency toward hollowing out. At the other extreme of our most fuel-rich condition where OH seems to be the main oxidizing agent, there is a tendency to see less hollowing out. It is important here to mention the break-up phenomenon along with the hollowing out which has been asked about. In going over the range of conditions, giving an increased role of oxidation by O_2, we definitely see an increase in the amount of breakup occurring. As was pointed out, breakup always occurred under fuel-rich conditions but not under lean conditions.

I. Glassman *(Princeton University)*

Jack, in the heyday of the aerospace field we dealt with ram rockets, the afterburning of rockets. When carbon graphite was put in the rocket motor it was impossible to afterburn it. But when methylacetylene was decomposed in a mono-propellant rocket you could afterburn the soot formed quickly and readily. At that time I came to the conclusion that it was the presence of the hydrogen within the soot itself that played a role in the oxidation process. Shouldn't that come under consideration in the burning of these particles rather than just the hydroxyl attack on the carbon itself? I think it's something we may be overlooking — just the hydrogen content itself is playing a role in this oxidation phase in these particles.

Howard

Soot or other carbonaceous solids, such as chars and cokes, become less reactive as graphitization, and hence hydrogen elimination, proceeds. It is not clear that the extent of graphitization of soot in the short residence times available in flames is large enough to exert a significant effect on soot reactivity, but the question would merit attention.

K. Otto (Ford Motor Company)

If you calculate the oxidation rate per surface area, you find that soot from diesel engines, for example, is two to three orders more reactive than graphite, and this is not only because you may have some catalysts in there. In fact, as burnoff — or burnout, as you call it — proceeds, the reaction rate per surface unit goes down; so there is considerable complexity in defining the reaction rate for the carbon because the carbon seems to change as the combustion continues.

Howard

More reactive to what?

Otto

As burnout proceeds the soot moves toward graphite. I have observed, for example, that it increases by one order of magnitude in its reactivity calculated per square meter of surface area.

Howard

This is in a diesel engine?

Otto

No, these are deposits of soot which are derived from diesel engines.

Howard

If one is dealing with reactivity in situations where OH and O are large contributors, then the difference between different materials would not be as large as would occur if O_2 were the main oxidizing species. With regard to the decrease in reactivity per unit surface area as burnout proceeds, it is possible that an experiment with a captive sample in a furnace could allow the occurrence of graphitization and an associated reduction in reactivity.

SESSION III
FORMATION DURING
CONTINUOUS COMBUSTION

Session Chairman
J. P. LONGWELL
Massachusetts Institute of Technology
Cambridge, Massachusetts

SOOT FORMATION IN JET-STIRRED REACTORS

R. M. KOWALIK, L. A. RUTH and W. S. BLAZOWSKI

Exxon Research and Engineering Company
Linden, New Jersey

ABSTRACT

Initial soot production data from a liquid fuel spray injected jet-stirred reactor are described and compared with previous data obtained from gas phase stirred reactors. Significant sooting differences observed for the two methods of fuel injection appear to be attributable to the pyrolysis, vaporization, and burning of liquid droplets within the spray injected reactor. Overall results imply that both chemical (structure) and physical (injection characteristics) fuel parameters can significantly affect the production of soot in stirred reactors

INTRODUCTION

Exxon Research is conducting, for the U. S. Department of Energy, a research program to characterize the effects of alternate fuels in continuous combustion systems. A major part of this work has been to study soot formation in highly backmixed combustion. Our primary goal is to obtain engineering data for the development of models which can predict the degree of soot formation in gas turbine combustors burning alternate fuels [1].

The overall experimental approach of the program has been to obtain soot formation data from the combustion of various pure hydrocarbons and hydrocarbon blends in jet-stirred reactors. The hydrocarbons have been specifically selected to investigate the effects of the various fuel components expected in future broad specification and synthetic fuels. The jet-stirred reactor provides a highly backmixed combustion environment which simulates the primary zone of a gas turbine combustor, and, to a first approximation, it produces spatially uniform species' concentrations and temperature distributions which allows one to characterize the reactor with a single set of species' concentration and temperature measurements.

Previous soot formation experiments in stirred reactors have been conducted

References p. 294.

with prevaporized fuels, premixed with oxidizer (air). Practical combustors, however, generally operate with separately injected fuel and oxidizer, and the fuel is usually injected as a liquid spray. Consequently, we have recently employed a jet-stirred reactor designed to accommodate the injection of liquid fuel sprays to investigate the sooting differences between the two types of fuel injection in highly backmixed stirred reactors.

This paper reports and discusses results obtained from liquid fuel spray experiments. We first review previous results from this and other studies of soot formation in stirred reactors. The liquid fuel jet-stirred reactor is then described. Recent data from this combustor are presented next, and a discussion of these results, including comparisons with previous stirred reactor results using prevaporized, premixed fuels concludes the paper.

REVIEW OF PREVIOUS WORK

Soot formation was first studied in jet-stirred reactors by Wright [2, 3]. His results for prevaporized, premixed fuels indicated that the onset of soot formation, reported as a critical O/C ratio, was not very different for premixed flames and jet-stirred reactors; the critical O/C ratio was slightly smaller in the jet-stirred reactor. For some fuels, however, quantitative measurements of soot concentrations in a premixed flame and a jet-stirred reactor indicated that soot concentrations were much smaller in the stirred reactor. Wright claimed that this large reduction of soot emissions could be attributed to dilution effects associated with the backmixing of products and reactants in stirred reactors.

Blazowski [4, 5] obtained additional data in a similar jet-stirred reactor and classified hydrocarbons into three general sooting categories. Category 1 consisted of principally aliphatic fuels. With these fuels, the reactor experienced rich blow-out before measurable quantities of soot could be obtained. Significant concentrations of unburned hydrocarbons (>1% as methane) were also observed before the onset of soot formation. Category 2 contained principally single-ring aromatic fuels. These fuels produced measurable quantities of soot for equivalence ratios greater than or equal to the equivalence ratio at which appreciable concentrations of unburned hydrocarbons (>0.1% as methane) could be detected in the exhaust gases (hydrocarbon breakthrough). The magnitudes of both soot and unburned total hydrocarbon (THC) concentrations from these fuels were found to be larger for larger equivalence ratios, lower temperatures, and shorter average residence times within the reactor. The common dependence of soot and THC concentrations on these parameters suggested that the THC concentration might be a key factor in the prediction of soot concentrations from Category 2 fuels in gas phase stirred reactors. Edelman et al. [6] used this idea to develop quasi-global rate expressions for the production of soot from toluene in gas phase stirred reactors.

Blazowski also investigated the sooting characteristics of 1-methylnaphthalene (a double ring aromatic) and several isooctane/toluene blends. 1-methylnaphthalene had sooting characteristics which were similar to the sooting characteristics of Category 2 fuels; however, corresponding soot concentrations from 1-methylnaphthalene were significantly larger at similar reactor conditions. Blends of isooctane (Category 1) and toluene (Category 2) produced soot and THC concentrations which appeared to represent combinations of the Category 1 and Category 2 sooting characteristics. These data suggested that soot formation models for each category might be combined to predict the soot produced from various fuel blends.

EXPERIMENTAL APPARATUS

The device used to investigate the effects of liquid spray injection of fuel on soot formation in jet-stirred reactors was Exxon's Liquid Fuel Jet-Stirred Combustor (LFJSC). The combustor is shown schematically in Fig. 1. It contains a spherical jet-stirred zone and a cylindrical plug flow zone. The spherical portion consists of two castable refractory halves housed within a 15.2 cm diameter metal shell. The refractory material is Castable 141A, a product of Combustion Engineering Refractories, which is capable of withstanding temperatures up to 2140 K. Each section is cast separately with a 5.1 cm diameter hemispherical volume — the two sections together create the spherical reactor zone. The cylindrical plug flow section is composed of a 7.6 cm long by 2.2 cm diameter alumina tube.

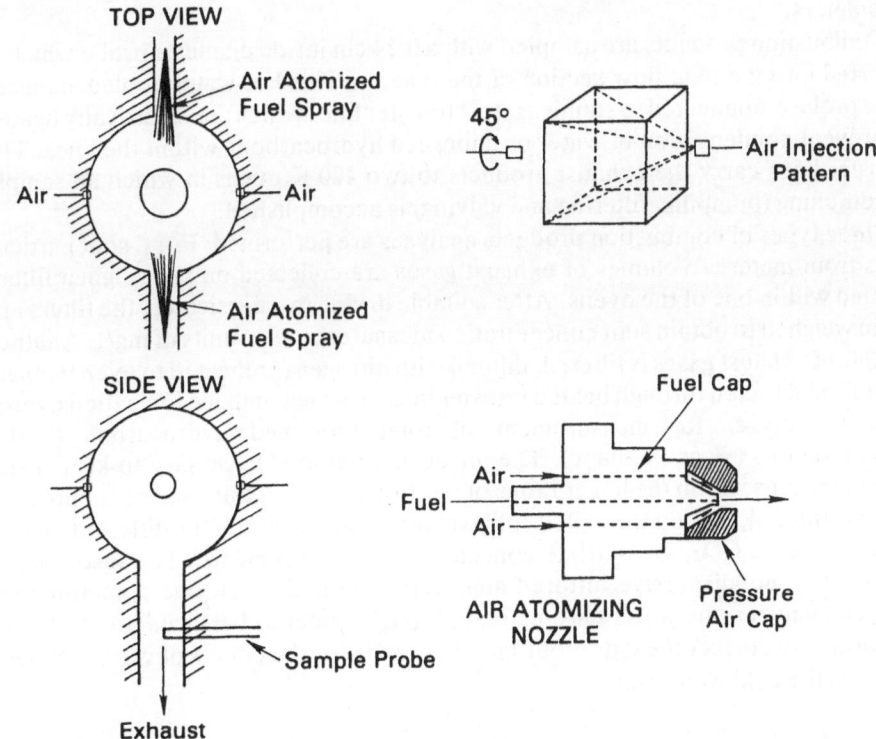

Fig. 1. Liquid fuel jet-stirred combustor.

Combustion air enters the spherical reaction zone through two sets of jets positioned 180° apart. Four 1.1 mm diameter air jets in each side of the reactor (total of eight) are aimed towards the corners of a cube imagined to sit within the spherical reactor (see Fig. 1). One set of air jets is rotated 45° with respect to the other to allow the opposing jets to mesh rather than collide. This approach provides for highly turbulent mixing and simulates the recirculating characteristics of gas turbine systems.

Fuel sprays enter the reactor at two positions 90° from each set of air jets. Air atomizing nozzles obtained from Spraying Systems Co. (1/4 J Series) produce the fuel sprays; nozzle assemblies include Model 1650 (0.04 cm inside diameter) fluid caps and Model 67147 (pressure) air caps. The nozzles are positioned approximately 7 cm from the center of the spherical section of the reactor; cylindrical cavities (2.3 cm diameter by 4.4 cm long) cast in the refractory contain the developing fuel sprays.

The LFJSC operates at atmospheric pressure. Estimates of the temperature within the reactor are obtained from a platinum-6% rhodium/platinum-30% rhodium thermocouple positioned at the center of the spherical reactor section; the temperature within the reactor is controlled by varying the concentration of oxygen in the main combustion "air" or main combustion and atomizing "air" flows. Metering of all air, oxygen and liquid fuel flows is accomplished using calibrated rotameters.

Combustion products are sampled with a 0.24 cm inside diameter probe which is inserted into the plug flow section of the reactor. This hot-water cooled stainless steel probe is connected to stainless steel transfer lines which are electrically heated to prevent condensation of water or unburned hydrocarbons within the lines. The transfer lines carry the exhaust products to two 420 K ovens in which all sample conditioning (pumping, filtering and valving) is accomplished.

Three types of combustion products analyses are performed. First, soot particulates from metered volumes of exhaust gases are collected on preweighed filters located within one of the ovens. After suitable drying in a dessicator, the filters are again weighed to obtain soot concentration measurements in units of mg/l. Another sample of exhaust gases is filtered, diluted with nitrogen (preheated to oven temperature) and directed through heated transfer lines to a heated flame ionization hydrocarbon analyzer for measurement of total unburned hydrocarbon (THC) concentrations (% as methane). The nitrogen dilution is necessary to keep THC concentrations within the linear range of our instrument. A third sample of product gases is filtered, passed through a cold water trap and sent to four different instruments for CO, CO_2, O_2 and H_2 concentration measurements. The instruments include two nondispersive infrared analyzers (CO and CO_2), one amperometric oxygen analyzer and a gas chromatograph (H_2). Material balances are routinely performed to correct the latter four measurements for the effects of water condensation in the cold water trap.

EXPERIMENTAL RESULTS

Two fuels have been studied in the LFJSC: isooctane (representing Category 1) and toluene (representing Category 2). Several sets of experiments have been run with each fuel to investigate the influence of fuel type, equivalence ratio, atomization air flow rate, and temperature on soot formation in the LFJSC. The average residence time of gases within the reactor (volume/volumetric flow rate at reactor temperature) was approximately 6 ms for all experiments; corresponding total "air" flow rates were approximately 200 g/min.

Fig. 2 displays soot concentrations obtained at various equivalence ratios. Atomization air flow rates (no added oxygen) and indicated reactor temperatures were held constant during these runs; temperature was controlled by varying the oxygen concentration in the main "air" jets. The error bars indicate the typical spread among repeated measurements. Toluene data obtained by Blazowski [7] with a different jet-stirred combustor that used prevaporized, premixed fuels are also included in the figure for comparison.

The effects of atomization air flow rates on soot production within the LFJSC are shown in Fig. 3. Data for each fuel were obtained at constant equivalence ratio, temperature, and total "air" (air + O$_2$) flow rates. Temperature was again controlled by varying the oxygen concentration in the main "air" jets; within each set of experiments (isooctane or toluene), overall oxygen concentrations were equal. No data are presently available to characterize the toluene and isooctane sprays at these flow conditions; however, some limited data from similar Jet A/nitrogen and

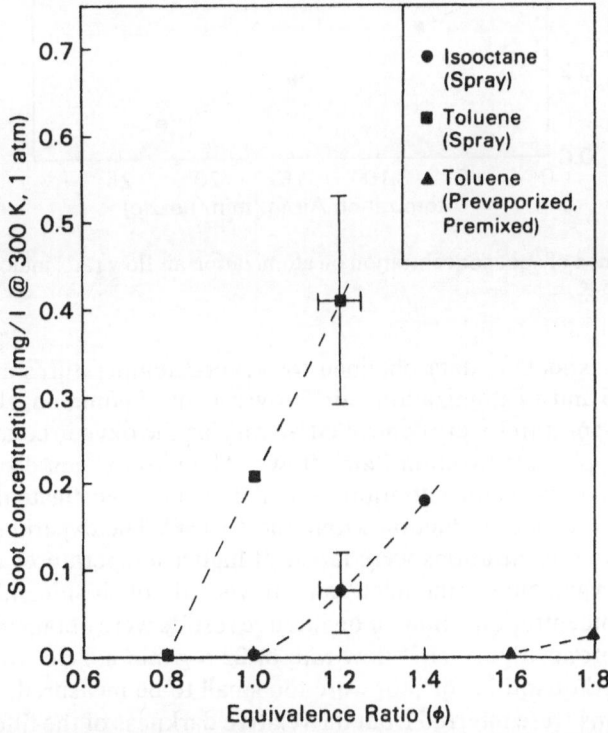

Fig. 2. Dependence of soot concentration on equivalence ratio; atomization air = 15 g/min/nozzle; indicated temperature = 1900 K $\pm$ 35 K.

References p. 294.

JP-10/nitrogen sprays in an approximately quiescent, room temperature atmosphere have been obtained with a Malvern Particle and Droplet Sizer (Model ST 1800). These data indicated that Sauter Mean Diameters (SMD) for the Jet A and JP-10 sprays were of the order of $30\mu m$ and that smaller diameters corresponded to larger atomization nitrogen flow rates (fixed fuel flow rate).

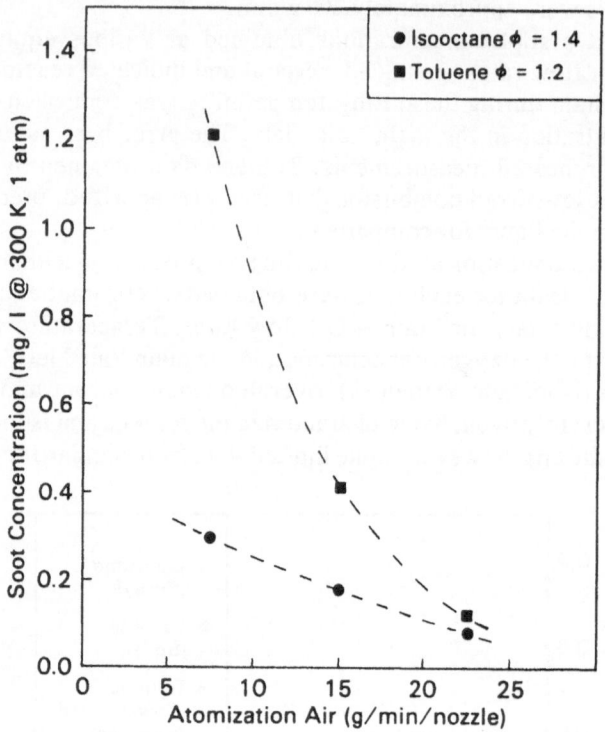

Fig. 3. Dependence of soot concentration on atomization air flow rate; indicated temperature = 1900 K ± 35 K.

Fig. 4 displays isooctane data obtained for several temperatures at an equivalence ratio of 1.45 and an atomization "air" flow rate of 15 g/min/nozzle. For these experiments, temperatures were controlled by varying the oxygen concentration in the main or main plus atomization "air" flows. The two oxygen distributions at each fixed overall oxygen concentration were included to investigate the effects of gas phase mixing on soot production within the LFJSC. The experimental results indicate that soot concentrations were larger at higher temperatures and that the distribution of oxygen among the inlet "air" flows did not significantly affect the measured soot concentrations. Similar qualitative results were obtained with isooctane at a higher atomization "air" flow rate of 22.5 g/min/nozzle. In these runs, however, collected quantities of soot were too small to be measured, and relative soot concentrations were inferred from the relative darkness of the filters.

Attempts to obtain variable temperature data for toluene sprays were not as successful. Large spreads in the data, similar to those shown in Fig. 2, were observed for these runs such that differences in the soot concentrations measured at several temperatures were not large compared to the differences observed among repeated measurements at fixed temperatures. After some toluene runs, coke deposits were also found in the reactor near the cylindrical surface where the spray cavities border the spherical zone of the reactor. Consequently, the effect of temperature on toluene soot production in the LFJSC could not be determined from the available experimental data.

One additional set of experiments was conducted in the LFJSC with gaseous ethylene fuel and air (no added oxygen). The objective of these experiments was to investigate the effects of gas phase mixing on combustion within the LFJSC. Mixing in the combustor was varied by either premixing ethylene with air entering through the main air jets or separately injecting ethylene through the atomization "air" passages of the fuel nozzles. At average residence times (τ) of approximately 7 ms, differences between measured major species' (CO, CO$_2$, O$_2$, H$_2$, THC) concentrations and indicated temperatures were small. Table 1 presents specific data for an equivalence ratio (ϕ) of 2.4, the largest equivalence ratio tested. At the conditions tested ($\phi \leqslant 2.4$, $\tau \approx 7$ ms), filters in the sampling system did not show any evidence of soot deposition for either method of ethylene injection.

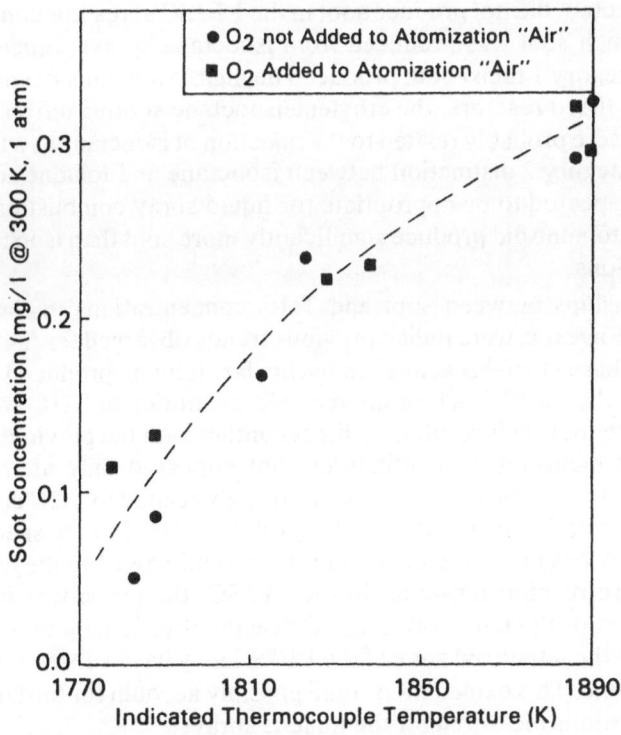

Fig. 4. Dependence of soot concentration on temperature for isooctane, $\phi = 1.45$, atomization "air" = 15 g/min/nozzle.

TABLE 1
Gas Concentrations for Ethylene/Air Mixing Experiments
in the Liquid Fuel Jet-Stirred Combustor.*

Configuration	Indicated Temperature	Concentrations (mole %)				
		CO	CO_2	O_2	H_2	THC (as CH_4)
Separately injected	1473	16.2	3.5	0.1	10.7	8.9
Premixed	1478	15.3	4.3	0.3	10.9	11.0

*Equivalence ratio = 2.4; total air flow rate = 200 g/min.

DISCUSSION

The initial data obtained from the the LFJSC suggest that the production of soot in jet-stirred reactors is very dependent upon the method of fuel injection, i.e., whether fuel enters prevaporized and premixed or as a liquid spray. Toluene and isooctane both produced significantly more soot when they were injected as liquid sprays, and the amount of soot produced by either fuel was strongly related to the fuel spray characteristics inferred from atomization "air" flow rates, i.e., more soot with larger diameter droplets. Also, gaseous ethylene, premixed with air or separately injected, did not produce soot in the LFJSC at reactor conditions similar to those at which soot was produced from isooctane sprays. Since ethylene and isooctane (Category 1 fuels) both produced negligible amounts of soot in premixed gas phase jet-stirred reactors, the ethylene/isooctane sooting differences obtained in the LFJSC were probably related to the injection of isooctane as a liquid spray. A Category 1/Category 2 distinction between isooctane and toluene sooting characteristics still appeared to be appropriate for liquid spray combustion in jet-stirred reactors since toluene did produce significantly more soot than isooctane at similar reactor conditions.

The relationships between soot and THC concentrations in the liquid spray experiments, however, were unlike previous trends observed for the two fuel categories in gas phase stirred reactors. In particular, toluene produced larger quantities of soot in the LFJSC before appreciable quantities of THC were measured ($>0.1\%$ as methane). This result is in direct conflict with the previous trends which suggested that measurable quantities of soot appeared only after THC breakthrough and that soot concentrations were strongly related to THC concentrations. The apparently conflicting trends may be partially related to the specific locations of the sample probes in each reactor. A probe was inserted into the jet-stirred zone of the gas phase reactor; however, in the LFJSC, the probe was inserted at the downstream end of the plug flow zone. Consequently, some unburned hydrocarbons present in the jet-stirred zone of the LFJSC may have subsequently burned in the plug flow zone. This explanation could possibly account for the large soot/small THC concentrations measured for the toluene sprays.

Isooctane produced significant ($>1.0\%$ as methane) THC concentrations for both types of fuel injection, but soot was measured only when isooctane was injected as a liquid spray. For the conditions tested in the LFJSC, soot concentrations appeared to be related to THC concentrations since both soot and THC concentrations increased with increasing equivalence ratio or decreasing atomization air flow rates (fixed temperature). In constant equivalence ratio and atomization "air" flow rate experiments, however, temperature did not appear to significantly affect the THC concentrations over the limited range investigated. This trend was unlike Blazowski's [7] findings that THC concentrations decreased with increasing temperature.

The effects of temperature on soot production in stirred reactors appeared to be exactly opposite for the two types of fuel injection. Blazowski's [7] previous temperature trend for prevaporized toluene/"air" mixtures was in agreement with expectations for premixed flames (less soot at higher temperatures), but the temperature trend reported here for isooctane sprays indicated that a significant portion of the fuel produced soot like a diffusion flame, i.e. more soot at higher temperatures [8, 9]. These trends suggested that fuel and oxidizer were not well-mixed in the LFJSC, despite the highly turbulent backmixing.

This unmixedness was probably related to the presence of liquid fuel droplets in the LFJSC. Data from the various atomization air flow rate conditions suggested that the characteristic droplet diameters of the sprays significantly affected both the soot and THC concentration levels, and results from the two sets of mixing experiments conducted in the LFJSC did not suggest any significant gas phase mixing effects at the conditions tested. In particular, variations of the oxygen distribution between main and atomization "air" flows and variations of the method of gaseous fuel injection did not appear to significantly alter the combustion product concentrations or the indicated temperatures. Consequently, gas phase mixing did not appear to significantly affect the soot production mechanisms within the LFJSC, and any dependence of soot production on unmixedness was probably related to the pyrolysis, vaporization, and burning of liquid fuel droplets within the combustor.

SUMMARY

The collection of results from the jet-stirred reactors suggest that both chemical (structure) and physical (injection characteristics) fuel parameters can significantly affect the production of soot within stirred reactors. When fuel is injected into a jet-stirred reactor as a liquid spray, fuel and oxidizer within the reactor may not be well-mixed. Corresponding sooting trends resemble those obtained in diffusion flames and are probably attributable to the pyrolysis, vaporization, and burning of liquid fuel droplets. Additional data, covering a wider range of fuels, spray characteristics, and reactor conditions, are highly desirable.

ACKNOWLEDGMENTS

The authors gratefully acknowledge the funding provided for this work by the U.S. Department of Energy under Contract DE-AC22-77-ET-11313 and the efforts of R. J. Kuriskin, who obtained the various experimental data reported in this paper.

References p. 294.

REFERENCES

1. R.C. Farmer, R. Edelman and E. Wong, *"Particulate Carbon-Formation During Combustion,"* Ed. D. C. Siegla and G. W. Smith, Plenum Press, New York, this volume, page 299.

2. F. J. Wright, *"Twelfth Symposium (International) on Combustion,"* The Combustion Institute, Pittsburgh, (1968), pp. 867-875.

3. F. J. Wright, Combust. Flame, Vol. 15 (1970), pp. 217-222.

4. W. S. Blazowski, Combust. Sci. Technol., Vol. 21 (1980), pp. 87-96.

5. W. S. Blazowski, ASME Paper No. 79-GT-155, (1979).

6. R. B. Edelman, A. Turan, P. T. Harsha, E. Wong and W. S. Blazowski, AGARD Conference Proceedings No. 275 (Combustor Modeling), (1979).

7. W. S. Blazowski, R. Edelman and E. Wong, Summary Technical Progress Report (15 August, 1978-31 January, 1980), U. S. Department of Energy Contract No. DE-ACO3-77-ET-11313, (February, 1980).

8. I. Glassman, Princeton University, Dept. of Mechanical and Aerospace Engineering Report No. 1450, Princeton, NJ, (July 1979).

9. I. Glassman and P. Yaccarino, *"Eighteenth Symposium (International) on Combustion,"* The Combustion Institute, Pittsburgh, to be published in 1981.

DISCUSSION

A. Levy *(Battelle Columbus Laboratories)*

To what extent are your droplets vaporized? Is their residence time long enough in the reactor?

Kowalik

The fuel sprays are recessed slightly from the reaction zone. We did some quick estimates and found that the droplets might reside in the recessed zone on the order of a millisecond and that smaller droplets may even vaporize before they enter the reaction zone. With a six-millisecond residence time in the reaction zone we have to have some fairly large droplets to remain in the liquid form so I think we are reasonably well vaporized.

I. Glassman *(Princeton University)*

Although I think this stirred reactor gives you a good sample of what happens in a gas turbine, I have to question whether it can give you results that will be related to other elements although they will look similar such as to predict the equivalence ratio to sooting. This is because I think we have learned at this meeting and others that the amount of soot that comes out of a combustion system is a function of two different variables, the rate soot is formed and the rate soot is oxidized. In this stirred reaction system, you are giving additional residence time within the combustion zone and therefore you have a different burnout time. I really doubt whether when you go to your bigger stirred reactor, that you will get the same correlation. If you use the parameter Mr. Longwell developed, m/Vp^2, and worry about the residence time of that fluid mass and vary that residence time by adjusting other parameters, I do not think you would get the exact same results.

Kowalik

Do I interpret your question to mean that mixing within the reactor might be suspect when you go to larger volumes?

Glassman

The time the soot particle stays within the reactor at reactor temperature.

Kowalik

It seems to me that if you increase the volume and if you increase your flow rate, you should be able to keep your residence times comparable.

Glassman

Well you can, that is what I am arguing. But if you change the residence time at a given equivalence ratio, you are going to get different soot concentrations.

Kowalik

That is correct. We have done some work showing that this is true in the premixed gas phase, and we just have not done it with the liquid phase.

Glassman

What is the significance of the sooting at a given equivalence ratio if it is dependent upon the other parameters of the system?

Kowalik

In the present set of experiments, the equivalence ratio is an overall equivalence ratio which describes the relative amounts of fuel and oxidizer within the reactor. It is not a fundamental parameter which can be directly compared to equivalence ratios in premixed laminar flames.

R. G. Gann (*National Bureau of Standards*)

In the small reactor, you mentioned that you could make one temperature measurement, and it was a pretty good one, that is in the gas-phase reactor. In this liquid-phase reactor do you have either significant time variation of temperature at the wall or is it much colder than the rest of the volume?

Kowalik

We have done some coarse profile measurements, and the temperature does not vary by much more than a hundred degrees Celsius as you traverse the reactor volume.

Gann

In the gas phase? But what about the wall zone?

Kowalik

I think the wall temperature is close to the gas phase temperature. We have a very good estimate of wall temperature.

Gann

And that stays steadily up there?

Kowalik

I am not sure what you mean by that — stays steadily?

Gann

There is not time during your experiment when condensation or a much, much lower temperature than this might exist in a small zone?

Kowalik

It is a steady-state system, and we preheat it usually with ethylene fuel.

Gann (*National Bureau of Standards*)

Then it is preheated? That is fine.

Kowalik

Yes.

N. A. Chigier (*University of Sheffield*)

In studying the effect of atomization you used an air-assist system, and you showed this very interesting reduction in soot concentration when you increased the atomizing air. There are, however, two effects of increasing the atomizing air. One is you reduce the diameter of the droplets, and the other is you are adding more atomizing air to the central region of the spray. One way of trying to separate these effects is by looking at a pressure jet spray which changes only the diameter of the droplets. I think it would be very interesting to see separately the effect of changing the droplet diameter on soot formation and the separate effect of adding air in the central region of the spray which changes the local air-fuel mixture ratio.

Kowalik

I agree. We have not been able to run those experiments as yet.

R. L. Bradow *(Environmental Protection Agency)*

It occurs to me that some of the value of the well-stirred reactor studies in the past has been a contribution to global kinetic properties, relative to the formation of soot. This would be a very valuable commodity simply from the vapor-phase studies. Are any of these published yet? Do you have global activiation energies or kinetic data for formation of soot in premixed systems?

Kowalik

I think there are some, and I think that question might better be directed toward the authors of the next paper since they have been collaborating with us on the modeling aspects of this work.

D. B. Olson *(AeroChem Laboratories)*

We have recently been trying to deduce and understand the effect of molecular structure on sooting, and we have analyzed a fairly extensive amount of literature data from diffusion flames and from premixed flames on soot points, smoke points and known critical equivalence ratios when soot is formed. In this regard I would like to make two comments. One is that in diffusion flames, dividing the fuels into families turns out to work very well. Under premixed conditions the classifications of fuels turns out to be much, much weaker. The vapor-phase, well-stirred reactor results turn out to correlate extremely well with the results from premixed flames. It gets very close to the same soot points as premixed flames. The second point is that I have analyzed some flame radiation data using that as an indication of the amount of soot using the nozzle-spray-fired, two-inch combustor data from Southwest Research Institute. Looking at the effect of the structure of the fuel on smoke, that is flame radiation, I find that by using diffusion flame smoke points the fuel data correlates the burner data much better than premixed flame data. This corresponds to what I think you are finding; that the spray fuel combustor performs much more like a diffusion flame than a premixed flame in its smoke effect.

Kowalik

I agree with that.

MODELING SOOT EMISSIONS IN COMBUSTION SYSTEMS

R. FARMER, R. EDELMAN and E. WONG

Science Applications, Inc.
Canoga Park, California

ABSTRACT

A comprehensive model for the characterization of the combustion kinetics of hydrocarbon fuels is described. The model framework is based upon the quasi-global concept and extensions appropriate for fuel rich oxidation. Soot formation and consumption processes are considered as an integral part of the model, and pyrolysis, partial oxidation, NO_x formation and detailed reaction steps for the path to completion of reaction are included. The model is being developed for wide ranges of application where both conventional and synfuels are of interest. Examples are given for toluene combustion and favorable comparisons between predicted and measured soot emissions demonstrate the utility of the model in dealing with the complexities of fuel-rich combustion.

INTRODUCTION

The requirement for more efficient utilization of conventional energy resources, the advent of synthetic fuels and the continually increasing demands for environmental compatibility have intensified the need for a more complete understanding of combustion processes in practical systems. The development of a model which would synthesize the many experimental observations of combustion phenomena into an ordered description of both the major and trace products of combustion and of the accompanying heat release rates would demonstrate the necessary understanding.

This paper describes soot emissions models in the context of the accompanying combustion processes. Lean, hot combustion does not produce soot; rich, cool combustion and pyrolysis does. Also, various fuels have markedly different sooting characteristics with aromatics being of particular concern because of the advent of synthetic fuels. These qualitative observations imply that rich combustion of hydrocarbon molecules typical of real fuels must be characterized before a practical soot formation model can be devised.

Interest in soot formation and consumption ranges from trace emissions in highly efficient combustors to carbon black production and from homogeneous nucleation in clean flames to coking of highly reactive catalytic surfaces. Stationary and mobile power plants including furnaces, boilers, gas turbines and diesel engines are specific examples of current interest. Besides the amount of soot formed, one also needs to know the temperature and the major chemical species present to determine the thermal efficiency of the combustor and other pollutant emissions such as NO_x.

Despite the large number of experiments and industrial observations, few methods have been developed which estimate the rate of soot formation. The major reason for this deficiency is that seldom have analytical and experimental studies been coordinated closely enough for all of the necessary experimental data to have been collected to fully determine the species, energy, and momentum transport of the system. Even fewer of the available methods are suitably structured to be included in an engineering analysis of the entire combustor system.

The purpose of this paper is to formulate a practical soot emissions model and to illustrate the application of the model to an adequately measured combustion process. Since net soot emission is controlled by chemical kinetics and by fluid mechanics, these effects must be studied separately before an overall model or understanding of their coupling can be obtained. This work deals only with the kinetics of the combustion process in general and of soot formation and consumption in particular.

THE STATE-OF-THE-ART

The vast literature on soot emissions has been ably reviewed on numerous occasions, most recently by Wagner [1]. Additional reviews by Glassman [2] and Tesner [3] should also be noted. Although many useful observations have been made, no generally applicable soot emissions model has yet been devised. A few correlation equations for the rate of soot formation and consumption have appeared, some of which will be referred to subsequently, but they are too experiment dependent and fuel specific to be generalized.

The concept of the model developed by Jensen [4] contains all of the necessary features to make it a useful point of departure. He suggested that active nuclei form chemically from combustion gases, these nuclei condense, grow, coagulate, and oxidize. These are the necessary steps to form and gasify soot; the difficulty is to quantify the steps. Jensen considered the relevant nuclei to be $C_2H\cdot$, $C_2\cdot$, and $C_3\cdot$ radicals in his study of methane combustion. An elementary set of pyrolysis reactions was used to generate these species. Tesner *et al.* [5, 6, 7] treated the nuclei as being an unspecified radical which is produced by a first order decomposition of the fuel. The choice of these active nuclei in these analyses was arbitrary; the species mentioned were present in concentrations high enough to exceed the critical supersaturation limit. Additional constraints are placed on their selection when the condensation process is described. Tesner [5] greatly simplified the nuclei formation step by assuming a global, branched-chain process which he described by

$$\frac{dn}{dt} = n_0 + (f\text{-}g)n - g_0 Nn \tag{1}$$

where n is the concentration of the nuclei forming radical,
  n_0 is the rate of production of this radical,
  f is the linear chain-branching coefficient,
  g is the linear chain-breaking coefficient,
  g_0 is the quadratic chain-breaking coefficient,
  N is the number density of soot particles.
Coupling to the gas phase combustion process was accomplished by including the concentration of the unburned fuel molecules (C_F) in a rate equation for nuclei formation,

$$n_0 = A \star C_F \exp \left\{ -E/RT \right\} \tag{2}$$

where A is the frequency of the oscillation of atoms along the bond being broken,
  C_F is the concentration of fuel molecules,
  E is an activation energy,
  R is the gas constant,
  T is the temperature.
  Because of the type, concentration, and size of the condensing nuclei, the size of the particle formed, and electrical charges which are present all strongly affect the condensation process, the type of species which condense have been very speculative. Prado and Howard [8] measured large positive hydrocarbon ions in flames and postulated that they were nuclei. Lahaye and Prado [9] suggested that the nuclei are macromolecules which condense to liquid droplets then form soot. Jinno et al. [10] included both of these ideas within their thermodynamic analysis of nucleation. The idea of the particles being liquid droplets has been criticized because not enough large molecules are believed to be present to cause condensation at flame conditions [3]. Jensen notes that the gas phase nuclei may condense directly to a solid soot particle. Tesner proposed a rate equation for the particle forming step and evaluated the parameters in his equations with experimental data for both acetylene [6] and aromatic [7] fuels, viz.

$$\frac{dN}{dt} = (a\text{-}bN)n \tag{3}$$

where N is the concentration of soot particles
  a, b are the constant parameters
This step is identical to one treated in the two-phase combustion problem with finite-rate condensation treated by Genovese, Edelman and Fortune [11].

Growth of the nuclei were estimated by Jensen with the model of active growth species sticking to nuclei. The growth species were arbitrarily specified to be the nuclei species and acetylene.

Narasimhan and Foster [12] reported a correlation equation for the rate at which the soot number density increased by adding methane in a methane/air flame. This rate depended upon the local temperature, methane concentration, soot concentration and soot number density.

Conglomeration was treated by a sticking together of particles; the extent to which these particles then coagulate and coalesce is not well known. Jensen accounted for agglomeration and growth by a discrete distribution of particle sizes. This use of discrete increments is the most unwieldy feature of this model. It makes the soot emissions combersome to use as an element of a more general combustor model.

Magnussen and Hjertager [13] used a soot formation model of the Tesner-type to analyze a turbulent combusting flow. Even though particle size effects were not treated, the model is complex. Furthermore, only the limiting case of infinitely fast reactions, with respect to the turbulent mixing, was treated.

Khan and Greeves [14] presented a soot correlation equation for high-pressure diesel engine data. Parameters in this model were local values of temperature, unburned hydrocarbon concentration, and the equivalence ratio. A corresponding combustion scheme was not suggested. Edelman *et al.* [15] showed that this model was not suitable for extrapolation to other conditions than those upon which it was based.

The oxidation of carbon has been modeled by several investigators. The models of Lee, Thring, and Beer [16] and Nagle and Strickland-Constable [17] are of particular interest because rate constants were determined from soot or soot-like combustion experiments. Since the model of Lee *et al.* can be deduced from the somewhat more general model of Nagle and Strickland-Constable, only the more general model will be discussed. The rate of soot oxidation per unit area of soot surface is as follows:

$$R_{ox} = 12 \left[\left(\frac{k_A P_{O_2}}{1 + k_Z P_{O_2}} \right) \chi + k_B P_{O_2} (1 - \chi) \right] \quad [gm/cm^2 \cdot sec] \qquad (4)$$

where

$$\chi = \left[1 + k_T \Big/ (k_B P_{O_2}) \right]^{-1}$$

$$k_A = 20 \exp(-30,000/RT) \qquad k_T = 1.51 \times 10^5 \exp(-97,000/RT)$$

$$k_B = 4.46 \times 10^{-3} \exp(-15,200/RT) \qquad k_Z = 21.3 \exp(4100/RT)$$

Although oxidation is the primary way in which soot is consumed, gasification by soot reactions with steam, CO_2, OH and other species is also possible.

In summary, the status of soot emissions models is outlined. The critical active species and reactions are not identified for any fuel, much less current commercial fuels or possible synfuels. The models reviewed are typically limited by the lack of coupling of the soot emissions processes to the combustion kinetics processes.

MODEL FORMULATION

A more fruitful approach to soot emissions modeling is to address soot formation and oxidation in the same manner as that by which other species involved in the combustion process are described. Since the combustion kinetics of even the smallest hydrocarbon molecules require such an excessively large number of elementary reactions that comprehensive engineering combustor models cannot utilize such reaction mechanisms, Edelman and Fortune [18] first introduced the ideal of a quasi-global model to obtain a reasonably sized kinetics scheme. The quasi-global scheme involves the partial oxidation of large hydrocarbon molecules by sub-global steps to form fragments which can be managed with elementary reactions. Converting the hydrocarbon to carbon monoxide and hydrogen in one step, followed by detailed steps to accomplish the final oxidation has proven to be a very useful idea. However, it should be noted that for a particular problem, any convenient combination of global and detailed steps deemed appropriate could be used, and the savings in computation time associated with using the abbreviated set of reactions would still be realized. This model has been developed and refined for fuel lean combustion [19], and its utility and limitations have been demonstrated. Further applications of this model which includes the use of oxygenated intermediates in the hydrocarbon combustion scheme are reported by Edelman and Fortune [18] and Roberts *et al.* [20]. Both considered aldehydes as intermediates.

To apply the concept of quasi-global kinetics to soot emissions modeling, the pyrolysis and oxidative pyrolysis of rich combustion must be included along with the steps which form and consume soot. The kind of reactions which would be required in such a model are shown generically in Table 1. Only those reactions and species which are necessary to analyze any particular problem would be used at any one time. The reactions given in Table 1 are meant to indicate types of reactions which might be important; they are not meant to exclude pertinent reactions for particular applications. For example, the entire class of sulfur oxidation reactions might be included, if SO_x formation is of concern. The hydroxy radical in the global steps gives the model great versatility. The $C_x H_y O_z$ in the oxidative pyrolysis steps represents not only stable products but also radicals such as peroxy or formal which might eventually prove beneficial. The major combustion products are specifically indicated in the partial oxidation steps, and the $C_x H_y O_z$ is retained to indicate partially oxygenated species, such as formaldehyde, whose importance is not yet established.

While the complexities of combustor modeling, which include coupled chemistry, fluid dynamics, and soot emissions are fully appreciated, it has long been the contention of these investigators that meaningful analyses may be peformed with the proper coupling of unit models which represent the elements of complex combustor flow fields (Edelman *et al.* [21]). Therefore, a combustion model is being developed which is based upon a quasi-global kinetics model concept that integrates soot formation and consumption processes with the chemistry of the heat release processes as outlined in Table 1.

TABLE 1
Generic Quasi-global Model

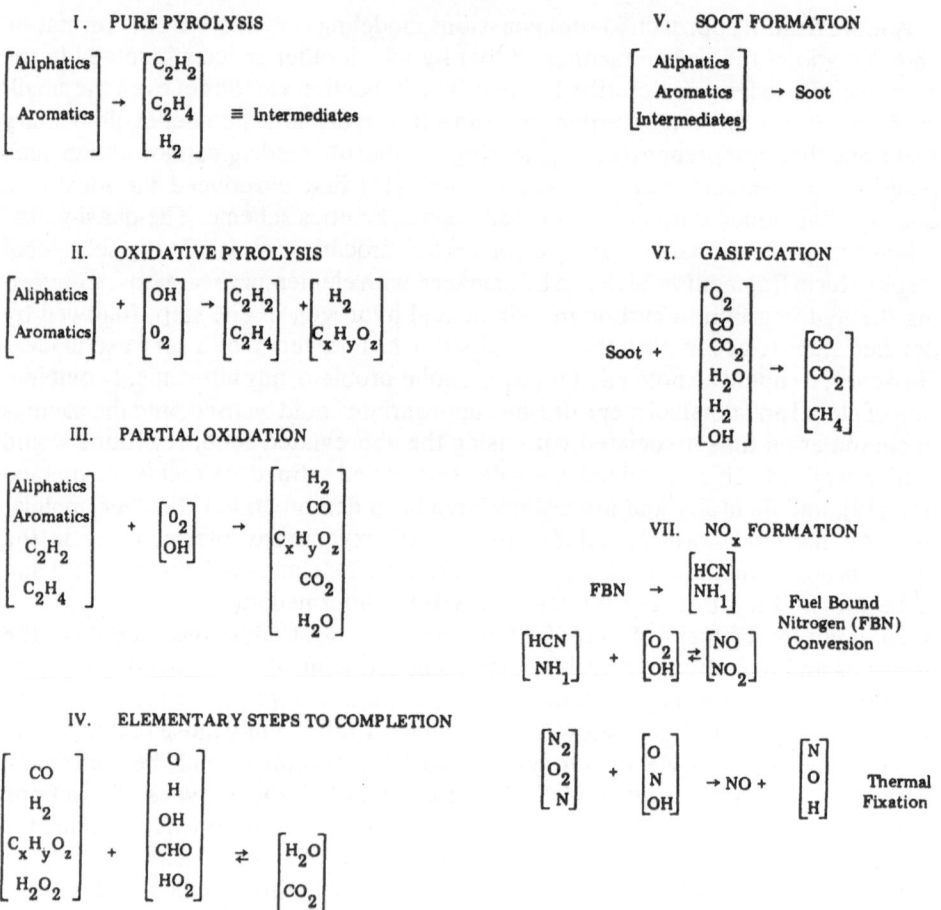

To apply the quasi-global model to sooting flames, a rather thorough set of jet-stirred reactor experiments which have been designed and performed by Exxon to study typical residence times, pressures, temperatures, fuels, and equivalence ratios for application to gas turbines relevant to the power industry (Blazowski *et al.* [22]) has been modeled. These experiments involve essentially premixed, near-adiabatic combustion with residence times in the range of 2 to 10 ms.

Utilizing the concept of a quasi-global model, Edelman *et al.* [15] developed an expression for the rate of formation of soot in a jet-stirred reactor. The soot oxidation rate used in this development was that of Nagle and Strickland-Constable [17]. The formation rate, R_F, was found to be represented by the following equation.

$$R_F = A\,T^a\,C_{HC}^{\ b}\,C_{O_2}^{\ c}\,\exp\left\{-E/RT\right\}$$ (5)

where A, a, b, c, and E are constants.

The temperature, hydrocarbon, and oxygen concentrations used in the correlation were those obtained from experiments.

Wang *et al.* [23] noted that between 1 and 4 ms in the reflected shock region of a shock tube a quasi-equilibrium was established between the apparent rate of soot formation and oxidation. Wang used the same global expressions as those in equations 4 and 5 to correlate his data. Temperature was inferred from measured shock velocities. The hydrocarbon and oxygen concentrations were calculated with a kinetics analysis using a mechanism suggested by McClain *et al.* [24]. Thirty-one species and 42 reactions were used. Since many of the molecules were very large, many of the reactions were more global than elementary for this small a ratio of species to reactions. Using exactly the same parameters obained by Edelman *et al.* [15], Wang obtained predicted sooting concentrations within a factor of four of those he measured. Considering the wide difference in experiments considered and that a published kinetics model was used without modification, this comparison is remarkable.

Lean combustion data from the Exxon experiments were appropriately modeled with the original quasi-global model. Predicted and measured CO_2 and CO concentrations produced by the combustion of iso-octane are shown in Figs. 1 and 2 [15]; the CO_2 agreement is excellent as is the O_2, unburned hydrocarbon, and temperature. The CO predictions differ in magnitude from the data, but the trend as a function of equivalence ratio is in general agreement with the data.

For the prediction of net soot emissions under well-stirred environments, soot reaction steps are coupled into the combustion analysis with the following rate expressions. The species conservation equation for soot can be expressed as follows:

$$\frac{\dot{m}}{\rho V}\,(C_s - C_{s_o}) = R_F - A_t R_{ox} \quad [\text{gm/cm}^3 \cdot \text{sec}]$$ (6)

where $\dot{m}$, ρ, V, C_s, C_{so}, are total mass flow rate, exhaust gas density, reactor volume, outlet and inlet soot concentrations. As a whole, $\dot{m}/\rho V$ represents the reciprocal of residence time. A_t which can be expressed in terms of C_s, soot density (ρ_s) and soot particle diameter (D) is the total surface area available for oxidation. Finally, R_F and R_{ox} are the reaction rates for the soot formation and oxidation steps.

References pp. 316-317.

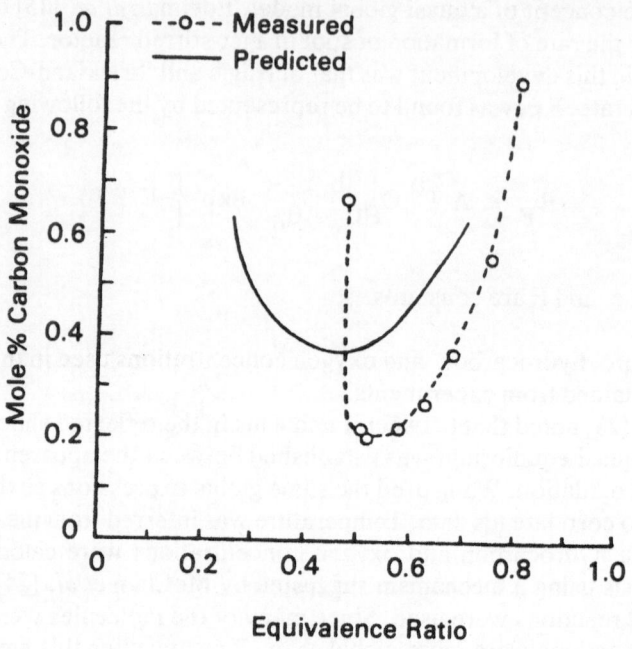

Fig. 1. Comparison of predicted and measured CO concentrations, isooctane.

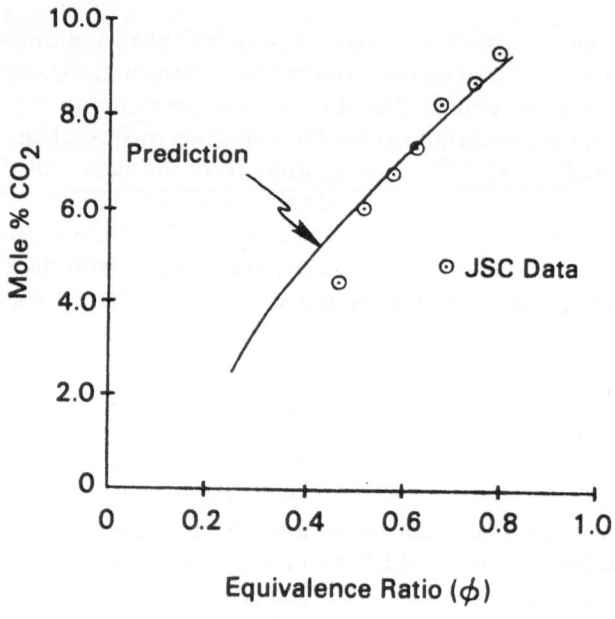

Fig. 2. Comparison of predicted and measured CO_2 concentrations, isooctane.

The formation step was given by equation 5. Tentative values for these constants have been determined by using jet-stirred reactor data [15]. These data include three residence times, 3, 4, and 7 ms. The 3 ms set of data which had air flow rates of 112.5 g/min was used to determine these rate constants. Experimental values of unburnt hydrocarbon as acetylene and oxygen concentrations were used to obtain the following set of rate constants.

$$A = 4.66 \times 10^{14}$$
$$a = -1.94$$
$$b = 1.81$$
$$c = -0.5$$
$$E = 32,000$$

The soot oxidation step was modeled with equation 4. An assumed mean soot size of 25 nm was used in this equation. Due to the relative magnitudes of the residence time in the combustor and the soot oxidation rate, assumed sizes in the range of 10 to 70 nm caused only modest changes in the net soot emissions.

Further comparisons of the same model with isothermal jet-stirred combustor experiments are shown in Figs. 3-6. Agreement is excellent.

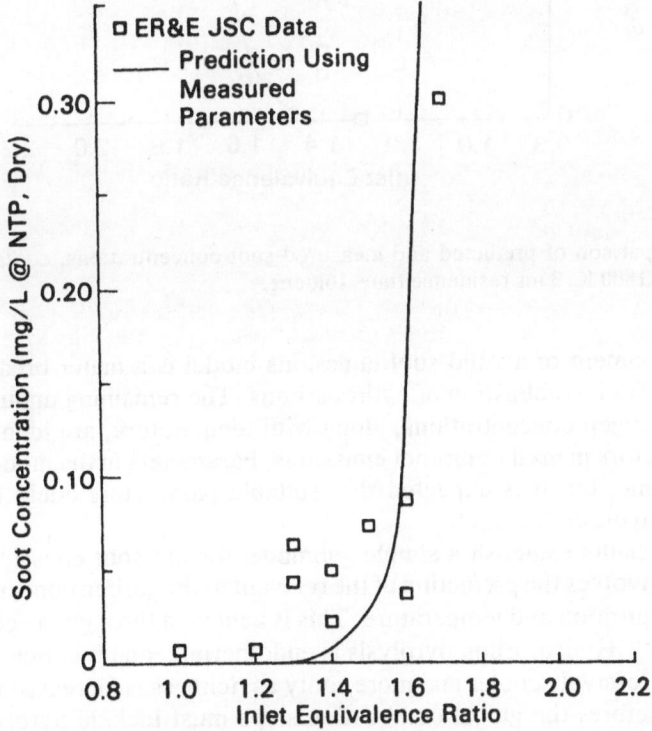

Fig. 3. Comparison of predicted and measured soot concentrations, controlled reactor temperature = 1700 K, 3 ms residence time, toluene.

References pp. 316-317.

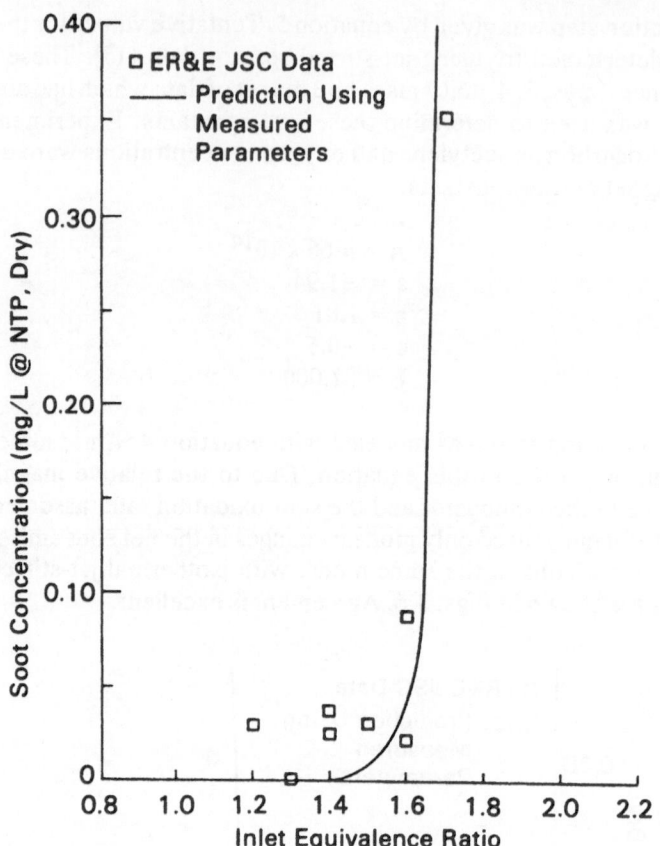

Fig. 4. Comparison of predicted and measured soot concentrations, controlled reactor temperature = 1800 K, 3 ms residence time, toluene.

This development of a valid soot emissions model is a major breakthrough in describing the rich combustion of hydrocarbons. The remaining unburned hydrocarbon and oxygen concentrations, along with temperature, are identified as the controlling factors in predicting soot emissions. Parameters in the model are evaluated for toluene, but it is expected that suitable parameters could be found to describe any hydrocarbon fuel.

While the results establish a simple submodel for net soot emissions, a major requirement involves the prediction of the relevant hydrocarbon concentration, the oxygen concentration and temperature. This is achieved through the coupling outlined in Table 1. Hydrocarbon pyrolysis is endothermic, and fuel rich combustion becomes progressively cooler and more sooty as richness is increased to the blowout limit; therefore, the global combustion steps must include pyrolysis. Strong sooters such as toluene (and other aromatics and some large straight chain compounds) might well be present as unreacted components of the original fuel. Acetylene is a strong sooter, and for the stirred reactor experiments in the 1700 to 2000K

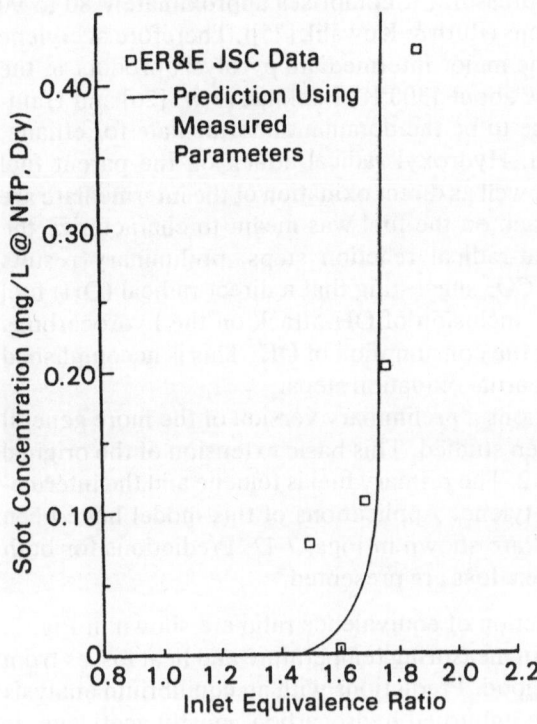

Fig. 5. Comparison of predicted and measured soot concentrations, controlled reactor temperature = 1900 K, 3 ms residence time, toluene.

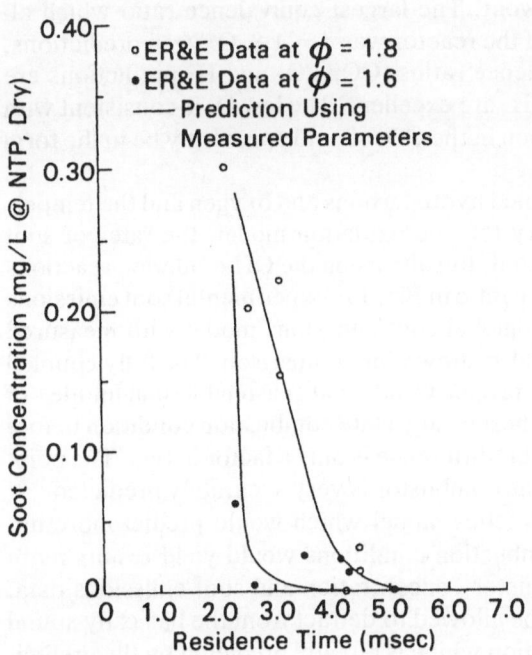

Fig. 6. Comparison of predicted and measured soot concentrations, controlled reactor temperature = 1900K, toluene.

temperature range at atmospheric pressure, it comprises approximately 80 to 90 mole % of the unburned hydrocarbons (Ruth & Kowalik [25]). Therefore acetylene is a likely candidate to represent the major intermediate pyrolysis product in the global steps. At temperatures below about 1300°K, Edelman et al. [26] and Hautman et al. [27] have shown ethylene to be the dominant intermediate for ethane, propane, and n-octane combustion. Hydroxyl radical attack on the parent fuel molecule and on the intermediate as well as direct oxidation of the intermediate are also included. Although the O_2 attack on the fuel was meant to characterize the intermediate hydrocarbon fragment-radical reaction steps, preliminary results have shown an over production of CO_2 suggesting that a direct radical (OH) fuel reaction is also required. Thus, the inclusion of OH attack on the hydrocarbons provides alternate parallel paths for the consumption of OH. This is accomplished in both the oxidative pyrolysis and partial oxidation steps.

To further investigate these reactions a preliminary version of the more general quasi-global model (Table 1) has been studied. This basic extension of the original quasi-global model is given in Table 2. The primary fuel is toluene and the intermediate hydrocarbon is limited to acetylene. Applications of this model have been made to jet-stirred reactor data and are shown in Figs. 7-12. Predictions for both specified reactor temperature and heat loss are presented.

Temperature predictions as a function of equivalence ratio are shown in Fig. 7. Considering the apparent difficulty in measuring temperature and heat losses from the reactor, these comparisons are good. Predictions with an equilibrium analysis are also shown for comparison. The unburned hydrocarbon, mostly acetylene, is shown in Fig. 8. The agreement with the equivalence ratio at which hydrocarbons first break through and appear as products is excellent. The level of hydrocarbon is predicted fairly well until near blowout. The largest equivalence ratio which allowed reasonably stable operation of the reactor was $\phi = 1.8$. Oxygen predictions, Fig. 9, are low at the higher equivalence ratios. CO, CO_2 and H_2 predictions are shown in Figs. 10 to 12. The CO and H_2 are excellent. The low O_2 is consistent with the high CO_2. Also, some of the carbon in the CO_2 should apparently be in the form of unburned hydrocarbon.

Using the concentrations of unburned hydrocarbons and oxygen and the temperature calculated with the preliminary toluene oxidation model, the rates of soot formation and oxidation were predicted. Results using the OH oxidation reactions are shown as a function of equivalence ratio in Fig. 13. Experimental soot emissions data points and results using the subglobal soot emissions model with measured concentrations and temperature are also shown for comparison. The fully coupled combustion kinetics model predicts proper trends and reasonable magnitudes of soot emissions. Predictions at the richest steady state combustion condition before blowout occurs differ the most, and this difference is only a factor of two. The point of hydrocarbon breakthrough from the combustor is very accurately predicted.

Refinements to the combustion kinetics model which would predict more unreacted oxygen under very rich combustion conditions would yield results more consistent with the jet-stirred combustor concentration and soot emissions data. However, such refinements cannot be allowed to detract from the basically sound combustion and heat release description which is already predicted by the prelimi-

TABLE 2
Quasi-global Model of Toluene Combustion

FUEL PYROLYSIS

$$C_7H_8 \rightarrow \frac{7}{2}C_2H_2 + \frac{1}{2}H_2$$

$$\frac{dC_{C_7H_8}}{dt} = -1.259 \times 10^{16} C_{C_7H_8} \exp(-98000/RT)$$

$$C_7H_8 + OH \rightarrow \frac{7}{2}C_2H_2 + H_2 + \frac{1}{2}O_2$$

$$\frac{dC_{C_7H_8}}{dt} = -1.0 \times 10^{12} C_{C_7H_8} C_{OH} \exp(-3000/RT)$$

FUEL OXIDATION

$$C_2H_2 + O_2 \rightarrow 2CO + H_2$$

$$C_2H_2 + 2OH \rightarrow 2H_2$$

$$\frac{dC_{C_2H_2}}{dt} = -1.664 \times 10^9 TP^{0.3} C_{C_2H_2}^{\frac{1}{2}} C_{O_2} \exp(-39000/RT)$$

$$\frac{dC_{C_2H_2}}{dt} = -7.929 \times 10^{18} C_{C_2H_2} C_{OH} \exp(-65000/RT)$$

SOOT FORMATION

$$C_2H_2 \text{ (plus } C_7H_8) \rightarrow \text{Soot}$$

See Text

SOOT OXIDATION

$$\text{Soot} + O_2 \rightarrow CO_2$$

Elementary H_2/CO Oxidation Mechanism: $K_f = AT^b \exp(-E/RT)$

REACTION	A	b	E/R
$CO + OH \rightleftarrows CO_2 + H$	4.000×10^{12}	0	4.030×10^3
$CO + O_2 \rightleftarrows CO_2 + O$	3.000×10^{12}	0	2.500×10^4
$H_2 + O_2 \rightleftarrows 2OH$	1.700×10^{13}	0	2.407×10^4
$OH + H_2 \rightleftarrows H + H_2O$	2.190×10^{13}	0	2.590×10^3
$2OH \rightleftarrows H_2O + O$	6.023×10^{12}	0	5.500×10^2
$O + H_2 \rightleftarrows OH + H$	1.800×10^{10}	1.00	4.480×10^3
$H + O_2 \rightleftarrows OH + O$	1.220×10^{17}	-0.91	8.369×10^3
$M + CO + O \rightleftarrows CO_2 + M$	6.000×10^{13}	0	0
$M + O + H \rightleftarrows OH + M$	1.000×10^{16}	0	0
$M + 2O \rightleftarrows O_2 + M$	9.380×10^{14}	0	0
$M + 2H \rightleftarrows H_2 + M$	5.000×10^{15}	0	0
$M + H + OH \rightleftarrows H_2O + M$	8.400×10^{21}	-2.00	0

$$[C] = \frac{gm\text{-mole}}{cc}, [T] = {}^oK, [P] = atm, [E] = \frac{kcal}{gm\text{-mole}}$$

References pp. 316-317.

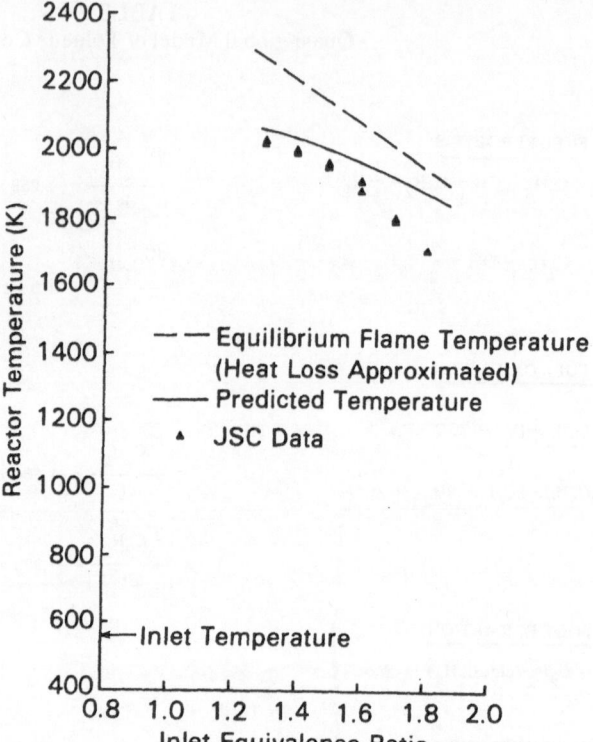

Fig. 7. Comparison of computed and measured temperature, toluene/air mixture, 3 ms residence time.

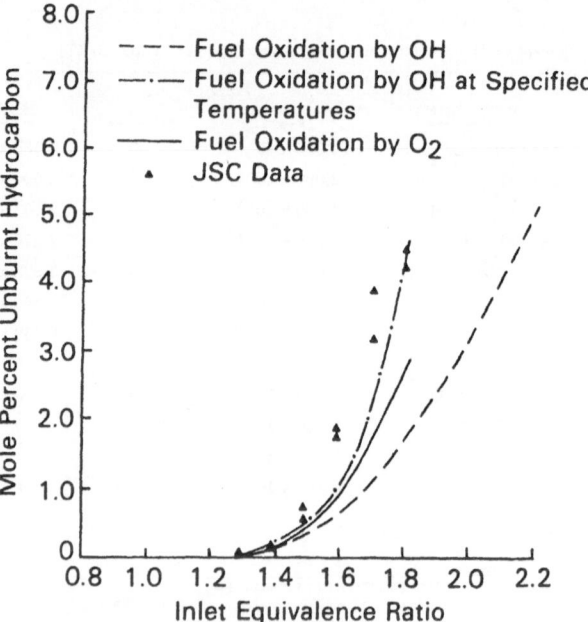

Fig. 8. Comparison of computed and measured unburnt hydrocarbon concentration, toluene/air mixture, 3 ms residence time.

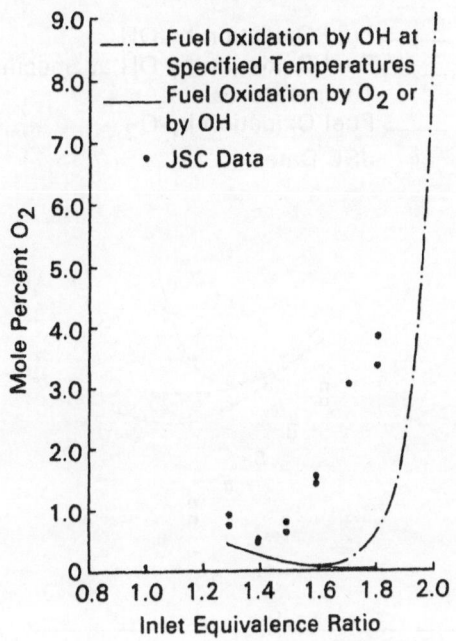

Fig. 9. Comparison of computed and measured O₂ concentration, toluene/air mixture, 3 ms residence time.

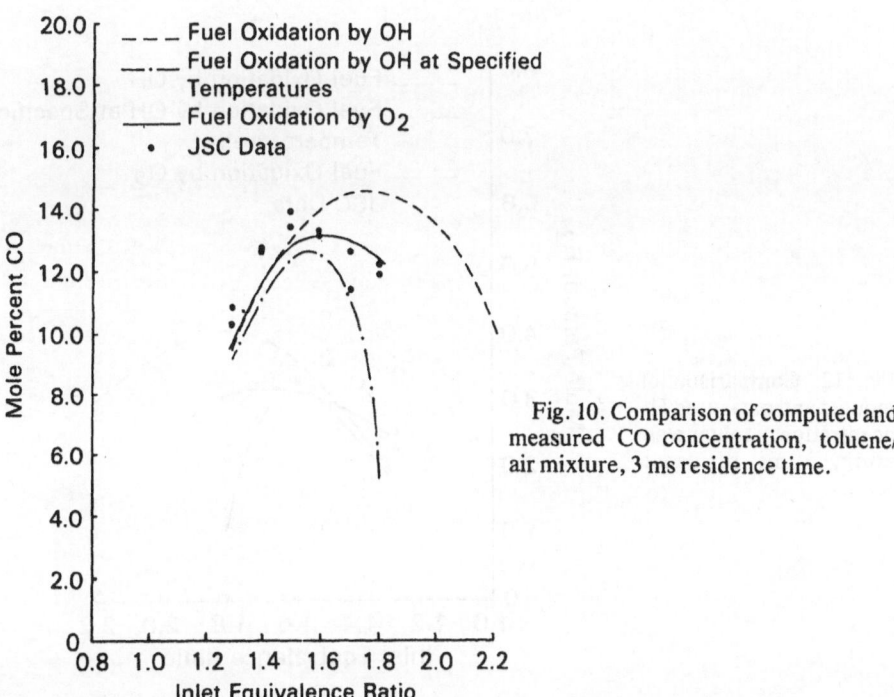

Fig. 10. Comparison of computed and measured CO concentration, toluene/air mixture, 3 ms residence time.

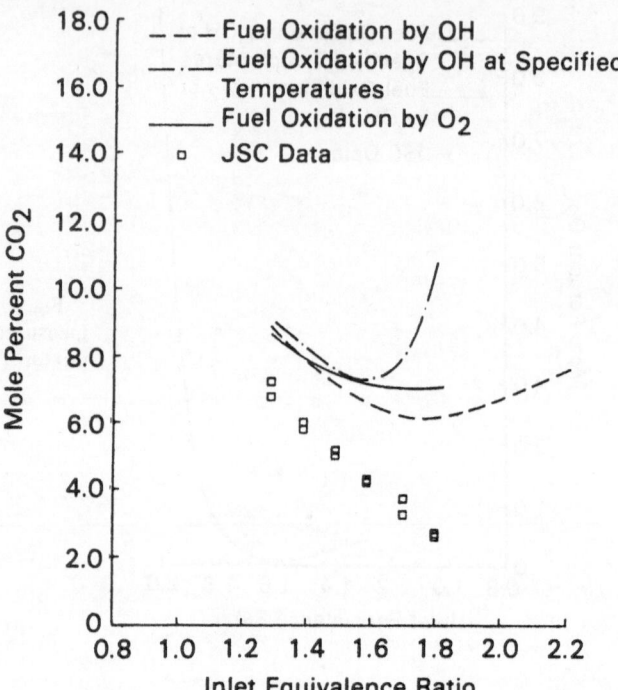

Fig. 11. Comparison of computed and measured CO₂ concentration, toluene/air mixture, 3 ms residence time.

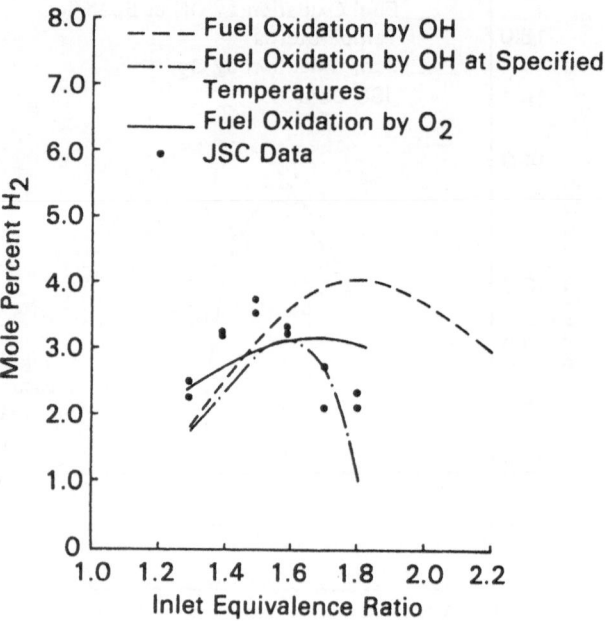

Fig. 12. Comparison of computed and measured H₂ concentration, toluene/air mixture, 3 ms residence time.

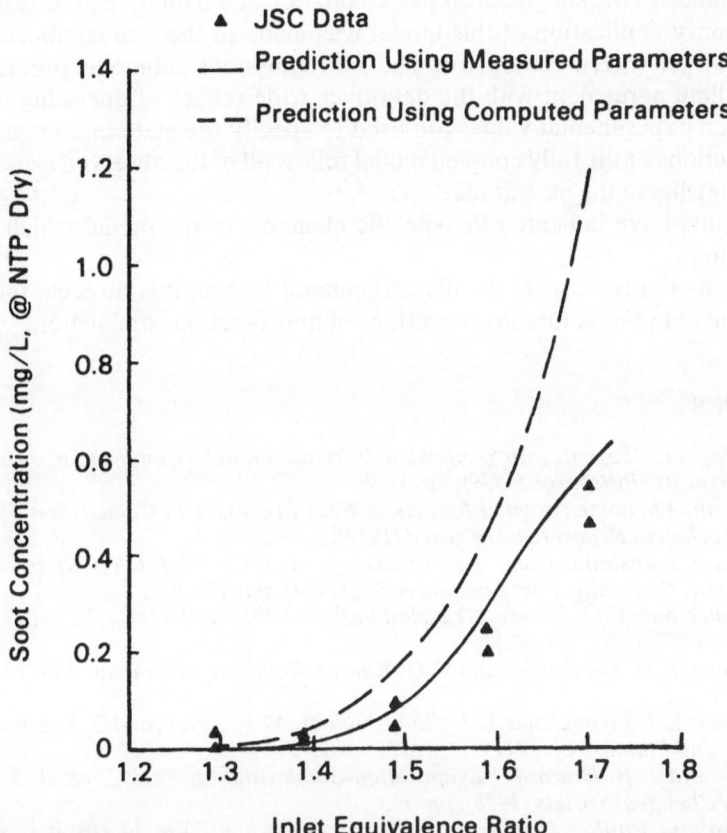

Fig. 13. Comparison of computed and measured soot emission, toluene/air mixture, 3 ms residence time.

nary toluene model. The concept of predicting soot emissions with a quasi-global kinetics model is clearly shown to be feasible with this preliminary toluene model.

CLOSURE

In summary, a fuel-sensitive combustion kinetics model which incudes soot emissions has been formulated.

- This model is based on the quasi-global concept and includes pure and oxidative pyrolysis, soot formation and oxidation, partial oxidation of the fuel and fuel fragments, and detailed kinetics for the completion of oxidation and NO_x formation.
- Soot formation and oxidation are shown to critically depend on the remaining unburned hydrocarbons and oxygen and temperature. The model accounts for unburned hydrocarbon as both primary fuel and intermediate fragments; an important feature with respect to describing the sooting ability of the particular hydrocarbon species.

References pp. 316-317.

- To describe pyrolysis and improve the relative rate of CO - CO_2 conversion, OH is included in the subglobal steps of oxidation and oxidative pyrolysis.
- A preliminary application of this model was made to the rich combustion of toluene in a jet-stirred combustor. The soot emissions submodel predictions give excellent agreement with the data over wide ranges of operating conditions, when experimental values are used to specify the state of combustion. The predictions of the fully coupled model follow all of the observed trends but differ somewhat in the magnitude.
- These results have indicated the specific elements of the model which need improvement.
- Since this model represents details and general trends, it is an ideal tool for guiding and planning future investigations of hydrocarbon combustion.

REFERENCES

1. *H. Gg. Wagner, "Seventeenth Symposium (International) on Combustion," The Combustion Institute, Pittsburgh (1979), pp. 3-19.*
2. *I. Glassman, Phenomenological Models of Soot Processes in Combustion Systems. AFOSR Technical Report to be released (1979).*
3. *P. A. Tesner, Translation from "Fizika Goreniya i Vzryva," Vol. 15 (1979), pp. 3-14.*
4. *D. E. Jensen, Proc. Roy. Soc. London, A-388 (1974), pp. 375-396.*
5. *P. A. Tesner and V.G. Knorre, Translation from "Fizika Goreniya Vzryra," Vol. 6 (1970), pp. 386-390.*
6. *P. A. Tesner, T. D. Snegiriova, and V. G. Knorre, Combust. and Flame, Vol. 17 (1971), pp. 253-260.*
7. *P. A. Tesner, E. I. Tsygnakova, L. P. Guilazetdinov, V. P. Zuyev, and G. V. Loshakova, Combust. and Flame, Vol. 17 (1971), pp. 279-285.*
8. *G. Prado, and J. B. Howard, "Evaporation-Combustion of Fuels," ed. J. T. Zung, American Chemical Society (1978), pp. 153-166.*
9. *J. Lahaye and G. Prado, "Chemistry and Physics of Carbon," Vol. 14, eds. P. L. Walker, Jr., and P. A. Thrower, Marcel Decker, Inc., NY (1978), pp. 167-294.*
10. *H. Jinno, S. Fukutani, and A. Takaya, "Sixteenth Symposium (International) on Combustion," The Combustion Institute, Pittsburgh, (1976), pp. 709-718.*
11. *J. Genovese, R. B. Edelman, and O. F. Fortune, AIAA Journal, Vol. 8 (1971), pp. 352-357.*
12. *K. S. Narasimhan and P. J. Foster, "Tenth Symposium (International) on Combustion," The Combustion Institute, Pittsburgh, (1965), pp. 253-257.*
13. *B. F. Magnussen and B. H. Hjertager, "Sixteenth Symposium (International) on Combustion," The Combustion Institute, Pittsburgh, (1976), pp. 719-729.*
14. *I. M. Khan and G. Greeves, "Heat Transfer in Flames," eds N. H. Afgan and J. M. Beer, Halsted Press, NY (1974), pp. 389-404.*
15. *R. B. Edelman, A. Turan, P. T. Harsha, E. Wong, and W. S. Blazowski, Combustor Modelling, AGARD-CP-275, (1979), pp. 13-1 to 13-14.*
16. *K. B. Lee, M. W. Thring, and J. M. Beer, Combust. and Flame, Vol. 6 (1962), pp. 137-145.*
17. *J. Nagle, and R. F. Strickland-Constable, "Proc. Fifth Carbon Conf.", Vol. 1 (1962), pp 154-164.*
18. *R. B. Edelman and O. F. Fortune, AIAA Paper No. 69-86, (1969).*
19. *R. B. Edelman and P. T. Harsha, Progress in Energy and Combustion Science, Vol. 4 (1978), pp. 1-62.*
20. *R. Roberts, L. D. Aceto, R. Kollrack, D.P. Teixeira and J. M. Bonnell, AIAA Journal, Vol. 10 (1972), pp. 820-826.*

21. R. B. Edelman, O. Fortune, and G. Weilerstein, "Emissions from Continuous Combustion Systems," eds. W. Cornelius and W. G. Agnew, Plenum Press, New York, (1972), pp. 55-90.
22. W. S. Blazowski, R. B. Edelman and E. Wong, Fundamental Effects in Continuous Combustion Systems, Summary Technical Progress Report for Period August 15, 1978 – January 31, 1980, DOE Contract DE-AC03-77-ET-11313.
23. T. S. Wang, R. A. Matula and R. C. Farmer, "Eighteenth Symposium (International) on Combustion," The Combustion Institute, Pittsburgh, to be published.
24. A. G. McLain, C. J. Jachimowski, and C. H. Wilson, NASA Technical Paper 1472 (1979).
25. L. A. Ruth and R. M. Kowalik, Fundamental Characterization of Alternate Fuel Effects in Continuous Combustion Systems, Technical Status Report for Period May 1, 1980 – May 31, 1980, DOE Contract AC22-77-ET-11313.
26. R. B. Edelman, R. J. Gelinas, C. W. Wilson, E. Y. Wong, Study of Net Soot Formation in Hydrocarbon Reforming for Hydrogen Fuel Cells, Quarterly Report for Period January, 1980 – March, 1980, DOE Contract DE-AC21-79 MC 12735.
27. D. J. Hautman, F. L. Dryer, K. P. Schug, and I. Glassman, A Multiple-Step Overall Kinetic Mechanism for the Oxidation of Hydrocarbons, submitted for publication (1980).

DISCUSSION

R. D. Kern (University of New Orleans)

Did you say that the global reactions for toluene pyrolysis was to the seven-halves power for acetylene and to the one-half power for hydrogen?

Farmer

No. The stoichiometric coefficients for acetylene and hydrogen are 7/2 and 1/2 respectively.

Kern

How did you get the global rate constants for them?

Farmer

The literature contains pyrolysis-rate information for toluene and a number of other hydrocarbons. The effect of oxygen is included in this literature. Initial estimates are derived from this information. Final rates are obtained by characterizing experimental data, in this case, the Exxon stirred-reactor data.

Kern

But there are numbers now that have been measured.

Farmer

That is correct. The one that I mentioned in comparison with the shock tube data was obtained from [24].

Kern

But to formulate that overall rate constant, you sifted through the other 24 elementary steps and came up with a composite.

Farmer

No, we postulated rates to establish the best representation of the Exxon data.

D. B. Olson *(AeroChem Laboratories)*

You showed a slide of oxygen concentration versus equivalence ratio. The data rises slightly when you get toward the higher equivalence ratio. I can almost understand that at the beginning of blow-off, but the computed curve rises also and I cannot understand what computational process makes the oxygen process go up. Can you shed some light on that?

Farmer

The model that we are applying to the Exxon data is a stirred reactor model which not only predicts steady-state operation, but blow-out on both the lean and rich end of the equivalence ratio spectrum. This includes the rise in unburned hydrocarbon and oxygen concentrations and the rapid dropoff in temperature as blow-out is approached. Thus the model is applicable to flame stabilization phenomena in systems using intense back-mixed regions typical of primary zones in combustors using flameholders, etc.

C. Hunter *(Ford Motor Company)*

In the plot you have of soot, you have it versus intake oxygen?

Farmer

Versus inlet equivalence ratio.

Hunter

Inlet equivalence ratio! Now the formation process is maybe a function of inlet oxygen, but the oxidation process you would think would be a function of the exhaust or end-gas oxygen?

Farmer

The oxygen concentrations, the hydrocarbon concentrations, and the temperatures that we put into the soot formation and oxidation model were those that existed in the stirred reactor and those that are the exhaust compositions. These concentrations are those that our model predicts.

C. T. Bowman *(Stanford University)*

As someone who has spent a good portion of his life modeling chemical kinetic systems, I have to make this comment about what I have heard here today. You have a model which apparently fits some Exxon data, but you have a model which is not a kinetics model based on fundamental interactions between molecules and realistic chemical kinetic rate coefficients. I do not want anyone in this room to think that you have constructed this model based on realistic chemical kinetics. I have to look at your results and say that you have been able to adjust parameters in a way where you have been able to fit a very limited set of experimental data but I do not want this model going out in the guise that it is based on fundamental chemistry because it certainly is not.

Farmer

Detailed kinetics mechanisms comprised of elementary reactions which describe fuel-rich combustion and soot emissions would certainly be valuable information to have and to use. If data for the fuel of interest were not available, physical insight could be derived from the study of other definitively characterized fuel systems. If the complete mechanism became too large for convenient use, mathematical simplifications could be developed.

Such elementary mechanisms will probably never be fully determined for commercial liquid fuels, and it is unlikely that they will be determined for molecules as large as toluene or octane. Current toluene studies have suggested some elementary pyrolysis reactions, but only global oxidation reactions have been postulated [24]. Not only would the number of elementary reactions necessary to model a real fuel be excessively large, but in any case the rates for each postulated reaction would still have to be determined from experimental data.

In the unlikely event that sufficient elementary kinetics data to describe real fuels ever did become available, it could not be conveniently used to analyze practical combustion systems. The computational time and cost requirements would have to be reduced by approximation to make the analysis practical.

The kinetics model reported in this paper can be experimentally determined and validated and can be conveniently used for computations. Combustion design information can be derived from experimental measurements with the quasi-global model for hydrocarbon fuels of commercial interest.

J. P. Longwell *(Massachusetts Institute of Technology)*

I think kinetics has two meanings, one to engineers and one to physical chemists.

B. S. Haynes *(Massachusetts Institute of Technology)*

As a chemical engineer who has dabbled in physical chemistry, yes, you have fitted some data for a combustion system and that is indeed a time-honored chemical engineering approach. However, do you not think it would be worth going to another combustion system, perhaps even another fuel, to check out how useful these correlations are. I can produce a series of premixed flat flames, can do time-resolved tests on this model and, with very small changes in inlet fuel-air ratio, I can give you four orders of magnitude change in soot concentration as opposed to a factor of five or ten in dealing with this model. So I am not saying that soot is going to be handled as a fundamental chemical reaction, but I do think we need some modeling of this and more fundamental studies.

Farmer

I really appreciate your leading me like this. I showed you jet-stirred combustor data that was modeled. In a combustion paper at the Waterloo (Combustion Institute) meeting, I showed the correlation for shock-tube data. We have ongoing programs with Pennsylvania State University, specifically to look at some flat-flame data, and we would dearly love to look at your flat-flame data, also. I feel that when you look at more than one kind of experiment and when the correlation-type equations are applied to these data, you get an idea of what the range of applicability is. As far as the Exxon data goes, we are looking at combustion times of the order of two to ten milliseconds. I cannot say that this is going to satisfy conditions outside of that range because it was a characterization of that data. Certainly looking at other applications is a way to determine how general or limited these correlations are.

MODELING OF REACTION PROCESSES
IN TURBULENT FLAMES
WITH SPECIAL EMPHASIS
ON SOOT FORMATION AND COMBUSTION

B. F. MAGNUSSEN

The Norwegian Institute of Technology
Trondheim, Norway

ABSTRACT

The present paper reviews features of the eddy-dissipation concept developed by the author for treating chemical reactions in turbulent flow.

An essential feature of this concept is that it takes into account the fact that the molecular mixing between reactants, which is associated with the dissipation of turbulence, takes place in concentrated, isolated regions that occupy only a small fraction of the total volume of the fluid.

The mass fraction occupied by the dissipative regions, as well as the mass transfer rate between these regions and the surrounding fluid, are determined from turbulence theory thus providing new general fluid mechanical information for the solution of reaction problems. This enables fast and accurate calculations of turbulent combustion phenomena.

The treatment of fast and slow chemical reactions in turbulent flow is discussed in relation to this concept. Comparison is made with experimental data.

Special attention is given to the modeling of soot formation and combustion in turbulent flames. A two-step model for the soot formation is applied, i.e., one rate equation for the formation of nuclei and one for particles. The interaction between the chemistry and the turbulence is modeled according to the eddy-dissipation concept. Comparison is made between experimental data and calculations for acetylene.

It is interesting to notice that when the same rate equations with the same constants are applied also for methane and propane, results are obtained which seem to be closely related to physical reality.

NOMENCLATURE

a	constant or flux absorption coefficient
a_0	constant
b	constant

References pp. 334-335.

c	concentration (kg/m^3)
c_p	specific heat
C_1, C_2, C_D	constants
D	nozzle diameter
E	activation energy or blackbody emissive power
F	flatness factor or radiation flux sum
f	mixture fraction or linear branching coefficient
g	linear termination coefficient or gravitation constant
g_0	coefficient of linear termination on soot particle
ΔH_R	reaction enthalpy difference
h	enthalpy
I	intensity of scattered light
k	turbulence kinetic energy
L^*	characteristic length of fine structures
$\dot{m}$	exchange rate of mass with fine structures
m	mass concentration (kg/kg)
m_p	mass of soot particle (kg/part)
N	concentration of soot particles ($part/m^3$)
n	nucleus concentration ($part/m^3$)
n_0	rate of spontaneous formation of nucleus ($part/m^3/s$)
p	pressure
Re	Reynolds number
R_{fu}	rate of fuel combustion ($kg/m^3/s$)
$R_{n, c}^-$	rate of nucleus combustion ($part/m^3/s$)
$R_{n, f}^-$	rate of nucleus formation
$R_{s, c}^-$	rate of soot combustion
$R_{s, f}^-$	rate of soot formation
R_Φ	source term
r	stoichiometric oxygen requirement to burn 1 kg fuel or soot
T	temperature (K)
ΔT	excess temperature of reacting fine structures
u^*	characteristic velocity of fine structures
u'	turbulence velocity
U	axial velocity
V	lateral velocity
x	axial coordinate
y	lateral coordinate
ρ	density
ϵ	rate of dissipation of turbulence kinetic energy
μ_t	effective turbulent viscosity
σ	turbulent Prandtl/Schmidt number
ν	kinematic viscosity
γ^*	mass fraction occupied by fine structures
χ	fraction of fine structures reacting
Φ	undefined quantity
Λ	integral scale of turbulence

Superscripts
-	time-mean value
*	fine structure
o	surrounding fluid

Subscripts
fu	fuel
n	nucleus
pr	product
s	soot
Φ	undefined quantity

INTRODUCTION

The great number of interacting processes taking part in turbulent combustion makes it difficult to readily judge the effect of a certain variable on the combustion result.

The strong impact of energy and environmental problems faces the engineer with the need for accurate design methods that can be applied to complex systems. The need for numerical methods for further development of turbulent combustion systems whose applications range from I.C. engines to large utility boilers is obvious.

Finite-difference methods have been developed to predict instantaneous and time-averaged states of the flow. Due to limitations of present computers these methods cannot be extended to include rigorous treatment of chemical kinetics and are unable to calculate all the correlations of fluctuations in scalar quantities needed for closure. Consequently the interaction between the turbulence and the chemical kinetics must be modeled.

A very interesting specie in flames is soot whose presence is called for in some cases and unwanted in others. Methods to predict soot behavior in combustion devices are therefore strongly needed. The complexity of the chemical kinetics of soot has for a long time paralyzed development of prediction methods.

A certain modeling approach which takes care of the interaction between turbulence and chemical kinetics will be discussed in the following with special emphasis on soot formation and combustion.

MODELING OF REACTION RATES IN TURBULENT FLAMES

Characteristics of the Molecular Mixing Process [1, 2] — Chemical reactions take place when reactants are mixed at a molecular scale at sufficiently high temperature. In real turbulent flows the rate-controlling phenomenon for reactant consumption is the molecular mixing. It is known that the microscale processes which are decisive for the molecular mixing as well as dissipation of turbulence energy into heat are severely intermittent, i.e. concentrated in isolated regions whose entire volume is a small fraction of the volume of the fluid.

These regions are occupied by fine structures whose dimensions are small in one or two directions, however not in the third. These fine structures are believed to be vortex tubes, sheets or slabs whose characteristic dimensions are of the same magnitude as the Kolmogorov microscale [3, 4, 5, 6, 7].

The fine structures are responsible for the dissipation of turbulence into heat. Within these structures one can therefore assume that reactants will be mixed at a molecular scale. These structures, thus, create the reaction space for non-uniformly distributed reactants.

In a modeling context one can assume that the reactants are homogeneously mixed within the fine structures. Thus, in order to be able to treat the reactions within this space, it is necessary to know the reaction volume and the mass transfer rate between the fine structures and the surrounding fluid.

In turbulent flow, energy is transferred from the bigger eddies to the fine structures where mechanical energy is dissipated into heat. On the basis of similarity considerations of this energy transfer one can postulate the following expression for the mass fraction of the fluid contained in the fine structures:

$$\gamma* = (u*/u')^3 \tag{1}$$

where u' is the turbulence velocity and $u*$ is the characteristic velocity of the fine structures.

The transfer of mass per unit mass of fluid and unit time between the fine structures and the surrounding can then be expressed:

$$\dot{m} = 2 \ \frac{u*}{L*} \ \gamma* \qquad (1/s) \tag{2}$$

where $L*$ is the characteristic length of the fine structures. The characteristic velocity and length can be expressed [8]:

$$u* = 1.74 \ (\nu \epsilon)^{1/4} \tag{3}$$

$$L* = 1.43 \ \nu^{3/4} / \epsilon^{1/4} \tag{4}$$

where, ν is the kinematic viscosity and ϵ the rate of turbulence energy dissipation per unit mass and unit time.

For nearly isotropic turbulence Eqs. (1) and (2) can be replaced by:

$$\gamma* = 9.7 \ (\frac{\nu \epsilon}{k^2})^{3/4} \tag{5}$$

and

$$\dot{m} = 23.6 \ (\frac{\nu \epsilon}{k^2})^{1/4} \frac{\epsilon}{k} \quad (1/s) \tag{6}$$

The mean mass transfer rate between a certain fraction, χ, of the fine structures and the rest of the fluid, R_i, can now for a certain specie, i, be expressed as follows:

$$R_i = \dot{m}\chi \ (\frac{c_i^o}{\rho^o} - \frac{c_i^*}{\rho^*}) \tag{7}$$

where * and o refer to conditions respectively in the fine structures and the surrounding.

The mass transfer rate can also be expressed per unit volume in the fine structure fraction, χ, as

$$R_i^* = \frac{\dot{m} \rho^*}{\gamma^*} \ (\frac{c_i^o}{\rho^o} - \frac{c_i^*}{\rho^*}) \tag{8}$$

Finally, the concentration of a specie, i, in the fraction, χ, of the fine structures and in the surrounding is related to the mean concentration by:

$$\frac{\bar{c}_i}{\rho} = \frac{c_i^*}{\rho^*} \gamma^* \chi + \frac{c_i^o}{\rho^o} \ (1 - \gamma^* \chi) \tag{9}$$

It is now possible to write balance equations for the reacting fine structures and the surrounding fluid including chemical kinetic rate expressions.

The Rate of Combustion — If the rate of reaction between fuel and oxygen is considered infinitely fast, the rate of reaction will be limited by the mass transfer between the bulk and the fine structures. In this case the concentration of fuel or oxygen will be very small within the reacting fine structures. If reaction took place in all the fine structures, the rate of combustion would be expressed by:

$$R_{fu} = \dot{m} \ \frac{\bar{c}_{min}}{1 - \gamma^*} \quad (kg/m^3/s) \tag{10}$$

where $\bar{c}_{min}$ is the smallest of $\bar{c}_{fu}$ and $\bar{c}_{O_2}/r_{fu}$, where $\bar{c}_{fu}$ and $\bar{c}_{O_2}$ are the local mean concentrations of fuel and oxygen, and r_{fu} the stochiometric oxygen requirement.

Not all the fine structures will be sufficiently heated to react. This is obviously the case in combustion of premixed gases where both fuel and oxygen are present in the fine structures.

References pp. 334-335.

The fraction of the fine structures which reacts can be assumed to be proportional to the ratio between the local concentration of reacted fuel and the total fuel concentration:

$$\chi = \frac{\bar{c}_{pr}/((1+r_{fu})\gamma^*)}{\bar{c}_{pr}/(1+r_{fu}) + \bar{c}_{fu}} \tag{11}$$

where $\bar{c}_{pr}$ is the local mean concentration of reaction products. Equation (11) implies that the reaction products are kept within the fine structure region until a concentration is reached which yields χ equal to unity.

By combination of Eqs. (10) and (11) the following general expression is obtained for the rate of combustion at infinite reaction rate between fuel and oxygen:

$$R_{fu} = \dot{m} \frac{\chi}{1 - \gamma^*\chi} \bar{c}_{min} \tag{12}$$

This expression is closely related to the previous expressions by Magnussen and Hjertager [8, 9, 10]. The reacting fine structures under these conditions will have a temperature, ΔT, in excess of the local mean temperature.

$$\Delta T = \frac{\Delta H_R \, \bar{c}_{min}}{\bar{\rho} \, c_p} \quad (K) \tag{13}$$

where ΔH_R is the heat of reaction (kJ/kg fuel) and c_p the local specific heat (kJ/kg/K). The temperature, T^*, of the reacting fine structures is consequently:

$$T^* = \bar{T} + \Delta T \quad (K) \tag{14}$$

and the surrounding temperature

$$T^o = \bar{T} - \Delta T \cdot \frac{\gamma^*\chi}{1 - \gamma^*\chi} \quad (K) \tag{15}$$

where $\bar{T}$ is the local time mean temperature.

MODELING OF SOOT FORMATION AND COMBUSTION IN TURBULENT FLAMES

The modeling of soot formation has been based on the two equations as used previously by Magnussen et al. [2, 9, 10, 11, 12]. An equation for the rate of formation of radical nuclei is expressed by

$$R_{n,f} = n_o + (f-g) n - g_o n N \quad (part/m^3/s) \tag{16}$$

where n is the concentration of radical nuclei, N is the concentration of soot particles and n_0 is the rate of spontaneous formation of radical nuclei,

$$n_o = a_o \bar{c}_{fu} \exp(-E/RT) \quad (part/m^3/s) \tag{17}$$

where $\bar{c}_{fu}$ is the mean concentration of fuel (kg/m^3), f, g, g_0, and a_0 are certain constants. The rate of soot particle formation is expressed by

$$R_{s,f} = m_p (a-bN) n \quad (kg/m^3/s) \tag{18}$$

where m_p is the mass of a soot particle (kg/part), and a and b are constants.

Soot was allowed to be formed both in the heated fine structures and in the surrounding fluid, and it was assumed that the mass fraction of the fuel contained in the heated fine structures was proportional to $\gamma^* \chi$. Local equilibrium with the surrounding fluid was assumed for the fine structures leading to the following equations:

$$R_{n*,f} - \frac{\dot{m} \rho^*}{\gamma^*} \left(\frac{n^*}{\rho^*} - \frac{n^o}{\rho^o} \right) = 0 \tag{19}$$

and

$$R_{s*,f} - \frac{\dot{m} \rho^*}{\gamma^*} \left(\frac{c_s^*}{\rho^*} - \frac{c_s^o}{\rho^o} \right) = 0 \tag{20}$$

The mean rates of formation could then be expressed by

$$R_{\bar{n},f} = n_{o,T*}\gamma^* \chi\bar{\rho}/\rho^* + n_{o,T}o(1-\gamma^*\chi)\bar{\rho}/\rho^o \tag{21}$$

$$+(f-g)\bar{n} - g_o n^* N^* \gamma^* \chi \bar{\rho}/\rho^* - g_o n_o N^o (1-\gamma^*\chi)\bar{\rho}/\rho^o$$

and

$$R_{\bar{s},f} = m_p (a-bN^*)n^*\gamma^*\chi\bar{\rho}/\rho^* + m_p (a-bN^o)n_o (1-\gamma^*\chi) \bar{\rho}/\rho^o \tag{22}$$

References pp. 334-335.

The mean rates of nucleus and soot particle combustion are expressed by Eq. (12) as follows:

$$R_{\tilde{n},c} = R_{fu}\bar{n}/\bar{c}_{fu} \qquad (part/m^3/s) \qquad (23)$$

and

$$R_{\bar{s},c} = R_{fu}\bar{c}_s/\bar{c}_{fu} \qquad (kg/m^3/s) \qquad (24)$$

CALCULATIONS

Calculations have been made for a hydrogen diffusion flame and diffusion flames from C_2H_2, CH_4 and C_2H_8. The calculations have been performed by the GENMIX program.

Basic Equations — The flows considered are of the boundary layer type and the various equations have been expressed as follows:

Mass:

$$\frac{\partial}{\partial x}\,\bar{\rho}\bar{U} + \frac{1}{y}\,\frac{\partial}{\partial y}\,(y\bar{\rho}\bar{V}) = 0 \qquad (26)$$

Momentum and scalar quantities:

$$\bar{\rho}\bar{U}\frac{\partial\phi}{\partial x} + \bar{\rho}\bar{V}\frac{\partial\phi}{\partial y} = \frac{1}{y}\,\frac{\partial}{\partial y}\,(y\,\frac{\mu_t}{\sigma_\phi}\,\frac{\partial\phi}{\partial y}) + R_\phi \qquad (27$$

The various variables, Prandtl/Schmidt numbers and source terms are listed in Table 1. The effective turbulent viscosity has been expressed:

$$\mu_t = \bar{\rho}C_D k^2/\epsilon \qquad (28)$$

The radiation flux sum, F, in the source term for the enthalpy equation has been determined by:

$$\frac{1}{y}\,\frac{\partial}{\partial y}\,(\frac{y^2}{1+ay}\,\frac{\partial F}{\partial y}) = a(F-E) \qquad (29)$$

where the flux absorption coefficient, a, is determined from the concentration of soot, water vapor and carbon dioxide.

TABLE 1
Turbulent Prandtl/Schmidt Numbers
and Source Terms for the Various Variables

Φ	σ_Φ	R_Φ	
$\bar{U}$	1	$g(\bar{\rho}_\infty - \bar{\rho})$	
$\bar{h}$	σ_h	$2a(F - E)$	
k	σ_k	$C_D \bar{\rho} k^2 / \epsilon \ (\partial \bar{U}/\partial y)^2 - \bar{\rho}\,\epsilon$	
ϵ	σ_ϵ	$C_1\, C_D \bar{\rho} k\, (\partial \bar{U}/\partial y)^2 - C_2 \bar{\rho}\, \epsilon^2/k$	
$\bar{m}_{fu}$	σ_{fu}	$-R_{fu}$	Eq. (12)
$\bar{f}$	σ_f	0	
$\bar{m}_s$	σ_s	$R_{\bar{n},\,f} - R_{\bar{s},\,c}$	Eqs. (22) and (24)
$(\bar{n}/\bar{p})$	σ_n	$R_{\bar{n},\,f} - R_{\bar{n},\,c}$	Eqs. (21) and (23)

Constants — Table 2 gives the constants in the turbulence model.

According to Tesner et al. [11], the following constants were used for the soot formation equations: $a = 10^5$, $f\text{-}g = 10^2$, $g_0 = 10^{-15}$, $b = 8 \cdot 10^{-14}$, $E/R = 9 \cdot 10^4$, and $\rho_S = 2$ (g/cm³). According to Ref. 2, a particle diameter $d_p = 178.5$ A and $a_0 = 12.5 \cdot 10^{36}$ part/kg/s have been applied.

TABLE 2
Constants in the Turbulence Model

C_D	C_1	C_2	σ_h	σ_k	σ_ϵ	σ_{fu}	σ_f	σ_s	σ_n
0.09	1.44	1.79	0.7	1.0	1.3	0.7	0.7	0.7	0.7

Results and Discussion — Fig. 1 shows comparison between experimental and calculated mean temperatures for a hydrogen diffusion flame. As can be observed from the present and previous [10] calculations by the method reviewed here, very accurate results are obtained. Included in Fig. 1 are also some preliminary results for NO, where super-equilibrium for the O-atom concentration in the fine structures and an ordinary O/O_2 equilibrium for the surrounding fluid have been assumed. These results are very promising.

Also included in the figure are calculated fine structure temperatures.

Fig. 2 shows the intermittent character of the soot behavior in turbulent diffusion flames. The overall features of this appearance is taken care of in the present model. Figs. 3, 4 and 5 show some previous experimental and calculated results [2] for soot concentration in C_2H_2 diffusion flames. It can be observed that global effects have strong influence on the soot concentration. This is also reflected in the calculated results.

References pp. 334-335.

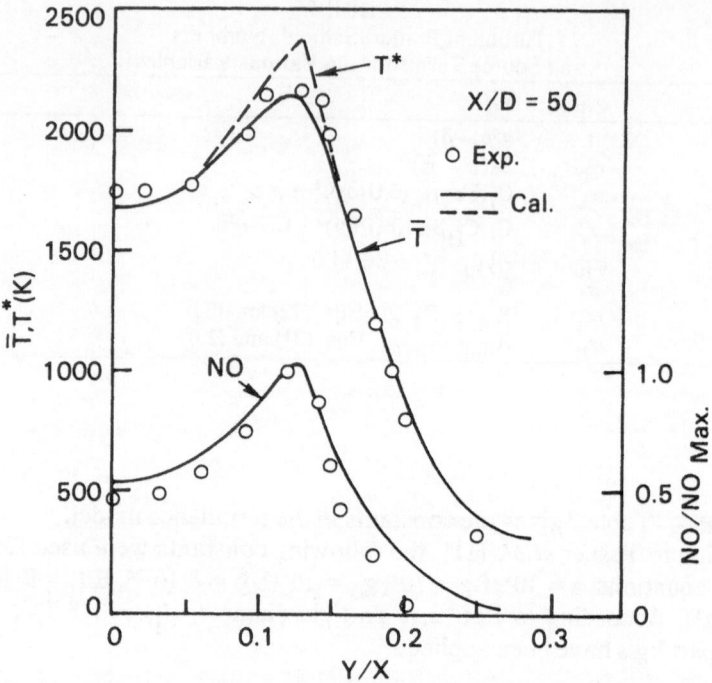

Fig. 1. Radial variation of temperature and nitrogen oxide in a hydrogen diffusion flame.

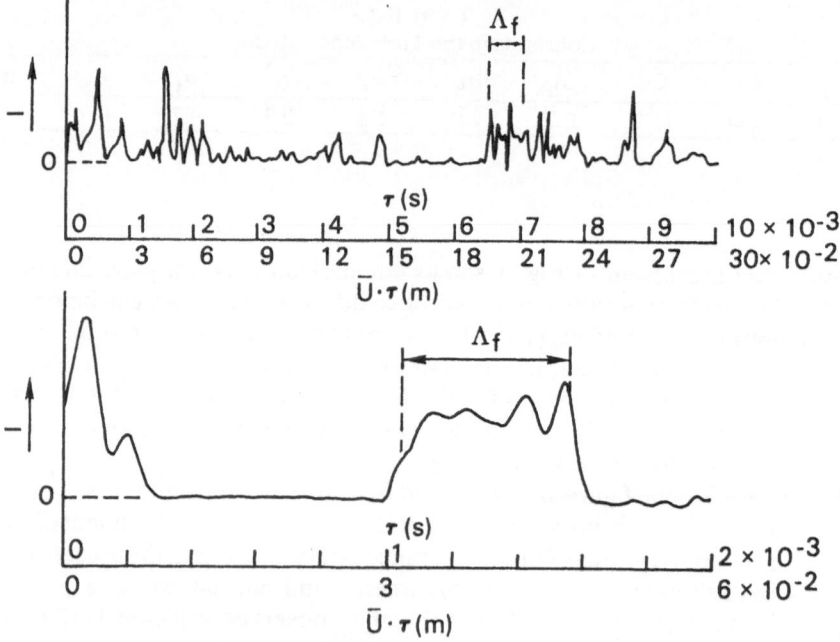

Fig. 2. Recordings of scattered light from soot particles in an acetylene turbulent diffusion flame showing the behavior of soot at the flame axis at X/D=120.

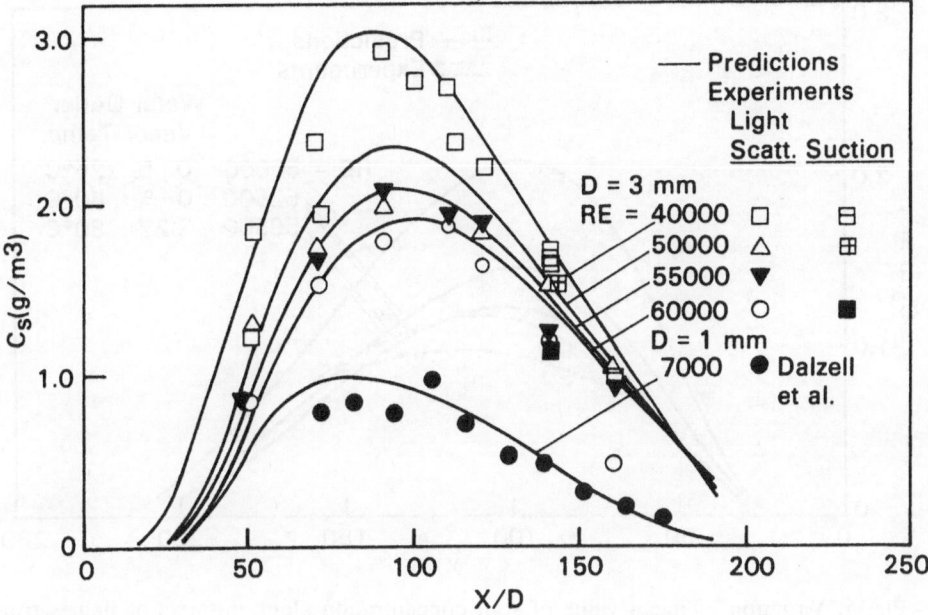

Fig. 3. Variation of mean value of soot concentration along the axis of flames from pure acetylene.

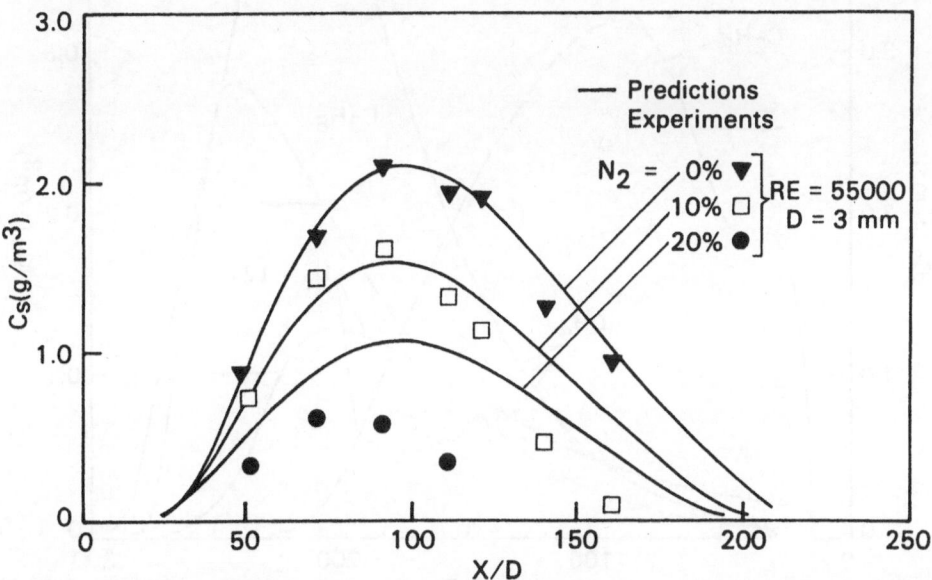

Fig. 4. Variation of mean value of soot concentration along the axis of flames from acetylene diluted with nitrogen.

References pp. 334-335.

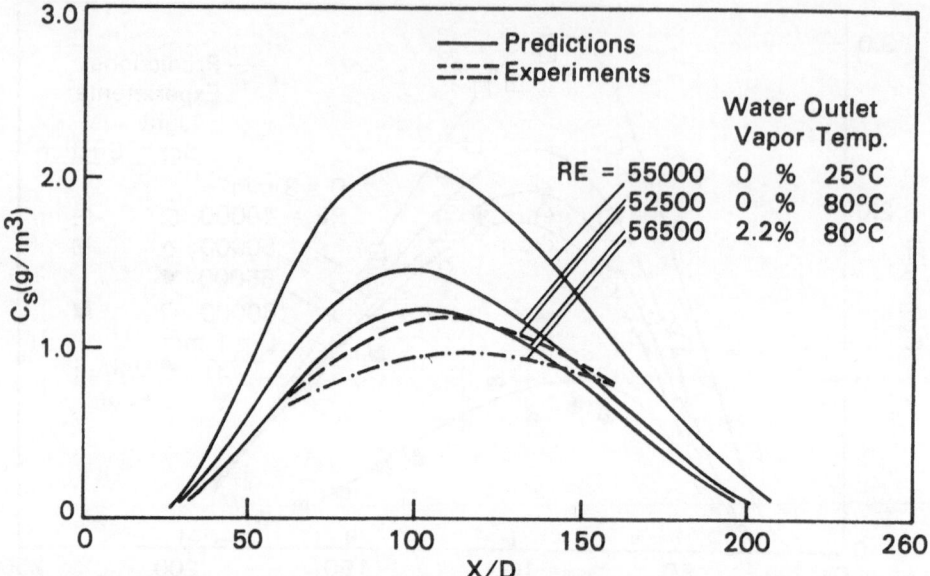

Fig. 5. Variation of mean value of soot concentration along the axis of flames from acetylene with different preheating and with water vapor dilution.

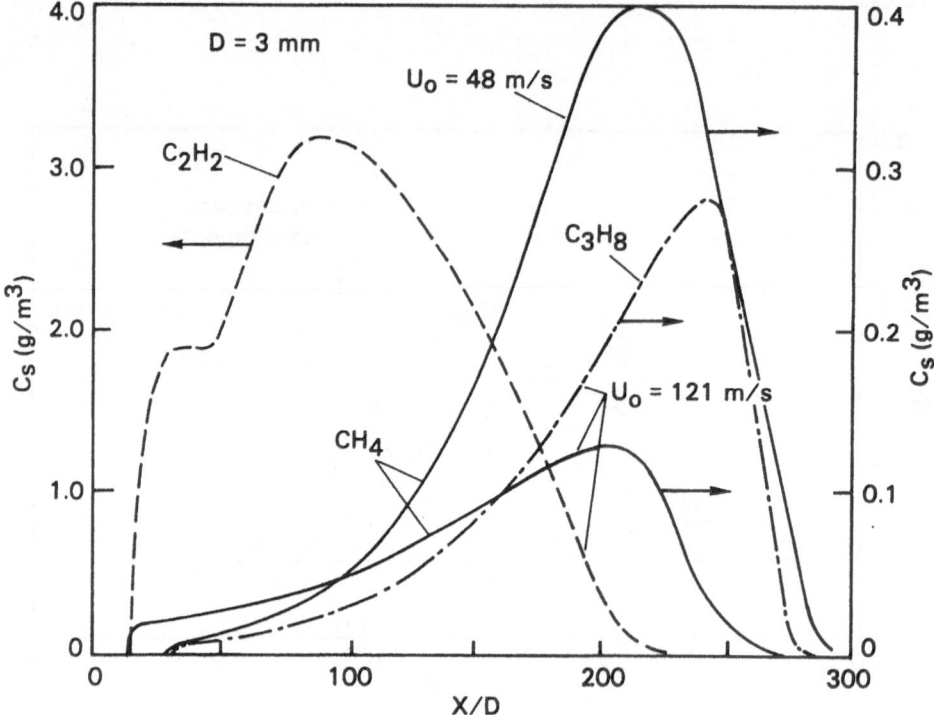

Fig. 6. Calculated soot concentration variation along the axis of flames from C_2H_2, CH_4, and C_3H_8.

Fig. 6 shows results for calculated soot concentration variations along the centerline of C_2H_2, CH_4 and C_3H_8 diffusion flames. It can be observed that the same rate equations with exactly the same constants give soot concentrations which are an order of magnitude lower for the CH_4- and C_3H_8-flames compared to the C_2H_2-flame. The results are in good agreement with what could be expected for these flames [14].

Figs. 7 and 8 show that the major reason for the great difference in soot concentration between the C_2H_2-flame and the CH_4 and C_3H_8 flames is the big difference in flame temperature.

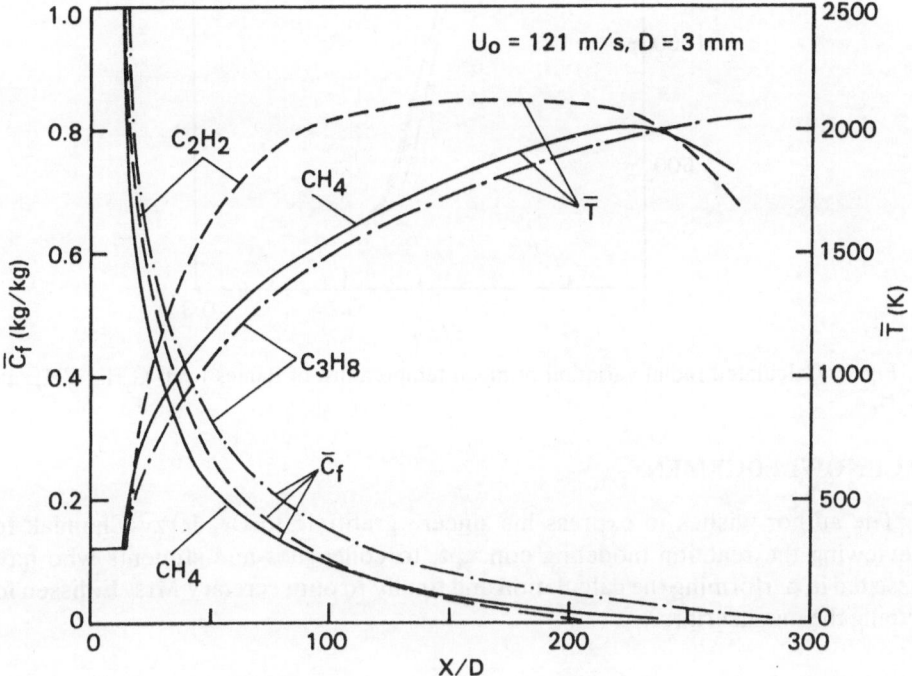

Fig. 7. Calculated variation of mean temperature and fuel concentration along the axis of flames from C_2H_2, CH_4, and C_3H_8.

CONCLUSION

The methods presented here can readily handle complex chemistry and at the same time take care of the turbulence interaction. The soot concentration calculations seem to indicate very promising results with respect to applicability of the method and equations to different fuels and environmental and flow conditions. The most striking advancement in the modeling of soot formation would be a decoupling of the soot formation rate equation for density of particles which would enable introduction of growing soot particles.

References pp. 334-335.

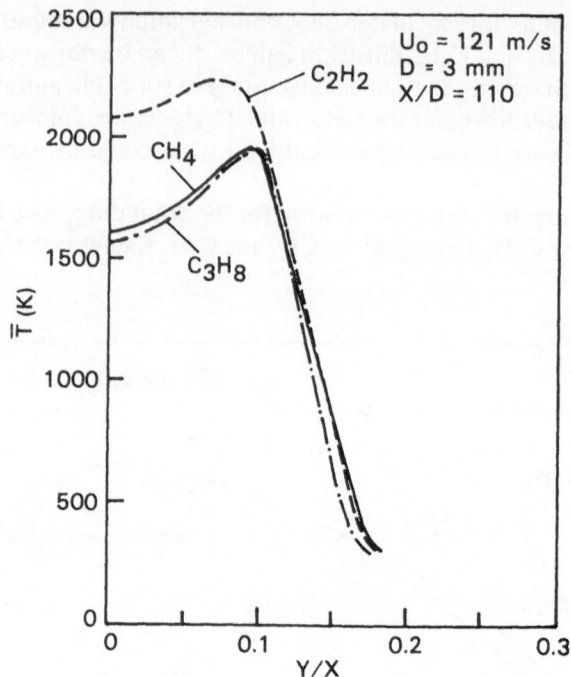

Fig. 8. Calculated radial variation of mean temperature of flames from C_2H_2, CH_4, and C_3H_8.

ACKNOWLEDGEMENT

The author wishes to express his sincere gratitude to Dr. Jerzy Chomiak for reviewing the reaction modeling concept, to colleagues and students who have assisted in performing the calculation and finally to our secretary Mrs. Eidissen for typing the manuscript.

REFERENCES

1. B. F. Magnussen, On the Structure of Turbulence and a Generalized Eddy – Dissipation Concept for Chemical Reaction in Turbulent Flow. Report NTH, (1978).
2. B. F. Magnussen, B. H. Hjertager, J. G. Olsen and D. Bhaduri, Seventeenth Symposium (International) on Combustion, The Combustion Institute, Pittsburgh, (1979), p. 1383.
3. A. N. Kolmogorov, J. Fluid Mech., Vol. 13 (1962), p. 82.
4. S. Corrsin, Phys. Fluids, Vol. 5 (1962), p. 1301.
5. H. Tennekes, Phys. Fluids, Vol. 11 (1968), p. 669.
6. A. Y. Kuo and S. Corrsin, J. Fluid Mech., Vol. 50 (1971), p. 285.
7. A. Y. Kuo and S. Corrsin, J. Fluid Mech., Vol. 56 (1972), p. 477.
8. B. F. Magnussen, Some Features of the Structure of a Mathematical Model of Shear Flow Turbulence, Report NTH, (1975).
9. B. F. Magnussen and B. H. Hjertager, Deuxieme Symposium Europeen sur la Combustion, Section Francaise du "Combustion Institute," (1975), p. 385.
10. B. F. Magnussen and B. H. Hjertager, Sixteenth Symposium (International) on Combustion, The Combustion Institute, Pittsburgh, (1976), p. 719.

11. *P. A. Tesner, T. D. Snegiriova and V. G. Knorre, Combustion and Flame, Vol. 17* (1971), p. 253.

12. *P. A. Tesner, E. I. Tsygankova, L. P. Guilazetdinov, V. P. Zuyev and G. V. Loshakova, Combustion and Flame, Vol. 17 (1971), p. 279.*

13. *T. Takagi, M. Ogasawara, K. Fuju and M. Daizo, Fifteenth Symposium (International) on Combustion, The Combustion Institute, Pittsburgh, (1974), p. 1051.*

14. *H. A. Becker and S. Yamazaki, Sixteenth Symposium (International) on Combustion, The Combustion Institute, Pittsburgh, (1976), p. 681.*

DISCUSSION

J. C. Dent *(University of Technology – Loughborough)*

I would like to discuss how I have utilized Professor Magnussen's concept of soot formation and burn-up in the dissipative eddies of a turbulent flow structure to correlate turbulent mixing rate with exhaust smoke levels and unburned hydrocarbon emissions from diesel engines.

Fig. 1 shows the results of some work conducted in conjunction with AVL on a direct injection diesel engine with intake air swirl, where the effects of turbulence intensity (upper diagram) and Taylor microscale (lower diagram) on exhaust smoke

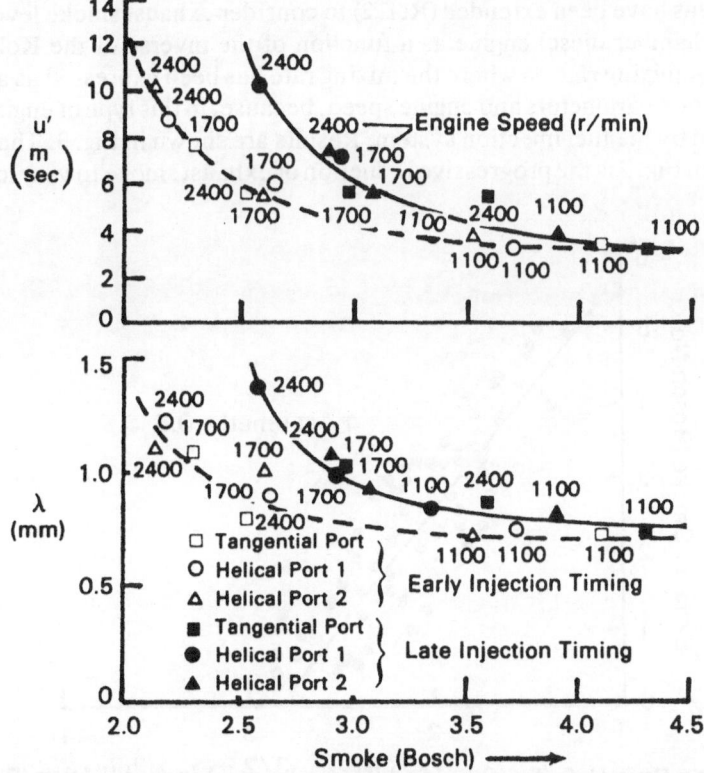

Fig. 1. Variation of exhaust smoke with r.m.s. of turbulence (u') and Taylor Microscale (λ). Data from Brandl et al. (1).

were studied (Ref. 1). This shows that increases in turbulence intensity and scale cause a reduction in exhaust smoke. Using Professor Magnussen's concept enables the replotting of the data in Fig. 1, to give exhaust smoke level as a function of a mixing rate which is the inverse of the Kolmogorov time scale. This is shown in Fig. 2.

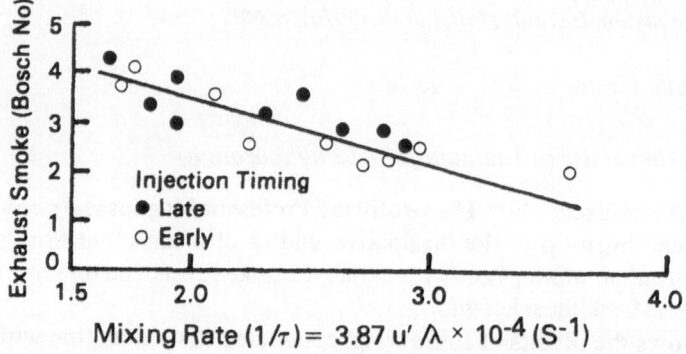

Fig. 2. Replot of data in Fig. 1 on basis of turbulent mixing rate. Figure from Ref. 2.

These ideas have been extended (Ref. 2) to consider exhaust smoke levels from a quiescent chamber diesel engine as a function of the inverse of the Kolmogorov time scale — mixing rate — where the mixing rate has been expressed as a function of fuel injection parameters and engine speed, because, in this type of engine mixing is controlled by the fuel injection system. Results are shown in Fig. 3. The interesting feature in Fig. 3 is the progressive reduction of exhaust smoke to very low values

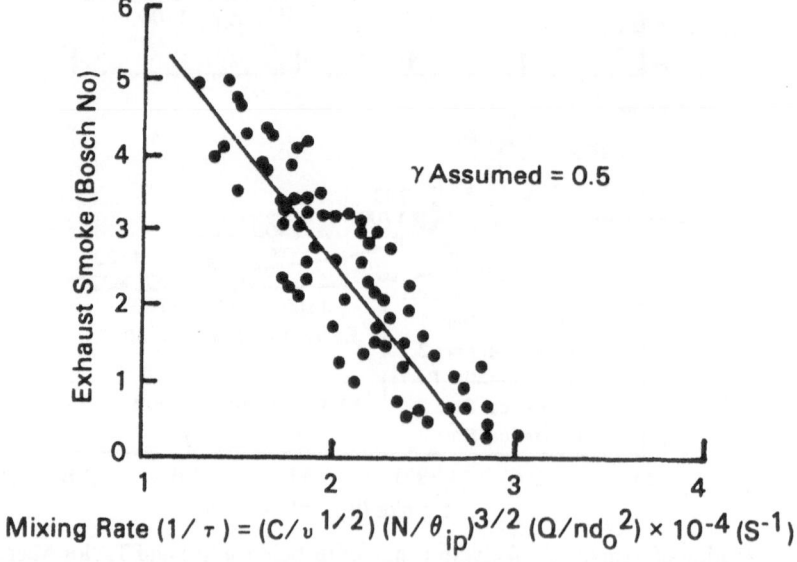

Fig. 3. Exhaust smoke against mixing rate-quiescent chamber engine. Figure from Ref. 2.

with corresponding increase in mixing rate. This poses a problem that at light load conditions a situation can exist where the mixing rates are too high, causing quenching of the combustion process and hence high unburned hydrocarbons. Data shown in Fig. 4 for increased unburned hydrocarbon emission with increasing mixing rate at light load support this idea.

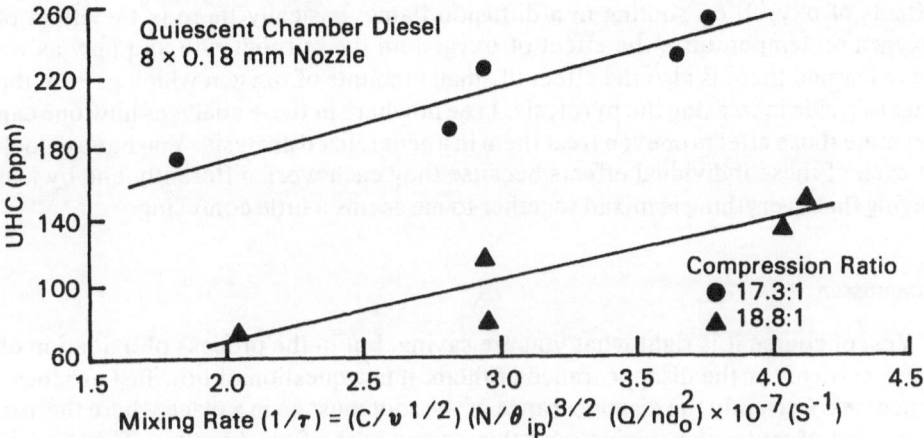

Fig. 4. Effect of mixing rate and compression ratio on unburned hydrocarbons.

REFERENCES

1. *F. Brandl, I. Reverencic, W. Cartellieri and J. C. Dent, S.A.E. Paper No. 790040, (1979).*
2. *J. C. Dent, S.A.E. Paper No. 800092, (1980).*

Magnussen

May I make a comment about this extinction. Certainly this is one of the things which can be treated and which we will treat. This is a discussion, of course, to what extent turbulence can extinguish or stop the reaction just by the cooling effect on the mass transfer. And there are, of course, some experiments which make me doubt whether it is a large scale effect or a fine structure effect. This is not very easy to say, but we are going to mathematically model similar cases just to see whether these fine structures can explain the extinction.

I. Glassman *(Princeton University)*

I am a little confused by Professor Dent's remarks and your paper in that I thought your analyses were for the premixed situation and Professor Dent is reporting data on diffusion flames.

Magnussen

The concept is valid irrespective of diffusion or premixed. We have calculated both premixed and diffusion flames.

Glassman

I am concerned if it is a diffusion flame in the following context. There are three effects of oxygen on sooting in a diffusion flame: basically there is the effect of oxygen on temperature, the effect of oxygen on the burnout rate and now as we have learned there is also the effect of small amounts of oxygen which gets on the fuel-rich side increasing the pyrolysis. I see nowhere in these analyses how one can separate those effects or even treat them in a generalized analysis. You have to look at each of these individual effects because they each work differently and by just saying that everything is mixed together to me seems a little confusing.

Magnussen

Yes, of course it is right what you are saying, but in the process of oxidation of these particles or the disappearance of them, it is a question of how fast are these processes. Certainly the disappearance of the soot must be in a place where there is some sort of molecular mixing, and this occurs in the fine structure. This then is rate-determining, and the question is what is the lifetime within this fine structure.

Glassman

It would seem then from the general discussion that in the process of considering the experiments just mentioned in your analyses that it must be the afterburning of the particles which is the dominating step in giving the correlation between experiment and theory because that is what Professor Dent sees in his experiments.

H. Gg. Wagner *(Universität Göttingen)*

As my first question I would like to ask this. You mentioned experiments involving large scale turbulence, nice large flames. How do they fit into the picture you have described here?

Magnussen

We are now calculating these large scale effects. We have been running large flame experiments on diffusion flames in cross winds in Southern France this summer, and what we have measured, of course, is the flame radiation. We have measured the turbulence characteristics of the air, but the whole tactic was to measure the radiation. We are now running calculations for these flames to predict the soot concentration. The visible flame length as predicted by these methods is in agreement with what we can see. We have a very accurate calculation of the radiation. Of course this is not proof for whether the prediction of soot concentration is correct, possibly in order of magnitude but not in detail. We know that if the scale is large enough, the flame becomes more or less black. But these flames were visible through, and from my feelings, I feel that the concentrations of soot we were calculating were of the same level.

N. A. Chigier *(University of Sheffield)*

In your discussion you concentrated very much on the small scale, fine structure of the turbulence. If one looks at the very interesting measurements which you have of the time histories of the soot concentrations, you can see what appears to be very clear evidence of the large-scale structures, and I wondered why you have placed so much emphasis on the fine structure when I am sure that you must feel, as I do, that the large eddy structures can give you a degree of unmixedness which could play a very dominant role in the whole mixing process.

Magnussen

I do not think that it is correct to say that we are treating just the fine structures and not the big structures. Of course, implicit to this solution is the consideration of large scale effects. Diffusion for instance is a macroscale or large scale effect. The reaction is taking place in the fine structure, of this I am quite certain. But the local concentrations for these fine structures are established by the large scale as you pointed out. So then it is a question what is the reason for what. In my understanding of turbulence, and I may be incorrect, the dissipation is something like a cataract which grows on into the flow where energy is then dissipated into heat and entrains other fluid into the flow. You can see from pictures how this dissipating eddy would develop in the fluid. This is more or less the fine structure character of these fine vortices or slabs or sheets which are developed. In fact, in my treatment it is a laminar flame. In a way I agree with you certainly.

J. P. Longwell *(Massachusetts Institute of Technology)*

I heard Professor Marble of Cal Tech talk about his way of treating the flame as dominated by very large structures. He treats the mixing process at the boundary of the large structures and he treats the rest as a laminar diffusion process. That is an extreme on the other end of the spectrum from where you are going.

Magnussen

No, it is not. It is the same applied in a different way. These laminar sheets he has around these big eddies are the fine structures. This is the location of the fine structures in the fluid. It is another way of trying to tell the same type of story.

Longwell

It did not fit my physical picture, which is very crude, of a Kolmogorov eddy.

Magnussen

Yes, but they are localized. They must appear in stretched regions, and consequently, they are localized more in the space between the bigger eddies. That is necessary. So it is a different approach but more or less of the same kind.

Chigier

To take this point further, it appears that in many of these turbulent flows when you have a diffusion system, you can get large pockets of fuel and air, and they are so large that they are of the order of the dimensions of the mixing layer. Under those conditions you can have virtually complete separation of the fuel and air for long distances down stream. The only place where you can have the fine scale interaction is at the interfaces of these large eddies. So in looking at the mixing process we have to look at both of them. We have to take into account that those large-scale eddies will determine the mixture ratios, and we have to look in a Lagrangian point-of-view at taking into account their variation in time and space. Once we have established what are these overall mixture ratios within the eddies, then we can use the increase of Reynolds number, which will reduce the fine-scale structure, and increase the rate of mixing within these large eddies.

A. F. Sarofin *(Massachusetts Institute of Technology)*

I noticed that your rate of mixing the mass, $\dot{m}$, has the units of reciprocal time. I would very much like to hear a number for that referred to the burning time of the soot particle. Your mixing time versus burnout time ratios would be very interesting to have.

Magnussen

The next time we do these calculations, we will take down the mixing times. It is not completely, $\dot{m}$.

C. Hunter *(Ford Motor Company)*

In thinking about the effect of temperature, if you consider a certain (engine) intake or exhaust oxygen value, if soot concentration is low, maybe a (Bosch) smoke number of 2.5 or something like that, and you raise the temperature, you might not see a measurable effect of temperature. However, if you somehow change other parameters such to create more smoke, you can see a large effect due to temperature. Then you might do some experiment in which you change the temperature, and you do not see an effect so you say the temperature has no effect.

But if you change the injection timing or use EGR to create more smoke and then change the temperature, you see a greater effect. So sometimes you are attenuating formation, and that is the dominant effect, and other times you are attenuating oxidation. With all these different kinds of burners we can do a lot of different things. I think you have to keep that in mind when you say this applies or that applies.

SOOT STUDIES IN GAS TURBINE COMBUSTORS
AND OTHER TURBULENT SPRAY FLAMES

A. M. MELLOR

Purdue University
West Lafayette, Indiana

ABSTRACT

Fuel properties, injector design, combustor inlet conditions and gas turbine combustor design are all known to affect soot emissions in continuous combustion. Although the details of the formation and oxidation processes are only poorly understood, emission control technology can be applied, but frequently at the expense of other performance parameters. After a discussion of turbine combustor results, work performed in simplified axisymmetric spray burners will also be reviewed.

INTRODUCTION

Particulate emissions from jet aircraft were obvious and the source of complaints when this new powerplant first appeared in commercial service in the early 1960's [1]. Although gas turbine particulates have been stated to be nontoxic [2], the current studies of Diesel particulates as bearers of carcinogens should most likely be extended to gas turbine engines. Total emission levels from the latter are quite low however when compared with expected penetration of the Diesel engine into the automotive market.

There are other reasons to control particulate emissions from turbine engines. Soot particles can form deposits within the combustor which can alter fuel and/or air flow distribution from that intended. Hot streaks can result and lead to reduced hot section durability. If these deposits spall, they can lead to significant mechanical erosion of the turbine blades downstream. Increased soot within the combustor can also increase the radiation heat transfer to the combustor walls. Finally, in military applications invisibility of the exhaust plume is highly desirable.

Although for aircraft specification fuels, soot emissions have lent themselves to control primarily through combustion modifications to be discussed in the following, more highly aromatic fuels are appearing [3]. Since some correlation of smoke emissions exists with aromatic content, present engines can be expected to eventually again emit visible smoke. Odgers and Fortier [4] note that whereas aircraft fuels in 1950 contained about 7 wt. % aromatics, today the level is approaching 20% on the average. Further, fuels of the future from alternative sources, notably coal, are expected to be less expensive to refine if allowed to exhibit greatly increased aromatic contents [5].

In one of several alternative fuels programs currently underway, Gleason et al. [6, 7] predicted combustor life reductions of over 60% if the hydrogen content in military aircraft fuels drops from the current 14.5 to 12.0 wt. %; as aromatic content increases, hydrogen content decreases due to the ring structure of the former.

Thus new technology may be required in gas turbine engines if future fuels are to be accommodated without increased engine maintenance and visible smoke emissions. This paper will review the knowledge of soot formation and oxidation in continuous turbulent spray diffusion flames as found in conventional gas turbine engines. Because combustion rather than fuel modifications are more desirable, we shall concentrate on the first approach. Thus the reader is referred to Howard and Kaush [8] for a summary of additive technology, and to Moses [9] or Jahani and Gollahalli [10] as examples of emulsion research.

In turbulent diffusion flames it is generally agreed that soot forms in rich eddies associated with the fuel jet which heat as they approach the reaction zone and subsequently oxidize in contact with air and/or combustion products, where this latter rate is mixing controlled [11, 12, 13]. One obtains a similar situation in the turbulent spray combustion of a volatile fuel [14, 15]; with very heavy fuels liquid phase reactions may be important as well [16]. As implied, the net soot emissions result from a competition between formation and oxidation. Fig. 1 adapted from Toone [17], shows the critical carbon formation and oxidation zones in a conventional combustor. One obvious control technique in such combustors is thus to lean out the region around and inside the fuel spray to decrease the number of fuel rich eddies created per unit fuel flow [7, 18].

A major difficulty in comparing results of various studies, particularly in the older literature, is that many differing soot quantification methods have been proposed. Odgers and Fortier [4] have attempted to relate the various measures, and Dodds et al. [14] and Moses [19] have discussed the relations available between SAE smoke number and particulate concentration in mass per unit (standard) volume. Only approximate relations exist in general, and this is one source for inconsistency in reported trends.

The following sections are divided into review of soot studies in gas turbine engines, in terms of cycle and combustor design, and fuel and injector quality. Other reviews concerned with some of these trends are readily available [20, 21, 22]. More accessible laboratory turbulent spray flames will then be discussed. Finally, a laminar spray flame will be mentioned, because, in moving toward the Diesel engine simulation, an intermittent spray was utilized.

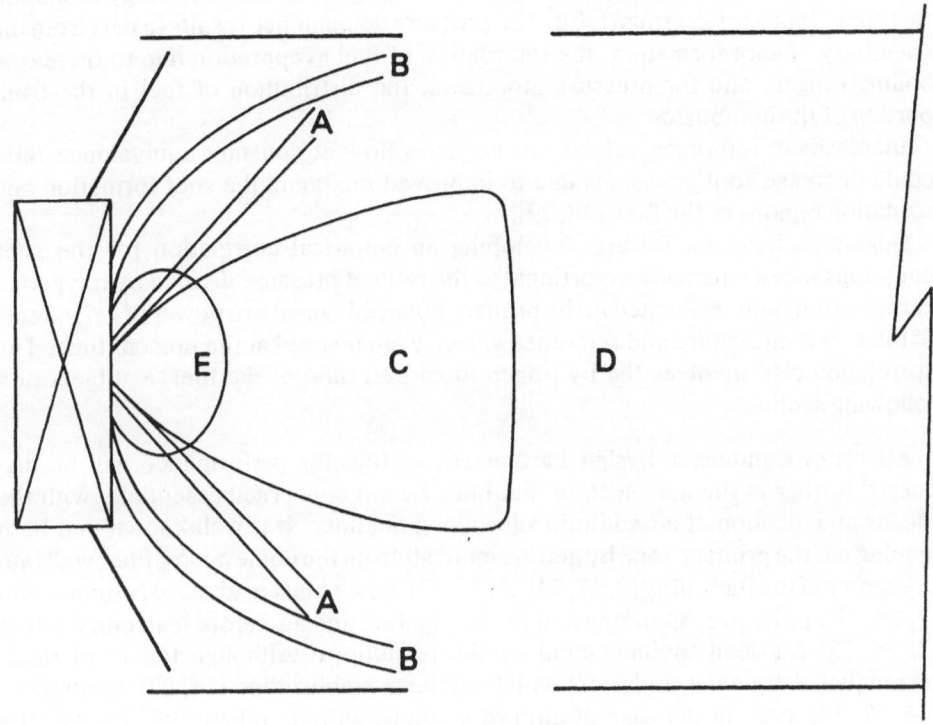

Fig. 1. Schematic of primary zone in conventional gas turbine combustor. Zone A — fuel spray; zone B — shear layer reaction zone; zone C — recirculation zone; zone D — dilution zone; zone E — region of soot formation. After Toone [17].

SOOT EMISSIONS FROM TURBINE ENGINES

In the following only selected citations are chosen to yield examples of trends; no effort has been made to provide a comprehensive review. Further, we emphasize studies in which direct soot measurements are reported, as opposed to those in which the indirect, but related flame radiation and liner wall temperature rise were monitored.

Effects of Cycle Design Parameters — Cycle design parameters are provided to the combustor designer because they are fixed by the type of engine and application. As technology in liner and blade cooling and materials advances, however, the designer generally encounters goals involving increases in inlet air temperature, pressure, and equivalence ratio in succeeding engine generations [23]. Increases in the latter two parameters are generally agreed to increase soot emissions from engines [2, 24-29], but the effect of increasing air temperature will lessen these trends somewhat [26, 27, 28]. Increasing equivalence ratio increases the number of rich, soot-forming eddies per unit mass fuel injected and thus increases net soot emissions. As increasing inlet air temperature generally increases both the rates of soot

References pp. 353-355.

formation and oxidation, the latter must have a higher activation energy to exhibit this effect (but see Glassman [30]). The pressure dependence results in part from the chemistry of soot formation, the retardation of fuel evaporation due to increased boiling ranges, and for pressure atomizers, the distribution of fuel in the front portion of the combustor.

Increases in reference velocity or air mass flow at constant equivalence ratio could decrease soot emissions due to improved mixing in the soot formation and oxidation regions of the flow [28, 29].

Odgers and Fortier [4] are developing an empirical correlation for the soot emissions index which is proportional to the ratio of pressure divided by oxygen to carbon atom ratio, estimated in the primary zone, all raised to a power of 2.7; effects of inlet air temperature and reference velocity are ignored at the present time. The correlation also involves the hydrogen to carbon ratio of the fuel, a subject of a following section.

Effects of Combustor Design Parameters — Injector performance will be discussed further in the next section, but here we are concerned essentially with the means and location of air addition sites along the liner. It is well known that both leaning out the primary zone by getting more air from the dome and/or liner wall into the center of the fuel spray [2, 17, 24, 25, 31, 32], by air blast or air assist atomization [2, 24, 27], or by prevaporizing and premixing fuel and air before lean combustion [2, 24, 32] can lead to substantial smoke reductions. Although for all of these techniques there exist trade-offs with lean flame stabilization [24, 27], ignitability [24, 27, 32], and, in the case of aircraft engines, altitude relight [24, 27, 28], the former two techniques have allowed compliance with smoke emission standards for aircraft engines operating on specification fuels.

These trade-offs arise because of the difficulty in tailoring simultaneously the equivalence ratio distribution in the primary zone for optimal smoke emissions, and flame stabilization and ignitability, over the entire engine operating envelope. Prevaporization and premixing prior to lean combustion reduce these other performance parameters to the point that it is doubtful airworthy engines can be designed, so even with alternative fuels the previous combustion modifications must suffice for aircraft engines. Obviously, automotive and stationary engines do not encounter equally stringent safety requirements; this distinction may ultimately define the quality of fuel to be used in each application (see for example Longwell [5]).

Effects of Fuel and Injector Properties — As noted above, pressure atomizers are not generally used in aircraft today; they have been largely replaced by airblast atomizers for their lower soot emissions. However, in the older literature there are several good discussions of how they interface with the smoke formation process (see for example [25, 27, 33, 34]). As can be imagined from Fig. 1, both fuel spray cone angle and penetration are quite important.

With regard to fuel properties, aromatic content is regarded as most significant [19, 29, 35-37]. In fact, several investigators report that smoke emissions increase linearly with decreasing wt. % hydrogen [6, 7, 26], but in specially blended fuels with identical hydrogen content Naegeli and Moses [38] find increases with increasing polycyclic aromatics. Odgers and Fortier [4] report a preliminary correlation of soot emissions index as a function of carbon atom to hydrogen atom ratio in the fuel with an exponent of 5.45.

Any effect of fuel viscosity or injector design through the Sauter mean diameter of the spray seems small or unclear [6, 7, 27]. Similarly, no effect of fuel volatility has been reported [6, 7] or of fuel end boiling point, at least to 675 K [6, 7, 19, 36-38].

Future Needs — Leaning out the primary zone by an appropriate air penetration jet scheme and/or by utilizing some type of airblast atomization can make gas turbine combustors insensitive to small variations in hydrogen content, at least for petroleum-derived fuels. One important research need is to ascertain the minimum degree of hydrotreating of coal liquids which is required for combustion without unacceptable smoke emissions in aircraft engines utilizing modern technology. Similarly, prevaporizing/premixing combustors for stationary and automotive turbine engines are expected to require somewhat less refined fuels, but the minimum refining is in need of specification. However, it must be realized that other considerations such as nitrogen, sulfur, and metal contents, or thermal stability may ultimately define the degree to which synfuels are hydrotreated and refined.

As has been implied, smoke emission suppression in gas turbine engines has been largely the result of expensive and empirical testing. More fundamental studies can hopefully contribute to the understanding of soot formation and oxidation processes in steady or intermittent spray flames; these are the topic of the following section.

SOOT EMISSIONS FROM LABORATORY SPRAY FLAMES

Continuous Combustion — Prado *et al.* [15] have examined soot (and polycyclic aromatic hydrocarbon) production in a quasi-steady swirl-stabilized spray flame at atmospheric pressure and temperature. Fuels examined were kerosene, benzene, and benzene with 0.5 wt. % anthracene, and air assist atomization was utilized. Dodds *et al.* [14] and Wyatt *et al* [39] were concerned with soot emissions from disc-stabilized spray flames at a (vitiated) air inlet temperature of 700 K and pressure of 304 kPa (3 atm). A simplex pressure atomizer mounted in the center of the disc was used for fuels including two Jet A's, JP 4, JP 5 refined from shale syncrude, DF 2, and a Jet A blended with 15% residual rubelene to increase the end boiling point of the fuel to 838 K. All of the studies used water-cooled probes and gravimetric analysis for local soot mass concentration measurements; Prado *et al.* [15] quenched their samples within the probe by direct water injection. None of the probes were isokinetic due to the method of probe insertion and presence of reversals in the flow, although Dodds *et al.* [14] addressed this issue.

High resolution electron microscopy was utilized by Prado *et al.* [15] to obtain arithmetic mean particle diameters of the spherical building units found in the soot chains, whereas Wyatt *et al.* [39] employed low resolution electron microscopy, photoextinction in Stokes flow to measure equivalent diameters of the agglomerates greater than 0.2 μm, and BET analysis for specific surface area. Dodds *et al.* [14] and Wyatt *et al.* [39] also examined the effects of sampling train design on their results.

Typical mass concentration measurements on the centerlines of the flames are shown in Figs. 2 through 4. At the lower pressure and with swirl opening the fuel spray, Prado *et al.* [15] were able to observe the soot formation region in Fig. 2, whereas Dodds *et al.* [14] and Wyatt *et al.* [39] found only the oxidation/dilution zone (Fig. 3). The latter result was also observed by Jahani and Gollahalli [10]. The measurements in Figs. 2 and 3 are comparable at the larger pressure drops across the air assist atomizer, and, as anticipated, this latter parameter has an important effect on net soot emissions.

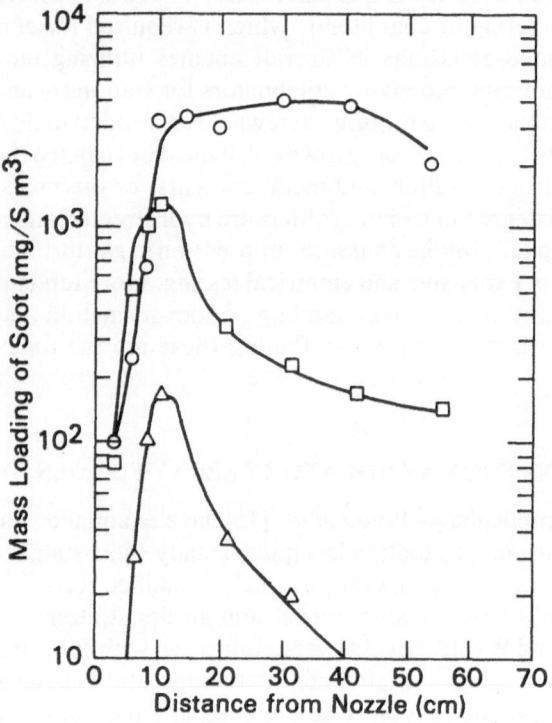

Fig. 2. Influence of atomizing-air pressure on axial profile of soot mass loading. Kerosene/air; equivalence ratio = 1.0; cold gas velocity = 0.96 m/s; inlet air temperature = 300 K; inlet air pressure = 101 kPa (1 atm); atomizing-air pressure = (o) 184 kPa (12 psig), (□) 205 kPa (15 psig), and (Δ) 239 kPa (20 psig). From Prado *et al* [15].

The results of fuel variations in poorly atomized sprays are displayed in Fig. 4 [15]. In general the sooting tendency in the flames of Figs. 3 and 4 increases with increasing aromatic content of the fuel (aromatic content of Jet A purchased in Indiana was considerably larger than that bought in Texas). However, Dodds *et al.* [14] found increased emissions attributed to effects of both front end volatility with JP 4 (not observed in the JT8D engine [1]) and end boiling point with Jet A blended. Note that this fuel has a final boiling temperature 160 K higher than the fuels tested to date in gas turbines discussed in the previous section, so that liquid phase chemistry [16] could be important in soot formation in this case [14].

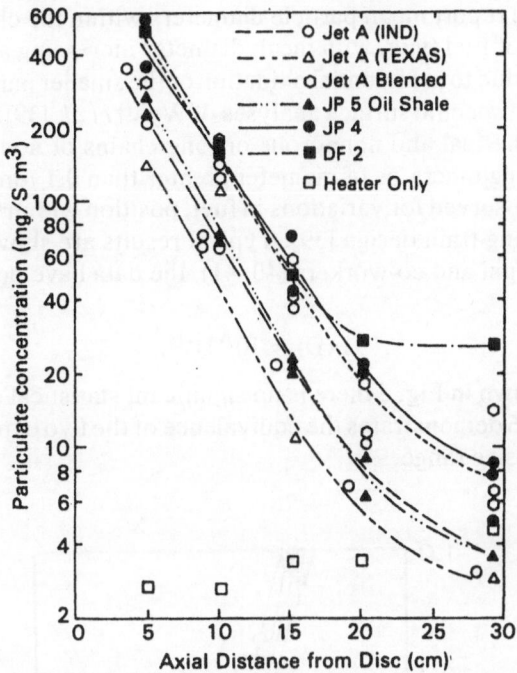

Fig. 3. Influence of fuel type on axial profile of soot mass loading. Equivalence ratio = 0.25; air approach velocity = 39.4 m/s; inlet air temperature = 700 K; inlet air pressure = 304 kPa (3 atm). From Wyatt *et al.* [39].

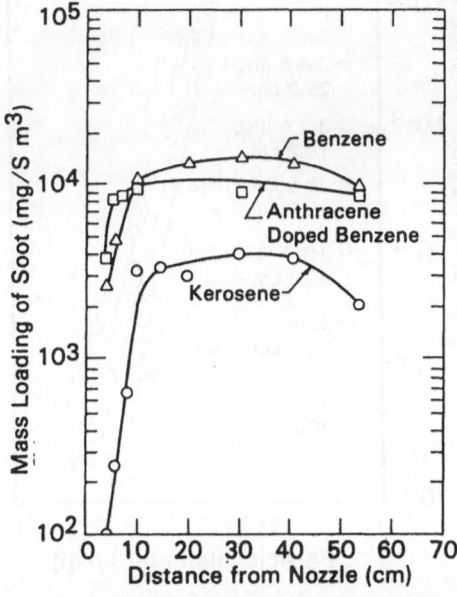

Fig. 4. Influence of fuel type on axial profile of soot mass loading. Fuel equivalence ratio = 1.0; air atomizing pressure = 184 kPa (12 psig); cold gas velocity = 0.96 m/s; inlet air temperature = 300 K; inlet air pressure 101 kPa (1 atm). From Prado *et al.* [15].

References pp. 353-355.

Prado *et al.* [15] report mean particle diameters within the chains of order 0.02 μm, independent of fuel type, with mean diameter increasing as the particles pass through the flame due to preferential oxidation of the smaller particles. This result is consistent with the specific surface analyses of Wyatt *et al.* [39] if their particles are assumed to be spherical and nonporous or long chains of nonporous spheres. In addition, for the agglomerates of diameter greater than 0.1 μm, no change in size distribution was observed for variations in fuel, position within the flame, counting method, or sampling train design [39]. Typical results are shown in Figs. 5 and 6; here, following Pagni and co-workers [40, 41], the data have been fit to a distribution law

$$F(D) = 10^A D^{-B}.$$

For the results shown in Fig. 5 there is no significant statistical difference between the fuels, and Fig. 6 demonstrates the equivalence of the two experimental methods in the overlapping size range.

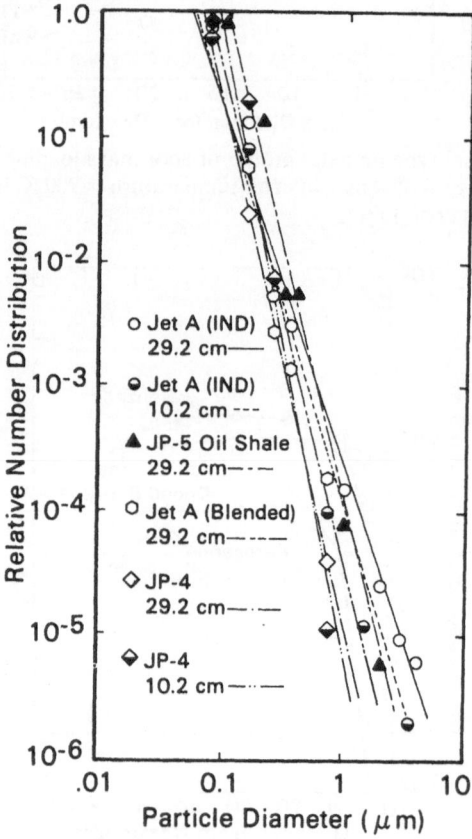

Fig. 5. Particle size distributions for a variety of fuels. Equivalence ratio = 0.25; air approach velocity = 39.4 m/s; inlet air temperature = 700 K; inlet air pressure = 304 kPa (3 atm). From Wyatt *et al.* [39].

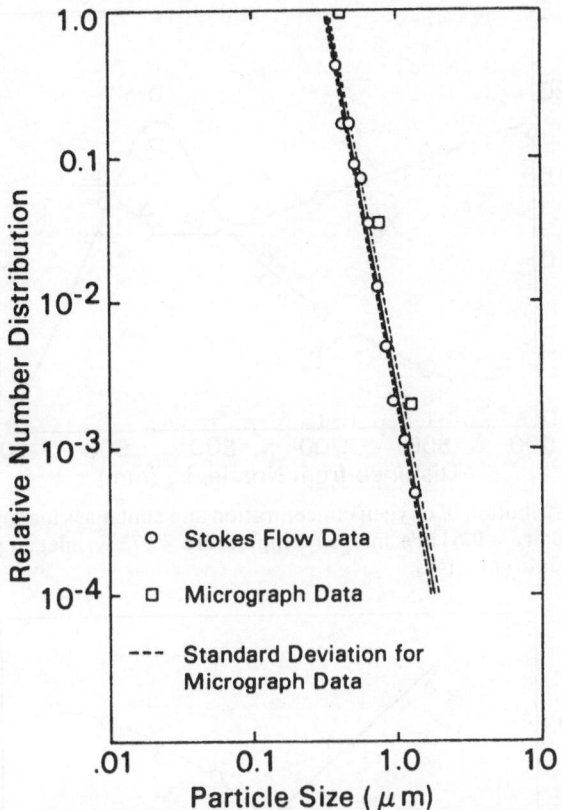

Fig. 6. Comparison of micrograph and Stokes-flow size distribution data for the 29.2 cm location (DF-2). Equivalence ratio = 0.25; air approach velocity = 39.4 m/sec; inlet air temperature = 700 K; inlet air pressure 304 kPa (3 atm). From Wyatt *et al.* [39].

This result for the agglomerates is not completely surprising. The great majority of particles emitted by flames of these types is known to have diameters less than 0.1 μm [42]. Further, using soot oxidation data based on the Nagle and Strictland-Constable correlation [22, 43], we estimate that at an overall equivalence ratio of 0.25 used by Wyatt *et al.* [39] the surface regression rate is only 6 x 10^{-4} μm/ms, too small to be detected. This of course is a kinetic rate, the maximum possible if the oxidation rate is mixing controlled in this turbulent diffusion flame.

Intermittent Combustion — It is of interest to compare results from the continuous flames with those obtained with a pulsed spray into a steady air flow. Kadota *et al.* [44] utilized a diesel fuel injection system to inject DF 2 into a co-flowing air stream at atmospheric pressure but heated to temperatures of 823 to 1073 K. Low air velocities were used such that the flow was laminar, and an inert gas quenching probe stationed at various positions downstream from the autoignition event collected the soot. Figs. 7 and 8 show typical results, where mass concentration was determined gravimetrically and mean diameter by electron microscopy. The number of particles, N, is calculated from the previous results. Here both regions of

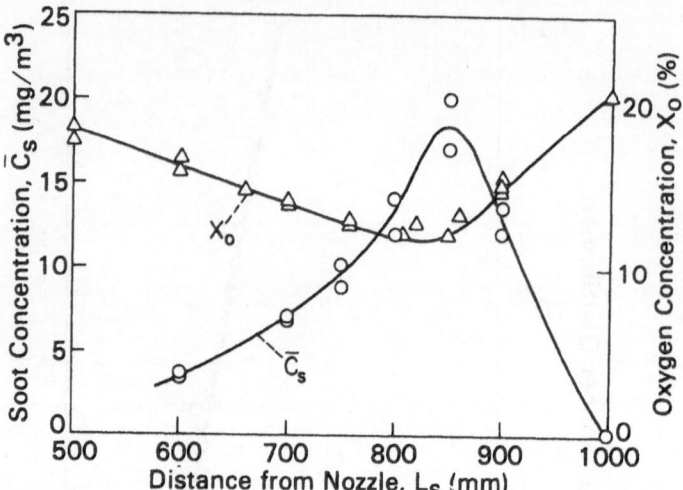

Fig. 7. Axial distribution of oxygen concentration and soot mass loading (DF 2). Intermittent spray; air velocity = 0.64 m/s; inlet air temperature = 973 K; inlet air pressure = 101 kPa (1 atm). From Kadota *et al.* [44].

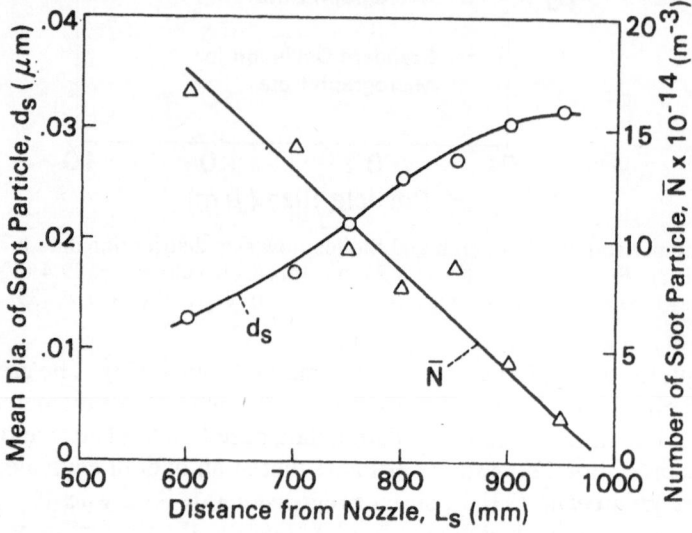

Fig. 8. Axial distribution of the mean diameter and number of soot particles (DF 2). Intermittent spray; air velocity = 0.64m/s; inlet air temperature = 973K; inlet air pressure = 101 kPa (1 atm). From Kadota *et al.* [44].

formation and oxidation are seen, soot loadings are considerably less than reported for the steady flames, and the surviving mean particle size is somewhat larger.

One possible reason for the lower particle concentration and more complete oxidation in the pulsed spray is suggested by the oxygen concentration profile in Fig. 7. Apparently each pulse is exposed to excess oxygen throughout its combustion, unlike the situation in the diesel [45]. The measured peak gas temperature is reported to be about 1500 K, much less than that expected in a diffusion flame.

Therefore perhaps each fuel packet burns as a "premixed" spray, with individual droplet diffusion flames precluded due to lowered oxygen concentrations and the large relative velocity between the fuel packets ($\simeq$ 40 mm/ms) and air flow (0.64 mm/ms). Wake flames [46] or lean homogeneous reactions would then be important. Increasing the air velocity is expected to further improve the mixing and reduce the soot emissions, as was observed by Kadota *et al.* [44].

Thus the combustion process in this experiment may be more typical of the first, premixed phase of diesel engine combustion than of the turbulent diffusion flame which follows surrounding the rest of the fuel which is injected. Based on the relative amounts of soot formed in the studies of Kadota *et al.* [44] and those in the turbulent spray diffusion flames discussed in the previous section, we conclude it is the second phase of diesel engine combustion which is responsible for the majority of soot emissions.

CONCLUSIONS AND SUGGESTED RESEARCH

If this last conclusion is correct, then perhaps the quasi-steady spray flames are a better simulation of the turbulent diffusive portion of diesel combustion, and the pulsed spray experiment of the premixed portion. Both warrant further study. It is of interest to extend the latter work to turbulent air flows to see how this affects the results. Although the gas turbine combustor results indicate how soot emissions vary with operating parameters, these dependencies should be verified in the laboratory flames, which have tended to emphasize fuel and injector effects. In general, results from the laboratory flames appear to be consistent. Both laboratory and engine studies should continue so that lower hydrogen content fuels can be accommodated in gas turbines.

The success in terms of soot emission control which the gas turbine combustor currently enjoys with specification fuels will not come easily for the diesel engine. The control techniques are the same, but their implementation in diesel hardware remains elusive and a challenge for the engineer.

ACKNOWLEDGEMENTS

Professor C. R. Ferguson of Purdue University provided valuable discussion of this paper. The author's current work on soot formation in spray flames is supported by the Army Tank-Automotive Research and Development Command (A. Jaeger and J. Lewakowski, technical monitors), the Army Mobility Equipment Research and Development Command (F. W. Schaekel, technical monitor), the Army Research Office (J. Murray and R. Singleton, technical monitors), and the Alternative Fuels Utilization Branch of the Department of Energy (R. Fleming, DOE and J. Clark, NASA-Lewis, technical monitors), to all of whom the author expresses his gratitude.

REFERENCES

1. R. E. George, J. A. Verssen and R. L. Chass, *Air Poll. Control Assoc.*, 19 (1969), pp. 847-855.
2. T. Durrant, *SAE Paper No. 680347*, (1968).

3. *P. P. Campbell, NASA Report No. CP 2146, (1980), pp. 11-14.*

4. *J. Odgers and P. Fortier, Report No. j.o. 91, Laval University, (1980).*

5. *J. P. Longwell, "Sixteenth Symposium (International) on Combustion," The Combustion Institute, Pittsburgh, (1977), pp. 1-15.*

6. *C. C. Gleason, T. L. Oller, M. W. Shayeson and D. W. Bahr, AFAPL-TR-79-2015, (1979).*

7. *C. C. Gleason, T. L. Oller, M. W. Shayeson and D. W. Bahr, AFAPL-TR-79-2018, (1979).*

8. *J. B. Howard and W. J. Kaush, Jr., ESL-TR-79-32, (1979).*

9. *C. A. Moses, WSS/CI Paper No. 76-34, Western States Section/Combustion Institute, (1976).*

10. *H. Jahani and S. R. Gollahalli, Combust. Flame, Vol. 37 (1980), pp.145-154.*

11. *W. H. Dalzell, G. C. Williams and H. C. Hottel, Combust. Flame, Vol. 14 (1970), pp. 161-170.*

12. *B. F. Magnussen, "Fifteenth Symposium (International) on Combustion," The Combustion Institute, Pittsburgh, (1975), pp. 1415-1425.*

13. *B. F. Magnussen and B. H. Hjertager, "Sixteenth Symposium (International) on Combustion," The Combustion Institute, Pittsburgh, (1977), pp. 719-729.*

14. *W. J. Dodds, J. E. Peters, M. B.Colket, III and A. M. Mellor, AIAA J. Energy, Vol. 1 (1977), pp. 115-120.*

15. *G. P. Prado, M. L. Lee, R. A. Hites, D. P. Hoult and J.B. Howard, "Sixteenth Symposium (International) on Combustion," The Combustion Institute, Pittsburgh, (1977), pp. 649-661.*

16. *A. Williams, "Energy and Combustion Science," Student Edition 1, Pergamon, Oxford, (1980), pp. 135-147.*

17. *B. Toone, "Combustion in Advanced Gas Turbine Systems," Pergamon, Oxford, (1968), pp. 271-296.*

18. *C. C. Gleason and D. W. Bahr, NASA CR-135413 (1979).*

19. *C. A. Moses, WSS/CI Paper No. 75-18, Western States Section/Combustion Institute, (1975).*

20. *J. B. Heywood, J. A. Fay and L. H. Linden, AIAA J., Vol. 9 (1971), pp. 841-850.*

21. *L. H. Linden and J. B. Heywood, Combust. Sci. Technol., Vol. 2 (1971), pp. 401-411.*

22. *S. W. Radcliffe and J. P. Appleton, Combust. Sci. Technol., Vol. 4 (1971), pp. 171-175.*

23. *R. E. Henderson and A. M. Mellor, Editors, AFAPL-TR-79-2115, (1979).*

24. *J. J., Faitani, SAE Paper No. 680348, (1968).*

25. *K. Gradon and S. C. Miller, "Combustion in Advanced Gas Turbine Systems," Pergamon, Oxford, (1968), pp. 45-76.*

26. *R. M. Schirmer, "Emissions from Continuous Combustion Systems," Ed. W. Cornelius and W. G. Agnew, Plenum Press, New York, (1972), pp. 189-210.*

27. *E. R. Norster and A. H. Lefebvre, "Emissions from Continuous Combustion Systems," Ed. W. Cornelius and W. G. Agnew, Plenum Press, New York, (1972), pp. 255-278.*

28. *J. S. Grobman, "Emissions from Continuous Combustion Systems," Ed. W. Cornelius and W. G. Agnew, Plenum Press, New York, (1972), pp. 279-303.*

29. *H. F. Butze and F. M. Humenik, NASA TM-79089 (1979).*

30. *I. Glassman, AFOSR TR 79-1147, (1979).*

31. *N. J. Friswell, "Emissions from Continuous Combustion Systems," Ed. W. Cornelius and W. G. Agnew, Plenum Press, New York, (1972), pp. 161-182.*

32. *D. W. Bahr, "Emissions from Continuous Combustion Systems," Ed. W. Cornelius and W. G. Agnew, Plenum Press, New York, (1972), pp. 345-373.*

33. *A. H. Lefebvre, "Combustion in Advanced Gas Turbine Systems," Pergamon, Oxford, (1968), pp. 3-19.*

34. *A. H. Lefebvre, "Combustion in Advanced Gas Turbine Systems," Pergamon, Oxford, (1968), pp. 211-226.*

35. *W. S. Blazowski, "Sixteenth Symposium (International) on Combustion," The Combustion Institute, Pittsburgh, (1977), pp. 1631-1639.*

36. C. A. Moses and D. W. Naegeli, ASME Paper No. 79-GT-178, (1979).
37. C. A. Moses and D. W. Naegeli, "Gas Turbine Combustor Design Problems," Hemi-
    sphere, Washington, (1980), pp. 39-69.
38. D. W. Naegeli and C. A. Moses, ASME Paper No. 80-GT-62, (1980).
39. W. R. Wyatt, J. A. Clark, J. E. Peters and M. A. Mellor, AIAA J. Energy, Vol. 3 (1979),
    pp. 285-290.
40. P. J. Pagni and L. Hughes, WSS/CI Paper 72-75, Western States Section/Combustion
    Institute, (1972).
41. P. J. Pagni, L. Hughes and T. Novakov, "Atmospheric Pollution by Aircraft Engines,"
    AGARD CP No. 125, (1973).
42. D. L. Fenton, E. H. Luebcke and E Norstrom, ASME Paper No. 79-GT-179, (1979).
43. C. Park and J. P. Appleton, Combust. Flame, Vol. 20 (1973), pp. 369-379.
44. T. Kadota, N. A. Henein and D. U. Lee, Paper No. CSS/CI-80-02, Central States
    Section/Combustion Institute, (1980).
45. Y. Aoyagi, T. Kamimoto, Y. Matsui and S. Matsuoka, SAE Paper No. 800254, (1980).
46. A. Sjogren, "Fourteenth Symposium (International) on Combustion," The Combustion
    Institute, Pittsburgh, (1973), pp. 919-927.

DISCUSSION

G. Prado (Massachusetts Institute of Technology)

I wanted to add a comment on the results you show from the particle size. I do not want anyone to think that the particle size increases during combustion. The size that you report is the average size of the population. We see that the small particles burn much faster than the large particles with the net result that the average size of the total population increases.

Mellor

Thank you Gilles. We have exactly the same interpretation.

F. L. Dryer (Princeton University)

I would like to augment a phrase that Prof. Mellor stated during his presentation concerning the importance of liquid-phase coking of material and the possibility for that mechanism to also produce carbon particulate. We have neglected discussion of that mechanism for the past two days based primarily on the premise that the morphology of the particulates that we see from diesels and from gas turbines look much like that that comes from premixed systems. We have compared that with the morphology of particulates that occurs in stationary combustion systems where we are certain coking occurs. This coking then results in very large particles, and those large particles result from the fact that spray size-distribution is very large in stationary systems in comparison to diesels. I would address the audience with the question that if one looks at the ratio of cenosphere size of this coke particulate to the initial droplet of the mean diameter for the stationary combustion system and also looks at the ratio of the spheroid size for gas turbines and diesels to the primary droplet size, you will find that those ratios are very similar. I wonder if we can really exclude the effects of liquid-phase coking from the diesel combustion problem. If one looks at some of the data which GM has produced on the effects of end points of the fuel, there is a definite trend of soot formation with end point. One should recognize that the fact that diesels operate at very high pressures means that the liquid phase achieves very high temperatures during the vaporization. In fact, temperatures are achieved that are equivalent to those that are achieved by stationary combustion of heavy fuels at atmospheric pressure.

Mellor

In the gas turbine literature which I reviewed in the paper, everyone says that there is no effect of end point on soot emission and therefore liquid-phase reactions are not important. The highest end-point that was examined in those studies was by Cliff Moses at Southwest Research, and it was a fuel which had an end point of 675 K. The particular blended fuel that we used where we thought that perhaps these liquid-phase reactions were becoming important at least in our flame had an end point 160 K higher. So that is 838 K. Whether diesel fuel, which was included both in the work at Southwest Research and in our own work, would exhibit this is unknown. However, of course, the flame is different in a diesel from that in a gas turbine, and it certainly is possible.

M. Brown (Exxon Research and Engineering)

If you were to do a calculation based on the number of droplets which are injected into the cylinder, you could determine the potential number of soot particles. If you do this you would find that there are about a million particles of soot for every droplet injected. Now that does not exclude the droplets as being a source of soot, but the relative importance of individual droplets turning into soot particles puts that in prespective.

Dryer

I would like to reply to that. There is something that is very interesting which happens to multi-component fuels during their combustion history as the concentration of the heavy components of the remaining fuel drop. Both Dr. Sangiovani, United Technologies Research Center, and ourselves at Princeton have noticed that there is a termination of combustion of such multi-component fuels. There is a very bright flash and that bright flash happens to be the dispersal of the remaining coke particle. So I would not want to attribute one single coke particle to each droplet. That is the first point. The second point is this area of nucleation initiation that Dr. Bauer has pointed out. I am wondering what the presence of solid coke particles from droplets does to his calculation.

Brown

Yes, I think you are right because if you do correlate the amount of soot produced per the fraction of the blend of the fuel that boils above the 90% volume point, you do find a tremendous increase, so there probably is some support.

A. D'Alessio (*University of Naples*)

My comments are of a general nature in the same sense as Prof. Glassman's yesterday. In regard to spray combustion, we are at best treating it as a two-dimensional problem, but actually it is a three-dimensional problem, so I am against a physical view from one dimension. You just showed what is happening along the axis of the spray, but the most important things may happen just on the border of the spray and outside. So what is going to happen along the axis is just a part of the story, and the rest of the story may happen in another place. We have evidence from light-scattering measurements that soot is formed on the border of the spray in high temperature regions since you need high temperatures in order to get soot. So in general when talking about spray flames, I would like to see both radial and axial distributions specified.

Mellor

I could not agree more. We are in the process of doing those experiments right now in our flames. The point of these measurements was to establish a ranking relative to diesel fuel which is the baseline for which we will be doing a detailed mapping.

A. S. Gordon (*California Air Resources Board*)

Fred Dryer's comment prompts me to ask the following question. In these fuels do we get to the point at high pressure that the temperature approaches the critical temperature and the critical pressure of these fuels, and what would be the consequence of that kind of situation?

Mellor

Not in the situation that I talked about here, nor in gas turbine applications, but in the diesel I think you might.

A. F. Sarofim (*Massachusetts Institute of Technology*)

Prof. Mellor, if I understood you right, you said that you could probably correlate your burn-out rate with the Nagle and Strickland-Constable relationship.

Mellor

I said I could correlate the lack of any observed burn-out.

Sarofim

The comment I would like to make is directed to both you and Prof. Magnussen as to whether you would expect the kinetic limitations or the mixing limitations to be controlling?

Mellor

Since these are turbulent diffusion flames, I would expect mixing to be controlling. I would simply estimate the maximum possible rate. Thank you for pointing that out. These flames were not at all different from gas-fed turbulent diffusion flames because in my opinion all these flames exhibit the case where the rate of fuel vaporization is very rapid compared to the rate of fuel-air mixing.

U. Bonne (Honeywell)

I wonder in your trends, have you ever tried changing or do you plan to change the inlet temperature of your injection jet as a form of a preheat above the boiling temperature (of the fuel) at the chamber pressure but still below the boiling temperature at your injection pressure. In this way you could get flash vaporization at the time of injection.

Mellor

Yes, we did.

Bonne

You would get smaller particles and greater mixing which would maybe suppress soot formation in that way.

Mellor

Yes, we did that experiment with propane because propane will flash evaporate under those conditions. It still forms sooty flames, and the only way that we could run this experiment with a practical multi-component fuel, superheated, would be to use a second air heater to heat the fuel. It requires a tremendous amount of energy even though it is not a very high flow rate. For diesel fuel, for example, we are talking about tremendous amounts of energy, so we have not done it. But in the sense that we did do it with pure propane which unfortunately is not a multi-component fuel, you still see yellow flames, so apparently it does not have that much effect.

Bonne

Your comment about yellow flames is interesting. We did some studies a couple of years ago. We wanted to see how much soot formation would be changed by preheating the fuel, and that was with a very simple plug-flow combustor with oil injection at one end. We were able to get essentially blue flames as opposed to the luminous normal oil combustion flames that one has in the original system.

Mellor

Yes, I would think that that might happen. The reason I was hedging is because I do not recall if a propane flame exhibits flame conditions which are yellow or not. We did not do soot measurements with propane although I would like to.

J. P. Longwell (Massachusetts Institute of Technology)

You found a pronounced effect of the end point of the fuel on your smoke formation. The nature of petroleum products is that if you look at the high boiling fractions, they tend to have an increasing number of polynuclear aromatics including benzene or even three-ring compounds as you go up in temperature. We heard this first paper today which pointed out that things like methyl napthalene are in a class by themselves. So I think there is a good chance there are some important chemical effects as well as the physical ones that we discussed, and I was wondering if you would have any hydrocarbon analyses that would bear this out.

Mellor

Yes, we do, but we do not happen to have them here. The blend was with 15% rubelene. We are also doing radiation measurements normal to these flames, and we do now have the PCAH analysis for the fuels. There is an effect, a very definite effect, of di- and tri-cyclic compounds. Cliff Moses of Southwest Research Institute has also seen this same effect although he is using blended fuels with extremely high concentrations, perhaps 20%, of the polycyclic hydrocarbons. We are using for most of our experiments fuel derived from petroleum so we are picking up the natural concentrations of 2-3% perhaps. But there is an effect on radiation, and, therefore, there is definitely an effect on soot.

SESSION IV
FORMATION DURING
INTERMITTENT COMBUSTION

Session Chairman
J. B. HEYWOOD
Massachusetts Institute of Technology
Cambridge, Massachusetts

A PHENOMENOLOGICAL DESCRIPTION OF PARTICULATE FORMATION DURING CONSTANT VOLUME COMBUSTION

T. M. DYER and W. L. FLOWER

Sandia National Laboratories
Livermore, California

ABSTRACT

Experiments to characterize soot particulate formation processes during the combustion of rich homogeneous mixtures have been conducted in a constant volume combustion bomb. Soot formation is observed to occur very rapidly in the flame front, and the particles are dispersed over the burned gas region as these gases are continually compressed. Particles sampled and analyzed by transmission electron microscopy exhibit three-dimensional chain-like structures. In the present premixed experiments, any parameter change that increases the flame zone temperature, such as reducing the diluent concentration, decreasing the diluent heat capacity, increasing the initial reactant temperature, or changing the fuel type, reduces the quantity of soot formed. Thus, temperature is shown to be a key parameter in determining the tendency of a mixture to soot. In many of these tests, burned gas temperature is nearly constant with time, suggesting that particulate blackbody radiation may be a stabilizing influence on the burned gas temperature. Coupling this radiation back into the chamber influences the particulate oxidation process and actually reduces the quantity of soot formed.

INTRODUCTION

Several authors [1-3] have reviewed soot formation phenomena in laboratory flames and/or engines. Glassman [3] observed that while soot formation in premixed flames does, of course, depend on equivalence ratio, it is perhaps more fundamentally related to the temperature of the burned gas. This observation was founded on the results of several rather exhaustive studies whose approach and pertinent conclusions are briefly described here. Perhaps the classic paper in this area is the work of Street and Thomas [4] who used a bunsen-type burner without secondary air entrainment. They found that:

   a. preheating the fuel/air mixture reduces the carbon-forming tendency, and

   b. dilution of a given fuel/air mixture with N_2 promotes carbon formation; enrichment with O_2 suppresses carbon formation.

They tested many different classes of fuel compounds and ordered them in terms of their relative tendency to form carbon based entirely on the equivalence ratio at the sooting threshold. It was not until the work of Millikan [5] that such an ordering was found to be inappropriate because of the different flame temperatures for the different types of fuels. Using flow velocity to control the heat loss from the reaction zone above porous metal burners, Millikan was able to show that temperature is one of the more important controlling parameters in premixed soot formation. The sooting order determined by Street and Thomas is nearly identical to the ordering of flame temperatures. Thus the flame temperature, and not directly the fuel type or equivalence ratio, is the variable which most fundamentally determines whether solid carbon will be liberated from a premixed flame. In a high-pressure laboratory flame MacFarlane et al. [6] have measured formation thresholds and soot quantity with premixed C_5 and C_6 hydrocarbon/air reactants. Pressure of the flame was varied up to 20 atmospheres, and the study considered the influence of equivalence ratio, pressure, and reaction temperature on the soot particulate and gaseous product concentrations. Several more practical experiments on operating diesel engines [7-10] have investigated the influence on exhausted smoke level of systematic variations in operating parameters such as injection period, injection nozzle design and timing, engine speed, intake air temperature, and compression ratio. In modeling soot formation phenomena, Amann et al. [11] have taken the approach of dividing the global engine processes into more easily modeled subscale processes. Similarly, a single-event constant volume combustion bomb provides a desirable compromise between the complex interactions within an operating engine and the simplicity of a flat flame burner.

Experiments to characterize and measure particulate formation processes are being conducted in a constant volume bomb. The objective is to develop insight into particulate formation, agglomeration, and oxidation in combustion environments representative of internal combustion engines. These experiments with rich, homogeneous mixtures may simulate the particulate formation and agglomeration processes which occur in inhomogeneous-charge engines where limited mixing results in locally rich mixtures. An understanding of high-pressure soot formation mechanism may point toward methods of in-chamber control of particulate emissions.

In these experiments the initial conditions, including pressure, temperature, equivalence ratio and diluent concentration, are parametrically varied, and their effect on soot formation is observed. Pressure history is measured in the chamber and high-speed shadowgraph movies are used for combustion visualization. The most informative technique is a coupled two-color emission/transmission pyrometer which simultaneously yields time-resolved soot volume fraction and temperature at a given position in the chamber.

This paper is divided into two major parts. The first part includes a brief outline of the facility and test conditions followed by a more thorough description of the diagnostic instrumentation. The second part deals with the combustion visualization, particle sizing measurements, particulate filtration results, and temperature and soot threshold measurements.

DESIGN OF THE EXPERIMENT

The Facility — The combustion bomb used in this work was designed and fabricated by Volkswagen Research and has been loaned to Sandia as part of an international cooperative program. The device is essentially a constant volume combustor in which the combustion processes occurring near top-dead-center in a conventional automobile engine are closely simulated. A schematic cross-section of the apparatus is shown in Fig. 1. The cylindrical combustion chamber is 80 mm in diameter by 29 mm thick. The 25 mm-thick glass windows on each end permitted full optical access to the combustion region. The premixed reactants were supplied to the intake valve assembly from a high pressure mixing reservoir. The mixer incorporated an explosion-proof, electrically driven fan for preparing the homogeneous mixtures of gaseous fuel and air. The walls of the bomb could be preheated by electrical resistance heaters, and for nearly all of the present tests were held at 420 K.

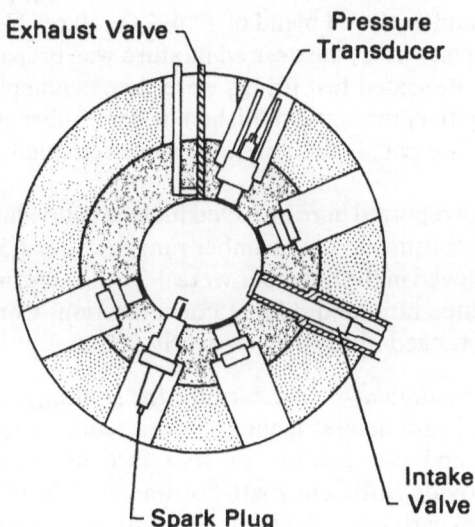

Fig. 1. Cross-section of combustion bomb.

During the normal sequence of events, the premixed fuel/air mixture was inducted into the initially evacuated chamber. The intake valve was held open long enough to equilibrate pressures in the bomb and the mixer. The valve then closed, sealing the chamber, followed by spark ignition at a predetermined time. Timing for valve operation, ignition, and diagnostic instrumentation was programmed in advance on digital counters, so that once the experiment was initiated, all events were automatically sequenced. During all of the testing reported here, the delay prior to ignition was sufficient to ensure that the mixture was quiescent and that the gas and wall temperatures were equilibrated. The facility and its operating characteristics are described in detail in Ref. [12].

In addition to the specialized diagnostic instrumentation discussed in a later section, standard facility instrumentation allowed transient recording of the pressure history within the combustion chamber using a piezoelectric pressure transducer (Kistler Model 601H) inserted in a water-cooled mount. Most of the diagnostic data were acquired and stored with an on-line PDP 11/34 minicomputer equipped with an A/D converter capable of digitizing several input channels.

Test Conditions — Some of the advantages afforded by a combustion bomb are the versatility and the control over the range of operating parameters. For this investigation, precise determination of equivalence ratio was essential. Accurate mixtures were relatively easy to prepare. Any desired volumetric fuel/air ratio was synthesized in the mixing reservoir by measuring the partial pressure of each component as it was added. The mixer fan stirred the gases thoroughly for at least 5 minutes before use. For most of these tests, the reactant gases were 99.5% pure propane and dry air whose composition was analyzed and found to be very close to standard atmospheric values. For the tests in which the diluent concentration was varied, the oxidizer mixture was synthesized from high purity N_2 or Ar and a specially prepared and analyzed blend of 49.4% O_2/50.6% N_2.

During a normal test series, the desired mixture was prepared at a pressure of 10 to 11 atmospheres. Repeated test firings were then accomplished out of the same reservoir with the initial pressure diminished on each subsequent firing. Generally, in these highly-sooting cases, the windows were removed for thorough cleaning between tests.

Most of the testing reported here involved fuel/air equivalence ratios between 1.7 and 2.6 and initial pressures in the chamber ranging from 4.5 to 10.0 atmospheres. Peak pressures achieved in the chamber were 4-7 times the initial pressure depending on the equivalence ratio and diluent concentration. Combustion duration for these rich mixtures ranged from 30 ms to nearly 500 ms.

Diagnostic Instrumentation — To determine the sooting characteristics as a function of the operating parameters, techniques are required for measuring the combustion properties and the quantity of soot formed. Gas-sampling techniques generally do not provide sufficient spatial or temporal resolution for the transient combustion bomb experiments. Instead, techniques based on the scattering, absorption, and emission of light were used to provide the spatial and temporal resolution required for following the soot formation process. Particulate filtration measurements were used to provide an independent check on the accuracy of the optical extinction technique, and soot samples were collected for analysis by transmission electron microscopy.

Light Extinction Techniques – Since the scattering and absorption of light by particles depend on the size, shape, composition, and quantity of those particles, measurements of scattering and absorption may in principle be used to obtain information regarding these properties. The intensity of light of wavelength λ transmitted through a pathlength L of particles with extinction cross-section C_{ext} is related to the initial intensity $I_{o\lambda}$ by [13]:

$$\frac{I_\lambda}{I_{o\lambda}} = \tau_\lambda = \exp\left[-C_{ext}\, NL\right] \quad , \tag{1}$$

where N is the number density of particles suspended in a transparent medium. The intensity ratio τ_λ is referred to as the transmission coefficient at wavelength λ. The extinction cross-section is the sum of the absorption and scattering cross-sections, which may be determined from the Mie theory [13, 14] for the case of plane waves incident on solid homogeneous spheres of arbitrary size.

Useful approximations for the cross-sections apply in the Rayleigh limit, when the Mie parameter $q = \pi d/\lambda$ is much less than unity. Here d is the particle diameter and λ is the wavelength of incident light. In the Rayleigh limit, the ratios of scattering and absorption cross-sections to the geometric particle cross-section are given by [13]:

$$\frac{C_{scat}}{\pi d^2/4} = \frac{8}{3} q^4 \left|\frac{m^2-1}{m^2+2}\right|^2 \tag{2}$$

$$\frac{C_{abs}}{\pi d^2/4} = -4q \; Im\left[\frac{m^2-1}{m^2+2}\right] \tag{3}$$

where m is the complex refractive index of the particles. Practically, the Rayleigh absorption cross-section is a good approximation for q less than about 0.1, and the Rayleigh approximation for scattering is good for q less than about 0.5. Even for q=0.3, the Rayleight absorption cross-section is within 5% of the cross-section predicted by the full Mie theory for values of m characteristic of soot particles.

In the Rayleight limit for absorbing particles, $C_{ext} \approx C_{abs}$ and Eqs. (1) and (3) may be combined to yield:

$$\tau_\lambda = \exp\left[\frac{6\pi L}{\lambda} \; Im\left(\frac{m^2-1}{m^2+2}\right) \left(\frac{\pi d^3}{6}\right) N\right] \tag{4}$$

$$= \exp\left[\frac{6\pi L}{\lambda} \; Im\left(\frac{m^2-1}{m^2+2}\right) f_v\right]$$

Here the product of the volume occupied by each particle, $\pi d^3/6$, times the particle number density N has been identified as the volume fraction f_v which is occupied by particles. Hence, measurement of the transmission coefficient for monochromatic light, τ_λ, provides a direct measure of the volume fraction occupied by the particles along the line of sight. It may be shown that f_v is proportional to $\ell n(\tau_\lambda)$ even if the particles have a distribution of sizes, provided that the Rayleigh approximation for C_{ext} still holds for all particles.

References pp. 384-385.

Detailed information on the quantity and size of particles can be obtained from simultaneous measurements of the transmission coefficients for light of two different wavelengths, as described, for example, in Ref. [15]. The application of the two-color extinction technique and a dual-polarization Mie scattering technique for making particle size and number density measurements in the combustion bomb has been described previously by the authors [16].

Combustion Visualization – When coupled with pressure histories, shadowgraph visualization can provide a highly informative qualitative assessment of fluid motion and combustion behavior inside the chamber. The most informative applications for such visualization, however, are when it is coupled with time-resolved measurements such as the line-of-sight optical measurements of temperature and soot quantity. In these cases the visualization allows the detailed behavior at a point to be related to phenomena occurring elsewhere in the chamber.

The application of high-speed laser shadowgraph cinematography to the combustion bomb has been described in Ref. [12]. The arrangement consisted of an argon-ion laser, a spatial filter to select the TEM_{00} mode and expand the beam, a 15 cm diameter spherically corrected achromatic collimating lens, and a Hycam camera speed-controlled at 5000 frames/s. In a transparent medium, the flame front is readily discernible as a dark line due to the strong gradient in the index of refraction. Fluid motion within the burned and unburned region is also made visible by density gradients arising from localized cooling of the gases in the vicinity of the walls. In the sooting environments studied here, the collimated beam is also attenuated directly by the soot particles, with the attenuation providing a measure of the soot volume fraction. Hence, the laser shadowgraph movies provide a visualization of soot formation during the combustion process in the bomb.

During some of the testing, quantitative simultaneous measurements were required of soot volume fraction at all locations across the bomb cross-section at any selected time. The optical arrangement used for these measurements is shown in Fig. 2. An expanded argon-ion laser beam passed through the combustion chamber, was down-collimated, and projected on a mylar screen. To minimize shadowgraph effects the screen was placed at the plane where the combustion chamber was imaged. A 35 mm camera with an electronically actuated shutter was used to photograph the image on the screen at a predetermined time during the burn. A narrow-bandpass filter between the camera and the screen transmitted the laser light and blocked essentially all radiation emitted by the soot particles. A microdensitometer was used to measure optical density of the processed film on a 64 x 64 point grid covering the image of the projected beam. The film density measurements were then used to determine the relative transmitted intensity of the laser beam at the grid points, hence mapping the soot volume fraction across the chamber cross-section.

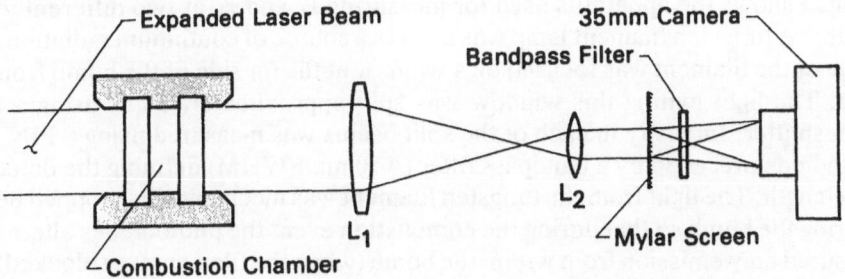

Fig. 2. Schematic of optical system for measurement of soot volume fraction.

Optical Pyrometry – The radiation emitted from an ideal blackbody at temperature T is given by Planck's law,

$$I_\lambda^B = \frac{2 c^2 h}{\lambda^5} \left[\frac{1}{\exp(hc/\lambda kT) - 1}\right], \tag{5}$$

where I_λ^B is the radiant power emitted by the blackbody per unit area per unit wavelength interval per steradian. For a non-blackbody, the radiation I_λ emitted is

$$I_\lambda = \epsilon_\lambda I_\lambda^B, \tag{6}$$

where ϵ_λ is the spectral emissivity of the radiation at wavelength λ. The spectral emissivity for a uniform distribution of radiators (suspended soot particles) is equal to the spectral absorptivity which is related to the transmissivity by [17]:

$$\epsilon_\lambda = 1 - \tau_\lambda \tag{7}$$

Combining Eqs. (5-7),

$$I_\lambda = (1-\tau_\lambda) \left(\frac{2c^2 h}{\lambda^5}\right) / [\exp(hc/\lambda kT) - 1]. \tag{8}$$

Hence, if I_λ and τ_λ are measured, the brightness temperature T of the particles may be calculated using Eq. (8). The characteristic times for conductive heat transfer for a 100-nm particle are on the order of nanoseconds [18, 19], which is short relative to the burn duration, the time scale on which the gas temperature changes. Hence in the absence of surface reactions it is reasonable to assume that the soot and gas temperatures are equal.

References pp. 384-385.

Fig. 3 shows the apparatus used for measuring I_λ and τ_λ at two different wavelengths. A tungsten filament lamp was used as a source of continuum radiation. The image of the filament was focused on a window at the far side of the bomb from the lamp. The light exiting this window was split approximately 50/50 using a cube beam splitter. Intensity in each of the split beams was measured using a PIN-8LC photodiode preceded by a bandpass filter ($\sim$10 nm FWHM) defining the detection wavelength. The light from the tungsten filament was mechanically chopped before entering the bomb so that during the combustion event the photodiodes alternately measured only emission from within the bomb (when the filament was blocked) and emission plus transmitted light from the lamp. The temperature T may then be inferred by inverting Eq. (8) using the measured values of I_λ (t) and τ_λ (t) at either detection wavelength obtained from the appropriate photodiode record. Since two wavelengths were used, redundant temperature measurements were made.

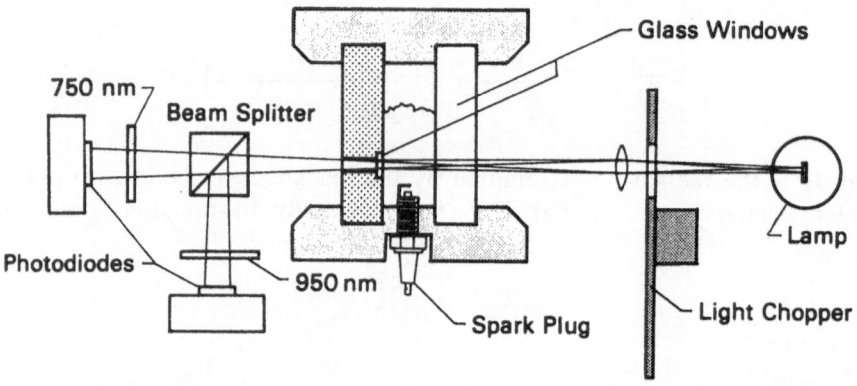

Fig. 3. Schematic of optical system for measurement of the soot particle temperature.

Interpretation of the photodiode response as a measure of the absolute intensity emitted by the soot required calibration of the response of the complete detection system to a source of known intensity. This was accomplished prior to the testing using a calibrated 30A tungsten ribbon lamp (with an Eppley standard lamp power supply) which was focused at the window used for detection of soot radiation. Uncertainties arising from the calibration may introduce as much as 25 K uncertainty in the soot temperature measurements. The excellent agreement between temperature measurements made with the two detection channels (usually better than 10-15 K) suggests that systematic errors, most significant of which are calibration uncertainties, are the primary source of uncertainty in the soot temperature measurements.

Fig. 4 shows a typical data record obtained from one channel of the two-wavelength emission/transmission pyrometer. For all of the temperature measurements reported here, 750 nm and 950 nm bandpass filters were used in the pyrometer. These wavelengths were chosen so that all measured radiation could be attributed to particulate emission (Swan band emission begins to interfere at wavelengths below 700 nm). The two channel pyrometer, of course, also permitted redundant measurements of particle volume fraction using the two transmission measurements, i.e., f_v can be calculated using Eq. (4) and either τ_1 or τ_2. As will be shown later, the particle size is small enough relative to 750 nm and 950 nm for the Rayleigh limit (implicit in Eq. (4)) to be a good approximation.

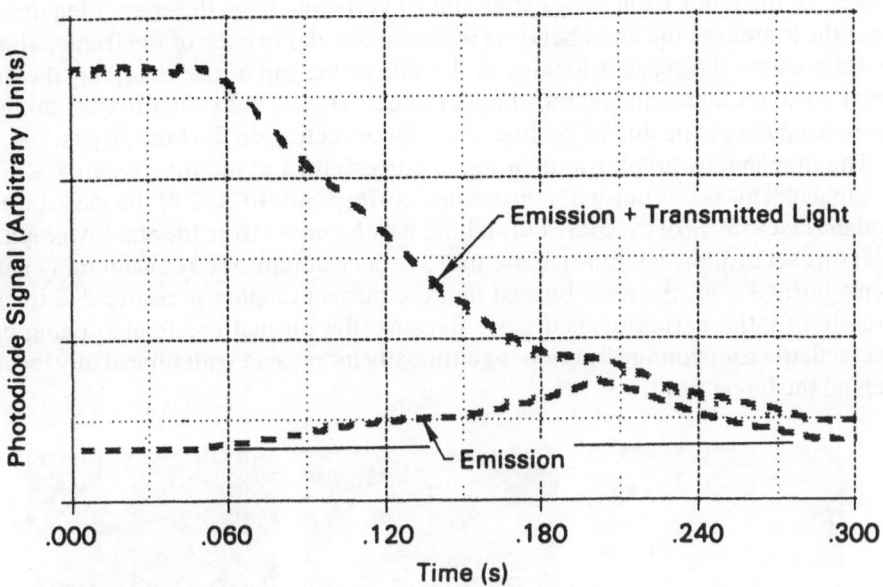

Fig. 4. Variation of photodiode signals with time during combustion for $\lambda = 950$ nm.

In a few short tests, a 35 mm camera was used to photographically record the spatial distribution of the soot radiation intensity. Infrared-sensitive film was used in these tests and a bandpass filter centered at 800 nm was placed in front of the camera lens. A microdensitometer was used to measure the optical density of the film on a 64 x 64 point grid covering the chamber image. These density measurements, combined with knowledge of the film response characteristics, were used to determine the variation of soot temperature across the bomb cross-section. In many cases the emissivity is nearly unity over the burned gas region, and the temperature distribution can be obtained by referencing each location in the chamber to the position where temperature measurements had been obtained with the optical pyrometer. Measurement of the temperature distribution in the general case where emissivity is not uniform requires simultaneous implementation of the technique for recording the spatial variation of soot transmission (Fig. 2). Point measurements with the optical pyrometer may, of course, be eliminated if the film response to a blackbody radiator of known temperature is calibrated.

EXPERIMENTAL RESULTS

Combustion Visualization — Fig. 5 shows a sequence of three frames from a laser-shadowgraph movie of the combustion event along with the measured pressure record for the same test. The flame front appears in all frames as a distinct line separating the burned and unburned gas. The burned gas appears as varying shades of gray, depending on the concentration of particles. The transmission of laser light in the burned gas provided a direct measure of the soot volume fraction (Eq. 4). Note that the flame front propagates almost vertically from the spark plug (located near the bottom of the chamber) and is nearly circular in each of the frames shown. In tests where the spark is located at the side or the top of the chamber, the flame front does not maintain such a smooth circular shape, but rather breaks up into a more irregular shape due to the buoyancy forces acting on the burned gas.

The first shadowgraph frame in Fig. 5 corresponds to the time when 5% of the original fuel mass has burned*, the second corresponds to 20% of the mass burned, and the last to 50% of the mass burned. As may be noted from the shadowgraph and pressure record, the last-burned mixture has been compressed considerably before being burned (i.e., the later-burned mixture burns at higher pressure due to compression by the earlier-burned gas). Because the burned gas is also being compressed, the soot volume fraction is continually increasing with time at any location behind the flame front.

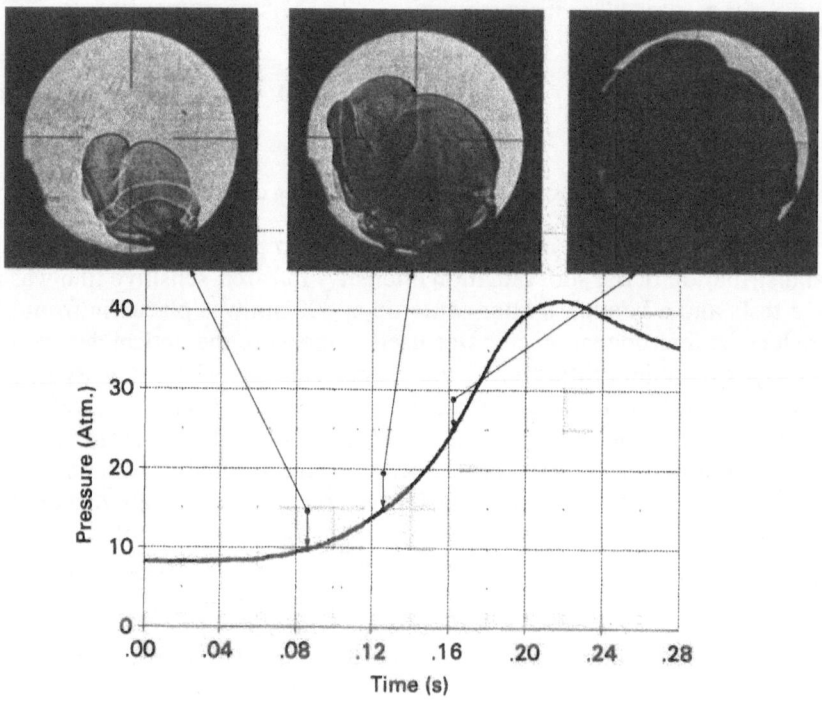

Fig. 5. Combustion development in bomb.

*To first order, mass fraction burned is proportional to the fraction of the total pressure rise in the chamber. See Lewis and von Elbe [20].

Shadowgraph visualizations of repeated tests with the same initial conditions reveal that the quantity of soot formed may vary significantly between apparently identical experiments. This nonreproducibility of burn conditions is also apparent in variations of the burn duration and measured peak pressure in the present rich sooting tests. Such variations are not evident in lean tests.

Fig. 6a shows the spatial distribution of the intensity of laser light transmitted through the combustion bomb recorded using the extinction mapping technique. Contours of constant film optical density, measured using a microdensitometer, are shown in Fig. 6b. Soot volume fraction may be related to film optical density using the relation for film response:

$$D = a \log (I\Delta t) + b, \tag{9}$$

where I is the intensity of light incident upon the film during exposure time Δt and D is the optical density of the developed film. The coefficients a and b, which depend on the type of film and the details of the developing process, were determined experimentally by measuring the film optical density resulting from exposure to a range of known incident intensities transmitted by a calibrated graybar. Inserting Eq. (4) into Eq. (9), it can be shown that the film optical density is linearly related to the soot volume fraction. The soot volume fractions corresponding to the contours of constant film density are indicated in the figure. Note that the highest soot density is found between the spark plug and the center of the bomb, the region containing the first-burned gas. The soot volume fraction in this region is approximately a factor of 2 greater than the average soot volume fraction in the burned gas.

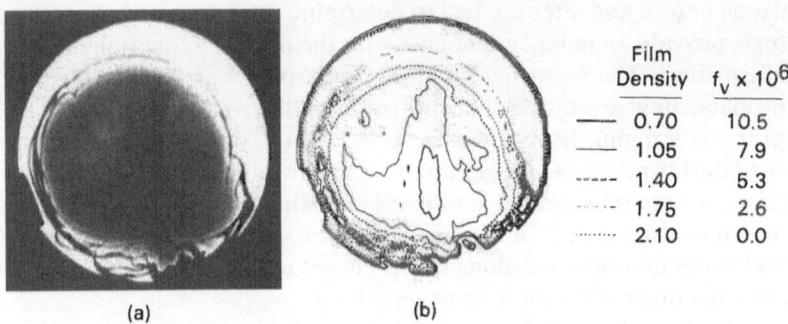

Film Density	$f_v \times 10^6$
—— 0.70	10.5
—— 1.05	7.9
----- 1.40	5.3
---·- 1.75	2.6
········ 2.10	0.0

(a) (b)

Fig. 6. (a) Photograph showing the attenuation of a collimated laser beam by soot in the burned gas region. (b) Contours of constant film density obtained by microdensitometer analysis of the negative.

Particle Sizing Results — Particle-size measurements made in the combustion bomb using the two-color extinction technique, described in detail previously [16], indicate that the particles are approximately 50 nm diameter with an uncertainty of approximately ± 50%. The relatively large uncertainty results from the fact that particle size is so small relative to the wavelengths of the probe lasers (488.0 and 632.8 nm) that the ratio of extinction cross-sections is relatively insensitive to particle size. Particle size does not vary significantly with time during the burn. Also, no dependence of size on equivalence ratio, initial pressure, or measurement location is observed within the sensitivity of the technique. The measurements do show, however, that particle number density and soot volume fraction increase with time during the burn for all test conditions. These increases appear to result from compression of the burned gases rather than from formation of additional particulates in the burned-gas region.

In some tests, thin SiO_2 substrates were mounted to the chamber wall for collecting soot samples for examination by transmission electron microscopy (TEM). Inspection of the deposit from any single test shows the soot to consist of three-dimensional chain-like structures composed of spheres of nearly uniform size. The diameters of individual spheres vary from 20 to 35 nm and appear to depend on initial pressure, with diameter smallest for the tests at the lowest initial pressure. It is not possible to deduce from examination of the samples whether the soot existed in such chain-like structures during the burn; however, the fact that the sizes of the individual spheres observed by the TEM technique are considerably smaller than the soot size measured *in situ* by the two-color extinction technique (near the uncertainty limits of the optical technique) suggests that some agglomeration occurs during the burn. Comparison of optical extinction measurements with the particle filtration results provides further support for this hypothesis.

Particulate Filtration Measurements — Measurement of the total mass of soot produced during a burn was accomplished following some tests by evacuating the chamber through a Pallflex teflon-coated glass-fiber filter or a Gelman type-A/E glass-fiber filter to collect the soot. The filters were dried and weighed on a precision balance before and after the test to determine the total mass of soot collected. These tests provide an independent check on the optical extinction measurements of soot formation. The test used for this comparison was the combustion of pre-mixed propane/air at an equivalence ratio of 2.1, 420 K initial temperature, and with initial pressure ranging from 4.4 atm to 8.9 atm. Total soot mass collected by filtration of the burned gases ranges up to 1.9 mg in individual tests. The uncertainty in the measurement of collected soot mass is relatively large, as it is the difference between two measurements of filter weight, each with an uncertainty of $\pm$ 0.2 mg. To minimize the potential weighing errors, in several test series a single filter was left in place to collect the soot from several tests.

If the soot volume fraction f_v measured optically at a single location is assumed to characterize the average soot loading throughout the burned gases after completion of combustion, then the total mass of soot M_s corresponding to the optical volume fraction measurement is

$$M_s = \rho_s f_v V_c \quad , \tag{10}$$

where V_c is the chamber volume and ρ_s is the density of the soot particles, assumed to be 2 g/cm³ [21]. The cumulative mass of soot collected on a single filter is on the average about a factor of 4.5 lower than the total mass obtained from the transmission measurements using Eq. (10). As previously noted, the soot loading in the burned gas is not uniform, and, in fact, the volume fraction measurement (made 11 mm from the spark plug) corresponds to the region of highest soot loading for these test conditions, as illustrated by Fig. 6. The volume fraction of soot here is about 2 times greater than the average over the burned gas region. Approximately half of the factor of 4.5 discrepancy between the optical-extinction and physical-sampling mass measurements must result from the use of a local volume fraction measurement in Eq. (10) rather than the mean value of f_v. Suggested explanations for the remaining portion of the discrepancy between measurements include: collection on the filters of only a fraction of the soot formed due to soot passing through the filter or being left on the chamber walls; absorption of light by gas-phase species in addition to particulates; higher absorption per unit mass by particulates than predicted by the Rayleigh theory; and uncertainties in the refractive index of the particles.

The amount of soot deposited on the walls of the bomb rather than collected on the filters can be estimated from the loss of window transmission after a test to be no more than about 10% of total soot measured optically. Changes in the rate at which the burned gas was evacuated (produced by changing the limiting flow-passage diameter) have no significant effect on the soot mass collected, indicating that the amount of soot deposited in the passage to the filter is not significant. The possibility that some soot passed through the filter was checked by placing two filters back-to-back in the filter holder in some tests and by measuring the optical absorption of the gases which passed through the filter in other experiments. There was no visible soot at all on the second filter when two filters were used, and the gases which passed through the filter were found to be completely transparent, indicating no significant particle loading.

Since the filtered combustion products were observed to be optically transparent, any absorbing gaseous species present would have to disappear before or during passage through the filter. Also, the excellent agreement (generally better than 2%) between the volume fractions measured at 750 nm and 950 nm when all absorption is attributed to particles in the Rayleigh size limit reinforces our confidence in this interpretation of the measurement — in general, absorption by gaseous species would not have the same $1/\lambda$ dependence as Rayleigh-limit particles. However, while the observed wavelength dependence of the absorption cross-section is consistent with the assumption that the particles are simple spheres in the Rayleigh size range, it does not confirm this model. Indeed, electron microscope pictures of sampled soot show series of interconnected spheres, each unit of which is in the Rayleigh size range. The optical technique may provide some measure of the volume of the agglomerated structure, rather than the volume occupied by condensed matter. The assumed soot density (2 g/cm³) would be inappropriate in this case, that is, if measured particle volume contains a large void fraction. While uncertainties in the refractive index of soot also affect the volume-fraction measurement, trial calculations with different suggested values of refractive index [22,

23] indicate that such uncertainties should contribute no more than ∽25% error.

The most plausible of the discussed explanations for the discrepancy between the optical-extinction and physical-sampling measurements of total soot formation is a combination of: (1) use of the local volume fraction measured in the most soot-laden region to characterize all of the burned gases; and (2) measurement of the volume of a relatively loosely packed structure by the optical technique (for which a lower density than 2 g/cm³ should be used). The spatial maps of optical extinction indicate that the contribution to the discrepancy caused by the use of a *local* volume fraction measurement should be a factor of about two. The remaining factor of about two would then most likely result from the measurement by the optical technique of the volume of a loose structure rather than the volume of the condensed phase. This interpretation of the remaining factor is consistent with the work of Greeves and Meehan [24] who measured attenuation coefficients per unit mass for diesel smoke three times greater than calculated, and with the results of Roessler and Faxvog [21] who measured similar attenuation coefficients in flames and diesel exhaust. In a recent paper by Janzen [25], spectral attenuation coefficients for colloidal suspensions of carbon black particles were typically a factor of 1.5 higher than those calculated from Rayleigh-limit expressions with the usual values of refractive index [23]. Janzen has chosen to resolve this discrepancy by modifying the refractive index and size distribution until suitable agreement is achieved.

Temperature Measurements — As detailed in a previous section, temperature measurements are relatively easy to make in sooting environments via the strong blackbody emission of particulates. The optical pyrometer used in this study measures brightness temperature at two wavelengths, using simultaneous time-resolved measurements of soot volume fraction to account for variations in emissivity. The *simultaneous* measurement of both temperature and soot concentration has proven to be of critical importance in view of the generally nonreproducible nature of the experiments.

Equilibrium calculations provide a well-defined reference point for comparison with the measurements. In general for most combustion environments, the approach to equilibrium is sufficiently fast that equilibrium predictions accurately describe the major species concentrations within the burned gas region. In addition, since heat losses are usually small for freely propagating flames, calculated equilibrium adiabatic flame temperatures provide realistic estimates of the temperature in the region immediately behind the flame front. For example, in previous work on the combustion bomb [12], the equilibrium code developed by Gordon and McBride [26] has been successfully used to predict flame temperatures determined from laser Rayleigh scattering measurements. The temperature history of any given mixture parcel can be obtained by first calculating the flame temperature directly from equilibrium considerations, then using isentropic relations to determine the evolution of the burned gas temperature with time.

Fig. 7 shows a comparison between temperature measurements (three repeat runs) at a point several millimeters from the spark plug and the calculated temperature history for the first burned mixture. The most striking observation regarding the measured temperature of the burned gas is the fact that it is nearly *constant* with time. This fact contrasts with the calculations, as well as with the usual observations in engines which show the temperature of the burned gas increasing with time because of compression during the burn [27]. Since the calculations assume adiabatic conditions, they do not account for either wall heat transfer or radiative heat transfer by the soot particles. These two factors are expected to contribute to the large difference between the measured and calculated equilibrium temperature histories. Typically for these test conditions and geometries, wall heat transfer accounts for approximately 10-15% of the total chemical heat release. Estimates using nominal values of soot concentration (emissivity), gas temperature and burn duration show that black-body radiation can account for another 20% of the total chemical energy release in the system.

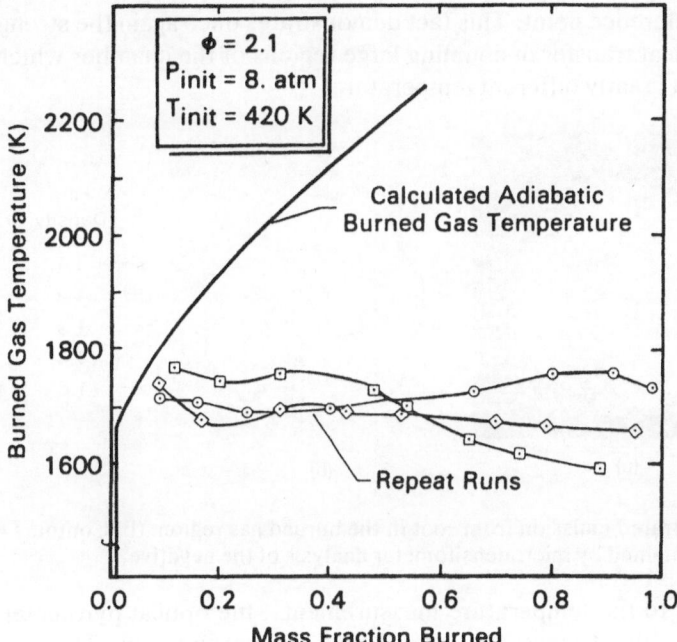

Fig. 7. Comparison of measured and calculated burned gas temperature versus mass fraction burned.

The observed constancy of temperature with time suggests the possibility that a dynamic energy balance is achieved within the burned gas. This behavior suggests that the temperature is radiatively-stabilized in this region with the σT^4 radiative heat transfer factor just balancing the constant compressive work done by combustion of the later-burned mixture.

The optical-pyrometer measurements of temperature history at a point in space show temperature to be nearly constant. To complement these results, the photographic technique described previously was implemented to determine the spatial distribution of temperature at a given time over the entire chamber. Figs. 8a and 8b show the infrared image of the distribution of radiating soot particles and the contours of constant film optical density respectively. The film density is related to temperature using the film response relation, Eq. (9), where I is the intensity of light emitted by the soot cloud in the 800 nm spectral region (Eq. (8)). In this case, since the emissivity is experimentally known to be nearly unity over the burned-gas region, the film density represents a linear measure of reciprocal temperature. An absolute calibration of the film was not performed, therefore the spatial variation in temperature is referenced to the point near the spark plug where the optical-pyrometric measurements were taken ($T_{ref} = 1700$ K). The contours indicate that temperature is nearly *uniform* over the burned gases ranging from -15 K to $+47$ K around the reference point. This fact demonstrates once again the strong influence of radiative heat transfer in coupling large regions of the chamber which normally would exhibit greatly different temperatures.

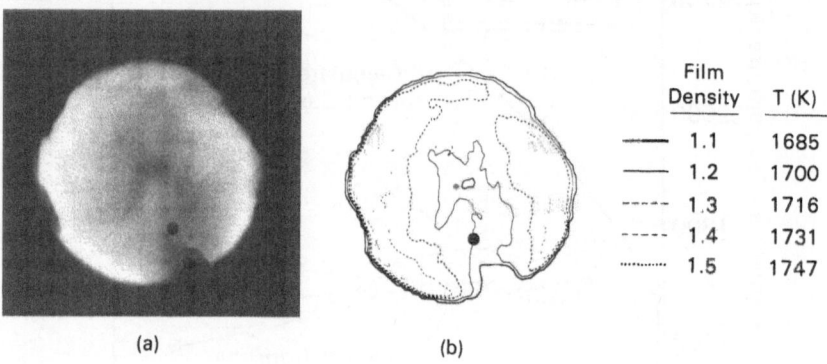

Film Density	T (K)
—— 1.1	1685
—— 1.2	1700
----- 1.3	1716
----- 1.4	1731
.... 1.5	1747

(a)   (b)

Fig. 8. (a) Infrared emission from soot in the burned gas region. (b) Contours of constant film density, obtained by microdensitometer analysis of the negative.

In addition to the temperature measurements, the optical pyrometer routinely yields time-resolved measurements of soot volume fraction. The authors have previously reported [16] that the soot formation ratio (mass of soot per unit mass of fuel burned) may decrease as the initial pressure in the combustion bomb is increased. Subsequent data obtained with the pyrometer and further testing with the mass filtering are too scattered to either substantiate or refute this trend. To date, most laboratory experiments with steady premixed burners show negligible influence of pressure on soot formation ratio. In contrast, the high-pressure experiments of MacFarlane et al. [6] show a dramatic increase in formation ratio with pressure and appear to be similar to the results obtained from steady diffusion flames. At present the pressure dependence of soot formation at high pressure is still an unresolved issue.

The Temperature Threshold for Soot Formation — The present work on the combustion bomb investigates the hypothesis that temperature is the dominant parameter governing soot formation at pressures similar to those experienced in engine environments. Three techniques have been adopted for changing the reaction-zone and/or burned-gas temperature independently of the fuel/air equivalence ratio:

a. changing the quantity of N_2 diluent in the premixed reactants,

b. varying the initial reactant temperature, and

c. trapping the particulate blackbody radiation within the chamber using gold mirrors.

In addition to this testing, the fuel type was varied from propane, a paraffin, to propylene, an olefin.

Varying the concentration of diluent in the reactant mixture is a convenient method of varying burned gas temperature while maintaining constant equivalence ratio. Except for third-body effects, N_2 diluent may be assumed to be inert, thus permitting parametric temperature variation without altering the heat release or carbon-forming kinetic mechanisms. Calculated equilibrium adiabatic flame temperatures are representative of the temperature of the first burned gases and are useful for comparing cases with different diluent concentrations. These flame temperatures are plotted in Fig. 9 as a function of equivalence ratio for various N_2/O_2 ratios. The decrease in temperature as equivalence ratio is increased results from the fact that excess fuel beyond stoichiometric conditions is broken down into H_2 and CO which do not completely oxidize and act as a thermal sink. Superimposed on these curves is the experimentally determined threshold for soot formation observed in propane tests on the bomb. For our particular case a sooting test is defined as one in which there is significant extinction of light passing through the 29

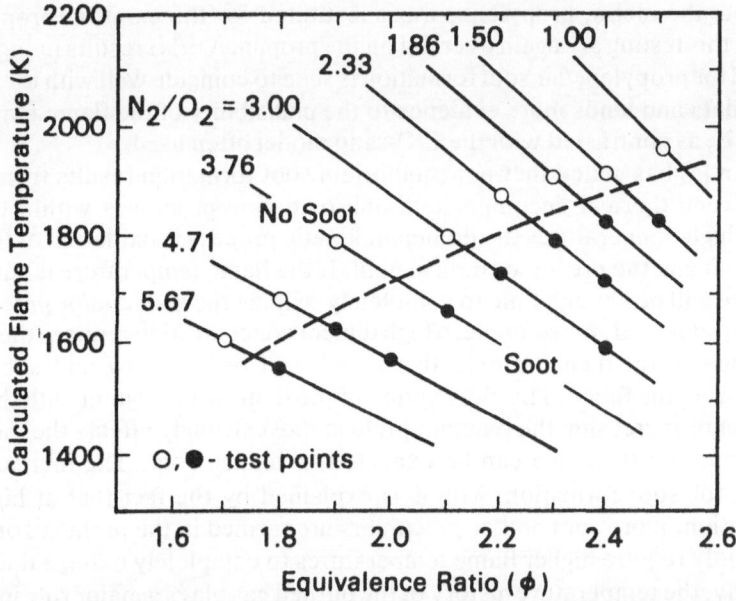

Fig. 9. Calculated adiabatic flame temperatures and measured sooting threshold for varying equivalence ratio and N_2/O_2 ratio.

References pp. 384-385.

mm depth of the chamber at the completion of combustion. In practice, this threshold is relatively easy to discern since in most cases a reduction of only 0.1 in equivalence ratio will shift the test from sooting to totally nonsooting (optically transparent). Clearly evident in the figure is the low temperature region below the line which yields soot and a non-sooting high temperature region above the line. Increasing quantities of soot are formed at conditions far below the line. When calculated flame temperatures are below about 1550 K, the mixtures generally cannot be spark-ignited and thus no data are obtainable. The fact that changing the diluent concentration alone can keep $\phi = 2.5$ mixtures from sooting, and conversely make $\phi = 1.8$ mixtures produce soot, strongly suggests that flame temperature is a critical parameter. Furthermore, identical test conditions with the initial reactant temperature increased from 420 K to 470 K generally result in a 50% reduction in soot volume fraction.

In another test series, argon was substituted for a portion of the N_2 diluent. The molar composition of the oxidizer blend was synthesized to be 25% O_2/25% N_2/50% Ar. Adiabatic flame temperature calculations for this particular blend overlay nearly exactly the temperatures for the $N_2/O_2 = 2.33$ blend as shown in Fig. 10. This, of course, results from the lower specific heat of argon compared to the nitrogen which it replaced. The experimental soot threshold is identical to that which was determined using N_2 as the diluent, demonstrating that the N_2 acts solely as a thermal sink and does not significantly participate in the carbon-forming reactions.

As mentioned previously, Glassman [3] has postulated that for premixed flames the fuel type is only important in soot formation insofar as it affects the reaction zone temperature. One simple experimental excursion in this parameter has been performed: the olefin, propylene, was substituted for the paraffin, propane. The results of this testing are again overlaid on the propane/O_2/N_2 results in Fig. 11. The threshold for propylene/air soot formation is seen to coincide well with the previous propane data and lends more evidence to the plausibility of the flame temperature hypothesis, as contrasted with the C/O-ratio model often used.

Millikan [5] has stated that nonequilibrium soot formation results from the balance between thermal decomposition and oxidation processes within the flame zone. At high temperatures the dominant kinetic process is oxidation of the newly formed soot and the precursor radical pool. If the flame temperature is sufficiently high, there will be enough time to completely oxidize the soot and/or precursors to gaseous products. If, for example, a high diluent concentration reduces the reaction zone temperature, then the oxidative attack will be too slow and soot will be liberated from the flame. This description of the flame is consistent with the testing on the bomb. Increasing the reactant preheat most strongly affects the flame zone temperature and thus also can be expected to reduce soot. The increase in the threshold for soot formation with ϕ is explained by the fact that at higher fuel concentration, more soot and/or precursors are formed in the preheat zone which subsequently require higher flame temperatures to completely oxidize the carbon.

Similarly, the temperature history of the burned gas plays a major role in controlling the oxidation of the soot particles liberated from the flame zone. In an equilibrium sense the burned gas region is still oxidative with respect to the solid carbon,

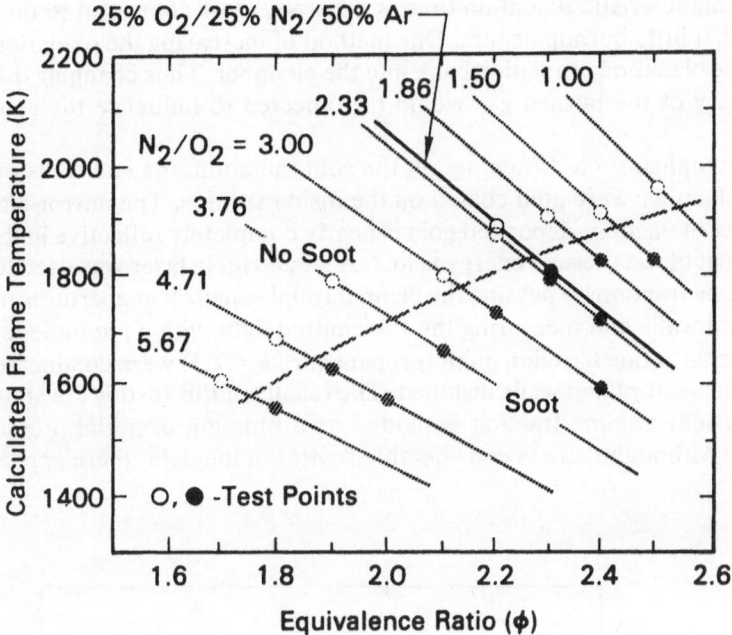

Fig. 10. Effect of diluent heat capacity on sooting threshold and calculated adiabatic flame temperature.

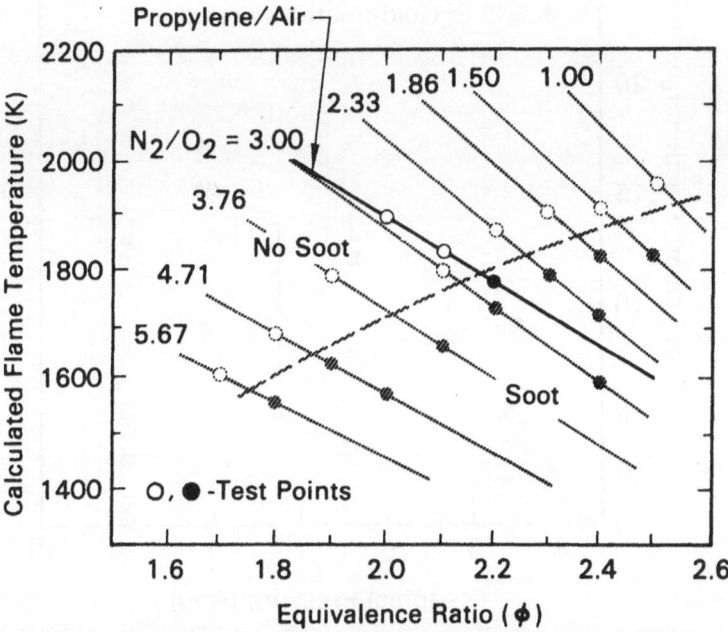

Fig. 11. Effect of substituting propylene for propane on sooting threshold and calculated adiabatic flame temprature.

References pp. 384-385.

but the characteristic oxidation time is generally slow compared to the residence time so that little burnup occurs. One method of increasing the oxidation rate is to couple the blackbody radiation back into the chamber. Thus changing the temperature history of the burned gas would be expected to influence the soot quantity evolved.

To accomplish such "trapping" of the soot radiation, the end walls of the combustion chamber were gold coated on the inside surface. The mirror-like 200-nm-thick layer of vacuum-deposited gold is nearly completely reflective in the infrared where most of the thermal energy is lost. A small HeNe laser was used to measure soot volume fraction by passing the beam through small clear apertures in the gold-coated end walls and measuring the transmitted light with a photodiode detector. Experiments at one baseline point (propane/air, $\phi = 2.1$) were conducted with and without the gold-plated walls installed. The results of this testing are shown in Fig. 12 where soot volume fraction is plotted as a function of initial pressure in the chamber. Although there is considerable scatter in the data, there appears to be a

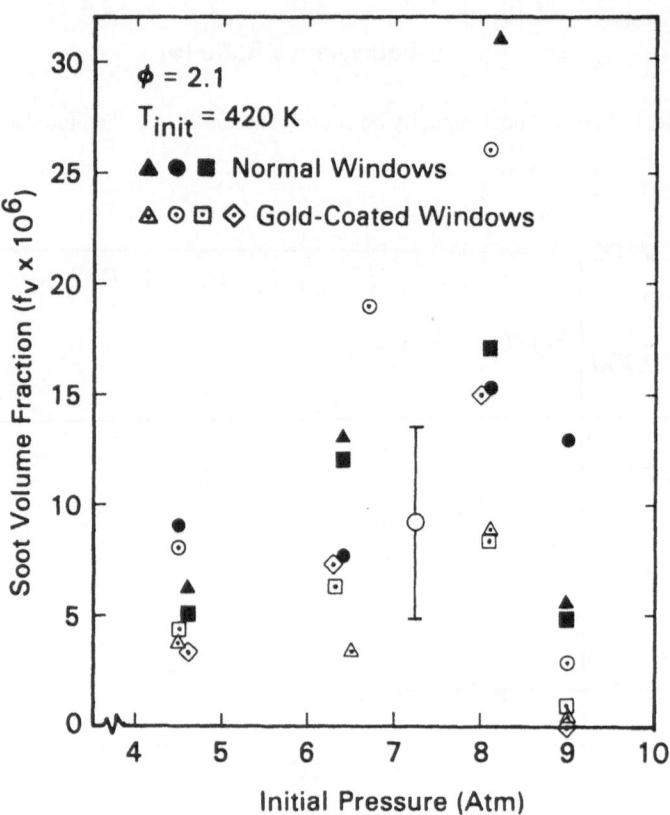

Fig. 12. Effect of initial pressure and gold-window-coating on soot volume fraction.

trend that less soot is present when the radiation is coupled back into the chamber by the gold walls. The data have been analyzed by a 28 degree-of-freedom statistical model which accounts for the major influences such as pressure and wall type (gold/ normal). The data exhibit better than 95% confidence level that the gold walls actually reduce the quantity of soot evolved. It is unlikely that the soot formation processes in the flame zone are affected by the windows since this zone radiates mainly to the opposite cold wall. Rather, the net reduction of soot is thought to be due to a long-term burnup of particles in the burned-gas region due to its increased temperature from radiative coupling. This explanation is consistent with the TEM analysis of soot particulates sampled with and without the gold walls. The average chain diameter is reduced by about one-third with the gold walls installed, indicating the probable effect of particle burnup.

Particle burnup is thus seen to be strongly affected by the temperature history of the burned gas. Moreover, because of the important role of radiative heat transfer from the particles, the quantity of soot present strongly influences the burned-gas temperature history.

SUMMARY AND PRACTICAL IMPLICATIONS

Soot formation processes have been characterized in a constant volume bomb. A variety of optical diagnostics have been used to gain insight into the sooting behavior of rich, premixed flames. Soot is formed very rapidly in the flame front in quantities far exceeding equilibrium predictions. The particles are then dispersed over the burned gas region as these gases are continually compressed.

Several major conclusions result from the present study:

1. Temperature is a key parameter in determining the tendency of a mixture to soot. In these premixed flames, it has been demonstrated that any change that increases the flame zone temperature such as reducing the diluent concentration, decreasing the diluent heat capacity, increasing the initial reactant temperature, or changing the fuel type, reduces the quantity of soot formed. This observation is consistent with previous work on laboratory burners.

2. Blackbody-type radiation from soot is an important energy transfer mechanism which can be a strong factor in determining the temperature history of the burned gas. In highly sooting tests, temperature is nearly constant in time, a result of the balance between radiative heat transfer from the soot and compressive work done by combustion of the later-burned gases. Radiative heat transfer also couples regions of the burned gas which would normally exhibit greatly different temperatures, resulting in a nearly uniform spatial distribution of temperature. The coupling between temperature and sooting behavior is demonstrated by the fact that the presence of soot strongly affects the temperature history, which in turn strongly affects the burnup-rate of those particles.

3. The soot mass obtained from optical extinction measurements is higher than the particulate filtration mass measurements by a factor of 2-3. This difference is thought to result from the three-dimensional chain-like structure of the

References pp. 384-385.

agglomerated soot particles as contrasted with spherical Rayleigh-limit absorbers assumed in the extinction model. The present study demonstrates the need for a more comprehensive assessment of the extinction technique and its dependence on particle shape as well as volume.

The process of soot formation during premixed constant volume combustion is much more straightforward to analyze than the complex interaction between fluid mechanics and chemistry which governs formation in real diesel engines and gas turbines. However, since the underlying physics is much the same, the general conclusions reached in this investigation may provide insight into the trends demonstrated in more practical environments. The present study has shown that soot formation and burnup is strongly dependent on temperature. NO_x formation would be enhanced in those high temperature regions which minimize soot. A common technique for controlling NO_x is exhaust gas recirculation which acts primarily to reduce the peak temperatures achieved in the chamber. This reduction in temperature decreases the precursor and particulate oxidation rates and thus results in generally higher particulate emissions. The dominant role of particulate radiation observed in these experiments may prove to have a similar infuence in real combustors.

ACKNOWLEDGEMENTS

The authors would like to acknowledge the support of Volkswagen Research, especially Drs. W. Lee and W. Brandstetter, for loan of the combustion bomb and their technical support of this program. Microdensitometer scans were performed by M. A. Palmer; the transmission electron microscopy analysis was accomplished by D. Look and G. J. Thomas. Many helpful discussions with the staff in the Combustion Sciences Department greatly enhanced the progress on this effort. This work was funded by the Divisions of Conservation and Basic Energy Sciences of the Department of Energy and by the Motor Vehicle Manufacturers Association.

REFERENCES

1. A. G. Gaydon and H. G. Wolfhard, "Flames," Chapman & Hall Ltd., London, (1970).
2. H. G. Wagner, "Seventeenth Symposium (Inernational) on Combustion," The Combustion Institute, Pittsburgh, (1979), pp. 3-19.
3. I. Glassman, Phenomenological Models of Soot Processes in Combustion Systems, Princeton University Dept. of Mech. and Aero. Engr. Rept. 1450 (1979); also AFOSR TR 79-1147.
4. J. C. Street and A. Thomas, Fuel, Vol. 34 (1955), pp. 4-36.
5. R. C. Millikan, J. Phys. Chem., Vol. 66 (1962), p. 794.
6. J. J. MacFarlane, F. H. Holderness and F. S. E. Whitcher, Combustion and Flame, Vol. 8 (1964), pp. 215-229.
7. I. M. Khan, Proc. Instn. Mech. Engrs., 184, Pt 3J (1969), pp. 36-43.
8. I. M. Khan, C. H. T. Wang and B. E. Langridge, Combustion and Flame, Vol. 17 (1971), pp. 409-419.
9. D. Broome and I. M. Khan, Proc. Instn. Mech. Engrs., C140/71 (1971), pp. 185-197.
10. D. F. Dolan and D. B. Kittelson, SAE Paper 780110 (1978).

11. *C. A. Amann, D. L. Stivender, S. L. Plee, and J. S. MacDonald, SAE Paper No. 800251 (1980).*
12. *T. M. Dyer, SAE Paper No. 790353 (1979).*
13. *M. Kerker, "The Scattering of Light," Academic Press, New York, (1969).*
14. *G. Mie, Ann. Physik, Vol. 25 (1908), p. 377.*
15. *P. A. Bonczyk, Combustion and Flame, Vol. 35 (1979), p. 191.*
16. *W. L. Flower and T. M. Dyer, Paper CSS/CI-01, Central States Meeting of the Combustion Institute, Baton Rouge, LA, March (1980).*
17. *S. S. Penner, "Quantitative Molecular Spectroscopy and Gas Emmissivities," Addison-Wesley Pub. Co. Inc., Reading, MA, (1959).*
18. *Y. Matsui, T. Kamimoto and S. Matsuoka, SAE Paper 790491 (1979).*
19. *W. L. Flower and J. A. Miller, Report No. SAND79-8607, Sandia Laboratories, Livermore, CA (1979).*
20. *B. Lewis and G. von Elbe, "Combustion, Flames and Explosions of Gases," Academic Press, New York (1961).*
21. *D. M. Roessler and F. R. Faxvog, J. Opt. Soc. Am., Vol. 70 (1980), p. 230.*
22. *H. Senftleben and E. Benedict, Ann. Physick, Vol. 54 (1918), p. 65.*
23. *W. H. Dalzell and A. F. Sarofim, J. Heat Transfer, Vol. 91 (1969), p. 100.*
24. *G. Greeves and J. O. Meehan, Proc. Instn. Mech. Engrs., C88/75 (1975).*
25. *J. Janzen, Appl. Optics, Vol. 19 (1980), p. 2977.*
26. *S. Gordon and B. J. McBride, NASA Lewis Research Center, NASA SP-273 (1971).*
27. *J. R. Smith, SAE Paper No. 800137 (1980).*

DISCUSSION

A. F. Sarofim *(Massachusetts Institute of Technology)*

Two comments. First, with regard to the statement you made concerning your results and Roessler's, I believe you said your optical measurements gave you soot concentrations which were three times too high.

Dyer

That is correct.

Sarofim

Roessler obtained an optical extinction measurement which was three times too high which would have given you an effect in the opposite direction.

Dyer

I am sure the results are consistent.

Sarofim

I just wanted to make the comment. The next one is that if indeed what you say about radiative exchange between different portions of your chamber is true, one would never see temperature gradients in flames. Have you calculated the rate of radiative exchange between different zones. Radiative transfer occurs in flames, certainly in highly sooting flames, where you have soot concentrations in excess of what you might even see in a pressurized bomb. It is not sufficient to equalize temperatures in the time frame of a few milliseconds.

Dyer

Perhaps another explanation for the phenomenon is that it is radiatively stabilized at exactly the same value over the entire burned gas region.

J. B. Howard *(Massachusetts Institute of Technology)*

I also have some question about the finding that the soot as measured optically exceeds the amount measured by filter, but I may not understand your technique well enough. I would guess that the amount of soot in that chamber was some fraction of a percent of the mass of the materials, so filtering it out with accuracy is not trivial, and an error in the collecting of the soot would tend to leave you with less soot observed than was in the chamber. A second point is that as I observed your system it seemed that the amount of soot present depended on both time and position in the chamber. Therefore, what you collect might not be what you see.

Dyer

That is correct.

Howard

There might be soot, for example, burning out before you collect it which would leave you with less soot than was present.

Dyer

What we see is that the extinction, the absorptivity of the medium, increases until the point at which peak pressure is achieved. It is then constant for many minutes thereafter. We use this extinction coefficient after allowing everything to become uniformly mixed within the chamber. So even though there are gradients during the burn, we allow for subsequent mixing to take place to establish a baseline. Typically, each shot yields on the order of one milligram of soot, and we agree very much with what you say. It is very difficult to collect all this mass and weigh it accurately. What we do is repeat several tests with the same filter until we can get a good measurement of total mass.

Howard

Do you remove particles collected on the wall of the vessel?

Dyer

We can account for that. We can measure the attenuation of our light beam after the contents of the chamber have been evacuated. The transmissivity through the windows (which is the major surface of the bomb) allows us to determine how much is left over in the chamber. It is typically 20%, and we have taken that into account.

H. Gg. Wagner *(Universität Göttingen)*

What you say about the temperature is known, but I would be a little careful about specifying absolute figures, because you said it is a calculated temperature.

Dyer

That is correct.

Wagner

On your pictures one could nicely see the influence of buoyancy on the flame and you also could see that you form polyhedral flames. You can reverse the temperature dependence when there are polyhedral flames, the soot formation goes the opposite way from that of a smooth flame. The flames when you operate are so slow. When you have such slow times, you get two things which act in opposite directions. Do you know what I mean?

Dyer

Perhaps.

Wagner

It is because polyhedral flames are very similar to a diffusion flame.

Dyer

Yes. On the movies that we showed, because the flames are so slow, they are significantly influenced by buoyancy. Vertical was to the right. If we had put the spark plug on the top of chamber, the flame shape would have been significantly different. By putting it on the bottom of the chamber, the buoyancy does help to make the flame propagate in a more symmetrical fashion.

Wagner

And the structure of the flame front?

Dyer

The structure of the flame front changes depending upon whether we have top ignition or bottom ignition.

Wagner

It is there of course. No, what I mean is your flame front is not a smooth one. If you see various lines passing through, these are characteristics for these cellular flames. When you look at these cellular flames, in the valleys of the cellular flames when you look from the side or the top, you can form soot.

Dyer

I see.

Wagner

I suggest you are doing that. There is one thing. When you go to turbulent mixtures, the whole situation could be different.

Dyer

I see what you mean. If we did produce excessive amounts of soot within the creases of the flame, then one would expect to see dark lines on the high-speed movies, because of the large amounts of soot that were formed in the crease. Do you think so or not?

Wagner

Did you look at it through the window? I mean you can see how this flame burns. Do you see streaks?

Dyer

The movies are the best diagnostic we have. The flashes are very fast, and so I have not noticed dark lines following the creases in the flames.

J. P. Longwell *(Massachusetts Institute of Technology)*

It might be interesting to use benzene as a fuel since it has a higher flame temperature and also produces more soot. I would doubt that it falls on your present correlation line.

Dyer

Yes, it might not.

Longwell

To follow up on Prof. Wagner's comment, the benzene flames of that nature tend to be much more cellular than propane flames so that you have that, in addition, to give you some more soot.

Dyer

Okay. Thank you.

A. D'Alessio *(University of Naples)*

Coming back to the comment made before about the shape effect. Is it in the Rayleigh regime? Probably, your particles may be in the Mie regime. Both considerations of shape effects take the calculation into the Mie regime. A look at the cross-section will explain the following to you. The question I want to ask is this. Do you think that it will be possible to study the influence of radiation on burning velocity? Is this a measurement you may make? The amount of soot that you have in this case will influence the burning velocity. I suggest that you study this.

Dyer

Thank you. We do not feel that the influence of radiation propagates upstream into the flame zone simply because there may not be a strong absorption within the flame zone of the long-wavelength infrared particulate emission. So we feel like the two processes of formation at the flame zone and radiation are pretty well decoupled.

TIME-RESOLVED SOOT PARTICULATES
IN DIESEL SPRAY COMBUSTION

T. KADOTA* and N. A. HENEIN

Wayne State University
Detroit, Michigan

ABSTRACT

Experiments have been made to study the soot particulates formed at different stages of combustion of a simulated diesel spray by applying a newly developed laboratory combustor in which the following two essential characteristics of diesel spray combustion are reproduced: intermittency and autoignition. The combustor is designed so that the fuel is injected by using a diesel fuel injection system into a steadily flowing high temperature air stream resulting in autoignition and combustion. Soot particulates have been collected at different stages of combustion by applying an inert gas quenching technique. The mass concentration, diameter, number and carbon-to-hydrogen ratio of soot particulates, oxygen concentration and temperature in the flame as a function of the time from the start of fuel injection have been obtained. Mutagenic activity of the time-resolved soot particulates has been determined by using the bacterial mutagenesis test. Thus, some factors affecting time-resolved characteristics of soot particulates have been studied. Theoretical analysis is also made of the physical mechanism for soot particulates formation.

The results indicate that this new technique is a useful tool to obtain the time-resolved characteristics of soot particulates in diesel spray combustion. The mass concentration of soot particulates increases during the early stage of combustion and reaches a maximum after which it decreases to very low values approaching zero near the end of the combustion process. The diameter and carbon-to-hydrogen ratio of the soot particulates increase with time while their number decreases. The mutagenic activity shows a maximum as a function of time. The rates of formation and gasification of soot particulates increase with the increase in flame temperature. Adding alcohol, indolene and water to diesel fuel has a definite effect on reducing the soot particulates in the flame.

Present address: Department of Mechanical Engineering,
  Hiroshima University, Hiroshima, Japan.

References p. 414.

INTRODUCTION

The superior fuel economy of the diesel engine over the gasoline engine has resulted in increasing its penetration in automotive applications and expanding its use from trucks to passenger cars. However, the use of the diesel engine in such vehicles might be ruled out if it cannot meet the expected future standards for the exhaust soot particulates. Therefore, experimental and theoretical studies are needed to understand the soot particulates formation mechanisms and to develop techniques for their control.

A review of the literature indicates that most of the previous research works on the soot particulates in the diesel engine have been limited to the measurements of their size distribution and the mass concentration in the exhaust. There is a surprising lack of information on the soot particulates formation at the different stages of the combustion process in the diesel engine. This has been caused by the complexity of the diesel combustion process and the difficulty of instrumentation needed to measure the time- and space-resolved soot particulates in the actual diesel engine. Due to the lack of experimental results on the time-resolved soot particulates in the engine, the current predictive analytical models in the literature have many uncertainties.

The primary objective of the present work is to obtain the time-resolved characteristics of soot particulates in diesel spray combustion under controlled and idealized conditions. To attain this, a laboratory combustor and a technique for collecting the soot particulates in the flame need to be newly developed. In the design of the laboratory combustor, the important characteristics of diesel spray combustion have to be reproduced. Intermittency and autoignition would be most essential. The sampling technique should be able to collect the soot particulates at the different stages of combustion for their physical, chemical and biological analysis. A sampling probe which collects continuously the soot particulates in the flame would be more appropriate than the rapid sampling system for the above objective. Optical techniques [1, 2] are effective in the study of the physical properties of the time-resolved soot particulates, but are incapable of determining their chemical and biological properties.

Prior to the measurements, it is necessary to define the characteristics of the fuel sprays and flames established in the laboratory combustor, and to obtain information on the performance of the newly developed sampling probe. To be measured at first are mass concentration and physical and chemical characteristics of the soot particulates which vary with time during combustion. Based on these data, a theoretical analysis is to be made of the physical mechanism for the soot particulates formation. The rates of formation and gasification of the soot particulates are also important information which have long been needed in diesel spray combustion.

There has been a deep concern about the health effects of the soot particulates. All the previous biological works on the soot particulates have been limited to study of the final products emitted in the exhaust. No work has yet been reported on the biological effects of the soot particulates formed during the combustion. Therefore, it is significant to study the mutagenic activity of the soot particulates collected at different stages of diesel spray combustion. To do this, bacterial mutagenesis tests would be available.

Since fuel modification is one of the techniques to reduce the soot particulates in the diesel engine, it would be important to study the effects of mixing diesel fuel with other available fuels of low smoking tendency and with water on the reduction of the soot particulates in and out of the flame. The air parameters are also expected to play an important role on the time-resolved soot particulates.

EXPERIMENTAL PROCEDURE

Instrumentation [3] — In the design of the experimental apparatus, it has been attempted to simulate two essential characteristics of diesel spray combustion, intermittency and autoignition. A single shot or successive intermittent sprays are injected using a modified diesel fuel injection system into a high temperature air stream resulting in the formation of liquid sprays, autoignition and combustion at constant pressure. A small portion of the resulting expanding and moving flame is quenched and filtered at different intervals from the beginning of fuel injection to the end of combustion.

A schematic diagram of the experimental apparatus is given in Fig. 1. It consists of an air supply system, an electric air heater with automatic temperature control, a modified diesel fuel injection system, a combustion chamber, a sampling system and an optical system. Referring to the figure, filtered shop air (1) is supplied through a valve (2) and a flow meter (3) into an electric air heater (42). The heater consists of an insulated stainless steel pipe of 200 mm inside diameter and 900 mm long. It contains five pieces of electric heating elements (43), each having a maximum output of 3.2 kw. The elements are connected separately to an automatic temperature controller (44). The air temperature can reach 1000°C with a fluctuation of $\pm 1\%$. The air temperature is measured by a chromel-alumel thermocouple (40) at the entrance of the combustion chamber (37). The thermocouple output is also used as an input to the automatic temperature controller. Fuel injection is made by a modified diesel pump-line-injector system. The injection nozzle (15) is located at the center of the combustion chamber, and the fuel is injected along the axis in the direction of the flowing air. The nozzle has a single orifice, 0.4 mm diameter. The needle opening pressure is 22.4 mPa. The nozzle as well as the fuel line are water-cooled to keep the fuel at a constant temperature. The injection pump (6) is driven by a stepless variable-speed electric motor (12). A disc with a narrow slit (13), a lamp (8) and a phototransistor (9) are used to measure the pump speed. The output from the phototransistor circuit is also used as a signal to trigger the display of the oscilloscope (19). A DC solenoid (7) can be used to produce a single shot of fuel injection. The solenoid is triggered by an electric circuit (10) which is actuated by the signal from the phototransistor (9). The solenoid controls the rack of the injection pump. The start of injection is detected by an optical system consisting of a phototransistor (16) and a lamp (39), which is located with its center line passing through the nozzle tip.

The combustion chamber (37) consists of an insulated stainless-steel tube of 100 mm inner diameter and 900 mm long. It has mica windows (36) to allow the observation and photographing of the flame. The wall temperature is measured with a thermocouple (35). The temperatue of the flame in the combustion chamber is measured by a Pt-PtRh (13%) bare thermocouple (21) having a diameter of 0.127 mm. The output of the thermocouple is displayed on oscilloscope screen (19).

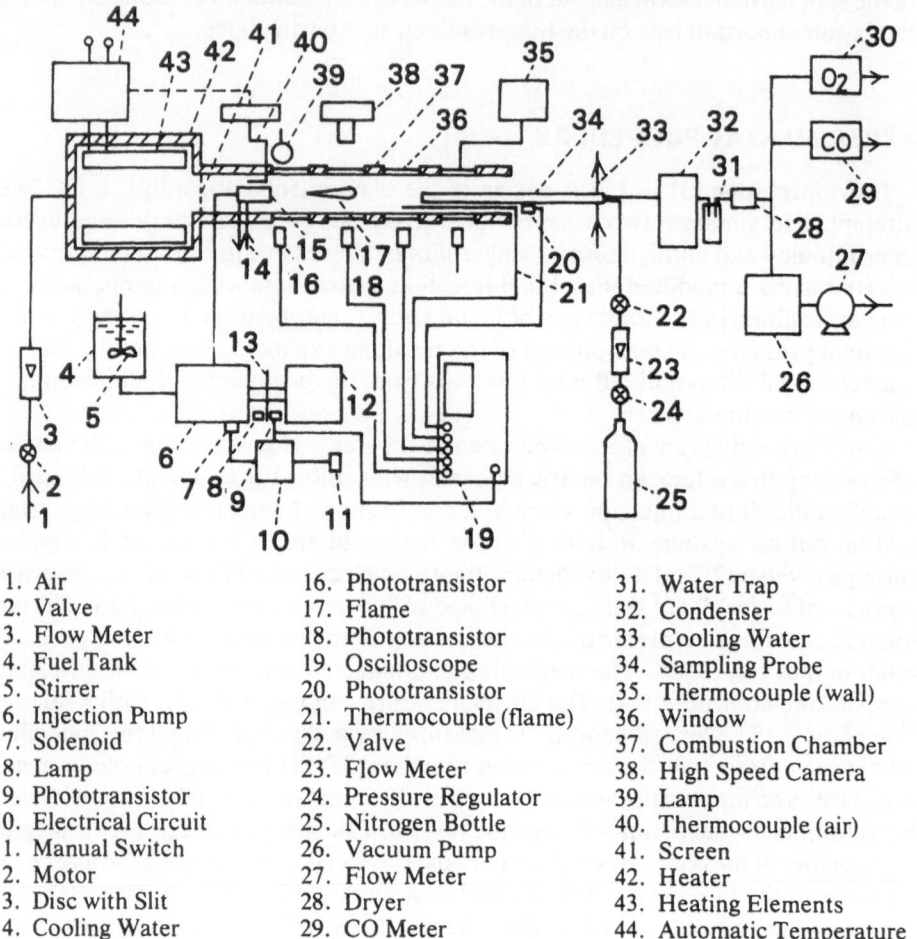

1. Air	16. Phototransistor	31. Water Trap
2. Valve	17. Flame	32. Condenser
3. Flow Meter	18. Phototransistor	33. Cooling Water
4. Fuel Tank	19. Oscilloscope	34. Sampling Probe
5. Stirrer	20. Phototransistor	35. Thermocouple (wall)
6. Injection Pump	21. Thermocouple (flame)	36. Window
7. Solenoid	22. Valve	37. Combustion Chamber
8. Lamp	23. Flow Meter	38. High Speed Camera
9. Phototransistor	24. Pressure Regulator	39. Lamp
10. Electrical Circuit	25. Nitrogen Bottle	40. Thermocouple (air)
11. Manual Switch	26. Vacuum Pump	41. Screen
12. Motor	27. Flow Meter	42. Heater
13. Disc with Slit	28. Dryer	43. Heating Elements
14. Cooling Water	29. CO Meter	44. Automatic Temperature
15. Injection Nozzle	30. O_2 Meter	Controller

Fig. 1. Experimental apparatus.

The soot particulates are collected by using a sampling probe (34) located at different positions in the combustion chamber. The sampling probe is schematically shown in Fig. 2. A water-cooled inert gas quenching technique has been newly developed for this probe. It is constructed with stainless steel pipes. The outer diameter of the probe is 30 mm. The water is allowed to flow through a concentric annular part. The probe has a hole of the diameter of 6.25 mm at its tip to suck flame gases. Gaseous nitrogen (F) supplied through a small pipe in the water jacket is discharged in an inward radial direction through a narrow slit (H) at the tip of the sampling probe. A small element of the flame gases (A) is withdrawn into the sampling probe under the effect of negative pressure caused by a vacuum pump (26). It is quickly quenched by the discharging gaseous nitrogen. The soot particulates are collected on a glass fiber filter paper (C) (24 mm in diameter) supported by a stainless steel mesh (D) at the end of the probe. If the quenching of the chemical reaction is complete, the materials collected on the filter paper could be representative of the soot particulates contained in the flame gases at the entrance of the probe. The mass of the soot particulates is obtained by weighing the filter paper before and after sampling. Prior to weighing, it is heated in a constant temperature drying oven at 130°C, and then kept in a desiccator to avoid the error which might be caused by the water and other materials contained in the filter paper, if any, as explained in detail in Reference [4].

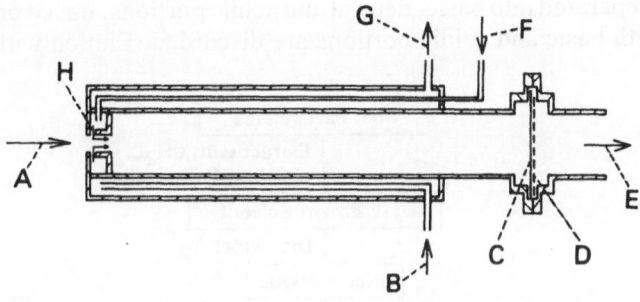

A. Flame Products	E. Sampled Gases Outlet
B. Water - In	F. Nitrogen Gas - In
C. Filter Paper	G. Water - Out
D. Stainless Steel Mesh	H. Nitrogen Gas - Out

Fig. 2. Sampling probe.

The analysis of the microstructure of the soot particulates is obtained by using an electron microscope. A formvar-coated grid (400 mesh) is installed on the glass fiber filter paper to sample the soot particulates. The ratio of carbon-to-hydrogen atoms (C/H ratio) in the soot particulates is also determined from measuring the mass of carbon dioxide and water produced by burning the collected soot particulates. After filtration, the sampled gas flows through a condenser (32), a water trap (31), a dryer (28), a vacuum pump (26) and a flow meter (27). It is also analyzed by using an oxygen meter (30) and a NDIR CO analyzer (29).

The characteristics of the fuel spray as well as the flame developed in this combustor have been obtained by using an optical system and a 16 mm high-speed movie camera (38). The optical system consists of a set of a lamp (39) and a phototransistor (16). The phototransistor (FPT 100) has a rise time of the order of microseconds and a maximum sensitivity at a wave length of 0.83 μm. To avoid the reflection from the wall of the combustion chamber, the phototransistor is located away from the walls, at the end of a cylinder which has two holes of 2 mm diameter each. Therefore, the radiation detected by the phototransistor is that produced by the flame in a location along the axis of the cylinder. This was checked by using a small lamp. Thus, the phototransistor signal is an indication of the instant of time when the spray or flame arrives at the location of the phototransistor. Another phototransistor (20) is installed at the end of the combustion chamber, in view of the injection nozzle, in order to detect the autoignition. All the signals from these phototransistors are displayed on the oscilloscope screen.

Bacterial Analysis [5] — The extraction of soluble organic materials in the soot particulates and their separation are made prior to the bacterial analysis. The flow diagram of the separation procedure is shown in Fig. 3. All the samples are kept under subdued light before their analysis. The soot particulates are extracted in a Soxhlet extractor with methylene chloride. The dried extract is redissolved in ether, being separated into basic, neutral and acidic portions. Based on the previous report [6], both basic and acidic portions are discarded. Elution with various sol-

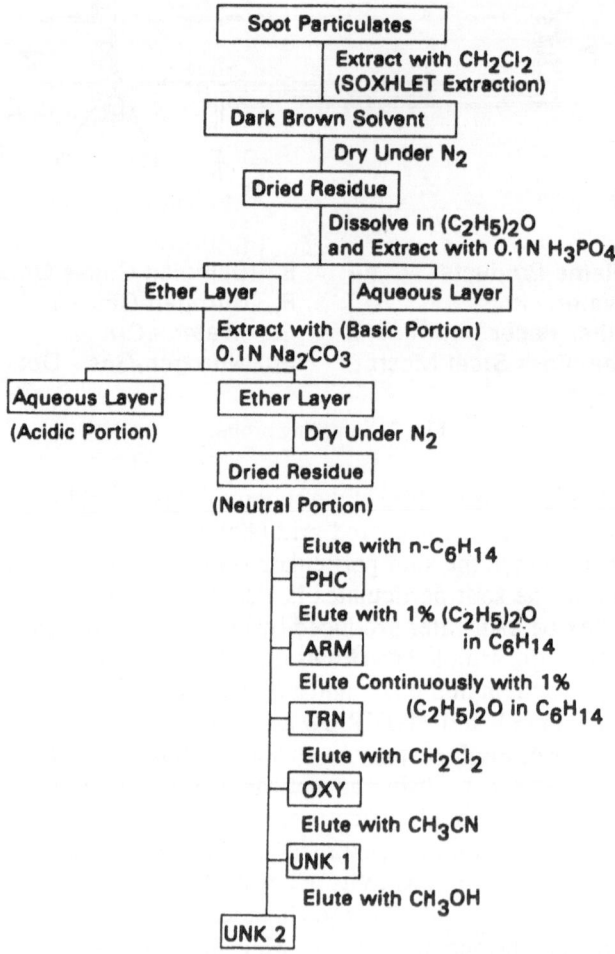

Fig. 3. Separation of organic materials.

vents on the neutral portion is done on a silica-gel column. The paraffinic hydrocarbons (PHC) are first collected by eluting the column with n-hexane. Then, the aromatic hydrocarbon (ARM), the transitional fraction (TRN) and the oxygenated fraction (OXY) are obtained. Thereafter, the column is eluted with acetonitrite and then methanol. These two last fractions are termed unknown 1 (UNK1) and unknown 2 (UNK2) respectively. Their chemical nature has not yet been characterized. They are expected to contain high molecular weight polynuclear derivatives with high polarity.

The mutagenic activity of each fraction thus derived is assayed in Salmonella typhimurium TA98 in the absence of external activation systems according to the technique by Ames [7]. The tested material is dissolved in dimethylsulfoxide, mixed with the bacterial suspension and a minimal agar (45°C) and then plated. The his^+ revertant colonies were enumerated after incubation in the dark for 3 days.

RESULTS OF MEASUREMENTS

Intermittent Spray and Flame Characteristics — Observations made on the combustor show that the fuel injected in the hot air stream autoignites and the flame tip moves toward the exit of the combustion chamber. The following points in time are detected by phototransistors installed at different locations outside of the combustion chamber and by a high speed movie camera: the start and end of fuel injection, the autoignition, and the arrival and the departure of the flame at different distances from the injection nozzle. Fig. 4 shows a sample of the traces for the time-resolved events of fuel injection, the start of illumination, and the arrival and departure of the flame at 555 mm, 685 mm and 785 mm from the injection nozzle. The results

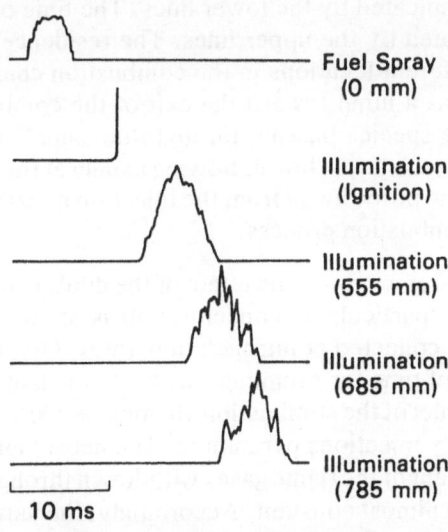

Fig. 4. Flame development detected by phototransistors.

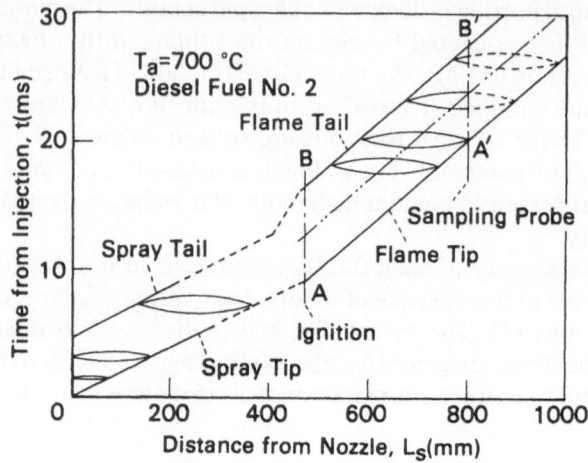

Fig. 5. The characteristics of flame and spray.

obtained for these events are plotted in Fig. 5. The solid lines are for measured results. Cycle-to-cycle variations in the combustion processes were observed. These are believed to be caused by aerodynamic factors because injection duration and injection timing did not show such variations. Sampling extended over a period of five minutes and the soot particulates from 600 consecutive flames were collected. This large number of sampled flames eliminates any errors which result from cycle-to-cycle variations common to combustion processes. The broken lines are extrapolations of the solid lines. The time of arrival of the spray tip and the flame at different locations is indicated by the lower lines. The time of arrival of the spray and flame tail is indicated by the upper lines. The residence time of the flame is almost the same at different locations in the combustion chamber. This indicates that the flame moves as a lump toward the exit of the combustion chamber at a constant velocity. The species passing through the sampling probe consist of a series of combustion products and hot air flowing axially at the location of the probe entrance. By moving the probe away from the injection nozzle, the sampling starts at a later time in the combustion process.

Effect of Sampling Parameters — The effect of the dilution of the products by the quenching gas on soot particulates concentration is shown in Fig. 6. The soot particulates have been collected at nitrogen flow rates, Q_N, ranging from zero to 120 cm³/s. The period of time for sampling was kept constant at five minutes. The air temperature at the inlet of the combustion chamber is 700°C. The injection rate is one-hundred and twenty injections per minute. The net sampling velocity, that is, the velocity of the element of the flame gases withdrawn through the entrance of the sampling probe, is kept almost constant. Accordingly, the variation in the nitrogen flow rate is proportional to the variation in the dilution ratio. It is seen from the figure that the soot particulates concentration increases with increasing nitrogen

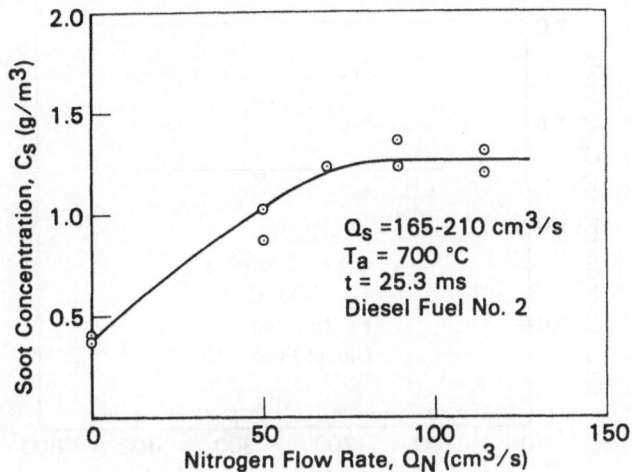

Fig. 6. Dilution effect on soot concentration.

flow rate from zero to about 80 cm³/s, after which it levels off. The remarkable change of the blackness of the soot particulates collected on a filter paper implys that the reduced soot particulates at the low dilution ratio could mainly be due to the incomplete quenching of the flame gases at the entrance of sampling probe. This indicates that the water-cooled sampling probe without dilution by discharging inert gases is not adequate for sampling the soot particulates in the flame. In all the experimental work in this study, the nitrogen flow rate has been kept constant at 90 cm³/s.

The effect of sampling velocity on the measured soot particulates concentration is shown in Fig. 7. The sampling velocity is varied by varying the negative pressure acting on the sampling probe. It is seen from this figure that, within a gas flow rate of 100 cm³/s to 410 cm³/s, there is no appreciable change in the measured concentration. In all the experiments, the sampled gas flow rate has been kept constant at 200 cm³/s.

It is very difficult to obtain the real flame temperature which shows temporal and spatial variations. If we assume that the flame temperature ranges from 1000 K to 2200 K, isokinetic sampling conditions correspond to the sampling rate of 120-270 cm³/s in Fig. 7. Therefore, it can be considered that the flame gases are isokinetically sampled in the present experiments.

Concentrations of Soot Particulates — Fig. 8 shows the time-resolved concentration of the soot particulates along the axis of the combustion chamber. The soot concentration increases with the lapse of time and reaches a maximum after which it decreases at a high rate to very low values approaching zero. The maximum occurs at 25.3 ms after the start of fuel injection. Also shown in this figure is the oxygen concentration and the flame temperature at different times. The oxygen

References p. 414.

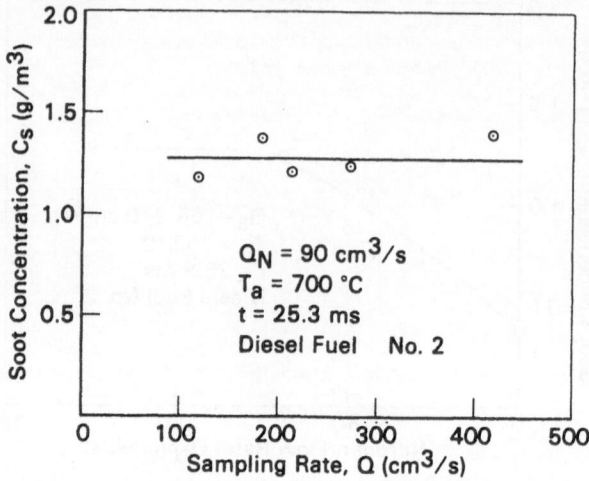

Fig. 7. Effect of sampling rate on soot concentration.

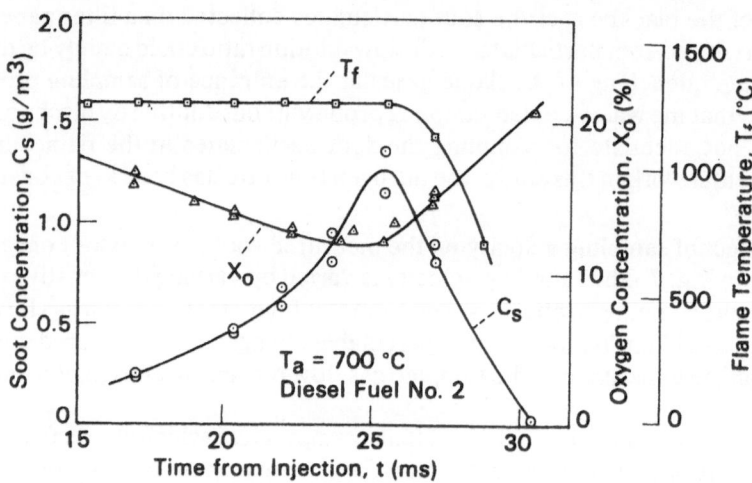

Fig. 8. Time-resolved concentration of oxygen and soot, and flame temperature.

concentration shows a minimum at the time close to that of maximum soot particulates concentration. The flame temperature measured at any time shows a periodical change with time, at a frequency equal to the one of the fuel injection. The average peak temperature is measured and plotted in the figure. It has a uniform profile except for the last stage of combustion.

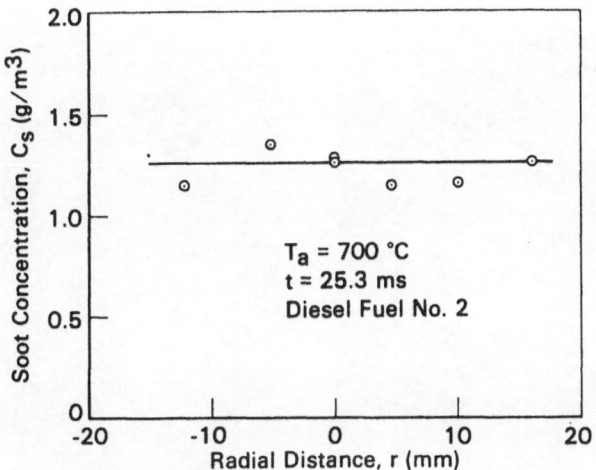

Fig. 9. Radial distribution of soot concentration.

Fig. 9 shows the radial distribution of soot particulates concentration at 25.3 ms. It is clear from this figure that the radial distribution of the soot particulates concentration near the center of the flame is almost uniform.

Physical and Chemical Properties of Soot Particulates — Fig. 10 shows a sample of the electron microscope photographs of the soot particulates, which are collected at different times after fuel injection. Several pictures are taken of the soot particulates at different randomly chosen positions in a grid. Total magnification is 120,000. These photographs indicate that there is no appreciable change in the structure of the soot particulates. But the size of the small soot particles which are almost spherical appears to vary with time. The mean diameter of these particles is determined and plotted in Fig. 11. The number of the small particles is calculated from their concentration in Fig. 8 and their mean diameter mentioned above. For their computation, the density of the soot particulates is assumed to be constant and equal to 2.0 g/cm^3. Also shown in Fig. 11 is the C/H ratio in the soot particulates collected at different times. The C/H ratio is defined as the number of carbon atoms divided by the number of hydrogen atoms. It is seen from the figure that both the mean diameter and the C/H ratio increase with time while the number of the soot particles per unit volume of flame gases decreases.

Mutagenic Activity — Fig. 12 shows the specific and total mutagenic activity, R_S and R_T, in the soot particulates formed at different stages of combustion. The specific mutagenic activity is defined as the revertants per unit mass of the soluble organic materials extracted from the soot particulates. The total mutagenic activity is defined as the revertants per total mass of the soluble materials derived from the soot particulates collected in the unit volume of flame gases. It is evident from the

References p. 414.

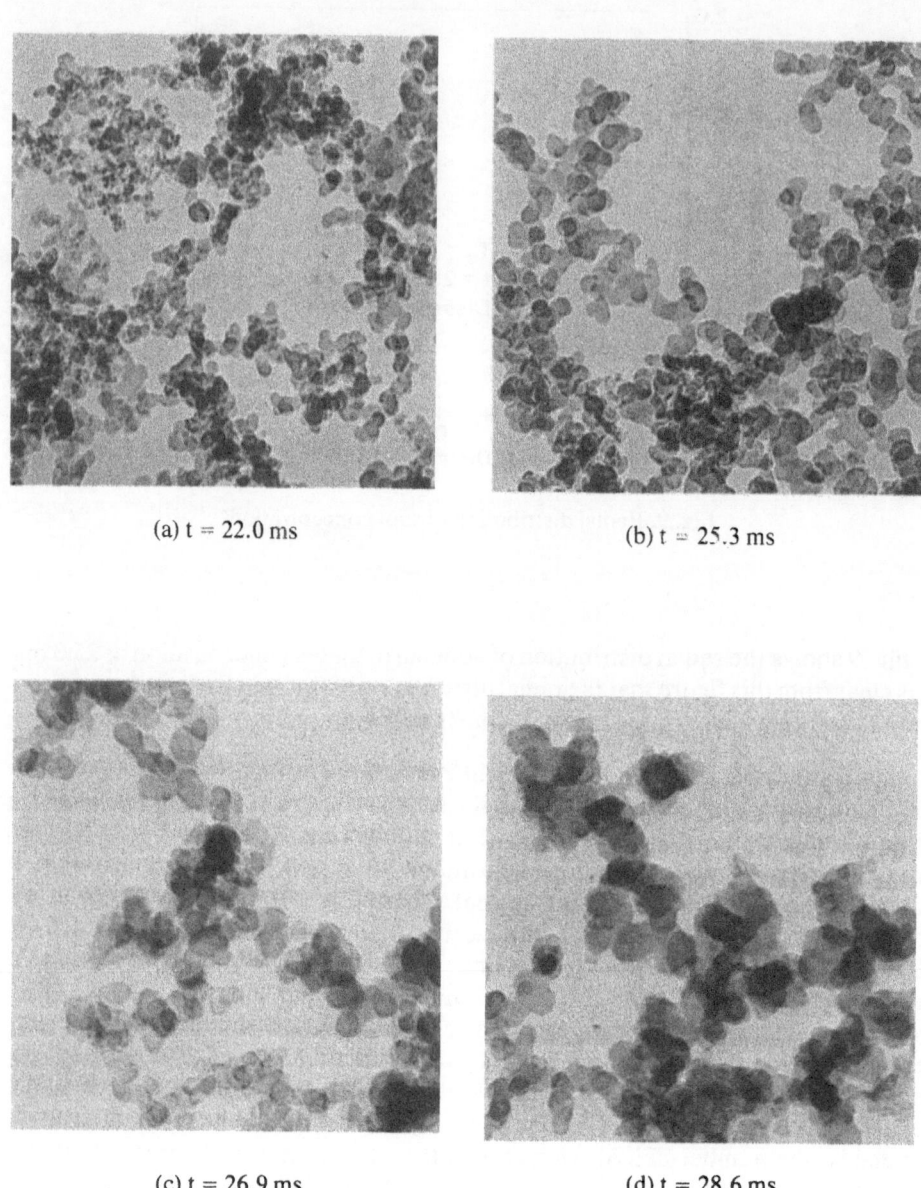

(a) t = 22.0 ms   (b) t = 25.3 ms

(c) t = 26.9 ms   (d) t = 28.6 ms

T_a=700°C, Diesel Fuel No. 2
Magnification: 120,000 X

Fig. 10. Electron microscope photograph of soot

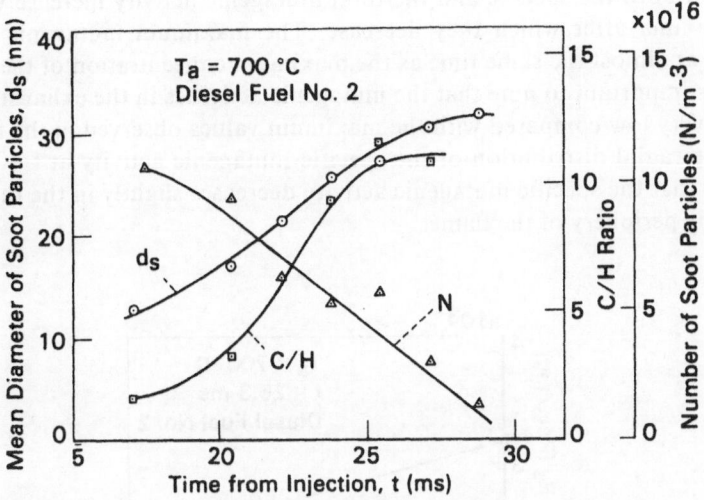

Fig. 11. Time-resolved mean diameter, number and C/H ratio of soot particulates.

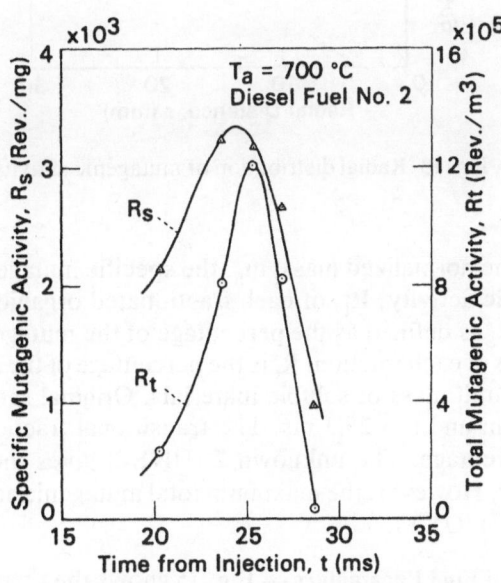

Fig. 12. Time-resolved mutagenic activity.

figure that both the specific and the total mutagenic activity increase with time and reach maxima after which they decrease. The maximum mutagenic activities are observed at almost the same time as the maximum concentration of the soot particulates. It is important to note that the mutagenic activities in the exhaust soot particulates are very low compared with the maximum values observed in the flame. Fig. 13 shows the radial distribution of the specific mutagenic activity at t = 25.3 ms. It is indicated that the specific mutagenic activity decreases slightly in the radial direction toward the periphery of the flame.

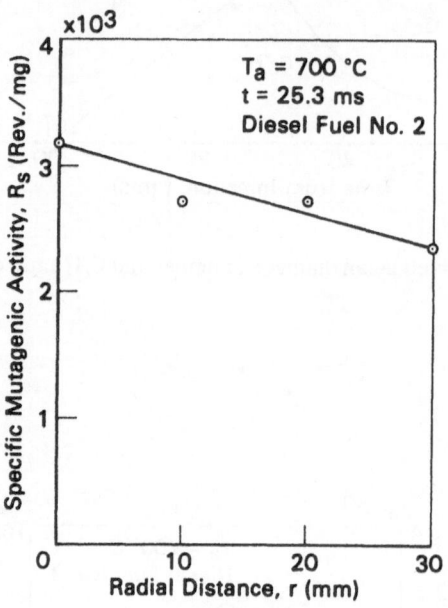

Fig. 13. Radial distribution of mutagenic activity.

Fig. 14 shows the normalized mass, m_S, the specific mutagenic activity, R_S^*, and the total mutagenic activity, R_t^*, of each fractionated organic material derived as shown in Fig. 3. R_S^* is defined as the percentage of the mutagenic activity contributed by a unit mass of each fraction. R_t^* is the percentage of the mutagenic activity of each fraction per unit mass of soluble materials. Original samples have been obtained at r = 0 mm and t = 25.3 ms. The transitional fraction (TRN) shows the highest mass percentage. The unknown 2 (UNK2) gives the maximum specific mutagenic activity. However, the maximum total mutagenic activity is given by the oxygenated fraction (OXY).

Effects of Air and Fuel Parameters — Fig. 15 shows the time-resolved soot particulates concentration at different inlet air temperatures. It is evident that the maximum soot concentration increases with the increase in the inlet air temperature. But the exhaust soot particulates are independent of the air temperature and approach

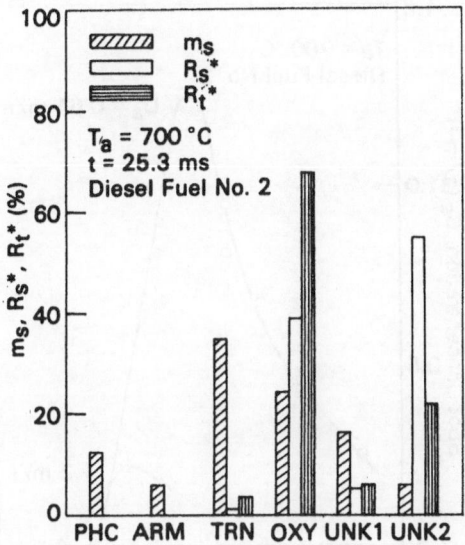

Fig. 14. Mutagenic activity of each fraction.

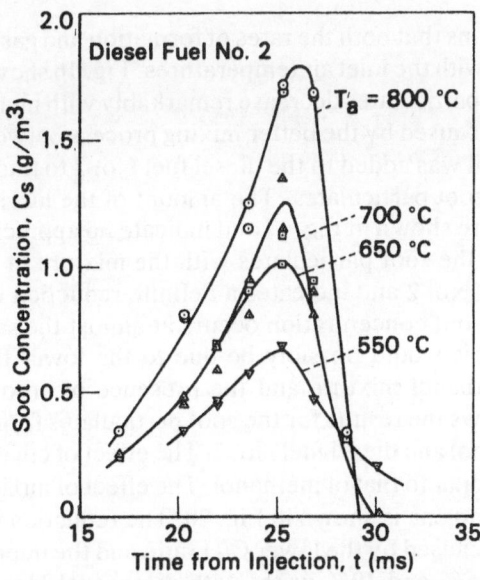

Fig. 15. Effect of inlet air temperature on soot concentration.

References p. 414.

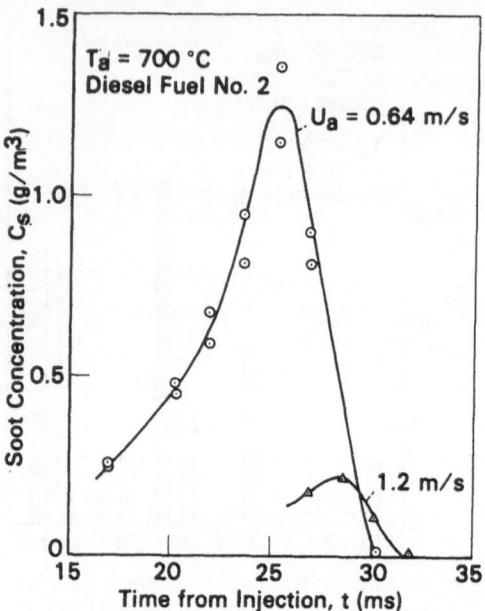

Fig. 16. Effect of air velocity on soot concentration.

the same value. It means that both the rates of formation and gasification of the soot particulates increase with the inlet air temperatures. Fig. 16 shows the effect of inlet air velocity. The soot particulates decrease remarkably with increasing air velocity. This is believed to be caused by the better mixing process between the air and fuel.

Fresh lubricating oil was added to the diesel fuel No. 2 to find its contribution to the formation of the soot particulates. The amount of the lubricating oil is 3% by volume. The results are shown in Fig. 17 and indicate no appreciable effect. Fig. 18 shows the results for the soot particulates with the mixture of 20% and 40% methanol and diesel fuel No. 2 and indicates a definite reduction in the soot particulates. The maximum soot concentration occurs at almost the same time, with and without methanol. This would possibly be due to the lower flame temperatures reached with the methanol mixture and the presence of an oxygen atom in the methanol. Fig. 19 shows the results for the soot particulates formed by combustion of the mixture of ethanol and diesel fuel No. 2. The effect of ethanol on reducing the soot particulates is similar to that of methanol. The effect of mixing diesel fuel No. 2 with 10% and 20% indolene is shown in Fig. 20. The reduction in the soot particulates is believed to be caused by the lower C/H ratio and the improved evaporation. The effect of mixing 5% and 10% water with diesel fuel No. 2 on reducing the concentration of the soot particulates is shown in Fig. 21. The reduction in soot particulates would primarily be due to the decreased flame temperature.

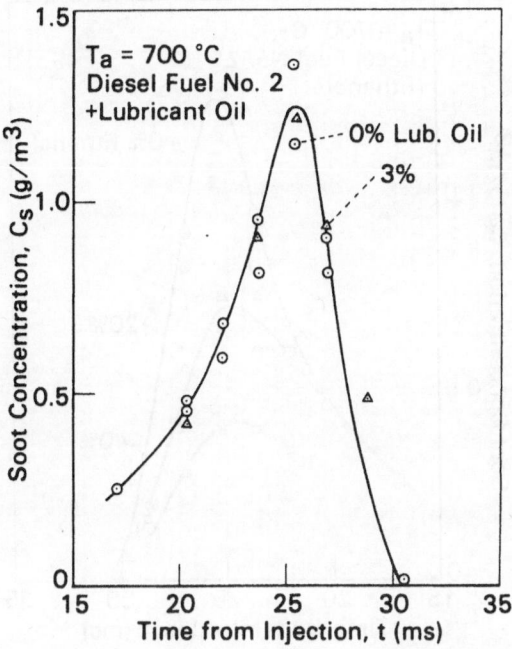

Fig. 17. Effect of lubricant oil added to diesel fuel No. 2.

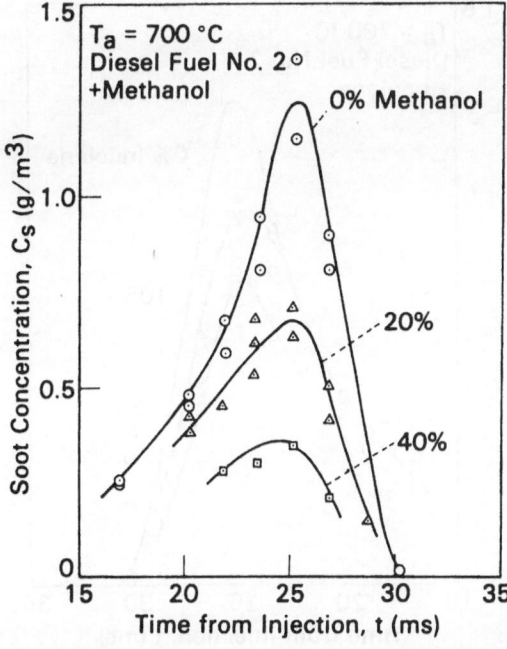

Fig. 18. Effect of methanol added to diesel fuel No. 2.

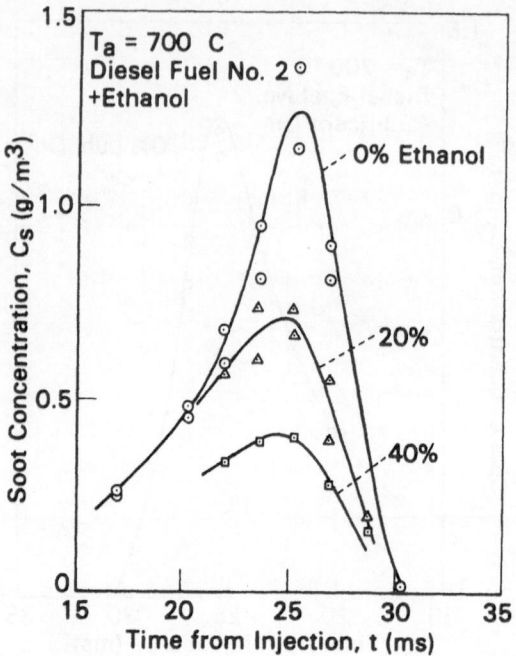

Fig. 19. Effect of ethanol added to diesel fuel No. 2.

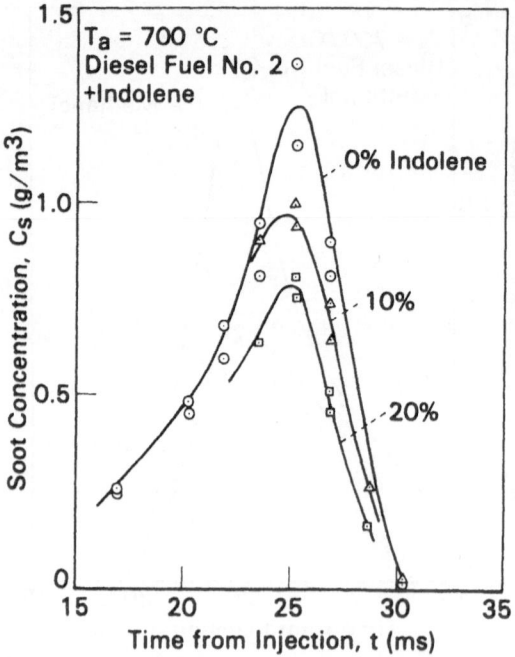

Fig. 20. Effect of indolene added to diesel fuel No. 2

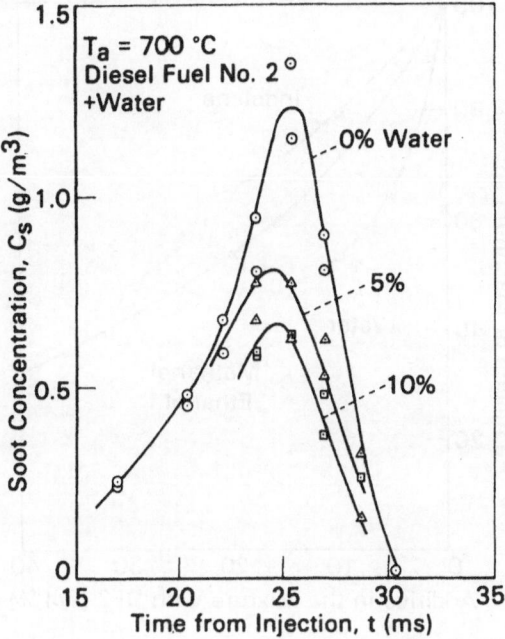

Fig. 21. Effect of water added to diesel fuel No. 2

Fig. 22 shows the effect of mixing different fuels or water with diesel fuel No. 2 on reducing the maximum concentration of the soot particulates in the flame. It is evident that mixing diesel fuel with water is the most effective, followed by alcohols and indolene. The present work indicates that the difference between the effect of methanol and ethanol on the maximum concentration of soot particulates is not substantial. The effects of air and fuel parameters on the soot particulates are summarized in Table 1 in terms of the mean rate of soot formation and gasification.

DISCUSSION

Physical Mechanism of Soot Formation and Gasification — Soot formation would be ruled by Eq. (1).

$$\dot{C}_s(t) = \dot{C}_f(t) + \dot{C}_c(t)$$

$$(1)$$

where $\dot{C}_s(t)$: net rate of soot formation

$\dot{C}_f(t)$: apparent rate of soot formation

$\dot{C}_c(t)$: rate of soot gasification

References p. 414.

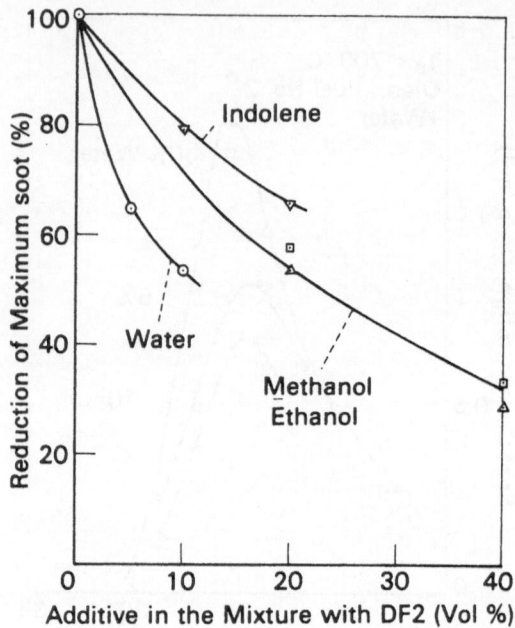

Fig. 22. Effect of additives on reducing the maximum soot concentration.

TABLE 1

The Mean Rate of Formation ($\dot{m}_f$) and Gasification ($\dot{m}_c$) of Soot Particulates

(a) Effect of Inlet Air Temperature

T_a(°C)	$\dot{m}_f$ (g/m³/ms)	$\dot{m}_c$ (g/m³/ms)
800	0.17	0.76
700	0.12	0.27
650	0.11	0.27
550	0.071	0.097

(b) Effect of Fuel Properties

Fuel	$\dot{m}_f$ (g/m³/ms)	$\dot{m}_c$ (g/m³/ms)
Diesel Fuel No. 2	0.12	0.27
Methanol 20%	0.057	0.17
Methanol 40%	0.036	0.092

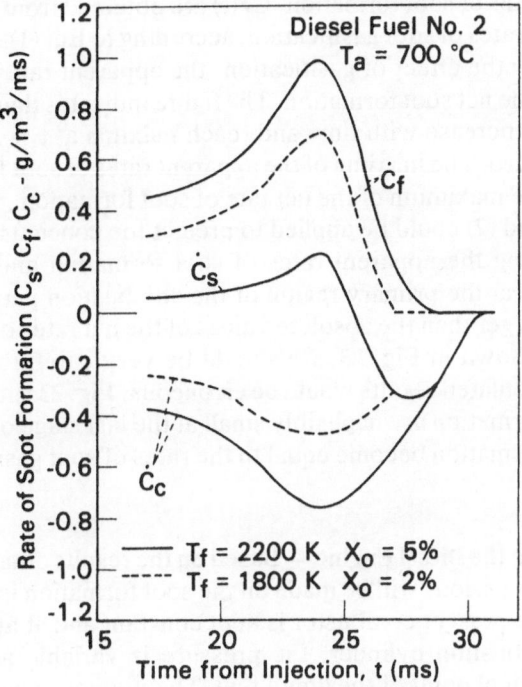

Fig. 23. The rate of soot formation.

The instantaneous concentration of the soot particulates $C_S(t)$ could be determined from Eq. (2).

$$C_S(t) = \int \dot{C}_S(t) \, dt \qquad (2)$$

An attempt has been made to obtain the values in Eq. (1). The results are shown in Fig. 23. The net rate of soot formation $\dot{C}_S(t)$ is derived from the time-resolved soot concentration as shown in Fig. 8. It increases with time and reaches a maximum at T ≈ 24 ms. After passing the maximum, it decreases to zero at $t \approx 25.3$ ms, where the apparent rate of soot formation balances with the rate of soot gasification. Thereafter, it shows a minimum after which it returns to zero again. Thus, the soot formation and gasification zones are clearly observed.

$\dot{C}_C(t)$ is the rate of soot gasification which has been calculated by using the results reported by Park and Appleton [8] as well as the physical properties of the soot particulates obtained in the present study. In the computation, two sets of the flame temperatures and oxygen concentrations in the flame gases are assumed based on the previous reports [9, 10]. It is likely that they would give the limits of the expected flame temperature and oxygen concentration in the present combustor. The calculated rates of soot gasification decrease with time and reach minima after which they increase to zero. The minima are observed at $t \approx 24$ ms.

References p. 414.

The apparent rates of soot formation $\dot{C}_f(t)$ are obtained from the net rate of soot formation and the rates of soot gasification, according to Eq. (1). If the soot particulates are not under the effect of gasification, the apparent rates of soot formation would determine the net soot formation. The figure indicates that the apparent rates of soot formation increase with time and reach maxima at $t \approx 24$ ms, after which they decrease to zero. The maxima of the apparent rates of soot formation are 3 to 5 times as high as the maximum of the net rate of soot formation.

Equations (1) and (2) could be applied to predict the concentration of soot particulates by assuming the apparent rates of soot formation and the rates of soot gasification. Since at the primary region of the combustion period, their absolute values are much larger than the absolute values of the net rates of soot formation to be predicted, as shown in Fig. 23, we should be very careful in assuming them. Otherwise, the calculated results would be erroneous. Fig. 23 shows that the apparent rates of soot formation are negligibly small at the last stage of combustion. The net rates of soot formation become equal to the rates of soot gasification.

Soot Formation in the Diesel Engine — Based on the results obtained in the present research, some discussions will be made on the soot formation in the diesel engine. The pressure in the present combustor is kept constant and at atmospheric. In the diesel engine combustion cylinder, the pressure is variable and reaches levels higher than the critical point of the liquid fuel. The previous studies on the laboratory flames [4, 11, 12] have reported that the increase in pressure causes the increase in the soot particulates mainly due to the increased thermal pyrolysis as a result of the decrease in molecular diffusion. Therefore, the soot particulates during combustion are expected to be higher in the diesel engine than in the present combustor.

The flame temperature in the diesel engine varies with time and would possibly be higher than in the present combustor during the primary portion of the combustion period. High temperature causes the increased soot particulates in the flame as shown in Fig. 15. During the expansion stroke in the engine, the temperature and the oxygen concentration drop below the values in the present combustor. This would result in the reduced rate of soot gasification [8], producing more soot particulates exhausted from the engine. This agrees with the experimental results that in the presence of sufficient oxygen the extremely high temperature causes the reduction of the exhaust soot particulates [13].

The time available for the gasification of the soot particulates is much less in the engine than in the present combustor. This would prevent the soot particulates from being gasified, resulting in much more soot particulates in the exhaust of the engine. These discussions are illustratively shown in Fig. 24.

As shown in Fig. 12, the mutagenic activity in the exhaust soot particulates is much lower than in the flame. This implies that increased mutagenic activity would be expected in the exhaust soot particulates if the flame is quenched.

These discussions lead to a conclusion that it is very important to gasify the soot particulates prior to their emission in the exhaust from the aesthetic and the toxic point of view.

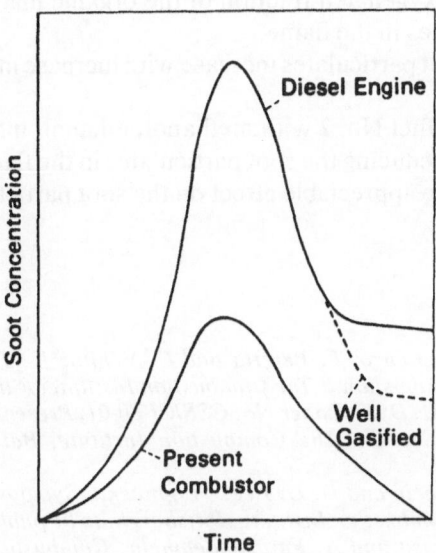

Fig. 24. Comparison between the process in the present combustor and the diesel engine.

SUMMARY AND CONCLUSIONS

Experiments have been made to study the soot particulates formed at different stages of combustion of a simulated diesel spray by applying a newly developed laboratory combustor in which the following two essential charateristics of diesel spray combustion are reproduced, intermittency and autoignition. The combustor is designed so that the fuel is injected by using a diesel fuel injection system into a steadily flowing high-temperature air stream resulting in autoignition and combustion. Soot particulates have been collected at different stages of combustion by applying an inert gas quenching technique. The mass concentration, diameter, number and carbon-to-hydrogen ratio of soot particulates, oxygen concentration and temperature in the flame as a function of the time from the start of fuel injection have been obtained. Mutagenic activity of time-resolved soot particulates has been determined by using the bacterial mutagenesis test. Thus, some factors affecting time-resolved characteristics of soot particulates have been studied. Theoretical analysis is also made of the physical mechanism of soot particulates formation.

The primary conclusions reached in the present study are as follows:

1. The concentration of the soot particulates increases with time and reaches a maximum after which it decreases to the low value approaching zero.
2. The diameter and C/H ratio of the soot particulates increase with time while their number per unit volume of flame gases decreases.
3. The mutagenic activity of the soot particulates shows a maximum as a function of time from the start of injection. The highest mutagenic activity is

observed in the oxygenated fraction of the organic materials extracted from the soot particulates in the flame.

4. The maximum soot particulates increase with increase in the inlet air temperature.

5. Mixing the diesel fuel No. 2 with methanol, ethanol, indolene or water has a definite effect on reducing the soot particulates in the flame. Fresh lubricating oil up to 3% has no appreciable effect on the soot particulates.

REFERENCES

1. A. D'Alessio, A. DiLorenzo, F. Beretta and C. Venitozzi, "Fourteenth Symposium (International) on Combustion," The Combustion Institute, Pittsburgh (1973), p. 941.
2. W. L. Flower and T. M. Dyer, Paper No. CSS/CI-80-01, Presented at the Spring Meeting, Central States Section, The Combustion Institute, Baton Rouge, Louisiana, March, 1980.
3. T. Kadota, N. A. Henein and D. U. Lee, "Eighteenth Symposium (International) on Combustion," The Combustion Institute, Pittsburgh, to be published in 1981.
4. T. Kadota, H. Hiroyasu and A. Farazandehmehr, Combustion and Flame, Vol. 29 (1977), p. 67.
5. T. Kadota, N. A. Henein, C. King, J. P. Horwitz, D. P. Lin and C. Wang, to be published.
6. J. Yergey, Paper No. CSS/CI-79-06, Presented at the Spring Meeting, Central States Section, The Combustion Institute, Columbus, Indiana, April, 1979.
7. B. N. Ames, "The Detection of Chemical Mutagens with Enteric Bacteria in Chemical Mutagens," A. Hollander, ed., Plenum Press, New York, NY (1971).
8. C. Park and J. P. Appleton, Combustion and Flame, Vol. 20 (1973), p. 369.
9. Y. Onuma and M. Ogasawara, "Fifteenth Symposium (International) on Combustion," The Combustion Institute, Pittsburgh (1975), p. 453.
10. Y. Matsui, T. Kamimoto and S. Matsuoka, SAE Paper No. 790491 (1979).
11. R. L. Schalla, T. P. Clark and G. E. McDonald, NACA Report 1186 (1954).
12. J. J. MacFarlane, F. H. Holderness and F. S. E. Whitcher, Combustion and Flame, Vol. 8 (1964), p. 215.
13. T. Kadota, K. Nakanishi and H. Hiroyasu, Transaction of the Japan Society of Mechanical Engineers, Vol. 46 (1980), p. 364.

DISCUSSION

S. M. Shahed (*Cummins Engine Company*)

My question is directed as much at Professor Mellor as it is to you. I understand that your sampling probe is sampling continuously, and the burning spray reaches it, is sampled and passes on. Then when the burning spray is not at the sampling point, you are sampling the ambient air.

Kadota

Yes.

Shahed

Therefore, the equivalence ratio is lean, the mean equivalence ratio as you measured it. So it is dangerous to then conclude that soot is formed under leaner conditions in your flames than in other conditions. Also, the peak may be simply because the peak is passing by your sampling point since your sampling is not intermittent as the spray is. Would you be able to extract the information out of your sampling data?

Kadota

You are talking about the air sampled between the two consecutive spray flames. It has been included in the data of the soot concentration reported in our previous publication, which was cited by Professor Mellor in the morning session. So, the soot concentration appeared to be low. However, the soot concentration in the spray flame formed in the present study is much higher than that, as I have shown here. The present data were derived by subtracting the air sampled between the flames. This has not been done in the data of the oxygen concentration. The oxygen concentration in the spray flame would be lower than the results presented here. It might not be appropriate to mention whether or not the soot is formed under leaner conditions in the present system.

Soot would show a peak as a function of time regardless of sampling methods. The technique developed in the present experiment would be sufficient to obtain the essential characteristics of the time-resolved soot particulates in the diesel spray combustion.

U. Bonne *(Honeywell)*

How do you eliminate the oxygen effect? How do you correct for that?

Kadota

The soot concentration in the flame is determined from the measured soot concentration by using the data of the residence time of the flame which is the only source of the soot particulates. It might be difficult to obtain the oxygen concentration in the flame. The results would give us some information on the time-resolved oxygen concentration in the flame.

C. T. Bowman *(Stanford University)*

You cannot conclude from that that the oxygen that you are measuring is playing an active role in the soot process. Is that correct? I think that is the question that you are trying to get to. What role is the oxygen playing?

Kadota

I am not trying to derive a conclusion concerning the role of oxygen on the soot formation. It is beyond the scope of the present investigation.

J. P. Longwell *(Massachusetts Institute of Technology)*

This is relevant to the last question. I think a lot of these considerations are much simpler if we report the fraction of the fuel that is converted into particulates. Milligrams per cubic meter is of interest to air pollution people but is not as useful in considering the chemical processes. A good bit of this concern on the last question would be eliminated if you were simply able to say that in your case roughly 0.1% of the fuel was converted to carbon. This has other values that I really want to mention sometime because everybody has used this milligrams per cubic meter. To give some insight as to possible mechanisms, I have seen in one of the runs reported by Professor Mellor on some MIT work that on the order of 20% of the fuel carbon ended up as particulate carbon. Now it is clear that that does not occur in a premixed flame because you are well beyond the flammability limits of a premixed flame, and so that could be taken as a diagnostic. It was made in something like a Cabot carbon-black manufacturing machine. It contains fuel vapor along with some gas that has already had the oxygen burned out of it. That is an important mechanism for soot formation and is not modeled and can give you things that have been reported so far. The flammability limits the data in premixed systems to hundredths or tenths of the fuel carbon going to soot, and this is much below what we see in diesel engines.

P. Hutchinson *(AERE Harwell)*

We recently have been doing some photographic studies of diesel spray injected into a nitrogen-filled bomb. What we see is that as you increase the pressure and temperature towards pressures and temperatures at the time-of-injection, there is no evidence at all that the droplets are there in the form of a spray for any significant time. If we choose a confined plume, it promptly flashes to vapor. Do you think that your results would still relate to a diesel engine if that happens to be true?

Kadota

The results obtained in the present experiments would be essential to the soot particulates in the diesel engine. For example, the soot particulates are expected to show a maximum as a function of time in the diesel engine combustion chamber, too.

J. B. Heywood *(Massachusetts Institute of Technology)*

I think they are very different conditions. I do not see how Professor Kadota can extrapolate over that range of pressure.

F. L. Dryer *(Princeton University)*

This has always been a very interesting question to me about whether or not there are droplets. What is your (Hutchinson's) evidence that suggest there are no droplets?

Hutchinson

There are not any droplets for a very significant length of time. We think that there is photographic evidence.

Dryer

How do you photograph something between 4 and 15 microns in a dense spray?

Hutchinson

There is a combination of photographic evidence, schlieren evidence and evidence from the use of an LDA in running diesel engines. The LDA works fine if you have particles that are of sub-micron size. We had severe difficulty looking down into the spray region in a running diesel engine to get good LDA signals. So we did the investigation in the high pressure cell mentioned previously.

N. A. Chigier *(University of Sheffield)*

I would just like to add that we looked at diesel sprays at atmospheric conditions, and we find that with high-magnification photography the spray consists of bulk liquid with a few small droplets at the outer periphery. Diesel nozzles provide very poor atomization.

Kadota

Nothing other than the diesel nozzle was applicable for the purpose of the present study.

J. C. Hilliard *(University of Michigan)*

A few years ago you published a paper based on single-droplet studies where you looked at soot emission rate as a function of oxygen concentration such that it rose through a maximum and then fell away to zero with increasing ambient oxygen partial pressure. Would you care to speculate on whether or not you might see similar effects in your system if you would actually enrich the air with oxygen.

Kadota

I expect similar effects of ambient oxygen partial pressure on the time-resolved soot particulates in the present system as I observed in the droplet studies.

S. Matsuoka *(Tokyo Institute of Technology)*

I would like to address the similarity of Professor Kadota's results with real diesel engine tests. In Fig. 1 taken from a paper presented at last February's SAE Congress* is shown a schematic illustration of the engine we have used. The bore diameter is 95 mm, and there is an observing window in the head. We measured flame temperature by the two-color method and simultaneously made gas sampling measurements. We measured both the flame temperature and the gas composition point-by-point in the combustion bowl. We also tried high-speed cinematography through the observing window. In Fig. 2 we compare the measurement of soot concentration by optical (two-color method) and gas sampling techniques. The gas sampling results show the highest average soot concentration at top dead center. However, the optical, two-color method shows its peak slightly later. But this is quite good agreement. I do not have time to explain the difference.

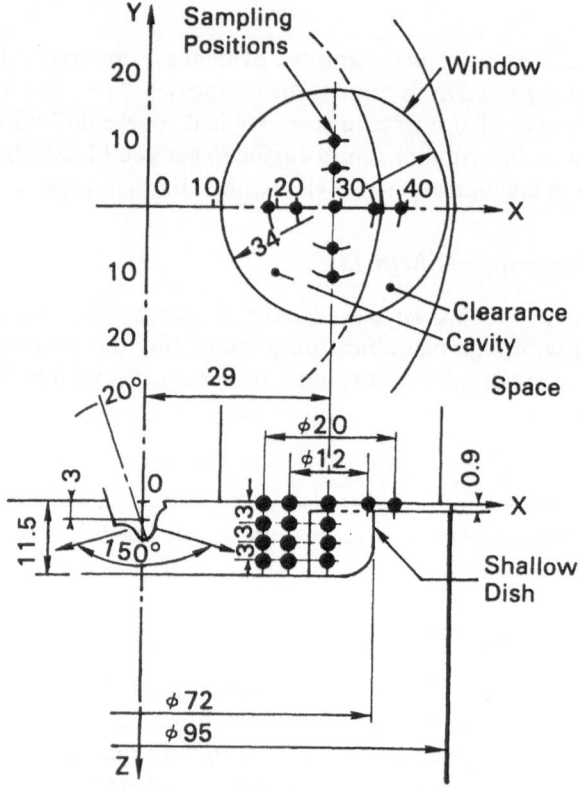

Fig. 1. Shallow dish combustion chamber and sampling positions.

---

*Y. Aoyagi, T. Kamimoto, Y. Matsui and S. Matsuoka, SAE Paper No. 800254 (1980).

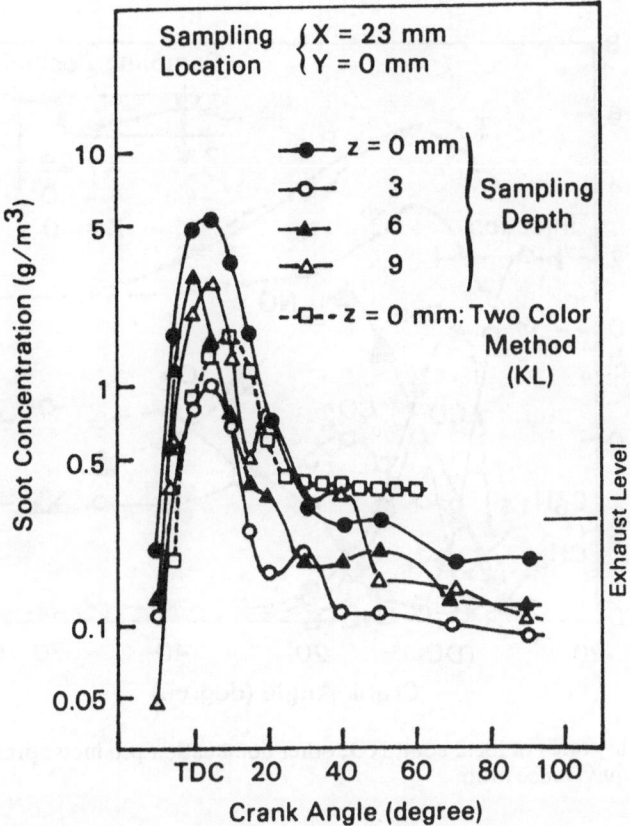

Fig. 2. Comparison of the local soot concentration obtained by two-color method and gas sampling method.

Fig. 3 shows the variation in concentrations, temperatures and pressures as a function of crank angle. The engine speed is 1200 r/min and the gas sampling interval is 1 ms which corresponds to 7°CA. The soot appears very rapidly during the injection process and also decreases very rapidly later on. In this case we measured a peak temperature of nearly 2200 K and from an analysis of combustion products calculated a maximum equivalence ratio of 1.8. The temperature is quite high so the soot generated is a function of equivalence ratio. When the equivalence ratio becomes leaner later in the engine cycle, the soot decreases. A very interesting point is that soot oxidation stops when the temperature drops to 1800-2000 K. We have made parametric studies changing many parameters. In all cases the soot oxidation stops at about 1800°K and maximum soot depends upon equivalence ratio. When the equivalence ratio is rich, soot is very high. The exhaust soot concentration, "Ex.", is shown at the right side of the graph. Its magnitude is comparable to that of the last gas sample measurement at 50°ATDC.

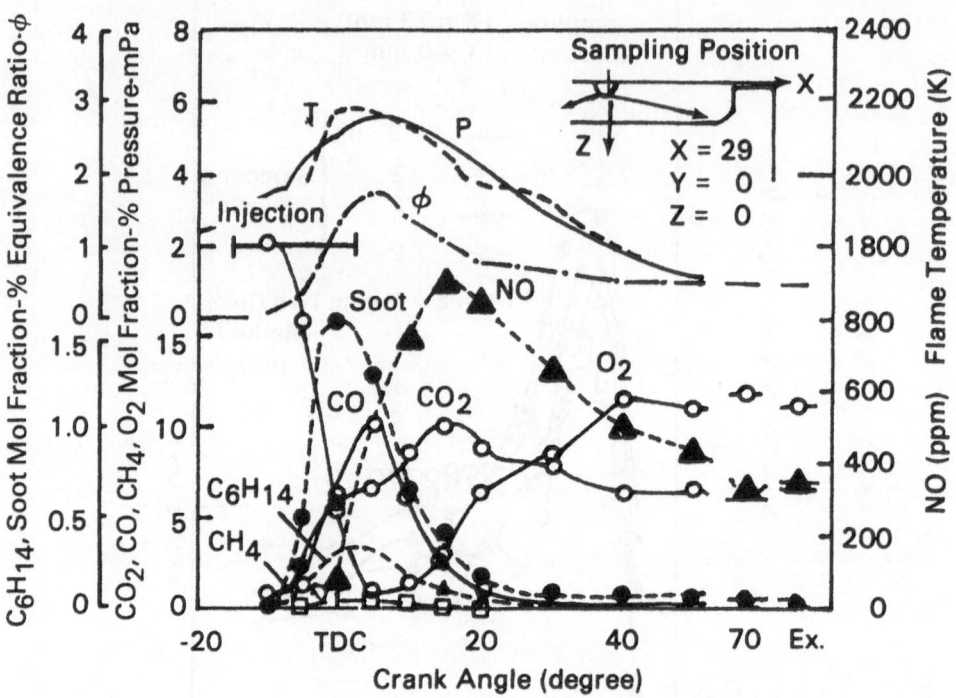

Fig. 3. Time histories of local soot, NO, other combustion products, pressure, temperature and local equivalence ratio.

We have made some parameter studies concerning change of injection timing. In Fig. 4 is shown the effect of injection timing on flame temperature, NO, soot and equivalence ratio. When the timing is advanced, $\Theta_i = 25°$ BTDC, the temperature reaches a peak of nearly 2400 K and then sharply decreases. When the timing is retarded, $\Theta_i = 5°$ BTDC, the temperature peak is reduced to 2100 K, and the temperature does not decrease as rapidly. Also to be noted is the very important influence of equivalence ratio. When the timing is advanced, the maximum equivalence ratio is somewhat leaner and decreases rapidly. When the timing is retarded, the maximum equivalence ratio is richer and does not diminish as rapidly. From data on the temperature and equivalence ratio, the formation of soot can be clearly explained. Soot is formed when the temperature is high enough, but its quantity depends on the equivalence ratio. Soot oxidation stops when the temperature decreases to 1800 K.

The big difference between Professor Kadota's data and this (engine) data is that in his case oxidation takes place in 5 to 7 ms whereas in the engine it is completed in about 1 ms.

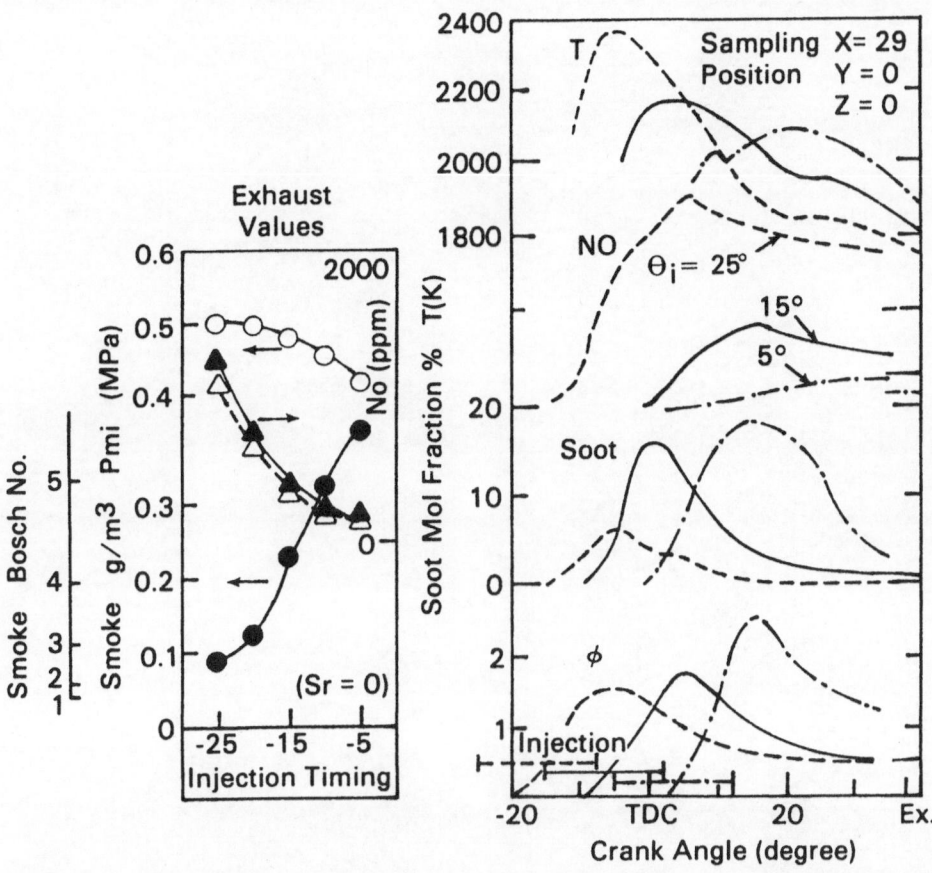

Fig. 4. Variations of local NO and soot concentration, T and ϕ, with changing injection timing.

EFFECTS OF FLAME TEMPERATURE AND AIR-FUEL MIXING ON EMISSION OF PARTICULATE CARBON FROM A DIVIDED-CHAMBER DIESEL ENGINE

S. L. PLEE, T. AHMAD, J. P. MYERS and D. C. SIEGLA

General Motors Research Laboratories
Warren, Michigan

ABSTRACT

The effect of flame temperature on particulate carbon emission from divided-chamber diesel engines was examined by adding various quantities of O_2 and N_2 to the intake air with the engines operating at several different loads and speeds. At a given operating condition for a fixed combustion chamber geometry, intake gas addition was expected to influence chemical kinetics without affecting the air-fuel mixing. Particulate carbon and CO were found to increase with N_2 addition and decrease with O_2 addition, whereas NOx emissions exhibited opposite trends.

Since the major portion of the combustion event in diesel engines is known to be diffusion controlled, changes in particulate carbon emissions due to O_2 or N_2 addition should be related to variations in the stoichiometric adiabatic flame temperature (T_f). Both soot and NOx emissions were correlated with calculated T_f^{-1}, yielding overall activation energies for the processes involved. The overall activation energy for each pollutant was found to be independent of engine speed and load for a given combustion chamber geometry, indicating that the kinetic mechanisms involved in the generation of these species were not altered significantly by these engine parameters. In addition, a one-to-one correspondence was observed in this study between soot and CO, which suggests some similarity between the kinetic and air-fuel mixing processes governing the formation and oxidation of both.

Air-fuel mixing effects were varied by changing the load and speed and by employing different combustion chamber geometries. Variations in particulate carbon emissions with load and speed were characterized by a mixing parameter involving average fuel flow rate, engine speed, and the quantity of fuel injected per cycle, while for NOx the mixing parameter depended on average fuel flow rate and engine speed.

References pp. 477-481.

NOMENCLATURE

A_{inj}	Nozzle opening area (Eq. 29)
a	Constant
B	Parameter (Table 2)
b	Constant
C, c	Constants
D	Combustion chamber diameter (Eq. 8)
D_s	Effective nozzle diameter (Eq. 4)
d	Nozzle orifice diameter (Eq. 8)
d_o	Injector hole diameter (Eq. 10 and Table 2)
E	Activation energy
EIC	Particulate carbon emission index
EINOx	NOx emission index
EICO	CO emission index
F	Quantity of fuel injected per cycle
F_s	Fraction of cylinder volume containing soot formation zones (Eq. 18)
G	Jet thrust (Table 2)
$\dot{G}_o$	Jet momentum flux (Eq. 2)
g	Gravitational constant
H	Depth of combustion chamber (Eq. 8)
J	Momentum flux ratio (Eq. 8)
K	Equilibrium constants
k	Reaction rate constants
L	Length scale
m_s	Mass of soot (Eq. 18)
$\dot{m}_f$	Fuel flow rate
$\dot{m}_{f_{ref}}$	Fuel flow rate at a reference condition
$\dot{m}_f^*$	Non-dimensional fuel flow rate
$\dot{m}_o$	Injector mass flow rate (Eq. 3)
N	Engine speed
N_{ref}	Engine speed at a reference condition
N^*	Non-dimensional engine speed
n	Number of injector holes (Eq. 10)
n	Constant (Eq. 18)
P	Pressure
p	Constant
P_o	Local partial pressure of O_2 (Eq. 19)
P_s	Partial pressure of unburned soot (Eq. 18)
Q	Fuel per cycle (Eq. 10)
q	Constant
R	Universal gas constant
r	Constant
Re	Reynolds number
Ri	Richardson number
S	Swirl number
T	Temperature
T_f	Flame temperature
T_o	Local temperature
T_ϕ	Burned gas temperature
U	Mean air velocity
U_s	Average velocity (Eq. 3)
U_{inj}	Mean injection velocity

u'	Turbulence velocity
W	Stirring factor (Table 2)
X_{O_2}	Mole fraction oxygen
Z_s	Pre-exponential factor (Eq. 18)
Z_o	Pre-exponential factor (Eq. 19)

Greek Symbols

α	Spray-plume area ratio (Eq. 8)
ϕ	Overall equivalence ratio
ϕ_s	Local equivalence ratio in the formation zone
λ	Taylor microscale (Eq. 9)
ν	Kinematic viscosity
ρ	Density
ψ	Mixing parameter (Eq. 8)
τ	Time scale
$\theta_{i.p.}$	Injection period in degrees crankangle

Subscripts

a	Air
D	Mixing delay (Table 2)
ign	Ignition
inj	Injection
∞	Ambient conditions
m	Mixing
s	Soot
s, f	Soot formation
s, o	Soot oxidation
SD	Mixedness (Table 2)
S	Residence (Table 2)

INTRODUCTION

Because of growing concern about dwindling petroleum supplies and spiraling gasoline prices, the past few years have seen a dramatic increase in the demand for fuel-efficient passenger cars. Concern over the environmental impact of the automobile has made the task of improving the fuel economy of today's spark-ignition (SI) engines a difficult one. In this situation, industry and government alike have been searching for an alternative to the conventional SI engine. For nearly half a century, the diesel engine has been a popular choice for heavy- and medium-duty applications due to its proven superiority in fuel economy over the SI engine. For the same reason, the diesel engine has recently emerged as an attractive alternative to the SI engine in light-duty passenger-car applications as well. Most diesel engines emit less unburned hydrocarbons (HC) and carbon monoxide than comparable SI engines. However, due to the heterogeneous nature of diesel combustion, they emit a much greater quantity of particulates.* The particulate composition is dominated by solid carbon generated during combustion. Concern over the environmental effects of diesel particulates could jeopardize the future of the diesel engine in passenger-car applications.

The concentration of particulates in diesel exhaust is the net result of their formation and oxidation in the combustion chamber. Following the pioneering

---

In this paper the terms soot, smoke, and particulates are used synonymously.

work of Khan and coworkers [1-5], various physical and chemical factors governing the formation and burn-up of carbonaceous soot in diesel engines have been studied extensively, as reviewed in References [6-8]. These factors, which will be discussed in some detail in a later section, interact in a complex manner. Therefore, it is difficult to isolate completely the effect of any one of the parameters on exhaust emissions.

In the present study, an attempt was made to separate the chemical-kinetic aspects of particulate emission from the fluid-mechanic aspects. At a given operating condition, addition of various quantities of O_2 and N_2 in the intake of the engine was expected to influence the flame temperature and chemical kinetics without affecting the air-fuel mixing. The air-fuel mixing in the combustion chamber was varied by operating the engine over a wide range of loads and speeds and by employing three different combustion chamber geometries. At this point, it is desirable to present a brief overview of the problem of particulate formation and oxidation in diesel engines, with emphasis on the effects of intake gas addition and air-fuel mixing on particulate emissions from various combustion systems.

PARTICULATE CARBON FORMATION
AND OXIDATION IN DIESEL ENGINES

Several reviews are available on the effects of pressure, temperature, molecular structure of the fuel, equivalence ratio, diluents, metal additives and electric fields on the emission of carbon from premixed and diffusion flames; see for example references [6, 9-18]. In particular, soot formation at high pressures is reviewed by McArragher and Tan [14], while references [6, 9, 16-18] contain discussions of the carbon formation process in diesel engines. In the present study, the effect of flame temperature and air-fuel mixing on the emission of particulate carbon from diesel engines will be emphasized.

The amount of soot emitted by a flame depends on the competition between the formation and oxidation reactions (oxidation of both soot particles and soot precursors) [19]. When the oxidation rate exceeds the formation rate, the flame may be luminous, indicating the presence of soot, but the flame will not smoke. However, when the rate of formation exceeds the rate of oxidation, soot will be present in the exhaust gases in sufficient quantity to cause visible smoke [19].

As discussed by Wagner [9] in his recent review paper, the formation of soot is controlled by the local equivalence ratio, the local temperature, and the residence time of the particles and gases in the soot-forming region. Increasing any of these parameters independently will, in general, increase the quantity of soot produced in the flame. Similarly, the amount of soot particles and soot precursors oxidized in the flame will increase with increasing local temperature and residence time and with decreasing local equivalence ratio. Thus, high equivalence ratios enhance soot emission by accelerating the formation process and at the same time inhibiting oxidation reactions. However, the net effect of increasing temperature on soot emission is not that straightforward because of the competition between formation and oxidation. To date, no study has been done which clearly defines how flame temperature affects soot emission [16]. No doubt the temperature dependence varies with local equivalence ratio and residence time in the formation and oxidation zones. These parameters are particularly difficult to define in diesel engines because of the nature of diesel combustion.

Combustion in Diesel Engines — Unlike spark-ignition engines, which burn an essentially homogeneous charge, the fuel and air are not premixed initially in diesel engines. As a result, the diesel combustion process is an extremely complex phenomenon which involves the burning of a liquid fuel spray under transient operating conditions. Combustion is governed by interrelated effects of fluid mechanics, chemical kinetics, and heterogeneity of the fuel spray.

Lyn [20] investigated the nature of the combustion process in diesel engines by examining the burning rates and concluded that combustion occurs in two phases. Initially, chemical kinetics are rate controlling and the premixed portion of the fuel burns with a nonluminous flame. However, during the heterogeneous combustion period, in which most of the energy is released, variables such as injection rate, engine speed, and chamber design influence the combustion process. Lyn's conclusions indicate that the flame is mixing (turbulent diffusion) controlled in this regime [20]. Although most investigators [1, 6, 7, 20, 21] generally agree that the main combustion event is diffusion controlled, the detailed mechanisms are not well understood at the present time [6, 7]. The diffusion flame can take two basic forms: the classical envelope flame surrounding each individual fuel droplet or a turbulent diffusion flame surrounding the entire spray.

Chigier [22] has reviewed the spray combustion literature in detail. In steady-flow combustion devices, no evidence has been found to support the individual droplet-combustion theories. Instead, the rate of burning is found to be controlled by the turbulent transport of fuel and air at the boundary of the fuel spray, and thus the entire process resembles a gaseous turbulent-diffusion flame. Chigier [22] concluded that droplet combustion is unlikely in dense sprays because evaporation ocurs in the fuel-rich central core of the fuel spray, which is beyond the rich flammability limits of the mixture.

In a direct comparison of temperature, velocity, and concentration profiles (O_2, CO, CO_2, HC and NOx) obtained from a gaseous propane flame, a kerosene spray flame, and a heavy-oil spray flame, Onuma and co-workers [23, 24] found that the structure of the liquid-spray flames closely resembled that of the gaseous propane flame. Under no conditions was there any experimental evidence which indicated the burning of individual droplets, even for the heavy-oil spray flame. Based on these results, Onuma and Ogasawara [23] divided the flame structure into three zones. In the central core of the fuel spray (zone 1), both liquid droplets and fuel vapor co-exist. Thus, zone 1 is the primary soot formation region because of the high local equivalence ratios which are present as a result of fuel droplet evaporation. Surrounding the core of the fuel spray is another rich zone (zone 2) which contains only a few droplets but very high CO concentrations. On the oustide of the fuel spray is an intermittent-burning regime (zone 3) in which combustion is completed.

The phenomenological model of diesel combustion proposed by Khan and co-

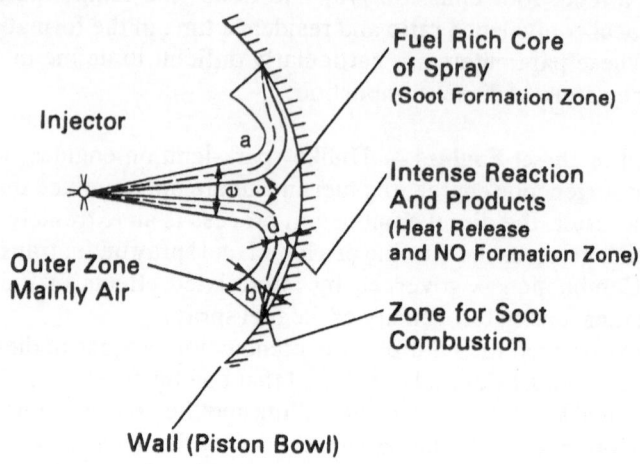

Fig. 1. A phenomenological model of diesel combustion from Reference [25].

workers [3, 4, 25] (Fig. 1) is similar to this steady-flow model. The fuel spray and air-entrainment region (zone e) is divided into the fuel-rich zone (c), where soot formation occurs, the intense reaction zone (d), which contains the majority of the heat release and NO formation, and the outer zone (b), where soot combustion takes place. Unfortunately, direct comparisons of gaseous and spray flames, like the ones by Onuma and co-workers [23, 24] in steady-flow burners, have not been attempted in transient-spray flames. Thus, one is forced to speculate as to the nature of the diesel combustion process. However, since diesel combustion occurs at very high air pressures and temperatures, as well as high fuel injection pressures (smaller initial droplet diameter) relative to steady-flow combustors, the character-istic droplet evaporation time is expected to be much shorter. This suggests that the concept of a gaseous diffusion flame surrounding the entire spray is also applicable to diesel combustion. In-cylinder sampling measurements by Duggal *et al.* [26] and theoretical calculations by Chiu and Liu [27] tend to support this general pheno-menological model for diesel combustion.

The influences of engine variables, inlet conditions, fuel type, and injection characteristics on diesel particulate emissions have been studied extensively in the past using direct-injection and divided-chamber diesel engines. Some of these studies are listed in Table 1. The effects of these variables on exhaust emissions are discussed in a review paper by Henein [6]. Unfortunately, trends are usually a function of several parameters in the system which are difficult or impossible to uncouple.

As illustrated by the table, changes in engine variables, inlet conditions, fuel type and injection characteristics can alter the chemical-kinetic, heterogeneous and fluid-mechanic aspects of the combustion process. Here, heterogeneous and fluid-mechanic effects include atomization and vaporization of the liquid fuel and mixing of the resulting fuel vapor with oxidizer, whereas chemical-kinetic effects include

TABLE I
Diesel Exhaust Smoke and Particulate Studies
Effect on Combustion

Variable	Chemical Kinetics	Heterogeneous Fluid Mechanic	Reference
Compression Ratio	X	X	7, 39
Overall Air-Fuel Ratio		X	7, 30, 31, 37, 39
Speed		X	3, 7, 31
Timing	X	X	3, 7, 29, 30, 31, 36, 37, 39
Swirl Ratio		X	3, 7, 39
EGR/Gas Addition	X		7, 30, 31, 37, 38, 39, 44
Water Injection	X	X*	7, 31, 33, 35, 37, 39
Inlet Temperature	X	X	7, 32, 37
Turbocharging	X	X	7, 30, 31, 38, 39
Humidity	X		32, 39
Fuel Injector Orifice Diameter		X	3, 7, 29, 36
Fuel Injection Rate		X	3, 7, 29, 36, 37
Spray Position		X	31
Cetane Number	X	X	34, 37, 43
Fuel Temperature		X	28
Additives	X		40, 41, 42, 43

*For emulsions

the combustion of the fuel and oxidizer once they have been mixed on a molecular level. Changes in compression ratio, inlet temperature, and turbocharging influence kinetic, heterogeneous, and mixing processes because initial pressure and temperature affect atomization, evaporation, and fuel penetration, as well as chemical reaction rates. Timing variations alter the initial pressure, temperature, and air motion to which the fuel is exposed at injection, while cetane number is expected to influence fuel burning and fuel evaporation characteristics. Overall air-fuel ratio is listed under fluid-mechanic effects rather than chemical kinetics because in diffusion flames it is the local stoichiometry rather than the overall stoichiometry which is important in combustion reactions [17, 19, 45]; the local equivalence ratio is determined by the mixing characteristics of the system.

In this study, a range of air-fuel ratios, engine speeds, fuel injection rates and intake gas compositions have been examined. Intake gas addition is expected to alter the chemical kinetics without affecting the air-fuel mixing process. On the other hand, variations in engine speed, overall air-fuel ratio, and rate of injection for a given oxidizer composition should influence primarily the mixing process and the distribution of fuel in the combustion chamber. The following sections review some of the earlier experimental work done in these areas and some of the existing correlation techniques.

References pp. 477-481.

Effect of Adding Oxygen or a Diluent — The influence of gas addition on soot emission has been studied extensively in the past in both premixed and gaseous diffusion flames. The addition of either a diluent or oxygen alters both the flame temperature and oxidant concentration and thus the competing soot formation and oxidation rates.

In premixed flames, Street and Thomas [11] found that the addition of nitrogen to the air increased the critical oxygen-to-carbon ratio (O/C) for carbon formation, while enrichment of the air with oxygen decreased this critical ratio. On the other hand, experiments by Wright [46] using a well-stirred reactor show a more complex response of critical O/C to gas addition. Wright [46] observed different effects with dilution depending on the type of fuel, and he concluded that transport properties, in addition to reaction kinetics, are also important in determining the point of incipient carbon formation.

Macfarlane *et al.* [47] separated out the effect of flame temperature in premixed flames by replacing N_2 with Ar at a given oxygen concentration. For mixtures richer than the soot formation threshold, they observed an increase in the rate of soot emission with an increase in flame temperature. In contrast, Street and Thomas [11] found that increasing flame temperature suppressed the initial formation of soot. Thus, it appears that near the threshold of soot formation, temperature influences the oxidation process more than formation, while in very rich mixtures, soot formation reactions are favored [13].

In diffusion flames, oxygen and diluents have been added to both the fuel [48-54] and the oxidizer [48, 49, 55-57]. In general, the addition of nitrogen to the fuel suppresses soot formation [48-50, 54], while the effect of premixing oxygen in the fuel varies depending on the oxygen concentration and fuel type. When small quantities of oxygen are added to the fuel, the sooting tendency generally increases [48, 50-54]. However, Dearden and Long [48] found that the sooting rate decreased when oxygen was added to propane, while Jones and Rosenfeld [51] observed no effect of oxygen addition to propane, propylene and butane. Wright [52] found that the level of soot increased when up to 10 percent oxygen was premixed with the fuel, but after about 10-15 percent oxygen addition, soot emission rapidly decreased and was eventually eliminated. Chakraborty and Long [53] also noticed that soot emission reached a maximum at a critical oxygen/fuel ratio.

Recently, Glassman and Yaccarino [54] have shown that the addition of diluents and small quantities of oxygen to the fuel primarily affect the soot formation process. They found that the tendency of a fuel to soot depends primarily on the diffusion flame temperature (stoichiometric flame temperature) and the fuel structure, rather than fuel partial pressure, which was found to be only of secondary importance. Obviously, as more oxygen is added to the fuel, the balance between soot formation and oxidation will change causing the trends to reverse, as observed by Wright [52] and Chakraborty and Long [53].

Of more importance to this study is the addition of oxygen and diluents to the combustion air. The effect of gas addition to the inlet air also depends on the balance between the soot formation and oxidation reactions. In general, dilution of the air by inert gases leads to an increase in the smoke point (decrease in sooting tendency) and a decrease in luminosity, but contradictory results have also been obtained [49].

Dearden and Long [48] varied the oxygen index of the air (mole fraction of oxygen) by adding pure oxygen and nitrogen, and they found that the sooting rate increased with increasing oxygen index. Chakraborty and Long [55] also found that soot emission increased with an increase in oxygen index from 0.18 to 0.26. However, after reaching a maximum value around 0.26, soot emission declined rapidly with further increases in oxygen concentration. Dearden and Long [48] attributed this difference to partial combustion of the soot particles and soot precursors in the Chakraborty and Long [55] burner. The addition of oxygen to the air increases flame temperature, which leads to more pyrolysis of the fuel, but it also leads to higher temperatures and increased O, O_2, and OH concentrations in the soot oxidation region. The net effect on total soot emission will depend upon which of these processes dominates. For the Wolfhard-Parker burner used by Dearden and Long, the soot particles formed in the flame are quenched before being exposed to the oxidizing medium, and thus only the soot formation process is affected [48].

The competition between soot formation and oxidation is further emphasized by Glassman and Yaccarino [57]. They determined the effect of oxygen index (varied from 0.09-0.5) on sooting tendency in a laminar diffusion flame by measuring sooting height. Using nitrogen as the inert gas, they found that the sooting tendency was minimum at an oxygen index of approximately 0.24 and increased on either side of this critical value, illustrating the importance of two competing mechanisms. The apparent discrepancy between the work of Glassman and Yaccarino [57] and Chakraborty and Long [55] is attributed to the different criteria used to establish sooting propensity. According to Glassman and Yaccarino [57], the effect of oxygen index on sooting tendency and soot emission should be different because they are governed by conditions in different parts of the flame.

Several investigators [55-57] have substituted argon for nitrogen to examine the effect of flame temperature without changing the initial oxygen concentration. The study by Schalla and McDonald [56] showed that for an oxygen-enriched mixture, the limiting fuel flow rate for soot formation was higher for the Ar-O_2 flame. Chakraborty and Long [55] observed that at the lower oxygen indices, approximately the same amount of soot was emitted by the Ar-O_2 flame as the N_2-O_2 flame. However, at higher oxygen indices, less soot was emitted by the Ar-O_2 flame, suggesting that combustion of soot particles and soot precursors had become more important than the formation reactions. Glassman and Yaccarino [57] have recently correlated the sooting tendency in both Ar-O_2 and N_2-O_2 flames with the stoichiometric adiabatic flame temperature.

McLintock [49] observed that the effect of oxygen index on the smoke point (maximum fuel flow at which soot initially leaves the flame) was even sensitive to the type of diluent present, conditions of dilution, and dimensions of the burner. Different results were obtained depending on whether the oxygen, the nitrogen, or the total flow was held constant as the oxygen index was varied. McLintock [49] further suggested that in addition to temperature and oxygen partial pressure, transport properties were important in determining the emission of soot particles from diffusion flames. However, Glassman and Yaccarino [57] have attributed these effects to the fact that McLintock's experiments were not performed with the proper overventilation.

Investigation of the effect of diluent and oxygen addition to the inlet air on the characteristics of soot formation in droplet and spray flames has been limited. Kadota et al. [58] examined droplet combustion and observed that soot emission reached a maximum at an oxygen index of 0.17 in O_2-N_2 mixtures and then decreased on either side of this critical value. Variations in soot-emission trends with increasing oxygen index have also been observed by Sjogren [59]. In spray flames, Cooper et al. [60] have shown that flames become nonluminous and soot concentration decreases when exhaust gases or additional diluents are added to the combustion air. A more recent study by Gollahalli [61] also suggests that the addition of diluents reduces soot emission.

The general tendency of diluents to reduce soot formation in diffusion flames is particularly interesting with regard to diesel engines since it suggests that exhaust gas recirculation (EGR) could be used to reduce NOx and particulate emissions simultaneously. Unfortunately, several studies indicate that although the addition of EGR reduces NOx, it also increases particulate and smoke emissions in both direct-injection and divided-chamber diesel engines [7, 30, 31, 37, 38, 39, 44].

The addition of exhaust gas to the intake air can influence the combustion process by:

1. decreasing flame temperature due to increased heat capacity of the intake charge.
2. lowering oxygen concentration due to dilution of the intake charge.
3. increasing inlet air temperature due to the introduction of heated combustion products.

In spark-ignition engines, the effect of EGR on NOx emissions has been related to the decrease in flame temperature [62]. In diesel engines, a recent study by Tsunemoto and Ishitani [44] concluded that the reduction in NOx with EGR was a result of lower oxygen concentrations in the intake charge rather than decreased flame temperature. However, in their analysis, Tsunemoto and Ishitani [44] assumed that combustion occurred at the overall equivalence ratio rather than at the stoichiometric equivalence ratio. Wilson et al.[7], have been able to predict the effect of EGR on NOx emissions using the peak flame temperature in a diffusion flame. In support of Wilson et al. [7], several studies have successfully correlated NOx emissions from steady-flow diffusion-flame combustors with the stoichiometric adiabatic flame temperature [63-67].

Very little information is available on why EGR increases smoke in diesel engines while the opposite effect is generally observed in gaseous diffusion flames and spray flames. Neither Khan and coworkers [3, 4, 25] nor Hiroyasu and coworkers [68, 69] have compared the model predictions of the effect of EGR on smoke with experimental data. In this study, O_2 and N_2 have been added to the intake charge rather than adding exhaust gas in order to eliminate the effects of inlet temperature, water vapor and carbon dioxide.

It should be obvious from this brief review that the effect of gas addition on soot is an extremely complicated process because it involves the competition between soot formation reactions and the oxidation reactions for soot particles and soot precursors. Furthermore, the time a fluid particle spends at a given temperature and equivalence ratio in either the soot formation or soot oxidation regions is extremely important and depends on the flow situation [9]. In turbulent diffusion flames, the air-fuel mixing process determines these time scales.

Effect of Air-Fuel Mixing — A large number of practical combustion systems use diffusion-type burning. Liquid or gaseous fuel and oxidizer enter the combustion chamber and mix together, giving rise to a diffusion- or a mixing-controlled combustion situation. Under these circumstances, fuel-rich pockets are formed where O/C ratio is above its critical value for sooting [9]. In the case of spray combustion systems, the mixing intensity and the degree of atomization critically influence the formation of soot. High-intensity turbulence can mix the fuel and oxidizer rapidly, enhancing the oxidation of soot precursors [19]. A complete characterization of turbulent mixing in combustion systems requires knowledge of the profiles of mean and fluctuating temperature, velocity, and concentration of species as well as the drop size distribution within the combustion space. The turbulence-modeling approaches reported by Patankar and Spalding [70], Gosman et al. [71], Hutchinson et al. [72], Launder and Spalding [73] and Bilger [74] represent a step towards the predictive capability of the profiles of these quantities. However, due to the complex nature of turbulence and the chemical reactions involved, these models utilize a large number of empirical constants that depend on the system under consideration. Therefore, the application of these models to practical combustion systems has not been widespread. A simpler approach to characterize the mixing process has been the use of non-dimensional parameters and characteristic time scales. These will be discussed here in some detail. For the sake of the discussion of the effect of mixing on soot emission from various combustion systems, it is convenient to categorize them as turbulent-jet diffusion flames and spray flames.

Open turbulent diffusion flames at atmospheric pressure have been utilized extensively to understand the mechanisms of soot formation and combustion in more complex systems. Magnussen [75] studied the effect of turbulence on the appearance of soot in a free C_2H_2 jet flame by performing a Fourier transform analysis of the scattered light signal. Magnussen's results indicated that, similar to the findings of Dalzell et al. [76], the center-line soot concentration reached a maximum prior to the flame tip and then decreased due to subsequent oxidation. He also observed that soot was contained in eddies which seemed to grow in size in the fuel-rich core of the jet in a manner similar to turbulent eddies. Near the edges of the jet and the flame tip, a dissipation of soot-containing eddies, similar to turbulent dissipation, was observed. On the basis of these observations, Magnussen and Hjertager [77] developed a simplified eddy-dissipation model for soot combustion, which when combined with a soot-formation model originally proposed by Tesner and co-workers [78, 79], was able to predict soot concentration in an acetylene jet with reasonable accuracy. In a later study, Magnussen et al. [80] studied the effect of variations of Reynolds number and diluent addition in a C_2H_2 flame. An increase in the Reynolds number of the fuel stream was accompanied by a significant decrease in the amount of soot emitted. Interestingly, the eddy-dissipation model was able to predict these trends with fair accuracy. Becker and Yamazaki [81] investigated experimentally the effect of Reynolds number and buoyancy parameters in free turbulent diffusion flames. Their results suggested that the soot concentration field was mixing controlled. They characterized mixing by a Reynolds number in the forced convection regime and by a Richardson number in the natural convection regime. These dimensionless parameters were defined as follows:

$$Re = \frac{U_s D_s}{\nu_\infty} \tag{1}$$

and

$$Ri = \frac{\pi}{4} g \rho_\infty \frac{D_s^3}{\dot{G}_o} \left(\frac{L}{D_s}\right)^3 \tag{2}$$

where ρ_∞ is the air density, L the flame length, $\dot{G}_o$ the jet momentum flux and ν_∞ the kinematic viscosity. Also

$$U_s = \frac{\dot{G}_o}{\dot{m}_o} \tag{3}$$

where $\dot{m}_o$ is the mass flow rate through the nozzle, and

$$D_s = \left(\frac{4 \dot{G}_o}{\pi \rho_\infty U_s^2}\right)^{1/2} \tag{4}$$

The soot concentration in the forced-convection regime, denoted by the condition $Ri \to 0$, was significantly lower than in the natural-convection regime. Furthermore, the location of the maximum soot concentration along the centerline was found to be much closer to the burner tip in the high-Re regime in the case of $Ri \to 0$.

In enclosed-jet turbulent flames, such as flames in industrial furnaces and back-mixed reactors, studies of soot emission in relation to mixing intensity are almost non-existent. Most of the studies focus on the influence of mixing on combustion efficiency without devoting much attention to soot emission. However, some of the schemes that various investigators have employed to characterize mixing in confined gas jets are summarized in Table 2, since these parameters could have some relevance to diesel combustion.

TABLE 2
Some Mixing Parameters for Enclosed Turbulent Flames

Type of Combustion System	References	Parameters Used to Characterize Mixing and Recirculation
1. Ducted turbulent-jet flames	82	Craya-Curtet parameter
2. Free and ducted turbulent-jet flames	83	Intermittency factor β (defined as the probability of finding the nozzle fluid at a certain point)
3. Swirled & unswirled jets in a confined space	84, 85, 86	Stirring factor W $$W = \frac{\tau_S}{\tau_D + \tau_S}$$ τ_S = residence time τ_D = mixing delay time
4. Partially stirred reactors	87, 88	Mixedness parameter τ_{SD} $$\tau_{SD} = \frac{\tau_S}{\tau_D} \propto \left[\frac{u'}{U}\right]_{max}$$ u' = turbulence velocity U = mean velocity
5. Enclosed turbulent-jet flames	89	B parameter $$B = \frac{1}{(\text{Craya-Curtet No.})^2} + 1/2$$ or $$B = \frac{1}{(\text{Thring-Newby No.})^2}$$ $S = T/(G\, d_o)$ = swirl number

A great majority of the practical combustion systems utilize heterogeneous burning of liquid or solid fuels in the form of droplets or particles respectively. Characterization of the flow field in spray flames is far more difficult than in gaseous flames. The intrusive probes often used are vulnerable to direct impingement of droplets and blockage of orifices by soot particles [90, 91, 92]. Chigier and co-workers [93, 94] have made significant contributions in developing non-intrusive optical techniques for detailed measurements in spray flames. However, there are limitations in terms of cost and applicability which will have to be overcome before their use becomes widespread [95] in practical combustion systems. Thus, the information available about the interaction between mixing and soot emissions (in sprays) is even more limited than for gaseous jets. Some of the important investigations that have considered this interaction are reviewed briefly.

Mellor and co-workers [63-67, 96-99] have employed a characteristic-time concept similar to the one developed by Swithenbank *et al.* [88] to correlate NO, CO, and HC emissions from gas turbine combustors. They characterized the process of spray combustion by time parameters representing droplet evaporation, mixing, and chemical reaction. These time scales are combined in the following manner to obtain correlations for NO and CO

$$\text{EINOx} \propto \frac{\tau_{m, \text{NO}}}{\tau_{\text{NO}}} \tag{5}$$

and

$$\text{EICO} \propto \frac{\tau_{\text{CO}}}{\tau_{m, \text{CO}}} \tag{6}$$

where   EINOx and EICO   = emission indices for NOx and CO respectively (g/kg fuel)

$\tau_{m, \text{NO}}$ and $\tau_{m, \text{CO}}$ = mixing times for NOx and CO respectively (ms)

τ_{NO} and τ_{CO} = kinetic times for NOx and CO respectively (ms)

Tuttle *et al.* [63, 97] demonstrated the success of these correlations by comparison of data for various combustor geometries, operating conditions and fuel types.

Dodds *et al.* [99] have proposed a similar qualitative scheme for correlating soot emissions. This correlation is as follows:

$$\text{EI}_{\text{smoke}} \propto \left(\frac{\tau_{m,f}}{\tau_{s,f}} \right) \cdot \left(\frac{\tau_{s,o}}{\tau_{m,o}} \right) \tag{7}$$

where EI_{smoke} = smoke emission index (g/kg fuel)

$\tau_{m,f}$ and $\tau_{m,o}$ = mixing times for soot formation and oxidation respectively (ms)

$\tau_{s,f}$ and $\tau_{s,o}$ = kinetic times for soot formation oxidation respectively (ms)

Dodds *et al.* [99] stress the oversimplified nature of this correlation whose validity, to date, has not been verified by experiments.

Kadota *et al.* [100] have reported soot concentration measurements in a spontaneously ignited intermittent-spray flame along the axis of a duct. Their measurements consist of soot concentration along the centerline of the duct for various air temperatures and velocities. Kadota *et al.* [100] do not report detailed measurements of the flow field in the duct. However, they observed a drastic reduction in the centerline soot concentration with increasing duct air velocity.

Among the practical systems utilizing spray combustion, the diesel engine represents the most difficult experimental and analytical problem [101] due to the heterogeneous, transient and multidimensional nature of diesel combustion, with possible impingement of fuel on the chamber walls. Unlike steady-flow combustors, the diesel spray flame is extremely difficult to probe because of the limited size of the combustion chamber and its inaccessibility. In addition, the quantities of interest vary spatially as well as temporally, which leads to serious measurement problems in obtaining a representative sample. Thus, only a few investigators [26, 91, 102-107] have attempted time- and space-resolved measurements in diesel engines. Even fewer studies have been performed to characterize air-fuel mixing and its effect on soot emission from diesel engines because, as shown in Table 3, air-fuel mixing in diesel engines is influenced by a large number of engine parameters.

TABLE 3
Air-Fuel Mixing Process

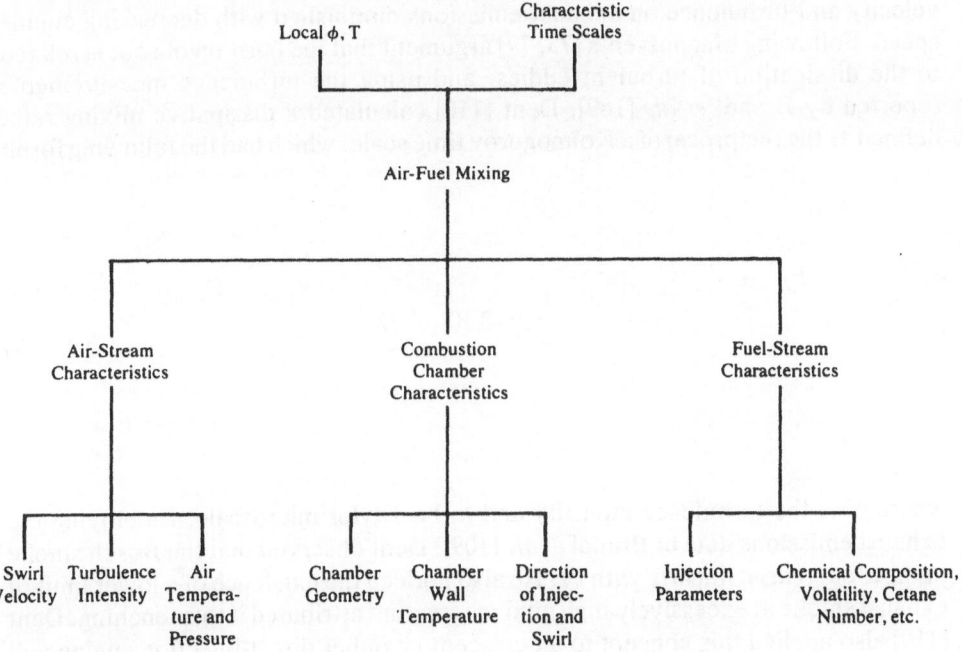

Morris and Dent[108] simulated successfully the fuel-air mixing process in a high-swirl, open-chamber diesel engine by a steady-state gas jet injected into a steady air-swirl flow. The simulation was based on a mixing parameter defined as follows:

$$\psi = \frac{\pi}{4} \frac{J}{\alpha} \frac{d^2}{D\,H}$$

(8)

where $J =$ jet momentum flux/cross-flow momentum flux

$\alpha = \dfrac{\text{area of the spray plume with air-fuel ratio} \leqslant 10:1}{\text{total area of the plume}}$

$d =$ nozzle orifice diameter (mm)
$D =$ combustion chamber diameter (mm)
$H =$ depth of combustion chamber (mm)

Morris and Dent [108] also reported some data showing the influence of ψ on exhaust emissions. The engine operating conditions associated with high values of ψ, signifying poor mixing, produced high concentrations of exhaust smoke and HC and lower NOx. Brandl *et al.* [109] measured mean and turbulent velocities in the combustion bowl of a motored direct-injection diesel engine. At high engine speeds, the exhaust smoke decreased with increasing magnitude of velocity and turbulence intensity in the bowl. Brandl *et al.* [109] also found that the effect of air velocity and turbulence on exhaust emissions diminished with decreasing engine speed. Following Magnussen's [75, 77] argument that the burn-up of soot is related to the dissipation of turbulent eddies, and using the turbulence measurements reported by Brandl *et al.* [109], Dent [110] calculated a dissipative mixing rate, defined as the reciprocal of a Kolmogorov time scale, which had the following form:

$$\frac{1}{\tau} = 3.87 \; u' \, /\lambda$$

(9)

where u' is the turbulence intensity and λ the Taylor microscale. Employing the exhaust emissions data of Brandl *et al.* [109], Dent observed that the Bosch smoke number decreased linearly with $1/\tau$. He also noticed the existence of a lower limit of exhaust smoke at excessively high mixing rates and attributed it to quenching. Dent [110] also applied this concept to a quiescent-chamber direct-injection engine and correlated the exhaust smoke for several speeds, loads, and injectors with the mixing rate defined as follows:

$$\frac{1}{\tau} = \left[\frac{C}{\nu^{1/2}}\right] \left[\frac{N}{\theta_{ip}}\right]^{3/2} \left[\frac{Q}{n\,d_o^{\,2}}\right] \tag{10}$$

where

$C =$ a constant
$\nu =$ kinematic viscosity
$N =$ engine r/min
$\theta_{ip} =$ injection period in degrees
$Q =$ fuel in mm³/cycle
$n =$ number of injector holes
$d_o =$ injector hole diameter

Greeves [111] studied the effect of the rate of fuel injection on exhaust smoke, NO_x and brake specific fuel consumption (BSFC) for several diesel engines. His results revealed that in non-impinging situations, an increase in the rate of injection (ROI) resulted in a decrease in smoke and BSFC and an increase in NOx. This observation led him to define a "maximum useful rate of injection" (MUR) beyond which no further improvement in smoke, CO, or BSFC was found. His results also indicated that the values of MUR for the engines with significant impingement were much lower than for those with no impingement.

In general, most of the previous efforts to relate mixing and exhaust particulates have focused on direct-injection diesel engines. In the absence of detailed in-cylinder measurements, the concepts developed by Dent [110] and Greeves [111] should be considered for application to divided-chamber diesel engines as well.

APPROACH

The present study undertakes to examine separately the effects of flame temperature and air-fuel mixing on exhaust particulate carbon emissions from a divided-chamber diesel engine. To examine the effects of flame temperature and chemical kinetics, various quantities of oxygen or nitrogen were added to the intake air in order to change the intake oxygen mole fraction. This is expected to alter both the flame temperature and the concentrations of the various oxidizing species without significantly affecting the air-fuel mixing process (see Table 1).

Since the majority of the combustion process in divided-chamber diesel engines is diffusion controlled, this suggests that diesel exhaust emissions might correlate in the same fashion as emissions from steady-flow diffusion-flame combustion systems. As discussed previously, both NOx emissions from gas-turbine combustors [63-67] and the sooting tendency in laminar diffusion flames [54, 57] have been successfully correlated with the characteristic diffusion flame temperature (stoichiometric flame temperature). In this study, stoichiometric flame temperature correlations are presented for both NOx and particulate carbon emissions.

References pp. 477-481.

NOx emissions are considered in this paper for two reasons. First, the kinetic reactions for NO formation from atmospheric nitrogen are much better understood than are the mechanisms for particulate carbon. Hopefully, as a result, some basic understanding can be obtained by examining NOx emissions before attempting to apply the flame temperature correlation to particulate carbon. The second reason, which will be discussed in more detail later in this paper, is that there also appears to be a strong coupling between NOx and particulate emissions from diesel engines.* That is, almost any design or change in operating conditions which lowers particulates, invariably increases NOx emissions. Thus, an important tradeoff always exists when designing automotive diesel engines. For similar reasons, CO emissions have also been reported and are discussed briefly in this paper. The fact that CO and particulate carbon emissions generally follow the same trends* indicates possible similarities between the processes governing the formation and oxidation of both.

The influence of air-fuel mixing on exhaust particulate carbon is examined by varying several engine parameters. These parameters include: engine speed, fuel flow rate, quantity of fuel injected per cycle, and prechamber geometry. These changes are expected to alter the characteristic mixing times in the system without significantly affecting the chemical kinetics (see Table 1). As discussed in Table 1 as well as by Glassman [19], overall air-fuel ratio does not determine reaction kinetics in a diffusion-controlled combustion process. Details of the experiments are provided in the following sections.

EXPERIMENTAL ASPECTS

To acquire the experimental data necessary for this investigation, a wide range of instrumentation and equipment was required, along with an extensive series of test conditions.

Apparatus — *Single-Cylinder Engine (0.72 L)* – A schematic of the test facility is shown in Fig. 2. Included in this schematic are the engine instrumentation, emission-measurement equipment, and gas- and air-metering systems. The first engine tested was a 5.7 L V8 modified for single-cylinder operation. A cross section of the active cylinder is shown in Fig. 3. An experimental Stanadyne injection pump was used, with seven of the eight injectors firing into a pressurized chamber. Fuel from these seven injectors was recirculated so that only fuel injected into the active cylinder was measured. The engine cylinder head was modified to accommodate pressure transducers in both the main chamber and the prechamber. CAV poppet injectors were used. The injector in the operating cylinder was equipped with a proximeter for monitoring the poppet lift. A line-pressure transducer was also installed in the fuel line about 50 mm ahead of the injector. For this engine, both a single-throat and a dual-throat hot plug were used; the two geometries are shown in Fig. 4. For some tests, the glow plug was removed and a flame luminosity probe

---

*See References listed in Table 1.

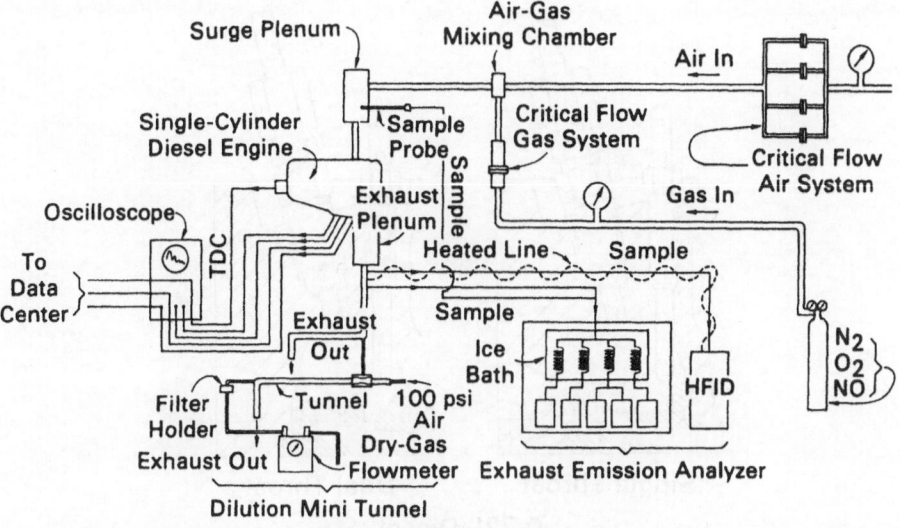

Fig. 2. Schematic of test facility.

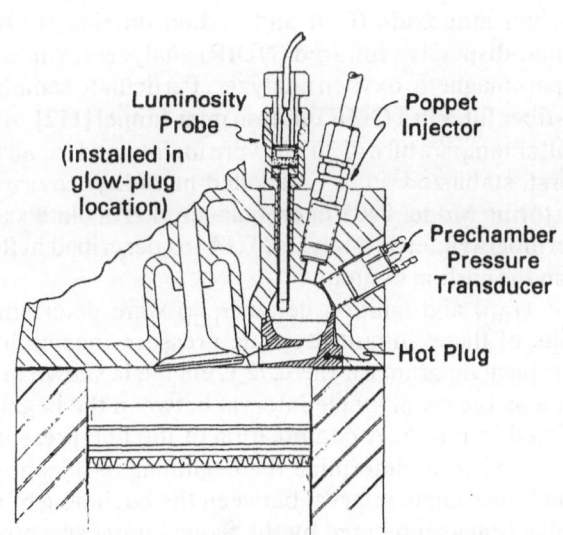

Fig. 3. Cross section of the 0.72 L engine combustion chamber.

(same size as the glow plug) was installed in its place. As indicated in Fig. 2, plenums were installed before the intake port and after the exhaust port for reducing pressure fluctuations and improving mixing. A summary of key engine parameters is given in Table A-1 of the Appendix.

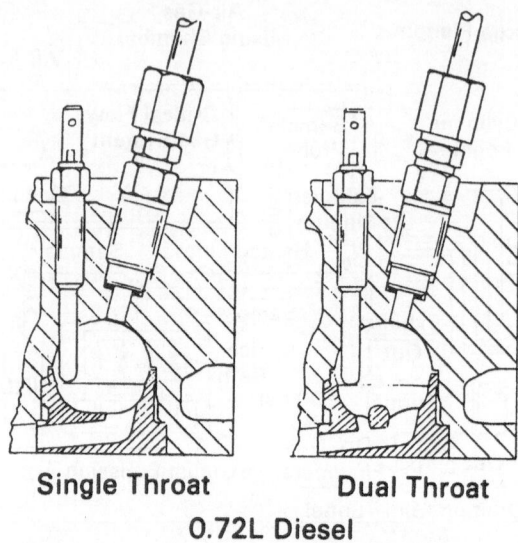

Single Throat   Dual Throat

0.72L Diesel

Fig. 4. Prechamber geometries used in the 0.72 L engine.

Both gaseous and particulate emissions were measured at all operating conditions. Oxides of nitrogen (NOx) and hydrocarbon (HC) emissions were measured with a chemiluminescent analyzer and a heated flame ionization detector (HFID) respectively. Carbon monoxide (CO) and carbon dioxide (CO_2) measurements were made with non-dispersive infrared (NDIR) analyzers, while oxygen (O_2) was measured with a paramagnetic oxygen analyzer. Particulate samples were collected on a 47-mm glass-fiber filter in a GMR dilution mini-tunnel [112]. A nominal dilution ratio of 10 and a filter temperature of 300 K were maintained for all tests. Particulate samples were first stabilized in a controlled-humidity environment and then weighed on a Sartorius Model 2405 microbalance. Particulate samples were then subjected to a thermogravimetric analysis (TGA) as described in Reference [113] in order to determine the carbon content.

Injection timing (τ_{inj}) and ignition delay (τ_{ign}) were determined from oscilloscope photographs of flame luminosity, line pressure, needle lift, and the TDC marker pulse. A typical diagram for the 0.72 L engine is shown in Fig. 5. Injection timing was defined as the crank-angle interval between the beginning of injection and TDC as depicted in Fig. 5. A combination of the line-pressure signal and the needle-lift signal was used to determine the beginning of injection. Ignition delay was defined as the crank-angle interval between the beginning of injection and the beginning of combustion as indicated by the flame luminosity probe. The average fuel flow rate for all test conditions was obtained from the measured fuel per cycle and the injection duration. The measurement of the injection duration from the needle-lift profile ignored the tail of the profile (see Fig. 5), which could be interpreted as secondary injection.

Single-Cylinder Engine (0.52 L) – A single-cylinder version of a 2.1 L diesel engine was also used to examine the effect of O_2 and N_2 addition (Fig. 6). This

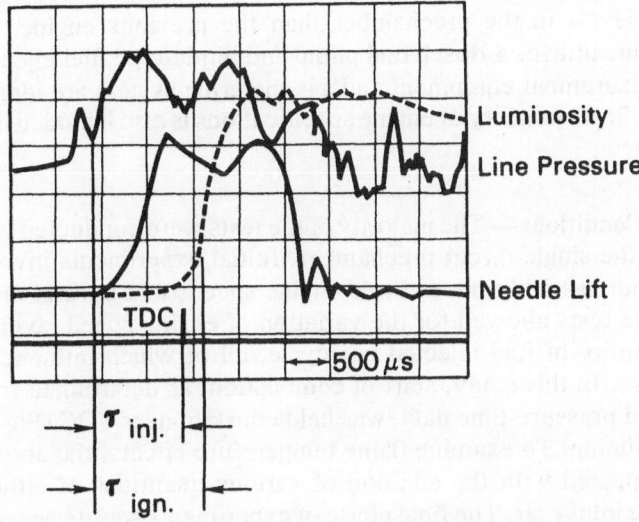

Fig. 5. Typical oscilloscope trace of flame luminosity, line pressure, and needle lift.

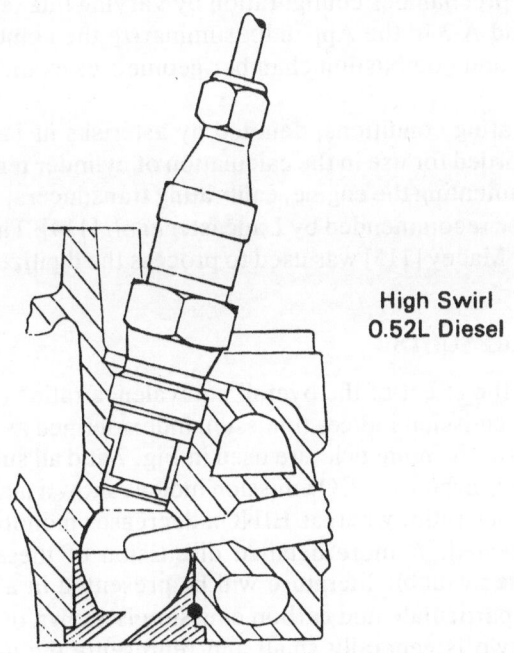

**High Swirl
0.52L Diesel**

Fig. 6. Prechamber geometry used in the 0.52 L engine.

engine, described completely in Reference [112], is a divided-chamber diesel having a Ricardo Mark Vb combustion chamber. This combustion chamber has much higher swirl levels in the prechamber than the previous engine. The Ricardo-chamber engine utilizes a Bosch fuel pump and pintle-type fuel injector. The basic emission measurement equipment and gas-metering system are identical to those shown in Fig. 2. A summary of engine specifications is also included in Table A-1 of the Appendix.

Operating Conditions — The majority of the tests were conducted with the 0.72 L engine using the single-throat prechamber. Initial experiments involved baseline testing, without gas addition, at three engine speeds and several different loads. These baseline tests allowed for the variation of engine speed, average fuel flow rate, and quantity of fuel injected per cycle, all of which influence the air-fuel mixing process. In this study, start of combustion, as determined from the flame luminosity and pressure-time data, was held constant near TDC (Fig. 5) by adjusting injection timing. To examine flame temperature effects, the above mentioned tests were repeated with the addition of various quantities of either oxygen or nitrogen to the intake air. The final phase of experiments was designed to study the effect of prechamber geometry using both the 0.72 L dual-throat combustion chamber and the 0.52 L high-swirl combustion chamber. In these tests, engine speed and load were held constant for a given engine, and the effect of gas addition was examined for each prechamber configuration by varying intake oxygen mole fraction. Tables A-2 and A-3 in the Appendix summarize the combinations of speed, load, gas addition, and combustion chamber geometries examined during this investigation.

At selected operating conditions, denoted by asterisks in Table A-2, pressure-time data were recorded for use in the calculation of cylinder temperatures. Procedures used in instrumenting the engine, calibrating transducers, and recording data were similar to those recommended by Lancaster et al. [114]. The technique developed by Fisher and Macey [115] was used to process the digitized data.

EXPERIMENTAL RESULTS

Fig. 7 illustrates the effect of the overall equivalence ratio* (ϕ) on exhaust carbon, CO, and NOx emission indices (emission index defined as grams of pollutant per kilogram of fuel). The nomenclature used in Fig. 7 and all subsequent figures is listed in Table 4. The carbon and CO emission indices show a marked increase with increasing equivalence ratio, whereas EINOx decreases monotonically with ϕ for all engine speeds tested. A more detailed discussion of these trends and their comparison with the available literature will be presented in a later section. The effect of ϕ on both particulate and carbon emission is shown in Fig. 8. The difference between the two is generally small and represents the amount of volatiles deposited on the exhaust particulate. The present study is concerned with particulate carbon, and therefore no attempt was made to characterize the physical or

---

* $\phi = \dfrac{Fuel/Oxidizer}{(Fuel/Oxidizer)_{stoichiometric}}$ $\left(\begin{array}{l} oxidizer\ includes\ O_2\ and\ N_2 \\ entering\ cylinder \end{array}\right)$

TABLE 4
Nomenclature Used in Figs. 7-34

Symbol	Speed (r/min)	Fuel (mm^3/cycle)	Combustion Chamber	Engine
▷	600	15	Single throat	0.72 L
△	600	22	Single throat	0.72 L
▽	600	36	Single throat	0.72 L
ბ	1200	16	Single throat	0.72 L
○	1200	22	Single throat	0.72 L
○-	1200	28	Single throat	0.72 L
-○	1200	32	Single throat	0.72 L
♀	1200	40	Single throat	0.72 L
◇	1800	14	Single throat	0.72 L
□	1800	22	Single throat	0.72 L
◯	1800	34	Single throat	0.72 L
⌂	1200	22	Dual throat	0.72 L
⊔	1500	21	High Swirl	0.52 L

Symbol	Gas Added
open	none
solid	nitrogen
half	oxygen

chemical nature of the volatiles. However, previous work [116, 117] indicates that the volatiles are mainly comprised of polyaromatic hydrocarbons with a carbon number ranging between 10 and 40.

The effect of changing the mole fraction of oxygen in the intake air (X_{O_2}) on carbon, CO, and NOx emissions is shown in Figs. 9-13. In Figs. 9, 10, and 11, variations in exhaust emissions with intake O_2 concentration are illustrated for three different speeds at a constant fueling level of 22 mm^3/ cycle. For all three speeds considered, the exhaust carbon and CO increase rapidly with decreasing oxygen concentration in the engine intake, while the NOx emissions show an opposite trend. Fig. 12 shows the effect of intake O_2 concentration on exhaust emissions for different fueling levels at 1200 r/min. The general trends observed in Fig. 10 are apparently valid for all fueling levels tested. The systematic variation in the exhaust emissions with fueling level primarily results from changes in mixing and will be discussed in detail in a later section. The various combustion chamber geometries tested also show similar emission trends with addition of O_2 and N_2, as illustrated in Fig. 13.

ANALYSIS AND INTERPRETATION

General Emission Characteristics — Before examining the effects of gas addition and air-fuel mixing in detail, some of the general emission tradeoffs between NOx, CO, and particulate carbon will be presented and discussed briefly. These general

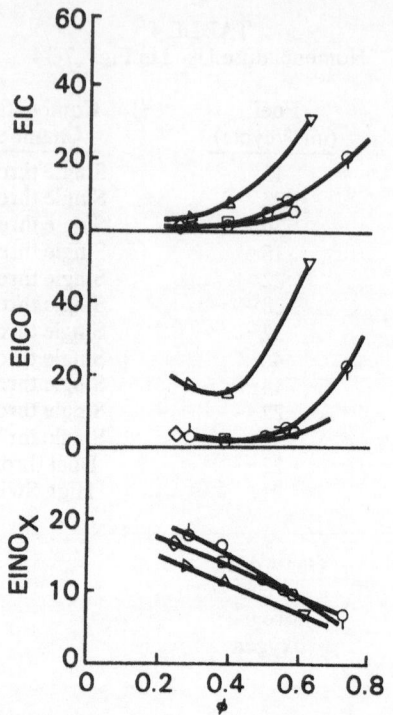

Fig. 7. Effect of overall equivalence ratio on exhaust emissions without gas addition.

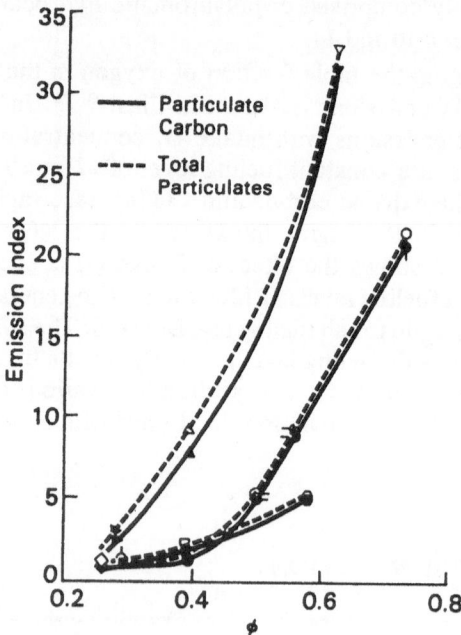

Fig. 8. Comparison of particulate carbon and total particulates as a function of overall equivalence ratio.

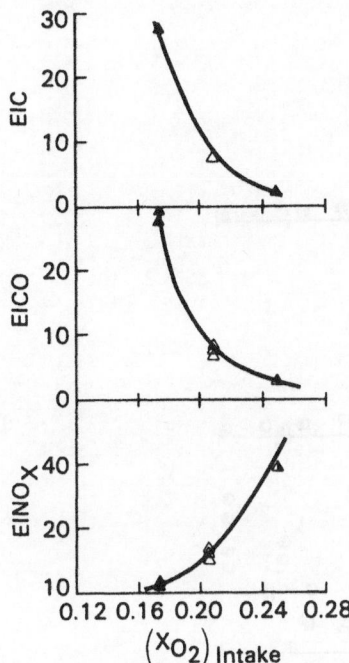

Fig. 9. Effect of intake O₂ mole fraction on exhaust emissions at 600 r/min.

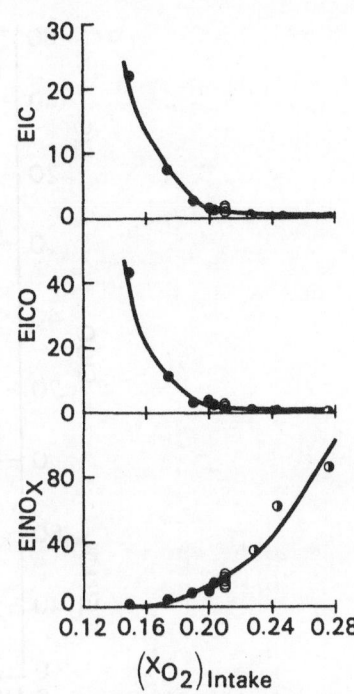

Fig. 10. Effect of intake O₂ mole fraction on exhaust emissions at 1200 r/min.

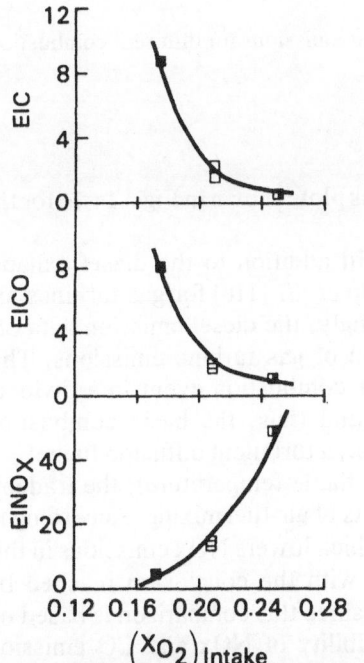

Fig. 11. Effect of intake O₂ mole fraction on exhaust emissions at 1800 r/min.

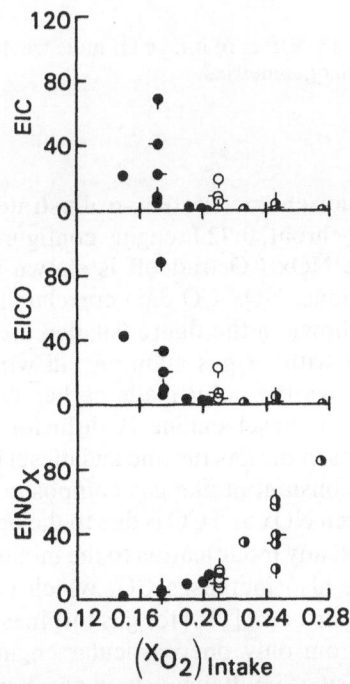

Fig. 12. Effect of intake O₂ mole fraction on exhaust emissions at different fueling levels.

References pp. 477-481.

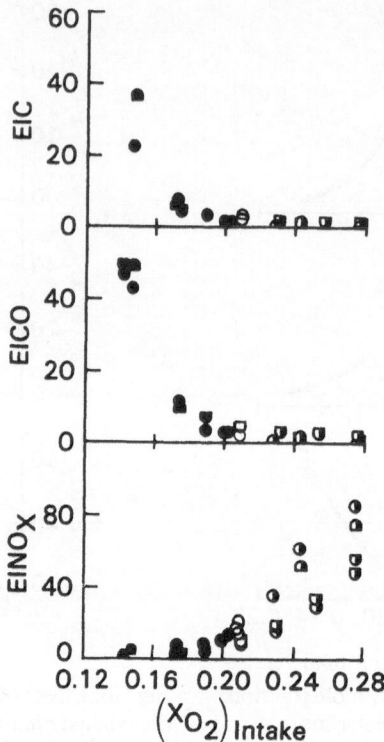

Fig. 13. Effect of intake O_2 mole fraction on exhaust emissions for different combustion-chamber geometries.

emission characteristics are illustrated in the cross plots shown in Figs. 14-16 for the single-throat, 0.72 L engine configuration.

The NOx-CO tradeoff is shown in Fig. 14. In addition to the diesel exhaust emissions, NOx-CO data correlated by Verkamp *et al.* [118] for gas turbines are also shown in the figure (shaded area). Surprisingly, the diesel emission data obtained without gas addition fall within the band of gas turbine emissions. This illustrates the point made earlier that the major combustion event in a divided-chamber diesel engine is diffusion controlled and thus, the basic combustion process in the gas turbine and diesel is similar, i.e., a turbulent diffusion flame.

At constant intake gas composition (constant flame temperature), the tradeoff between NOx and CO is due to the opposite effects of air-fuel mixing. According to Fig. 14, any modification to the mixing process which lowers NOx emissions in this engine also increases CO, which is consistent with the conclusion reached by Verkamp *et. al* [118] for gas turbines. However, since this comparison is based on data from only one particular engine, the possibility of NOx and CO emission increasing simultaneously in some situations cannot be ruled out. The addition of either oxygen or nitrogen shifts the NOx-CO tradeoff slightly, as shown in Fig. 14.

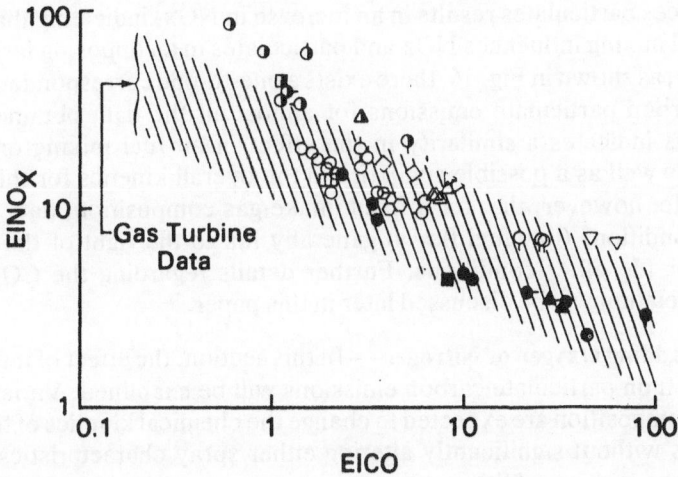

Fig. 14. Variation in EINOx with EICO for different engine speeds, fueling levels and intake oxygen mole fractions.

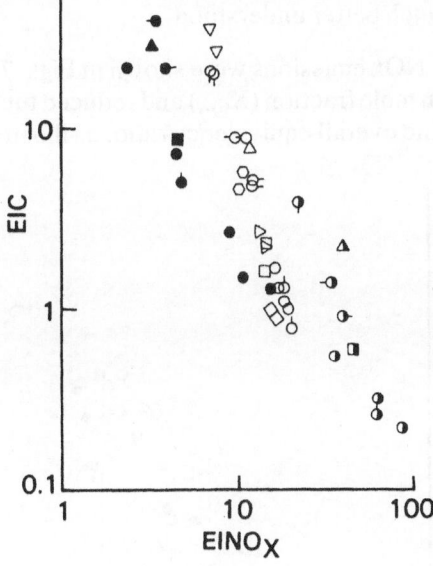

Fig. 15. Variation in EIC with EINOx for different engine speeds, fueling levels and intake oxygen mole fractions.

For a given CO emission level, NOx emissions are higher with O_2 addition than with no gas addition. The trend reverses with the addition of nitrogen to the intake air.

The NOx-particulate carbon tradeoff (Fig. 15) also shifts with gas addition. For a fixed NOx level, less particulate matter is obtained with the addition of N_2 to the intake air. For a fixed intake gas composition, any change in operating conditions which reduces particulates results in an increase in NOx, indicating that like NOx-CO, air-fuel mixing influences NOx and particulates in an opposing fashion.

However, as shown in Fig. 16, there exists a one-to-one correspondence between CO and carbon particulate emissions for almost all the data obtained with this engine. This indicates a similarity in the effect of air-fuel mixing on these two emissions as well as a possible similarity in the overall kinetics for this particular engine. Note, however, that for a fixed intake-gas composition, data obtained at fuel-lean conditions (low fuel flows) generally fall to the right of the correlation obtained for the other conditions. Further details regarding the CO-particulate carbon correlation will be discussed later in this paper.

Effect of Adding Oxygen or Nitrogen — In this section, the effect of intake oxygen mole fraction on particulate carbon emissions will be examined. Variations in the intake gas composition are expected to change the chemical kinetics of the combustion process without significantly altering either spray characteristics or air-fuel mixing (Table 1). The addition of oxygen or nitrogen directly affects the flame temperature and the oxygen concentration, which subsequently influences the concentrations of various oxidizing radicals such as O and OH. First, the effect of gas addition on NOx emissions will be considered, since the kinetic mechanisms for NO formation are much better understood.

NOx Emissions – NOx emissions were shown in Figs. 7 and 9-12 to increase with higher intake oxygen mole fraction (X_{O_2}) and reduced fueling (mm³/cycle). Both of these effects lower the overall equivalence ratio, as illustrated in Fig. 17 for a given

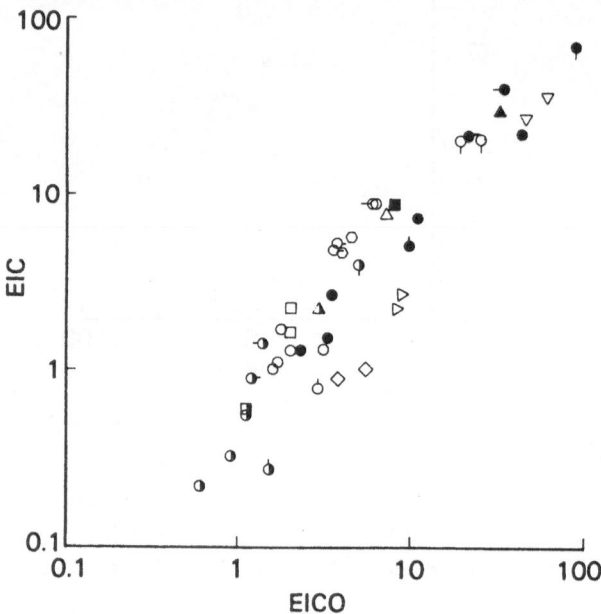

Fig. 16. Variation in EIC with EICO for different engine speeds, fueling levels and intake oxygen mole fractions.

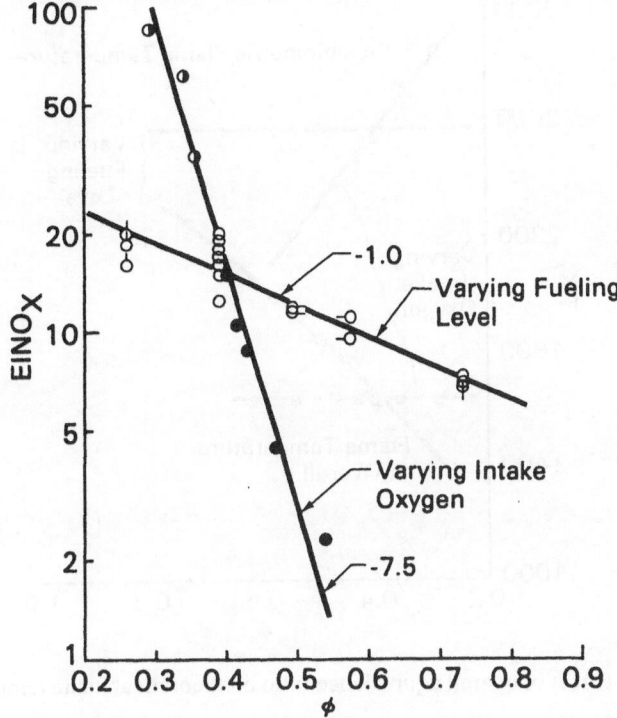

Fig. 17. Effect of overall equivalence ratio on EINOx for different fueling levels and intake O₂ mole fractions.

engine speed. However, the variation in NOx with ϕ when O_2 or N_2 are added to the intake air is much more dramatic than when the fueling level is changed at a given intake gas composition. On a graph of log EINOx versus log $(X_{O_2})_{intake}$, a slope of approximately six is obtained, which indicates an unreasonably high reaction order for NOx formation. This suggests that, in the case of gas addition, temperature effects are also important.

Fig. 18 illustrates how the flame temperature varies with equivalence ratio for the conditions of Fig. 17. Both the adiabatic flame temperature at stoichiometric equivalence ratio (T_f) and the flame temperature based on the overall equivalence ratio (T_ϕ) are shown. Here flame temperatures have been calculated using a version of the NASA equilibrium program [119], which includes effects of dissociation. In divided-chamber diesel engines, the peak pressure during combustion is usually not much higher than the TDC pressure for motoring conditions, since very little pre-mixed combustion occurs. Therefore the cycle pressure and temperature at top-dead-center (TDC), obtained from the analysis of pressure-time data taken under motoring conditions, were used in the flame temperature calculations (Table 5). Measurements with the flame luminosity probe indicated that combustion began near TDC for all operating conditions examined.

Comparing the flame temperature trends in Fig. 18 with the NOx trends in Fig. 17 illustrates that neither the stoichiometric flame temperature nor the burned gas

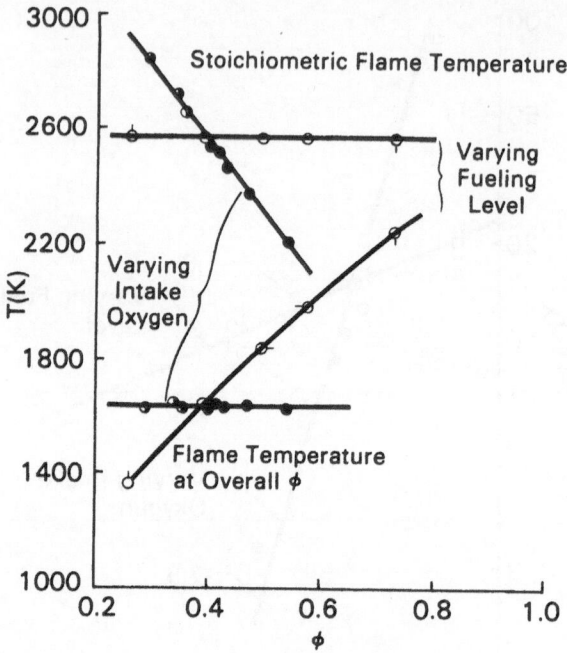

Fig. 18. Effect of overall equivalence ratio on theoretical flame temperatures.

TABLE 5
Prechamber Gas Pressure and Temperature
at Top-Dead-Center

Speed (r/min)	P (atm)	T (K)
600	45.6	690
1200	49.3	747
1800	52.5	758

temperature (flame temperature based on the overall equivalence ratio) can explain the decrease in NOx observed with increased fueling. The stoichiometric flame temperature remains constant for a given intake gas composition, while the burned gas temperature for the overall mixture increases with ϕ. Since NOx is known to increase with temperature, the observed reduction in NOx with ϕ in Fig. 17 is attributed to changes in the air-fuel mixing process rather than chemical kinetics.

However, the effect of gas addition can be interpreted via changes in the stoichiometric flame temperature since both NOx and T_f decrease rapidly with a reduction in intake oxygen concentration, whereas the burned gas temperature is relatively unaffected by gas addition. This should not be too surprising since the major portion of the combustion event in diesel engines is diffusion controlled. As discussed earlier, the ability of T_f to correlate NOx emissions from steady-flow combustors has already been demonstrated [63-67].

The formation of NO from atmospheric nitrogen proceeds via the extended Zeldovich mechanism, which includes [120]:

1. $O + N_2 \rightleftharpoons NO + N$
2. $N + O_2 \rightleftharpoons NO + O$
3. $N + OH \rightleftharpoons NO + H$

In steady-flow combustors, the NO decomposition reactions are usually omitted since the formation process is quenched long before equilibrium is approached [63, 64, 121]. By neglecting the reverse reactions and invoking the N-atom steady-state approximation, the NO formation rate simplifies to

$$\frac{d\,[NO]}{dt} = 2\,k_{1f}\,[O]\,[N_2] \tag{11}$$

where k_{1f} is the forward rate coefficient for reaction (1). If the O-atom is assumed to be in chemical equilibrium via

4. $O + O + M \rightleftharpoons O_2 + M$

the NO formation rate becomes

$$\frac{d\,[NO]}{dt} = \frac{2\,k_{1f}\,[O_2]^{1/2}\,[N_2]}{K_4} \tag{12}$$

where K_4 is the equilibrium constant for reaction (4). Combining the activation energies in k_{1f} and K_4 [122] and integrating Equation (12) yields

$$[NO] \propto [O_2]^{1/2}\,[N_2]\,\exp\,(-134\,630/RT) \tag{13}$$

Sawyer et al. [64] argued that in gas turbine combustors, NOx production occurs primarily in the stoichiometric regions of the flow. As a result, NOx emissions should correlate with the stoichiometric adiabatic flame temperature (T_f). Sawyer et al. [64] plotted ℓn (EINOx) versus T_f^{-1} and obtained an apparent activation energy of 136 000 cal/mole, which is practically identical to the theoretical prediction in Equation (13). More recently, Washam and Mellor [65] have shown that ambient effects on NOx such as humidity are also correlated with T_f^{-1}.

Mellor and coworkers have expanded on this basic concept to correlate NOx data from simplified steady-flow combustors [63] and gas turbine combustors [66, 67]. They found that NOx emissions scale linearly with a Damkohler number, defined as the ratio of a turbulent mixing time to a NO formation time (Equation 5). The model considers that NO forms in the high-temperature regions of the combustor which are near stoichiometric. The characteristic time for NO formation has been taken from the theoretical expression for NO formation (Equation 13), while the mixing time in Equation (5) specifies the lifetime of these NOx-forming zones. Using this simplified approach, NOx emissions from GT-309 (automotive), T-63 (helicopter) and JT-9D (aircraft-jumbojet) gas turbine combustors have been correlated simultaneously [66, 67].

Since in diesel engines the nature of the heterogeneous combustion event, in which most of the energy is released, is similar to that in steady-flow combustors, NOx emissions in gas turbines and diesels should correlate in a similar fashion. The majority of the NO production in diesels is also expected to occur near the stoichiometric reaction zone. Fig. 19 illustrates the correlation obtained between the NOx emission index (EINOx) and T_f^{-1} at 1200 r/min and constant fueling level. The linear correlation of ℓn EINOx with the inverse of the stoichiometric flame temperature indicates that the effect of gas addition on NOx is a flame temperature effect, in agreement with Wilson et al. [7], rather than a result of changes in oxygen concentration as proposed by Tsunemoto and Ishitani [44]. The interpretation by Tsunemoto and Ishitani [44] requires an unreasonably high reaction order for NO formation, as discussed earlier, and fails to consider the fact that NOx formation occurs primarily in stoichiometric reaction zones.

According to Fig. 19, the effect of O_2 and N_2 addition to the intake air can be correlated with the simple expression

$$NOx \propto \exp\left(-38\ 700/T_f\right) \tag{14}$$

yielding an overall activation energy of -76 858 cal/mole (E/R = -38 700). The apparent activation energy obtained in this study is considerably lower than the activation energy derived for steady-flow combustors (Equation 13), but larger than the one predicted by NO equilibrium. Assuming that reactions (1) and (2) of the Zeldovich mechanism are equilibrated,

$$[NO]_{equil} = \left(K_1\ K_2\ [O_2]\ [N_2]\right)^{1/2} \tag{15}$$

(K_1 and K_2 are equilibrium constants for reactions (1) and (2) respectively). Substituting for K_1 and K_2 from Reference [122] yields

$$[NO]_{equil} \propto [O_2]^{1/2}\ [N_2]^{1/2}\ \exp\left(-21\ 627/RT\right) \tag{16}$$

There are several possible reasons for the difference in apparent activation energy between Equation (13) and the diesel correlation (Equation 14). These include

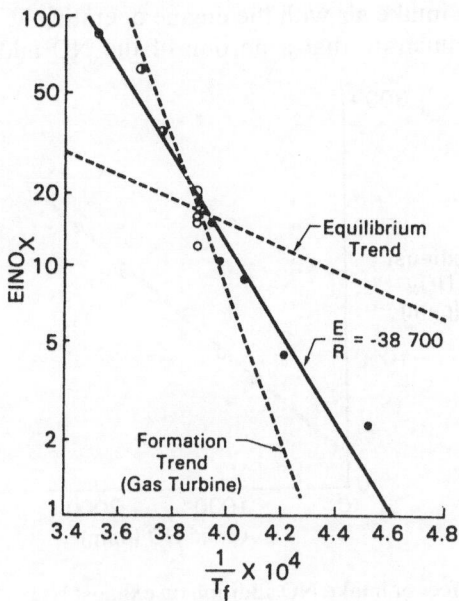

Fig. 19. Correlation of EINOx with stoichiometric flame temperature at 1200 r/min, 22 mm³/cycle.

the following:

1. The O-atom is not in equilibrium as assumed in Equation (13).
2. The computed flame temperature is not representative of the actual process.
3. NO decomposition is important.

The assumption of O-atom equilibrium via reaction (4) is generally accepted to be valid in spark-ignition engines [123, 124] and steady-flow combustors [63-67, 121] and therefore should also be a good assumption in diesel engines.

The use of the stoichiometric adiabatic flame temperature with initial conditions evaluated at TDC is obviously an approximation for the entire NO formation process in diesel engines. Calculation of the stoichiometric adiabatic flame temperature based on conditions at TDC cannot alone explain the known variation of NOx level with changes in injection timing. However, in this study, injection timing was varied to maintain start-of-combustion near TDC. Although the flame temperature calculation includes dissociation, it does not consider heat losses due to radiation, turbulent transport, etc. Lower flame temperatures are also expected if the fuel impinges and burns off the walls of the combustion chamber. In addition to heat-transfer effects, the NO reaction rate should be integrated over the entire combustion duration rather than approximated at TDC conditions. However, Blumberg and Kummer [124] have shown that in spark-ignition engines, the major contribution to the overall NOx level comes from elements which burn first. Based on this work, it seems that NOx emission trends should correlate with the maximum flame temperature obtained during the initial stages of combustion.

The importance of NO decomposition has been demonstrated in SI engines [124, 125]. To examine this effect in diesel engines, various quantities of pure NO were introduced into the intake air with the engine operating at 1200 r/min. The results shown in Fig. 20 illustrate that a portion of the NO added to the intake air is

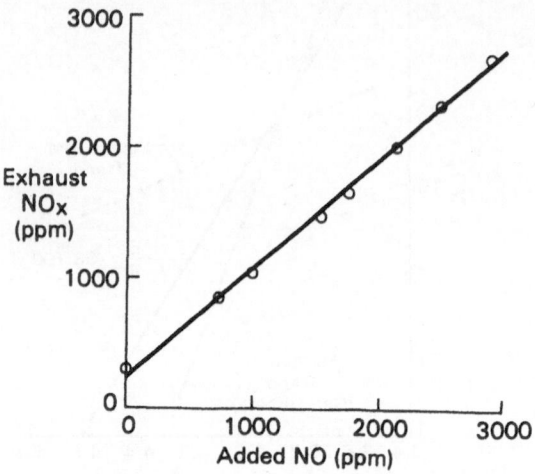

Fig. 20. Effect of intake NO addition on exhaust NOx concentration.

destroyed in the engine. The amount of NO emitted by the exhaust is described by the equation

$$(NOx)_{ex} = (NOx)_o + 0.84 \, (NO)_a \tag{17}$$

where

$$(NOx)_{ex} = \text{NOx measured in the exhaust (ppm)}$$
$$(NOx)_o = \text{NOx measured in the exhaust with no gas}$$
$$\text{added to the intake air (ppm)}$$
$$(NO)_a = \text{NO added to intake air (ppm)}$$

Thus, NO decomposition seems to explain why the apparent activation energy for NO from diesel engines is between the equilibrium value (Equation 16) and the overall activation energy for NO formation (Equation 13).

NOx emissions obtained at different engine speeds and fuel flow rates are plotted in Fig. 21 as a function of the stoichiometric flame temperature. Examination of the data in Fig. 21 indicates that approximately the same apparent activation energy is obtained for all engine operating conditions. This point is more clearly demonstrated in Fig. 22, where NOx emissions at a given operating condition have been normalized by the NOx level obtained without gas addition. For all practical purposes, changes in engine speed and overall air-fuel ratio do not alter the basic kinetic processes involved in the production of NOx. Any deviation in NOx emissions from the line in Fig. 19 is interpreted as a result of changes in the air-fuel

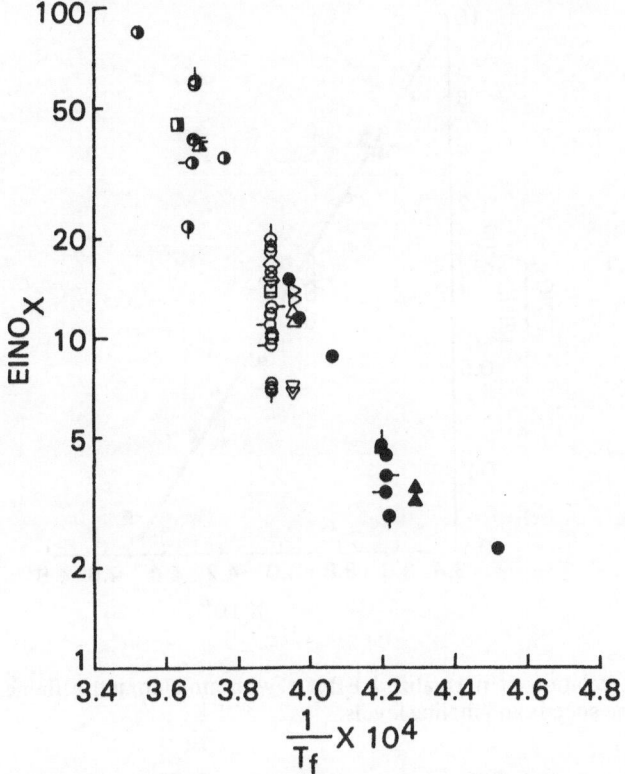

Fig. 21. Effect of stoichiometric flame temperature on EINOx at different engine speeds and fueling levels.

mixing process. Although engine speed influences the initial pressure and temperature at TDC (Table 5), and thus the calculation of T_f, this effect is only secondary compared to the effect of speed on the air-fuel mixing process.

The influence of combustion chamber geometry on NOx is demonstrated in Fig. 23. For the three combustion chambers examined in this, there is no appreciable change in the apparent activation energy representing the formation and destruction of NOx. Thus, it appears that NOx kinetics are basically the same for these three geometries.

Thus, in summary, the effect of intake oxygen or nitrogen addition on NOx emissions is described by the simple expression

$$NOx \propto \exp(-38\ 700/T_f) \tag{14}$$

Equation (14) is valid for a given engine and operating condition and is solely a function of the stoichiometric adiabatic flame temperature. Since NOx kinetics are not significantly affected by engine speed, overall air-fuel ratio, and combustion chamber geometry in the configurations of this study, variations in NOx emissions with these engine parameters can be described by parameters characterizing the

References pp. 477-481.

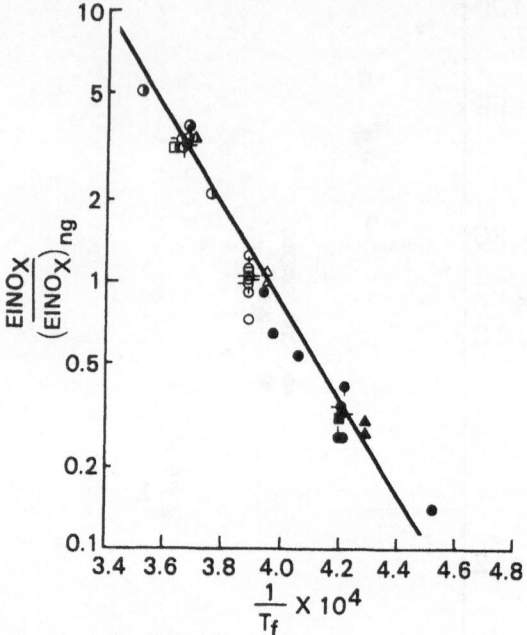

,Fig. 22. Correlation of normalized EINOx with stoichiometric flame temperature for different engine speeds and fueling levels.

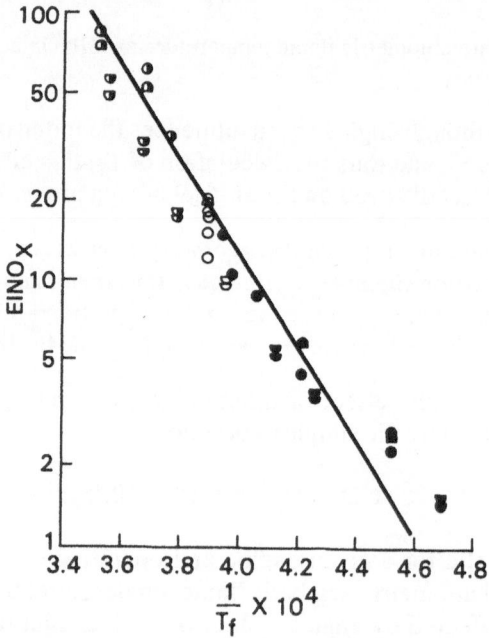

Fig. 23. Correlation of EINOx with stoichiometric flame temperature for different combustion chamber geometries.

air-fuel mixing process, which should appear as the proportionality factor in Equation (14).

Particulate Carbon – As shown in Figs. 9-12, the effect of decreasing oxygen concentration in the intake air is to increase the emission of particulate carbon from the engine. This differs from the general tendency found in diffusion flames for diluent addition to reduce soot emission. Khan [1] discusses the basic differences which exist between the classic diffusion flame and the combustion process occurring in a diesel engine. According to Khan [1], the diesel combustion process consists of a confined turbulent diffusion flame with intense recirculation. Once soot particles are released from the flame, they can be recirculated back through the hot reaction zones, as well as being consumed during the expansion process. Hence, carbon formation and oxidation occur simultaneously during the heat release period, and the mixing process is extremely important in determining the tradeoff between soot formation and oxidation.

At the present time, detailed mathematical descriptions of the mixing process and soot kinetics in diesels are not available. In fact, the mechanisms of soot formation and oxidation are far from understood even in the simplest combustion systems. To describe soot kinetics in diesels, most investigators utilize simple Arrhenius relationships to characterize formation and oxidation, with the pre-exponential factor, activation energy and concentration dependence determined empirically [4, 5, 17, 25, 68, 69].

Khan and co-workers [4, 5, 17, 25] describe the soot formation process by

$$\frac{dm_s}{dt} = Z_s F_s \phi_s^n P_s \exp\left[-E_s/RT_s\right] \tag{18}$$

where Z_s = pre-exponential factor
  F_s = fraction of cylinder volume containing soot formation zones
  ϕ_s = local equivalence ratio in formation zone
  P_s = partial pressure of unburned fuel in formation zone
  E_s = apparent activation energy for soot formation
  R = gas constant
  T_s = temperature in soot-forming zone
  n = constant

The above overall expression for soot formation includes the formation and oxidation of soot precursors, nucleation, and heterogeneous particle growth. No attempt is made to describe each of these processes individually.

Following formation, the particles undergo coagulation in accordance with the Whytlaw-Gray equation [126], and surface oxidation according to the global equation [4, 5, 17, 25].

References pp. 477-481.

$$-\frac{dm_s}{dt} = Z_0 \, \pi \, d^2 \, T_0^{-0.5} \, P_0 \, \exp\left[-E_0/RT_0\right] \qquad (19)$$

where Z_0 = pre-exponential factor
             d = soot-particle diameter
             T_0 = local temperature
             P_0 = local partial pressure of oxygen
             E_0 = apparent activation energy for soot oxidation

Equation (19) was obtained from the work of Lee et al. [127] in laminar flames, which assumes an activation energy of 39 300 cal/mole. However, as with soot formation, considerable controversy exists over the exact mechanism of soot combustion. The oxidation of soot particles in flames has been explained on the basis of reactions with oxygen [127-130], O-atom [131], OH-radical [132, 133] and carbon dioxide [134]. Appleton [135] has reviewed these studies and concluded that in practical combustion systems, the semi-empirical expression proposed by Nagle and Strickland-Constable [130] provides the best method for estimating soot oxidation rates. However, a recent study by Neoh et al. [133] found that the Nagle and Strickland-Constable expression drastically underestimated soot burnout rates under fuel-rich and near stoichiometric conditions. Neoh et al [133] observed that under their conditions, the OH-radical was the primary oxidant of soot, in agreement with Fenimore and Jones [132].

The trends shown in the experimental data (Figs. 9-12) indicate that the addition of oxygen or nitrogen to the intake air has a dramatic influence on the amount of particulate carbon emitted by the engine. This effect is further illustrated by Fig. 24, where EIC, the emission index for particulate carbon determined by TGA analysis, is shown as a function of overall equivalence ratio for data obtained at an engine speed of 1200 r/min. The strong dependence of EIC on ϕ with gas addition again implies a flame temperature effect. Based on the recent work of Glassman and Yaccarino [54, 57], it is reasonable to assume that changes in soot emission with addition of either O_2 or N_2 should also correlate with the stoichiometric flame temperature, since the main combustion event is diffusion controlled. The influence of fueling level on EIC (no gas addition) will be discussed in the section on air-fuel-mixing.

As illustrated in Fig. 25, the correlation of ℓn EIC with T_f^{-1} is linear with an apparent activation energy of 97 115 cal/mole (E/R = 48 900). Although higher temperatures increase the rates of both formation and oxidation, the general trend of EIC with flame temperature (Fig. 25) implies that the oxidation process is more strongly affected by gas addition than is soot formation. Thus, in contrast to classical diffusion flames in which diluent addition suppresses the net production of soot, the addition of nitrogen to a diesel engine dramatically increases particulate carbon, primarily because of changes in the oxidation mechanism.

However, the apparent activation energy in Fig. 25 is much higher than expected for the combustion of carbon via oxygen mechanisms [127-130]. This could be

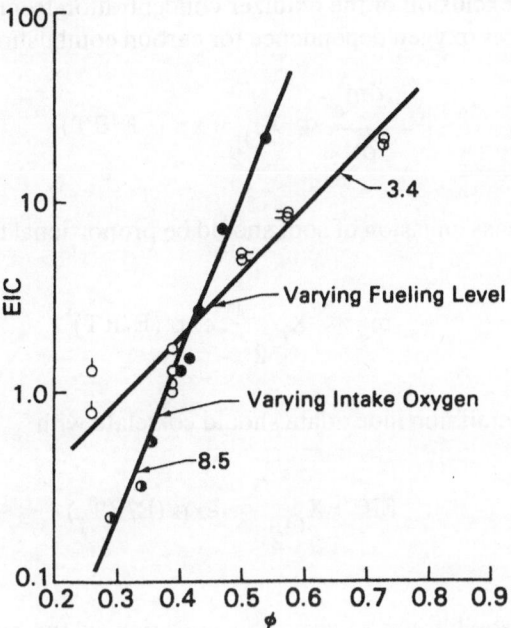

Fig. 24. Effect of overall equivalence ratio on EIC for different fueling levels and intake O_2 mole fractions.

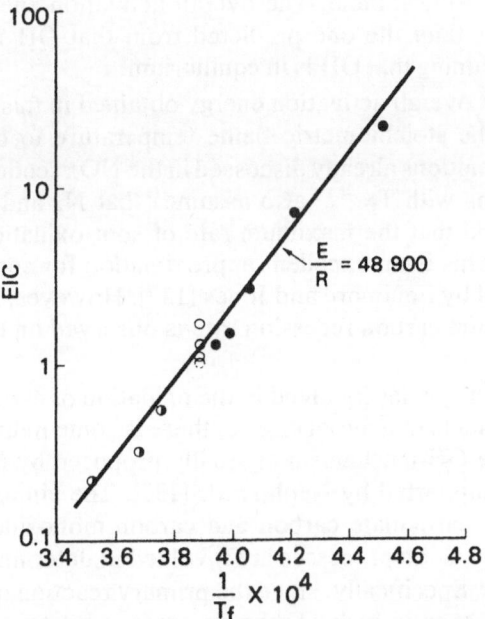

Fig. 25. Correlation of EIC with stoichiometric flame temperature at 1200 r/min, 22 mm³/cycle.

attributed to the exclusion of the oxidizer concentration from the correlation. Assuming a first-order oxygen dependence for carbon combustion [127, 128],

$$-\frac{dm_s}{dt} \propto X_{O_2} \exp(-E/RT) \qquad (20)$$

implies that the mass emission of soot should be proportional to

$$m_s \propto X_{O_2}^{-1} \exp(E/RT) \qquad (21)$$

Thus, the carbon emission index data should correlate with

$$EIC \cdot X_{O_2} \propto \exp(E/RT_f) \qquad (22)$$

Inclusion of the equilibrium oxygen concentration in the correlation drops the apparent activation energy from 97 115 to 51 835 cal/mole, although it is still slightly higher than values reported in the literature. In a recent review paper, Laurendeau [136] noted that for a first-order oxygen dependence, the overall activation energy is usually less than 40 kcal/mole. The overall activation energy obtained in this study is also higher than the one predicted from that OH mechanism for soot oxidation [132], assuming that OH is in equilibrium.

The high apparent overall activation energy obtained in this study could also be due to the use of the stoichiometric flame temperature to correlate results. In addition to the assumptions already discussed in the NOx section, the correlation of particulate emissions with T_f^{-1} also assumes that N_2 and O_2 addition affects mainly oxidation and that the maximum rate of soot oxidation exists near unity equivalence ratio. This is an excellent approximation for soot oxidation via OH radicals as proposed by Fenimore and Jones [132]. However, for soot combustion with O_2, the maximum carbon recession rate is observed on the lean side of stoichiometric [135].

Although the exact species involved in the oxidation of carbon cannot be determined from experiments in a diesel engine, there is some indirect evidence in this study to support the OH-mechanism originally proposed by Fenimore and Jones [132] and recently supported by Neoh et al. [133]. The almost one-to-one correspondence between particulate carbon and carbon monoxide (Fig. 16) tends to indicate that similar kinetic processes are involved in determining the emission of these two pollutants. Specifically, since the primary reaction of CO to CO_2 at high temperatures is almost entirely due to the elementary reaction [19, 137],

$$5. \quad CO + OH \rightleftarrows CO_2 + H$$

the direct correlation of particulate carbon with CO suggests that the OH-radical may also be important in the oxidation of particulate carbon.

In addition to the oxidation of carbon particles, oxidation of soot precursors may also be important in the diesel engine, which could account for the strong flame temperature dependence shown in Fig. 25. Thus, soot formation could also be affected by gas addition. Considering the diesel combustion process to be a confined turbulent diffusion flame with intense recirculation, as proposed by Khan [1], implies that both soot particles and soot precursors are continually recirculated through hot combustion zones and are thus affected by variations in the stoichiometric flame temperature. This is indeed a plausible explanation for the relatively high apparent activation energy observed in this study.

However, one can only speculate about the mechanisms involved at the present time because it is difficult to separate soot formation from soot oxidation in the diesel environment. Furthermore, flame structure can also be affected by gas addition. Obviously, before these processes can be completely understood in a diesel engine, the reactions of importance in simple laboratory flames must be defined. At the present time, the correlation of EIC with T_f^{-1} represents an adequate method of determining the overall effects of O_2 and N_2 addition on diesel particulate emissions. The success of the correlation using T_f evaluated at TDC conditions is probably due to the fact that in divided-chamber diesel engines, the peak pressure is not much above the motoring pressure at TDC. However, the calculation of the stoichiometric adiabatic flame temperature based on conditions at TDC cannot alone explain the known variation of particulate level with changes in injection timing.

The flame-temperature correlations for the other loads and speeds are presented in Fig. 26. Although a large variation exists in the carbon emission index at a given flame temperature, examination of the data in more detail indicates that the apparent activation energy for the other operating conditions is the same as in Fig. 25. This is more clearly demonstrated in Fig. 27, where the gas addition points have been normalized by the no-gas addition data obtained at the same operating conditions. Therefore, the kinetics of soot formation and oxidation, like those of NOx, are not significantly affected by variations in engine speed and overall equivalence ratio, as discussed in connection with Table 1. Instead, variations in EIC at a given flame temperature (Fig. 26) are interpreted as resulting from changes in the air-fuel mixing process.

For the 0.72 L engine, the effect of O_2 and N_2 addition to the intake air is described by a simple function of the stoichiometric flame temperature

$$EIC \propto exp (48\ 900/T_f) \tag{23}$$

Since the contribution of volatiles to the particulate mass is relatively small (Fig. 8), total particulate emissions would also correlate reasonably well with T_f^{-1}.

The effect of combustion chamber geometry on particulate carbon is shown in Fig. 28. Unlike the NOx correlation, the apparent activation energy for particulate carbon is found to be a function of engine type. The high-swirl combustion chamber

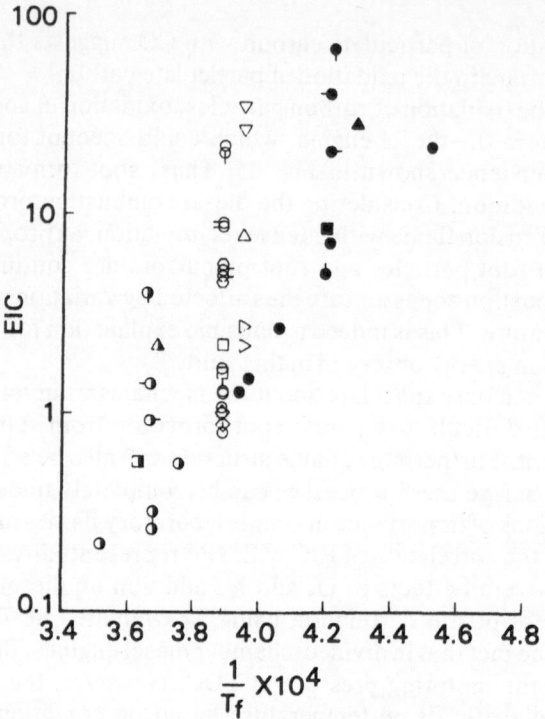

Fig. 26. Effect of stoichiometric flame temperature on EIC for different engine speeds and fueling levels.

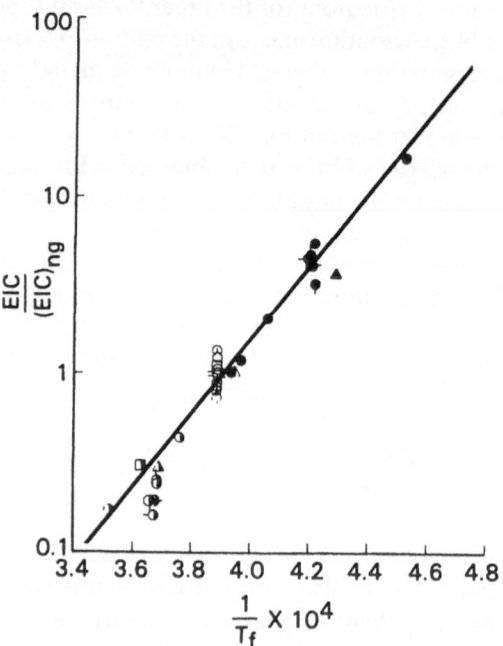

Fig. 27. Correlation of normalized EIC with stoichiometric flame temperature for different engine speeds and fueling levels.

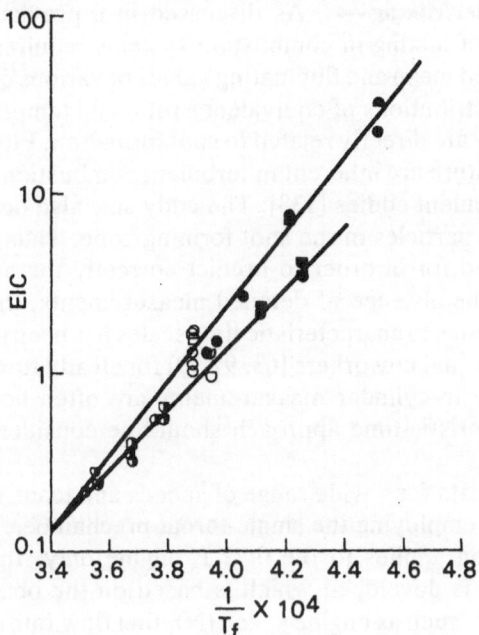

Fig. 28. Effect of stoichiometric flame temperature on EIC for different combustion-chamber geometries.

(the 0.52 L diesel) has a slightly lower overall activation energy than the 0.72 L diesel, which is interpreted as a change in the competition between soot formation and oxidation mechanisms. This is not unexpected considering the variation in trends observed in classical diffusion flames when O_2 or N_2 is added to the inlet air (see earlier discussion). In these simple flames, the emission of soot depends on the balance between soot formation and oxidation, which is strongly influenced by geometry [9].

Thus for a given engine the effect of O_2 and N_2 addition on soot emission is described by empirical expressions which are functions of the stoichiometric flame temperature. This is not to say that oxidizer concentrations are unimportant because these concentrations are directly related to the flame temperature. For the modified 0.72 L diesel engine, the correlation is

$$EIC \propto \exp{(48\,900/T_f)} \tag{23}$$

while for the Ricardo-chamber engine (0.52 L engine)

$$EIC \propto \exp{(38\,300/T_f)} \tag{24}$$

Equations (23) and (24) actually include processes of soot precursor formation and oxidation, nucleation, particle growth, and particle oxidation occurring in the diesel engine.

References pp. 477-481.

Effect of Air-Fuel Mixing —   As discussed in a previous section, a complete characterization of mixing in combustion systems requires a knowledge of time- and space-resolved mean and fluctuating values of various quantities. Among these quantities, the distributions of equivalence ratio and temperature are of particular interest since they are directly related to soot formation. Fluctuations in concentrations and temperature are inherent in turbulent combustion and are closely related to the size of turbulent eddies [138]. The eddy size also determines the residence time of gases and particles in the soot forming zone. Thus, turbulent fluctuations must be accounted for in order to predict correctly the exhaust emissions from flames [139]. In the absence of detailed measurements, the approach of relating turbulent fluctuations to characteristic time scales has been used by Swithenbank *et al.* [88] and Mellor and coworkers [63, 97-99] for steady flow combustors. Since in diesel engines the in-cylinder measurements are often not readily possible, the empirical characteristic-time approach should be considered until such data become available.

The emissions data for a wide range of speeds and loads were available only for the 0.72 L engine employing the single-throat prechamber. Therefore, the discussion in this section applies to the 0.72 L engine only. In the present study an empirical scheme is developed which is based on the observed effect of simple engine parameters, such as engine speed (N), fuel flow rate $(\dot{m}_f)^*$, and the quantity of fuel injected per cycle (F) on exhaust emissions. To formulate the basis for this scheme, the effects of N, $\dot{m}_f$, and F on particulate carbon emissions are first considered separately.

Effect of Speed   – The effect of engine speed on particulate carbon emissions is illustrated in Fig. 29. For all load conditions tested, the exhaust carbon decreases monotonically with the engine speed. For a fixed engine geometry, the engine speed influences directly the mean air velocity (U) and turbulence intensity (u′) in the combustion space. The mean velocity affects the large-scale convective transport of air to the fuel-rich zones and of fuel to the air-rich zones. It also influences the spray trajectory and penetration distance. The turbulence intensity enhances the molecular diffusion between fuel and air by straining the eddies containing both. Wrinkling of the eddy surfaces increases the contact area and the concentration gradients across the isoscalar surfaces. An increase in U and u′ is likely to have opposite effects on formation and oxidation of soot. It would decrease formation by leaning out the soot formation zones and at the same time enhance oxidation by increasing O_2 concentration in the combustion zones. An increase in U and u' would also tend to decrease the characteristic mixing times for both formation and oxidation of soot. The engine speed is also expected to affect the mean cycle temperature because an increase in speed could enhance the heat transfer to the combustion chamber walls. However, in this case, the calculations indicated a relatively small change in the cycle temperature over a wide speed range, as shown

---

*In this study

$$\dot{m}_f = \frac{volume\ of\ fuel}{injection\ time} \left(\frac{mm^3}{ms}\right) = \quad \begin{array}{l} average\ fuel\ flow\ rate\ during \\ the\ actual\ injection\ process \end{array}$$

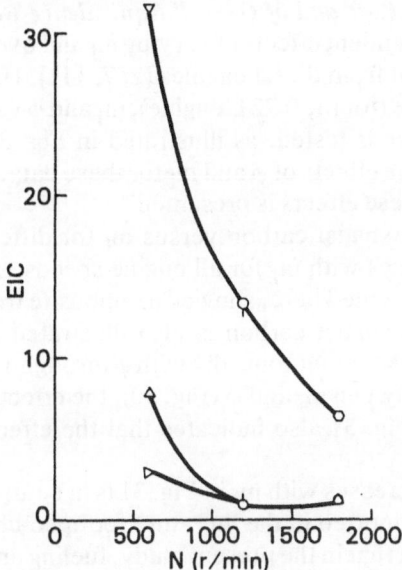

Fig. 29. Effect of engine speed on EIC at different overall equivalence ratios (no gas addition).

in Table 5. Thus, the effect of engine speed on exhaust soot for a particular engine geometry could be described as a combination of the effects of fuel dispersion and the characteristic mixing-time scales for formation and oxidation.

In a theoretical model for an open-chamber engine developed by Khan *et al.* [4], parameters representing diffusivity and entrainment of air into the fuel jet were used to describe the rate of heat release and fuel dispersion. Khan *et al.* found that it was necessary to increase the values of diffusivity and entrainment parameters to predict correctly the experimental trends, which showed a decrease in exhaust smoke with increasing speed. Their calculations indicated a reverse trend with speed for the exhaust NO. Henein [6] presented some data regarding the effect of speed on smoke emissions for several open- and divided-chamber diesel engines. The exhaust smoke for all the engines decreased with increasing speed for speeds up to 4000 r/min. Results for a divided-chamber diesel engine reported by Wilson *et al.* [7] indicate a load dependence on the effect of speed on exhaust soot. At equivalence ratios greater than about 0.45, the exhaust smoke decreased with increasing speed, but at lighter loads a reverse trend was observed. Furthermore, the trends for NO reported by Wilson *et al.* [7] agreed qualitatively with the findings of the present study as well as of Khan *et al.* [4] and Henein [6]. The experimental results for exhaust particulates reported by Hiroyasu and coworkers [68, 69] are somewhat in disagreement with the speed effects reported here. However, one of their theoretical models [69] does predict a decrease in smoke with speed. It should be pointed out that most of the literature data on exhaust particulates are based on potentially less accurate opacity measurements instead of total particulate mass measurements.

References pp. 477-481.

Effects of Fuel Flow Rate and of Overall Equivalence Ratio* — Several workers have studied the independent effects of varying $\dot{m}_f$ and overall equivalence ratio (ϕ) on the net exhaust soot from diesel engines [2, 7, 111]. However, for the injection system considered here (for the 0.72 L engine), $\dot{m}_f$ and ϕ varied simultaneously with load for all engine speeds tested, as illustrated in Fig. 30. Therefore, it was not possible to uncouple the effects of ϕ and $\dot{m}_f$ for these data. In the following sections a brief discussion of these effects is presented.

Fig. 31 is a plot of exhaust carbon versus $\dot{m}_f$ for different engine speeds. The exhaust carbon increases with $\dot{m}_f$ for all engine speeds examined. The results for NOx, although not presented here, showed an opposite trend. The effect of overall equivalence ratio on exhaust carbon is also illustrated in Fig. 31. The exhaust carbon emission increases monotonically with ϕ throughout the speed range tested. Due to the coupling between $\dot{m}_f$ and ϕ (Fig. 30), the effects of ϕ and $\dot{m}_f$ on EIC are qualitatively similar. Fig. 31 also indicates that the effect of ϕ on EIC is speed-dependent.

The fact that EIC increases with $\dot{m}_f$ in Fig. 31 is in contradiction with most of the results of the studies reported in the literature for open-chamber engines [3, 7, 29, 110, 111] due to the fact that in the present study, fueling and $\dot{m}_f$ changed simultaneously. Notably, in these studies, the rate of injection was varied maintaining constant fuel-per-cycle. Wilson et al. [7] have also reported that for a divided-chamber diesel engine the exhaust particulates increased by 120% when the fuel rate (mm³/° crankangle) was increased by 100%. They varied the rate of injection by changing the nozzle orifice size, which would affect the droplet size and the spray penetration and dispersion in the prechamber. Furthermore, it is not clear whether or not the total fuel injected per cycle was held constant in their experiments. Greeves [111] studied the effect of $\dot{m}_f$ maintaining a constant fueling level for both DI and IDI engines. His results indicate a decrease in the exhaust smoke with increasing $\dot{m}_f$ up to a certain injection rate, beyond which there was no further reduction in exhaust smoke. In contrast to Greeves results, a drastic increase in exhaust carbon with increasing $\dot{m}_f$ is reported here, which is clearly due to a simultaneous increase in the fueling level. Predictions of a mathematical model developed by Hiroyasu and Kadota [69] also indicate an increase in the exhaust soot with a simultaneous increase in $\dot{m}_f$ and ϕ. Fig. 31 also indicates that the effect of $\dot{m}_f$ is more significant at lower speeds. These trends could be explained as a result of excessive fuel impingement on the prechamber walls, which could lead to a poor fuel dispersion. Assuming that air-swirl momentum decreases faster than fuel-jet momentum with decreasing speed, impingement could be more significant due to a weaker swirl, resulting in a sharper rise in exhaust soot with $\dot{m}_f$ at lower engine speeds.

There could be several other contributing factors besides poor fuel dispersion due to impingement. For example, a variation in the nozzle opening area with $\dot{m}_f$ could affect spray fuel atomization and evaporation characteristics. In the present study, an outwardly opening poppet injector was used, for which no significant

---

**For no gas addition, overall equivalence ratio (ϕ) and the quantity of fuel injected per cycle are directly related and the terms are used interchangeably in the literature. However, these two will not be directly related when O_2 or N_2 are added to the intake air.*

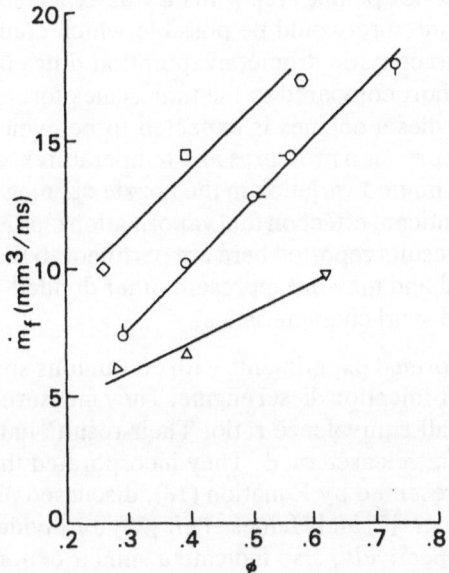

Fig. 30. Coupling of average fuel flow rate and overall equivalence ratio for different engine speeds (no gas addition).

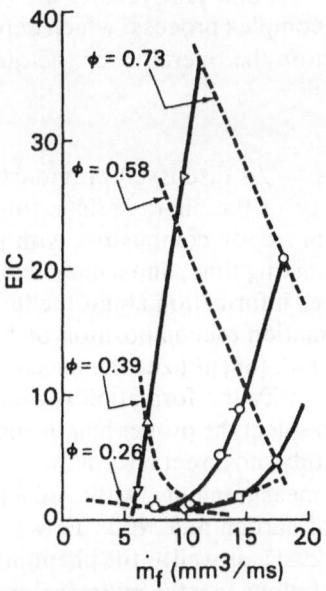

Fig. 31. Effect of average fuel flow rate and overall equivalence ratio on EIC for different engine speeds (no gas addition).

References pp. 477-481.

dependence of the nozzle opening area with $\dot{m}_f$ was expected. However, variations in the injection line pressure would be possible which could affect the drop-size distribution. The characteristic droplet evaporation times in gas turbine combustors are considered short compared to the time scales for mixing [63, 97, 99]. The evaporation time for diesel engines is expected to be even shorter, since diesels operate at higher compression pressures and temperatures, as well as higher injection pressures. Thus, limited variation in the nozzle opening area and line pressure should not have a significant effect on fuel vaporization characteristics. It should be emphasized that the results reported here are pertinent to the engine and the injection system employed and may not represent other divided-chamber engines with different injection and swirl characteristics.

Khan et al. [4] performed experiments with continuous sprays burning kerosene as well as with a direct-injection diesel engine. They measured the net soot released as a function of overall equivalence ratio. Their results indicated an exponential dependence of net soot released on ϕ. They incorporated these results in the soot formation model represented by Equation (18), discussed previously. The results reported by Wilson et al. [7] and Hames et al. [29] for divided-chamber and open-chamber engines, respectively, also indicate a similar exponential dependence of exhaust soot on ϕ. Although Hiroyasu and Kadota utilized a slightly different soot formation model than the one represented by Equation (18), the trends predicted by their calculations also agree with those of Khan et al. [4]. However, recent calculations of Hiroyasu and Arai [68] predict a leveling off of the exhaust smoke with increasing load which is in contradiction with most of the experimental data reported in the literature. It should be emphasized, however, that the soot formation correlations of Khan et al. [4] and Hiroyasu et al. [68, 69] represent a global oversimplification of a very complex process, which depends on local ϕ in the soot formation zones rather than on the overall ϕ or the quantity of fuel injected per cycle.

Correlations for Emissions — As discussed previously, Mellor and co-workers have demonstrated the utility of the characteristic-time approach in correlating NOx and CO emissions from steady combustors with parameters representing a droplet evaporation time, a mixing time, and a chemical kinetic time. The mixing-time characterization requires information about the length and velocity scales in the regions of pollutant formation (decomposition of NO is usually neglected in steady combustors). Tuttle et al. [63] utilized the measured concentration distributions for NOx and CO to identify the formation regions for these species. They employed this information to select the proper length and velocity scales for different species. In the present study, no direct information, such as high-speed movies or in-cylinder concentration measurements, was available to identify the regions of formation and destruction of carbon and NOx. However from the trends represented by Figs. 7, 14-16, and 29-31, as well as the phenomenological models of Khan et al. [4] and Henein [6], the following information emerges:

1. NO formation in diesel engines, as well as in gas turbines, occurs mainly in the reaction zone, i.e., zone b of Fig. 1 [63, 65-67, 97-99]. In contrast to gas

turbines, the decomposition of NO could be important in diesel engines, as the low overall activation energy suggests.

2. Soot formation occurs primarily in the fuel-rich region of the spray [2-4, 6, 68, 69]. Soot oxidation mainly occurs in the reaction zone, as suggested by the correlation between exhaust carbon and the flame temperature.

3. The processes of formation and oxidation of soot and CO are somewhat similar in the range of engine stoichiometries of interest in this study (Fig. 16).

NOx Emissions – In order to account for the decomposition of NO in diesel engines, the characteristic-time correlation of Tuttle *et al.* [63], namely Equation (5), should be modified by multiplying the formation term by a decomposition factor as follows:

$$\text{EINOx} \propto \frac{\left(\tau_{m, NO}\right)_f}{\left(\tau_{NO}\right)_f} \frac{\left(\tau_{NO}\right)_d}{\left(\tau_{m, NO}\right)_d} \tag{25}$$

where subscripts f and d denote formation and decomposition respectively. The kinetic-time scales $(\tau_{NO})_f$ and $(\tau_{NO})_d$ can be combined as an overall kinetic-time scale, τ_{NO}, which would be a function of the flame temperature, as shown previously. Thus, following Tuttle *et al.* [63], from Equation (14).

$$\tau_{NO} \propto \exp\left(38\ 700/T_f\right) \tag{26}$$

Due to the complex nature of the diesel flame and the lack of in-cylinder measurements, it is difficult to select the appropriate mixing-time scales for formation and destruction. Therefore, for simplicity, the mixing-time scales $(\tau_{m,NO})_f$ and $(\tau_{m,NO})_d$ are combined as an overall mixing-time scale, $\tau_{m,NOd}$. Thus, Equation (25) can be rewritten as follows:

$$\text{EINO}_x \propto \frac{\tau_{m,NO}}{\tau_{NO}} \propto \tau_{m,NO} \exp\left(-38\ 700/T_f\right) \tag{27}$$

In order to model $\tau_{m,\ NO}$ for a diesel engine, proper integral turbulence-length scales and turbulence-velocity scales must be selected. For a fixed combustion chamber geometry and injection system, the turbulence intensity, u', and the fuel injection velocity U_{inj} represent possible choices for the desired velocity scales. Measurements reported by Lancaster *et al.* [140] for a spark-ignition engine and Brandl [109] for a diesel engine show that the turbulence velocity can be approximated as a fraction of the mean velocity in the combustion chamber, which is roughly proportional to the engine speed for a fixed geometry [109, 141].

Thus, these velocity scales can be approximately related to simple engine parameters as follows:

$$u' \propto a N \tag{28}$$

and

$$U_{inj} \propto \frac{\dot{m}_f}{b A_{inj}} \tag{29}$$

where $\dot{m}_f$ is the average fuel flow rate during injection (mm³/ms) and A_{inj} is the nozzle opening area. All numerical constants have been lumped in a and b. The time scales based on the two velocities would be

$$(\tau_{m,NO})_1 \propto \frac{L_1}{a N} \tag{30}$$

and

$$(\tau_{m,NO})_2 \propto \frac{L_2 b A_{inj}}{\dot{m}_f} \tag{31}$$

L_1 and L_2 being the characteristic-length scales for the chamber and the injector respectively. NOx emissions from both direct-injection and indirect-injection diesel engines are significantly affected by N and $\dot{m}_f$, as discussed earlier. However, in many cases, these variables are coupled together. For example in the engine considered here, for a fixed amount of fuel delivered per cycle, the total injection duration was found to be somewhat dependent on the engine speed. Analysis of the data indicated that the effects of speed and fuel flow rate could be represented collectively by the parameter $N/\dot{m}_f$. Therefore, this parameter is possibly related to the overall mixing-time scale. Combining Equations (30) and (31), the following expression is obtained.

$$\tau_{m,NO} \propto \frac{N}{\dot{m}_f} \frac{C_2}{C_1} \tag{32}$$

where

$$C_2 = L_2 b A_{inj} \tag{33}$$

and

$$C_1 = \frac{L_1}{a} \tag{34}$$

Substitution of Equation (32) in Equation (27) yields the following correlation

$$EINO\,x = \frac{C_2}{C_1} \; \frac{N}{\dot{m}_f} \exp\left(-38\,700/T_f\right) \tag{35}$$

Fig. 32 illustrates the correlation for all load and speed conditions for the single-throat prechamber operating at a fixed start-of-combustion timing. For this engine and operating conditions, Equation (35) becomes Equation (36)

$$EINOx = C_{NOx} \; \frac{N^*}{\dot{m}_f^*} \exp\left(-38\,700/T_f\right) \tag{36}$$

where $C_{NOx} = 3.44 \times 10^6$ (g/kg fuel)

$$N^* = \frac{N}{N_{ref}}$$

$$\dot{m}_f^* = \frac{\dot{m}_f}{\dot{m}_{f_{ref}}}$$

and the normalizing parameters N_{ref} and $\dot{m}_{ref}$ are the engine speed and fuel flow rate for a reference condition (1200 r/min and 22 mm^3/ cycle).

Particulate Carbon – Equation (7) represents the characteristic-time model proposed by Dodds *et al.* [99] for smoke emission from gas turbines. The model suggests a linear relationship between the exhaust smoke and the times of interest. However, Figs. 29-31 indicate that it is improbable that such a linear relationship exists between EIC and the time scales, which in this case are directly related to N and $\dot{m}_f$. Furthermore, the engine data reported in the literature, as well as the findings of the present study, indicate an exponential dependence of exhaust soot on the quantity of fuel injected per cycle (F) or the overall equivalence ratio (Fig. 31). Therefore, empirical correlations of the exhaust soot should include the quantity of fuel injected per cycle also. For the particular engine and injection system of this study operating at a fixed start-of-combustion timing, the exhaust carbon can be represented empirically as follows:

$$EIC \propto N^p \; \dot{m}_f^{\;q} \; F^r \; \exp\left(E/RT\right) \tag{37}$$

where E is the overall activation energy for carbon emissions for the engine.

References pp. 477-481.

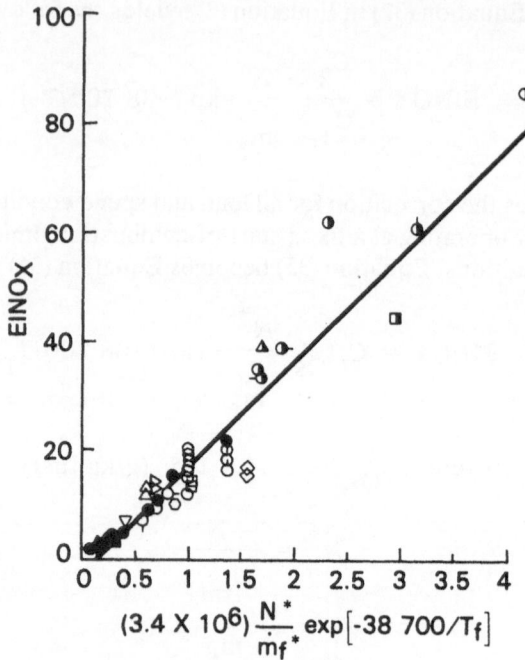

Fig. 32. Correlation of EINOx with flame temperature and air-fuel mixing for different engine speeds, fueling levels and intake O_2 mole fractions.

Analysis of the data presented in Figs. 29-31 indicated that the effects of speed (N), average fuel flow rate ($\dot{m}_f$) and the quantity of fuel injected per cycle (F) could be represented collectively by an empirical mixing parameter such that

$$\text{EIC} \propto \frac{\dot{m}_f^* \, F^{*3}}{N^{*2}} \tag{38}$$

when no gas is added to the intake air (values normalized to the same reference condition as in Equation 36). Note that the quantity of fuel injected per cycle has been used in the correlation rather than the overall equivalence ratio since ϕ does not remain constant when either O_2 or N_2 is added to the intake air. The correlation between the mixing parameter for particulate carbon and EIC (Equation 38) is illustrated in Fig. 33.

It should be noted that the parameter N^*/m_f^* was used to correlate the NO_x data. Fig. 15, which is a cross-plot between NO_x and exhaust carbon, indicates an approximately inverse relationship between the two. Thus, it is reasonable that the correlating parameter for exhaust carbon contains $\dot{m}_f^*/N^*$, which is the inverse of the parameter used for NO_x. Fig. 33 also indicates that EIC varies as the third power of F. Notably, Khan and Wang [2] also report an exponential dependence of exhaust smoke on ϕ for a direct-injection diesel engine operating at a fixed speed.

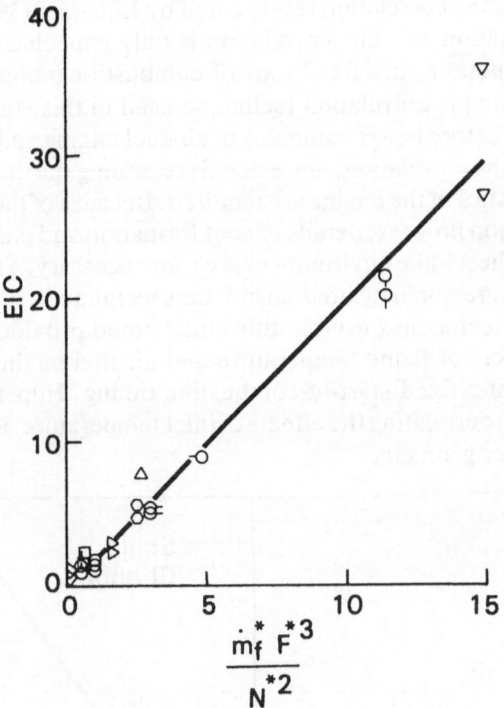

Fig. 33. Correlation of EIC with mixing parameter for different engine speeds and fueling levels.

Furthermore, Khan *et al.* [4] assigned the exponent of ϕ in Equation (18) a value of 3 to match the experimental results. However, soot formation and oxidation are local phenomena, and the correlation of the net exhaust soot with overall conditions in the combustion chamber in Equation (37) represents only a global-type empirical approach. Also, the success of the F^3 correlation depends, in part, on the validity of the choice of N and $\dot{m}_f$ as principal variables affecting EIC. In the present case, the use of a mixing parameter that includes N and $\dot{m}_f$ is suggested by the experimental observations for one particular engine. In general, however, it appears reasonable to argue that the exhaust soot for a particular engine and injection system operating at a fixed start-of-combustion timing could be represented empirically by Equation (37).

For the 0.72 L engine considered here

$$EIC = C_s \frac{\dot{m}_f^*}{N^{*2}} F^{*3} \exp\left(48\,900/T_f\right) \tag{39}$$

where $C = 5.50 \times 10^{-9}$ (g/kg fuel)

Fig. 34 illustrates the correlation represented by Equation (39). Like the correlation for NOx (Equation 36), this correlation is only expected to be valid for this particular engine operating at a fixed start-of-combustion timing.

Obviously, the simple correlation technique used in this study must be further refined. However, before better estimates of air-fuel mixing and flame temperature can be included in the correlation, more details regarding the fluid motion and heat-transfer characteristics of the engine are required. Because of the complex nature of the diesel combustion process, details of soot formation and oxidation kinetics in a more controllable diesel-like environment are also necessary. This information will allow the use of more sophisticated correlation techniques. For the interim, the simple correlation techniques used in this study could provide valuable tools for estimating the effects of flame temperature and air-fuel mixing on emissions for engines operating at a fixed start-of-combustion timing. Hopefully this approach will also be useful in correlating the effects of inlet temperature, injection timing and combustion chamber geometry.

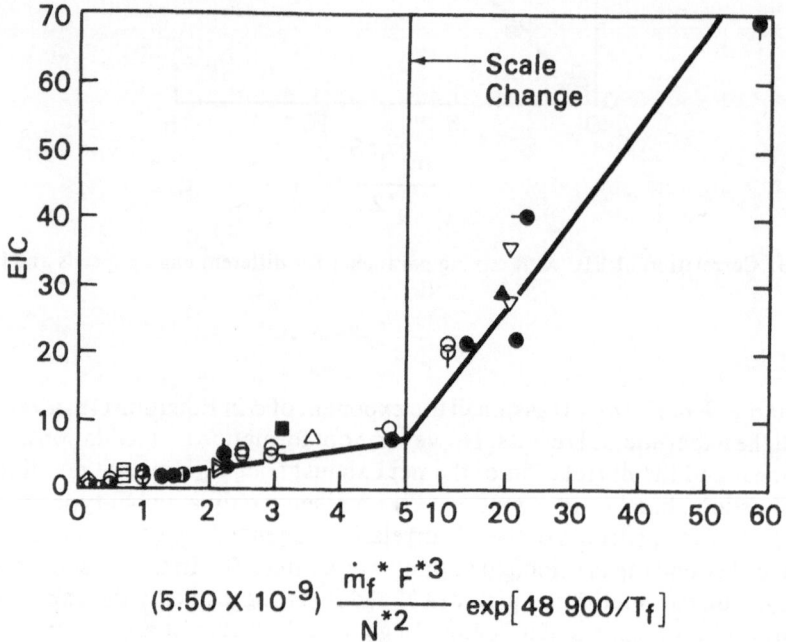

$$(5.50 \times 10^{-9}) \; \frac{\dot{m}_f{}^* \, F^{*3}}{N^{*2}} \; \exp\left[48\,900/T_f\right]$$

Fig. 34. Correlation of EIC with flame temperature and air-fuel mixing for different engine speeds, fueling levels and intake O_2 mole fractions.

SUMMARY

Exhaust emissions were obtained from three single-cylinder diesel engines operating at various intake-oxygen mole fractions (15 to 28 percent), engine speeds (600-1800 r/min), and fueling levels (14-41 mm³/cycle).

1. At all engine speeds and loads, both particulate carbon and CO emissions were found to decrease with an increase in oxygen mole fraction in the intake

air. However, NOx emissions exhibited opposite trends with intake-gas addition.

2. Under constant mixing conditions, the emissions for particulate carbon and NOx have been correlated with the stoichiometric adiabatic flame temperature. This correlation and the similarity demonstrated in the NOx-CO tradeoff between gas turbines and diesels provides new evidence that the basic combustion process in divided-chamber diesel engines is dominated by turbulent diffusion.

3. The flame-temperature correlation for particulate carbon indicates that the variation of intake-gas oxygen mole fraction primarily influences the oxidation of carbon particles and precursors. Furthermore, the one-to-one correspondence observed between particulate carbon and CO in this study provides indirect evidence to support the OH mechanism for particulate carbon oxidation. The apparent activation energy derived from the NOx correlation indicates that both NOx formation and destruction are important in diesel engines. This is in contrast to steady-flow combustors, where only NOx formation is important.

4. The apparent activation energies for particulate carbon and NOx were found to be independent of engine speed and overall air-fuel ratio, indicating that the kinetic mechanisms involved in the generation of these species were not altered significantly by these engine parameters.

5. Variations in emissions with engine load and speed are interpreted as changes in the air-fuel mixing process, since kinetics are unaffected by these parameters. In addition, the NOx-particulate carbon cross-plot has shown that air-fuel mixing affects particulate carbon and NOx in opposite directions. Particulate carbon emissions are correlated with an empirical mixing parameter involving average fuel flow rate, engine speed and the quantity of fuel injected per cycle. The mixing parameter for NOx emissions is related to average fuel flow rate and engine speed.

ACKNOWLEDGMENTS

The authors gratefully acknowledge the contributions of S. G. Wilson, S. R. McEwen and L. S. Jerzy (Engine Research Department) in the acquisition of the experimental data and of S. J. Swarin and W. R. Lee (Analytical Chemistry Department) for performing the TGA analyses. Special thanks are extended to Professor G. M. Faeth (The Pennsylvania State University) and Professor A. M. Mellor (Purdue University) for their valuable suggestions during the course of this investigation. The authors are also indebted to L. D. Sroda, R. J. Bayer, T. L. Card, A. H. Cook, and R. Stark for their contributions in preparing the manuscript.

REFERENCES

1. I. M. Khan, Proc. Instn. Mech. Engrs., Vol. 184, Pt. 3J (1969-70), pp. 36-43.
2. I. M. Khan and C. H. T. Wang, Proc. Instn. Mech. Engrs., C151/71 (1971), pp. 293-303.

3. I. M. Khan, G. Greeves and C. H. T. Wang, SAE Paper No. 730169, Part I, (1973).
4. I. M. Khan, G. Greeves and D. M. Probert, Proc. Instn. Mech. Engrs., C151/71 (1971), pp. 205-217.
5. I. M. Khan and G. Greeves, SAE Paper No. 730169, Part II, (1973).
6. N. A. Henein, Prog. Energy Combust. Sci., Vol. 1 (1976), pp. 165-207.
7. R. P. Wilson, E. B. Muir and F. A. Pellicciotti, SAE Paper No. 740123, (1974).
8. P. N. Blumberg, G. A. Lavoie and R. J. Tabaczynski, Prog. Energy Combust. Sci., Vol. 5 (1979), pp. 123-167.
9. H. Gg. Wagner, "Seventeenth Symposium (International) on Combustion," The Combustion Institute, Pittsburgh, (1979), pp. 3-19.
10. K. H. Homann, Combust. Flame, Vol. 11 (1967), pp. 265-287.
11. J. C. Street and A. Thomas, Fuel, 34 (1955), pp. 4-36.
12. H. B. Palmer and C. F. Cullis, "Chemistry and Physics of Carbon," ed. P. L. Walker, Jr., Marcel-Dekker Inc., New York, (1965), pp. 265-325.
13. A. G. Gaydon and H. F. Wolfhard, "Flames – Their Structure, Radiation and Temperature," Chapman and Hall Ltd., London, (1970).
14. J. S. McArragher and K. J. Tan, Combust. Sci. Tech., Vol. 5 (1972), pp. 257-261.
15. J. Lahaye and G. Prado, "Chemistry and Physics of Carbon," ed. P. L. Walker, Jr., and P. A. Thrower, Marcel-Dekker Inc., New York, (1978), pp. 167-295.
16. D. Anderton, U.S. Dept. of Transportation Report No. DOT-TSC-OST-76-57, (1977).
17. D. Broome and I. M. Khan, Proc. Instn. Mech. Engrs., C140/71 (1971), pp. 185-197.
18. C. A. Amann, D. L. Stivender, S. L. Plee and J. S. MacDonald, SAE Paper No. 800251, (1980).
19. I. Glassman, "Combustion," Academic Press, New York, (1977).
20. W. T. Lyn "Ninth Symposium (International) on Combustion," Academic Press, New York, (1963), pp. 1069-1082.
21. J. Rife and J. B. Heywood, SAE Paper No. 740948, (1974).
22. N. A. Chigier, Prog. Energy Combust. Sci., Vol. 2 (1976), pp. 97-114.
23. Y. Onuma and M. Ogasawara, "Fifteenth Symposium (International) on Combustion," The Combustion Institute, Pittsburgh, (1975), pp. 453-465.
24. Y. Onuma, M. Ogasawara and T. Inoue, "Sixteenth Symposium (International) on Combustion," The Combustion Institute, Pittsburgh, (1977), pp. 561-567.
25. I. M. Khan and G. Greeves, "Heat Transfer in Flames," ed. N. H. Afgan and J. M. Beer, John Wiley and Sons, New York, (1975), pp. 391-404.
26. V. K. Duggal, T. Priede and I. M. Khan, SAE Paper No. 780227, (1978).
27. H. H. Chiu and T. M. Liu, Combust. Sci. Tech., Vol. 17 (1977), pp. 127-142.
28. R. W. Temple-Pediani, Proc. Instn. Mech. Engrs., Vol. 187 (1973), pp. 395-404.
29. R. J. Hames, D. F. Merrion and H. S. Ford, SAE Paper No. 710671, (1971).
30. R. Pischinger and W. Cartellieri, SAE Paper No. 720756, (1972).
31. C. J. Walder, SAE Paper No. 730214, (1973).
32. S. R. Krause, D. F. Merrion and G. L. Green, SAE Paper No. 730213, (1973).
33. R. B. Melton, S. J. Lestz, R. D. Quillian and E. J. Rambie, "Fifteenth Symposium (International) on Combustion," The Combustion Institute, Pittsburgh, (1975), pp. 1389-1399.
34. G. P. Gross and K. E. Murphy, ASME Paper No. 78-DGP-26, (1978).
35. G. Greeves, I. M. Khan and G. Onion, "Sixteenth Symposium (International) on Combustion," The Combustion Institute, Pittsburgh, (1977), pp. 321-336.
36. G. Greeves, SAE Paper No. 790037, (1979).
37. W. F. Marshall and R. D. Fleming SAE Paper No. 710836, (1971).
38. R. F. Parker and J. W. Walker, SAE Paper No. 720755, (1972).
39. R. C. Bascom, L. C. Broering and D. E. Wulfhorst, SAE Paper No. 710484, (1971).
40. S. K. Ray and R. Long, Combust. Flame, Vol. 8 (1964), pp. 139-151.
41. D. W. Golothan, SAE Paper No. 670092, (1967).
42. C. O. Miller, SAE Paper No. 670093, (1967).
43. G. McConnell and H. E. Howells, SAE Paper No. 670091, (1967).

44. H. Tsunemoto and H. Ishitani, SAE Paper No. 800030, (1980).
45. U. Bonne, K. H. Homann and H. Gg. Wagner, "Tenth Symposium (International) on Combustion," The Combustion Institute, Pittsburgh, (1965), pp. 503-512.
46. F. J. Wright, "Twelfth Symposium (International) on Combustion," The Combustion Institute, Pittsburgh, (1969), pp. 867-875.
47. J. J. Macfarlane, F. H. Holderness and F. S. E. Whitcher, Combust. Flame, Vol. 8 (1964), pp. 215-229.
48. P. Dearden and R. Long, J. Appl. Chem., Vol. 18 (1968), pp. 243-251.
49. I. S. McLintock, Combust. Flame, Vol. 12 (1968), pp. 217-225.
50. K. P. Schug, Y. Manheimer-Timnat, P. Yaccarino and I. Glassman, Combust. Sci. Tech. Vol. 22 (1980), pp. 235-250
51. J. M. Jones and J. L. J. Rosenfeld, Combust. Flame, Vol. 19 (1972), pp. 427-434.
52. F. J. Wright, Fuel, Vol. 53 (1974), pp. 232-235.
53. B. B. Chakraborty and R. Long, Combust. Flame, Vol. 12 (1968), pp. 469-476.
54. I. Glassman and P. Yaccarino, "Eighteenth Symposium (International) on Combustion," The Combustion Institute, Pittsburgh, to be published (1981).
55. B. B. Chakraborty and R. Long, Combust. Flame, Vol. 12 (1968), pp. 226-242.
56. R. L. Schalla and G. E. McDonald, "Fifth Symposium (International) on Combustion," Reinhold, New York, (1955), pp. 316-324.
57. I. Glassman and P. Yaccarino, Combust. Sci. Tech., to be published (1981).
58. T. Kadota, H. Hiroyasu and A. Furazandehmehr, Combust. Flame, Vol. 29 (1977), pp. 67-75.
59. A. Sjogren, "Fourteenth Symposium (International) on Combustion," The Combustion Institute, Pittsburgh, (1973), pp. 919-927.
60. P. W. Cooper, R. Kamo, C. J. Marek and C. W. Solbrig, Amer. Petrol. Inst. Publ. No. 1723, (1964).
61. S. R. Gollahalli, Combust. Flame, Vol. 34 (1979), pp. 141-151.
62. A. A. Quader, SAE Paper No. 710009, (1971).
63. J. H. Tuttle, M. B. Colket, R. W. Bilger and A. M. Mellor, "Sixteenth Symposium (International) on Combustion," The Combustion Institute, (1977), pp. 209-219.
64. R. F. Sawyer, N. P. Cernansky and A. K. Oppenheim, "Atomospheric Pollution by Aircraft Engines," AGARD CP-125, Paper No. 22, (1973).
65. R. M. Washam and A. M. Mellor, AIAA J. Aircraft, Vol. 16 (1979), pp. 626-631.
66. A. M. Mellor, "Semi-Empirical Correlations for Gas Turbine Emissions, Ignition and Flame Stablization," Paper Presented at AGARD Propulsion and Energetics Panel, 54th Specialists' Meeting on Combustion Modeling, (1979).
67. A. M. Mellor and R. M. Washam, ASME Paper No. 79-GT-194, (1979).
68. H. Hiroyasu and M. Arai, "Development and Use of Spray Combustion Modeling to Predict Diesel Engine Efficiency and Pollutant Emissions," Paper Presented at Fifth International Automotive Propulsion Systems Symposium, (1980).
69. H. Hiroyasu and T. Kadota, SAE Paper No. 760129, (1976).
70. S. V. Patankar and D. B. Spalding, "Fourteenth Symposium (International) on Combustion," The Combustion Institute, Pittsburgh, (1973), pp. 605-614.
71. A. D. Gosman, W. M. Pun, A. K. Runchal, D. B. Spalding and R. Wolfshtein, "Heat and Mass Transfer in Recirculating Flows," Academic Press, London, (1969).
72. P. Hutchinson, E. E. Khalil, J. H. Whitelaw and G. Wigley, Journal of Heat Transfer, ASME, C98 (1976), pp. 276-283.
73. B. E. Launder and D. B. Spalding, "Mathematical Models of Turbulence," Academic Press, London, (1972).
74. R. W. Bilger, Prog. Energy Combust. Sci., Vol. 1 (1976), pp. 87-109.
75. B. F. Magnussen, "Fifteenth Symposium (International) on Combustion," The Combustion Institute, Pittsburgh, (1975), pp. 1415-1425.
76. W. H. Dalzell, G. C. Williams and H. C. Hottel, Combust. Flame, Vol. 14 (1970), pp. 161-170.

77. B. F. Magnussen and B. J. Hjertager, "Sixteenth Symposium (International) on Combustion," The Combustion Institute, Pittsburgh, (1977), pp. 719-729.

78. P. A. Tesner, T. D. Snegiriova and V. G. Knorre, Combust. Flame, Vol. 17 (1971), pp. 253-260.

79. P. A. Tesner, E. I. Tsygankova, L. P. Guilazetdinor, V. P. Zuyer and G. V. Loshakova, Combust. Flame, Vol. 17 (1971), pp. 279-285.

80. B. F. Magnussen, B. H. Hjertager, J. G. Olsen and D. Bhaduri, "Seventeenth Symposium (International) on Combustion," The Combustion Institute, Pittsburgh, (1979), pp. 1383-1393.

81. H. A. Becker and S. Yamazaki, "Sixteenth Symposium (International) on Combustion," The Combustion Institute, Pittsburgh, (1977), pp. 681-691.

82. H. A. Becker, H.C. Hottel and G. C. Williams, "Ninth Symposium (International) on Combustion," Academic Press, New York, (1963), pp. 7-20.

83. H. A. Becker, H. C. Hottel and G. C. Williams, "Tenth Symposium (International) on Combustion," The Combustion Institute, Pittsburgh, (1965), pp. 1253-1263.

84. S. T. R. Rao and R. J. Essenhigh, "Thirteenth Symposium (International) on Combustion," The Combustion Institute, Pittsburgh, (1971), pp. 603-615.

85. M. A. Zeinalov, M. Kuwata and R. J. Essenhigh, "Fourteenth Symposium (International) on Combustion," The Combustion Institute, Pittsburgh, (1973), pp. 575-583.

86. R. J. Essenhigh, "Combustion Technology: Some Modern Developments," Academic Press, New York, (1974), pp. 373-415.

87. L. A. Vulis, "Thermal Regimes of Combustion," McGraw-Hill, New York, (1960).

88. J. Swithenbank, I. Poll, M. W. Vincent and D. D. Wright, "Fourteenth Symposium (International) on Combustion," The Combustion Institute, Pittsburgh, (1973), pp. 627-638.

89. B. F. Magnussen, "Fourteenth Symposium (International) on Combustion," The Combustion Institute, Pittsburgh, (1973), pp. 553-566.

90. N. A. Chigier, Prog. Energy Combust. Sci., 3 (1977), pp. 175-189.

91. J. E. Bennethum, J. N. Mattavi and R. R. Toepel, SAE Paper No. 750849, (1975).

92. N. A. Chigier, "Experimental Diagnostics in Gas Phase Combustion Systems," AIAA, New York, (1977), pp. 337-356.

93. A. Ungut, A. J. Yule, D. S. Taylor and N. A. Chigier, AIAA J. Energy, Vol. 2 (1978), pp. 330-336.

94. N. A. Chigier, A. Ungut and A. J. Yule, "Seventeenth Symposium (International) on Combustion," The Combustion Institute, Pittsburgh, (1979), pp. 315-324.

95. D. L. Hartley, "Experimental Diagnostics in Gas Phase Combustion Systems," AIAA, New York, (1977), pp. 467-477.

96. J. H. Tuttle, R. A. Shisler, R. W. Bilger and A. M. Mellor, Purdue University Combustion Laboratory Report PURDU-CL-75-04, (1975).

97. J. H. Tuttle, M. B. Colket and A. M. Mellor, Purdue University Combustion Laboratory Report PURDU-CL-76-05, (1976).

98. A. M. Mellor, Prog. Energy Combust. Sci., Vol. 1 (1976), pp. 111-113.

99. W. J. Dodds, M. B. Colket and A. M. Mellor, Purdue University Combustion Laboratory Report PURDU-CL-76-06, (1976).

100. T. Kadota, N. A. Henein and D. U. Lee, "A New Approach to Study the Formation of Soot Particulates in Diesel Sprays," Paper Presented at the Spring Technical Meeting, Combustion Institute Central States Section, (1980).

101. G. M. Faeth, Prog. Energy Combust. Sci., Vol. 3 (1977), pp. 191-224.

102. K. T. Rhee, P. S. Myers and O. A. Uyehara, SAE Paper No. 780226, (1978).

103. D. R. Nightingale, SAE Paper No. 750848, (1975).

104. W. D. Whitehouse, E. Clough and P. S. Roberts, SAE Paper No. 770409, (1977).

105. Y. Matsui, T. Kamimoto and S. Matsuoka, SAE Paper No. 790491, (1979).

106. Y. Matsui, T. Kamimoto and S. Matsuoka, SAE Paper No. 800970, (1980).

107. S. Matsuoka, T. Kamimoto, Y. Matsui and Y. Aogi, "An Introduction to the Research on Combustion Mechanism and Formation-Extinction Processes of NOx and Soot-

Particulates in a Direct Injection Diesel Engine," Paper Presented at the Thirteenth International Congress of Automobile Technology, (1980).

108. C. J. Morris and J. C. Dent, Proc. Inst. Mech. Engrs., 190 47/76, (1976), pp. 503-513.

109. F. Brandl, I. Reverencic, W. Cartellieri and J. C. Dent, SAE Paper No. 790040, (1979).

110. J. C. Dent, SAE Paper No. 800092, (1980).

111. G. Greeves, SAE Paper No. 790037, (1979).

112. J. S. MacDonald, S. L. Plee, J. B. D'Arcy and R. M. Schreck, SAE Paper No. 800185, (1980).

113. R. L. Williams and D. P. Chock, *"Characterization of Diesel Particulate Exposure,"* Paper Presented to the International Symposium on Health Effects of Diesel Engine Emissions, December 3, Cincinnati, Ohio, (1979).

114. D. R. Lancaster, R. B. Krieger and J. H. Lienesch, SAE Paper No. 750026, (1975).

115. R. V. Fischer and J. P. Macey, SAE Paper No. 750028, (1975).

116. L. E. Frisch, J. H. Johnson and D. G. Leddy, SAE Paper No. 790417, (1979).

117. F. Black and L. High, SAE Paper No. 790422, (1979).

118. F. J. Verkamp, A. J. Verdouw and J. G. Tomlinson, AIAA J. Aircraft, Vol. 11 (1974), pp. 340-344.

119. S. Gordon and B. J. McBride, NASA Report No. SP-273, (1971).

120. L. S. Caretto, Prog. Energy Combust. Sci., Vol. 1 (1976), pp. 47-71.

121. J. P. Appleton and J. B. Heywood, *"Fourteenth Symposium (International) on Combustion,"* The Combustion Institute, Pittsburgh, (1973), pp. 777-786.

122. D. E. Jensen and G. A. Jones, Combust. Flame, Vol. 32 (1978), pp. 1-34.

123. J. B. Heywood, *"Fifteenth Symposium (International) on Combustion,"* The Combustion Institute, Pittsburgh, (1975), pp. 1191-1211.

124. P. Blumberg and J. T. Kummer, Combust. Sci. Tech., Vol. 4 (1971), pp. 73-95.

125. W. R. Aiman, *"Fourteenth Symposium (International) on Combustion,"* The Combustion Institute, Pittsburgh, (1973), pp. 861-868.

126. H. L. Green and W. R. Lane, *"Particulate Clouds: Dusts, Smokes and Mists,"* D. VanNostrand Co., Princeton, New Jersey, (1957).

127. K. B. Lee, M. W. Thring and J. M. Beer, Combust. Flame, Vol. 6 (1962), pp. 137-145.

128. P. A. Tesner and A. M. Tsibulevsky, Combust., Expl. and Shock Waves, 3 (1969), pp. 163-167.

129. C. Park and J. P. Appleton, Combust. Flame, Vol. 20 (1973), pp. 369-379.

130. J. Nagle and R. F. Strickland-Constable, *"Proceedings of the Fifth Carbon Conference,"* 1 (1962), pp. 154-164.

131. F. J. Wright, *"Fifteenth Symposium (International) on Combustion,"* The Combustion Institute, Pittsburgh, (1975), pp. 1449-1460.

132. C. P. Fenimore and G. W. Jones, J. Phys. Chem., Vol. 71 (1967), pp. 593-597.

133. K. G. Neoh, J. B. Howard and A. F. Sarofim, *"Particulate Carbon – Formation During Combustion,"* Plenum Press, New York, this volume, p. 261.

134. P. A. Tesner and A. M. Tsibulevsky, Combust. Flame, Vol. 11 (1967), pp. 227-233.

135. J. P. Appleton, *"Atmospheric Pollution by Aircraft Engines,"* AGARD CP-125, Paper No. 20, (1973).

136. N. M. Laurendeau, Prog. Energy Combust. Sci., Vol. 4 (1978), pp. 221-270.

137. R. M. Fristrom and A. A. Westenberg, *"Flame Structure,"* McGraw-Hill, New York, (1965).

138. H. Tennekes and J. L. Lumley, *"A First Course in Turbulence,"* The MIT Press, Cambridge, Massachusetts, (1972).

139. J. H. Kent and R. W. Bilger, *"Fourteenth Symposium (International) on Combustion,"* The Combustion Institute, Pittsburgh, (1973), pp. 615-625.

140. D. R. Lancaster, SAE Paper No. 760159, (1976).

141. S. Ohigashi, Y. Hamamoto and S. Tanabe, Proc. Instn. Mech. Engrs., C134/71, (1971), pp. 129-136.

APPENDIX

TABLE A-1

	0.72 L Engine	0.52 L Engine
Compression Ratio	21.4	22.0
Bore (mm)	103.1	88.0
Stroke (mm)	85.7	85.0
Displacement (mL)	717.0	517.0
Prechamber Volume (mL)	15.0	11.8

TABLE A-2
Operating Conditions for the 0.72 L Engine

Fuel Rate (mm³/cycle)	Speed r/min	ϕ^a	Inj Timing DBTDC	Gas Type	% Gas Added[b] (Nominal)	IMEP (kPa)	Geometry 1T/2T[c]	P-t Data
36.0	600	0.62	10	—	—	320	1T	*
22.3	600	0.39	7	—	—	440	1T	*
15.2	600	0.28	5	—	—	560	1T	*
21.9	600	0.46	7	N₂	15	420	1T	
23.5	600	0.34	7	O₂	6	445	1T	
40.4	1200	0.73	9	—	—	720	1T	*
22.3	1200	0.39	7	—	—	510	1T/2T	*
15.9	1200	0.29	6	—	—	340	1T	*
22.3	1200	0.40	7	N₂	3	510	1T	
22.3	1200	0.41	7	N₂	6	510	1T	
22.3	1200	0.43	7	N₂	9	510	1T	
40.7	1200	0.84	9	N₂	15	660	1T	
22.0	1200	0.46	7	N₂	15	500	1T/2T	*
14.3	1200	0.30	6	N₂	15	350	1T	
22.7	1200	0.55	7	N₂	30	490	1T/2T	
32.3	1200	0.56	7	—	—	650	1T	
28.4	1200	0.50	7	—	—	610	1T	
32.3	1200	0.66	7	N₂	15	640	1T	
28.7	1200	0.58	7	N₂	15	610	1T	
22.5	1200	0.36	7	O₂	3	500	1T	
22.7	1200	0.32	7	O₂	5	500	1T/2T	*
22.5	1200	0.30	7	O₂	9	500	1T	
40.9	1200	0.62	9	O₂	5	750	1T	
15.4	1200	0.21	6	O₂	5	350	1T	
32.6	1200	0.43	8	O₂	5	660	1T	
28.6	1200	0.39	7	O₂	5	610	1T	
22.3	1800	0.39	8	—	—	520	1T	
34.3	1800	0.58	9	—	—	700	1T	
14.8	1800	0.27	7	—	—	370	1T	
22.1	1800	0.49	8	N₂	15	550	1T	
22.6	1800	0.32	7	O₂	6	530	1T	

^{a}*The equivalence ratio is defined as:*

$$\phi = \frac{\text{Fuel/Oxidizer}}{(\text{Fuel/Oxidizer})_{\text{stoichiometric}}} \quad \left(\begin{array}{l}\text{oxidizer includes } O_2 \text{ and } N_2 \\ \text{entering cylinder}\end{array}\right)$$

^{b}*The fraction of gas added to the intake air is defined as:*

$$\% \text{ added} = \frac{\dot{m}_{\text{gas added}}}{\dot{m}_{\text{gas added}} + \dot{m}_{\text{air}}} \times 100$$

where $\dot{m}$ = mass flow rate

^{c}*1T and 2T denote single-throat and 2-throat prechamber geometries, respectively.*

*Denotes operating conditions at which P-t data were recorded (single-throat geometry only).

TABLE A-3
Operating Conditions for the 0.52 L Engine

Fuel Rate (mm³/cycle)	Speed r/min	ϕ^a	Inj Timing DBTDC	Gas Type	% Gas Added^b (Nominal)	IMEP (kPa)
21.2	1500	0.56	4	—	—	597
21.2	1500	0.61	4	N_2	9	594
21.2	1500	0.66	4	N_2	15	592
21.2	1500	0.80	4	N_2	30	544
21.2	1500	0.51	4	O_2	3	590
21.2	1500	0.47	4	O_2	6	587
21.2	1500	0.43	4	O_2	9	581

^{a}*The equivalence ratio is defined as:*

$$\phi = \frac{\text{Fuel/Oxidizer}}{(\text{Fuel/Oxidizer})_{\text{stoichiometric}}} \quad \left(\begin{array}{l}\text{oxidizer includes } O_2 \text{ and } N_2 \\ \text{entering cylinder}\end{array}\right)$$

^{b}*The percent of gas added to the intake air is defined as:*

$$\% \text{ added} = \frac{\dot{m}_{\text{gas added}}}{\dot{m}_{\text{gas added}} + \dot{m}_{\text{air}}} \times 100$$

where $\dot{m}$ = mass flow rate

DISCUSSION

I. Glassman *(Princeton University)*

First, I would like to congratulate you for taking an engine-type paper and trying to make as fundamental a correlation as you can possibly make of it. I think we should all be trying to do that. I would like to comment further that I don't think it's necessary to conclude that when you add nitrogen that the carbon necessarily has to go up. Your correlations only occur with the type of experiment that you have done because basically if you add nitrogen as you so nicely concluded it was affecting the burnout time. If somehow we could get the nitrogen in the fuel pockets, then indeed you could alter the temperature in the pyrolysis zone and also decrease NOx and carbon simultaneously. Now you can't do it with nitrogen, but if you could do as Fred Dryer is doing with his emulsified fuels and use an emulsified fuel, you could put the inert component in the fuel itself. Then when the fuel pocket forms, it would alter the flame temperature, and the other correlation we talked about would occur and therefore decrease both NOx and carbon. We haven't done it. No one has done it in the engine yet. But, so you then agree with that?

Plee

Yes, we agree with that. It's always a competition between formation and oxidation, and that's controlled by the mixing and where you put in these different diluents.

C. T. Bowman *(Stanford University)*

Steve, I want to explore this mixing parameter which you used to correlate the NOx emissions. I wrote it down, and I think I have it correctly. You said that you were going to correlate the emissions index in terms of N/m_f.

Plee

Right.

Bowman

I think that may be a truism because the definition of emissions index is $\dot{m}_{NO}/\dot{m}_f$ and is proportional to N/m_f. Then the m_f's cancel out, and we come to the conclusion that the mass flow rate of NO which is being produced is proportional to the amount of mass that you're pushing through the engine or, in other words, the mole fraction is going to be a function of the temperature only. So, is that really a mixing parameter?

Plee

Well it's really the quantity of fuel injected per degree crank angle. It depends upon how fast you put the fuel in. The duration of injection is also important and that's in there too.

C. A. Amann *(General Motors Research Laboratories)*

I wanted to respond to Professor Glassman's comment about adding water. We have indeed added water to our fuel, and it does decrease particulates and NOx. But it does one thing that those of us who have to worry about selling engines must be concerned about. It also increased hydrocarbon emissions to a degree that's intolerable in our experiments so far.

Glassman

It is still good because it shows that the concept we have is probably correct. I would agree; the hydrocarbons would probably increase.

F. L. Dryer *(Princeton University)*

Mr. Amann, is the engine a two-stroke or a four-stroke engine?

Amann

Four-stroke, indirect injection. The high swirl engine in the paper just presented.

J. C. Hilliard *(University of Michigan)*

When you are doing these experiments, do you measure the brake specific fuel consumption at the same time? I'm particularly interested in mechanisms that affect soot emissions and how they modify the thermal efficiency of the engine, the radiative losses, convective losses, pumping losses, etc.

Plee

We do monitor the thermal efficiency but not the others.

T. Ahmad *(General Motors Research Laboratories)*

The thermal efficiency does not change much.

A. S. Gordon *(California Air Resources Board)*

I'd like to ask you one thing. In a gas turbine, I think you said, NO activation energy correlated very well with stoichiometric flame temperature.

Plee

Right.

Gordon

But in your case it did not?

Plee

No, it still correlates with flame temperature, but you get a different slope.

Gordon

So you have a different energy of activation?

Plee

Yes, because what is happening is NO decomposition is also important.

Gordon

So then if you are getting decomposition, that is a thermal process. Then you would also be getting NO formation by a thermal process?

Plee

The apparent activation energy has both formation and decomposition in it. That is why it is between the pure formation trend and the equilibrium trend.

Gordon

Then it should depend on the amount of oxygen that is present during that period.

Plee

You can account for the oxygen concentration in there, but it doesn't change the slope significantly. The temperature dependence is much more important than the oxygen dependence.

Dryer

I'd like to make some comments about the question of water addition to diesels. You people have mentioned that that increases hydrocarbon emissions substantially, and that is true. It does that because water does more than just dilute the interior of the fuel-rich region. It also has a very drastic affect on the ignition delay of the fuel spray. I believe that if one chemically modifies the ignition delay or chemically modifies the fuel so that the ignition delay is similar to that of the pure fuel, you will find that you cannot only reduce NOx but also simultaneously reduce particulates and not see substantial increases in HC and CO.

Amann

I don't know as this adds or subtracts from what you said. In our particular experiments we varied injection timing to always develop maximum engine power. That's important because many people run this experiment without changing engine timing. The addition of water does affect ignition delay, and then when they do that, they are looking at the effect of two things simultaneously, timing and water injection.

Dryer

You were also looking at something simultaneously, in a sense, as you varied the injection timing to correct for the change of ignition delay. You now have a long period of time for mixing between the start-of-injection and the time of maximum pressure which will cause the highest temperature regions to last for the longest period of time. That automatically enhances NO.

Amann

Yes, but if we tried to alter the fuel chemically with an additive, which I think is what you are getting at, we are also changing two things at once.

Dryer

That's correct.

Amann

I'm not saying you shouldn't do that. You just can't change one thing at once.

SYMPOSIUM SUMMARY

H. B. PALMER
Pennsylvania State University
University Park, Pennsylvania

To begin this look at what has taken place these two days, I think the most profound thing to offer is an expression of gratitude to George Smith and Don Siegla on behalf of all of us here for organizing this most excellent symposium. The presentations and the discussions have been very stimulating, and the arrangements and facilities have been superb. Since one of their colleagues likes to quote Yogi Berra, I would say to Don and George that we all thank you for making this day necessary.

We have heard an impressive array of subjects related to soot formation, beginning with a perceptive and extremely helpful overview from Professor Wagner. The first topical session was on characterization. One of the interesting debates that it generated was, in the words of Dr. Mellor, the controversy over what we call what. This was in reference to whether carbon blacks and soot are really very much alike. Some people who make and study carbon blacks think they are not. There surely is more to be explored on that; however, my own persuasion is to go along with Dr. Lahaye, who thinks that the similarities are sufficiently extensive that they can provide added insight into the formation and properties of soot.

We also heard for the better part of a day papers on what were termed fundamentals, basics, basic mechanisms, and kinetics, with a number of fine presentations that I shan't summarize. All of this is vivid enough in our minds. Let me just remind you that presentations covered a spectrum from nucleation to growth to oxidation, and raised some important questions. I do intend to say a word or two about some of those questions.

Today has been devoted to the more practical aspects of soot and sooting, but not losing sight of the fundamentals. We discussed soot formation during continuous combustion this morning and then this afternoon the milieu was intermittent combustors. Overall, there has been an appropriately broad coverage of the wide-ranging subject of this symposium. And, also appropriately, this meeting has brought together people from numerous disciplines. It appears to me that this has resulted in some very effective cross-fertilization.

About fifteen years ago, Charles Cullis and I wrote a review of carbon formation in which we suggested that a useful goal in this field would be to reach the point where one could describe the process of carbon formation from a selected chemical compound (e.g., CH_4) under well-defined aerodynamic conditions, from start to finish. Clearly, we are not there yet, but progress is impressive. There has been excellent recent progress in both theory (by which I mean primarily detailed analytical modeling) and experiments — particularly the latter. The theory is perhaps lagging a bit at this point. The theorists, however, will justly remark that their prime need is better experimental data!

The pyrolysis and oxidation of hydrocarbons is a subject that was not discussed very much at this meeting but which is intimately part of this whole scheme of things. Progress there is good, in the case of small aliphatic species such as methane. A useful immediate goal might be to bring the understanding of benzene and toluene pyrolysis and oxidation to a comparable level. From this point one moves on to talking about formation of larger species. There is some very nice new work there that we heard about, concerning the formation of PAH or PCAH or whatever you want to call it (or them). The formation of these larger aromatic species that may move on then in the direction of forming soot is very complex, crucially important, and fascinating. The species must be intimately related to soot nuclei. Not as much is known about the nuclei (if one can identify them) as one might wish, and that is an area for work. But certainly in the realm of particle growth, agglomeration, and surface deposition there is good progress even if we are not sure what is depositing on the surface; at least some measurements are being made on how fast the surface grows.

Advances in measurements of particle and agglomerate characteristics are very, very promising. New techniques such as phase contrast microscopy have helped; also light scattering techniques and the introduction of laser systems in many optical diagnostics. All of these are leading in an exciting way toward new information.

There is further insight into the presence and the role of charged species in these sooting systems as epitomized in the paper of Olson and Calcote. It is good that we are still thinking openmindedly and perceptively but critically about this question of the relative roles of ions and neutral species in sooting.

Then there is a matter of burnout. Neoh, Howard, and Sarofim presented results of some very nice soot oxidation measurements that offer persuasive evidence that OH is the predominant soot oxidizer when systems are other than very O_2-rich. (Maybe it should be noted that in many practical systems, soot burnout does in fact take place under O_2-rich conditions.)

On the basis of this meeting in particular, and from other recent discussions, overall progress appears to be good, especially considering the modest level of funding of soot formation research. However, until that goal that I described is achieved with regard to the decomposition, oxidation, or oxidative pyrolysis, formation of nuclei, growth, agglomeration, and burnout — until that description is available, it seems to me that it would be difficult to couple an analytical model of soot chemistry and physics into a full model with any confidence; that is, couple the (soot) chemistry with gas dynamics and transport processes as is now being done in some homogeneous systems where the chemical mechanisms and kinetics are well understood.

There are several specific points that I might offer about what we have experienced here. They mainly concern research areas that seem to merit emphasis.

1. A detailed description of the nucleation process is very desirable. I am talking here about the chemistry: the detailed chemistry including rates and including identification of the species that are involved. Again, it may be wise to examine a particular system because "the process" undoubtedly is not identical in all systems. (It may not be identical in *two* systems!)

2. I have been very pleased to hear, especially today, interpretations of observations on practical systems such as engines in terms of fundamental knowledge, fundamental information about soot nucleation, growth, and burnout including the mixing and aerodynamics implicit in all this. Nevertheless, there is much room for further development in that direction. One of the things that should be facilitated by this kind of meeting is interaction that will advance that sort of interpretation. Interaction should occur among those who are, say, interested primarily in the physics and chemistry of these processes, those who are concerned about the gas dynamics, and those who are designing engines to run clean or designing furnaces.

3. Modeling: we have heard about some modeling efforts and there are differences of opinion about this that are stimulating. We should welcome detailed analytical modeling where and when the appropriate information is available and I am sure that should be a goal. However, I am also a strong supporter of "operational" kinetics or "global" kinetics, i.e. correlations that are useful over a well-defined range of parameters so that one can apply these correlations, these operational descriptions, in design work. We need to have more experimental information of that type; semi-empirical descriptions of sooting phenomena over a range of very carefully defined, very well controlled parameters such that these descriptions can be incorporated into mathematical models. In saying this I am in no way denigrating fundamental work; I strongly favor it. But we must "make do" with semi-empirical data in systems that are too complex for rigorous fundamental description at present. As an illustration, think of the early work of Lee, Beer, and Thring on the oxidation of soot, which has triggered much excellent work on a more basic level and at the same time has been useful in itself in matters such as furnace and engine design.

Another good example is the study of Narasimhan and Foster of soot growth rates that was published about fifteen years ago. It was essentially an empirical description in terms of readily measurable parameters and I am surprised that there has not been more research of that type. Always, one must define the conditions and limitations of the description very clearly. The work of Dr. Magnussen, reported today, is a good illustration of how one should go about modeling systems and making the appropriate experiments to establish the range of validity of the models.

4. Obviously there is a need for thorough study of the chemistry of aromatics in the gas phase. It seems to me that there are some things that could be done directly. It should be possible, for instance, to measure the rate of OH plus benzene in a discharge-flow system. I don't see any barrier to that. One should be able to follow the chemistry further and begin to sort it out. There should be considerable effort in that direction.

5. We need to identify the surface addition reactions that cause the growth of carbon particles. They are still in the realm of mystery.

6. We need to settle the controversy about the concentration, the absolute concentration, of ions in rich flames. It is essential to get the correct answer to that problem. I am not enough of an expert on that subject to be very critical about it, but surely it ought to be answerable. So should the question as to whether or not there are chemi-ions in the interior of diffusion flames. Are there chemi-ions in purely pyrolytic systems? That should be even more susceptible to study. There have been some studies of these questions by distinguished investigators such as Weinberg, but the problems still seem to me to be not entirely solved. I hope they receive further attention.

7. Why do particles stop growing? There are some general ideas about this but much study is needed. The modeling efforts in that direction are interesting, but I am sure that we need to know much more about the fundamental physical and chemical processes that occur as particles age. If one could delineate the kinetic and mechanistic processes involved, it should then be feasible to construct a model that describes the average size and the size distribution of particles as they grow old.

There are other things that one could discuss here, but let me finish by pointing out a couple of examples of research in this general area that failed to excite people at the time that they were published but which are now finding much applicability and interest. I refer first to the detailed studies by Ronald Long and his co-workers at Birmingham begun, I suppose, twenty years ago or more, on the polynuclear aromatic hydrocarbons or PCAH. Their careful studies in a variety of flame systems produced a vast array of information that fortunately was conscientiously put into the literature by Dr. Long and his colleagues. Researchers are only now beginning to fully appreciate that work. I think another example may be the very impressive study shown in a slide by Professor Wagner, the 1970 study by Kern and Spengler of a hexane-air diffusion flame which shows enormous and surely very valuable detail in the analytical results on that system.

The message here is simply this: that this is a field where developments have been very diverse, and one cannot be sure in advance of what may be crucially important. If you think that a careful experimental study of some combustion system might have a useful bearing on particulate formation, and if you think it is interesting but no one else gets excited about it or thinks it is worth doing but you *do*, chances are that you are right and you ought to do it. Your results may become very significant, more likely sooner than later. Predictions, as Mark Twain said, are difficult — especially when they involve the future. That is very much the case in this complicated and fascinating field.

PARTICIPANTS

Agarwal, P. D.
General Motors Research Laboratories
Warren, Michigan

Agnew, W. G.
General Motors Research Laboratories
Warren, Michigan

Ahmad, T.
General Motors Research Laboratories
Warren, Michigan

Alkidas, A. C.
General Motors Research Laboratories
Warren, Michigan

Amann, C. A.
General Motors Research Laboratories
Warren, Michigan

Ballal, D. R.
General Motors Research Laboratories
Warren, Michigan

Bauer, S. H.
Cornell University
Ithaca, New York

Beaman, R. T.
General Motors Research Laboratories
Warren, Michigan

Bechtel, J. H.
General Motors Research Laboratories
Warren, Michigan

Beer, J. M.
Massachusetts Institute of Technology
Cambridge, Massachusetts

Bidwell, J. B.
General Motors Research Laboratories
Warren, Michigan

Binder, K.
Daimler-Benz Ag
Stuttgart, West Germany

Bissett, E. J.
General Motors Research Laboratories
Warren, Michigan

Bittner, J. D.
Massachusetts Institute of Technology
Cambridge, Massachusetts

Blint, R. J.
General Motors Research Laboratories
Warren, Michigan

Bonne, U.
Honeywell Inc.
Bloomington, Minnesota

Borman, G. L.
University of Wisconsin
Madison, Wisconsin

Bowman, C. T.
Stanford University
Stanford, California

Bradow, R. L.
U.S. Environmental Protection Agency
Research Triangle Park, North Carolina

Brown, M.
Exxon Research and Engineering Company
Linden, New Jersey

Bryzik, W.
U.S. Army Tank-Automotive Command
Warren, Michigan

Burgess, A. J.
Lucas Industries, Inc.
Troy, Michigan

Burley, H. A.
Engineering Staff, GMC
Warren, Michigan

Butler, F. G.
Oldsmobile Division, GMC
Lansing, Michigan

Buzan, L. R.
General Motors Research Laboratories
Warren, Michigan

Cadle, S. H.
General Motors Research Laboratories
Warren, Michigan

Caplan, J. D.
General Motors Research Laboratories
Warren, Michigan

Cernasky, N. P.
Drexel University
Philadelphia, Pennsylvania

Chenea, P. F.
General Motors Research Laboratories
Warren, Michigan

Chigier, N. A.
University of Sheffield
Sheffield, England

Cole, R. M.
General Motors Research Laboratories
Warren, Michigan

Colucci, J. M.
General Motors Research Laboratories
Warren, Michigan

Countess, R. J.
General Motors Research Laboratories
Warren, Michigan

D'Alessio, A.
University of Naples
Naples, Italy

Dawson, J.
Oldsmobile Division, GMC
Lansing, Michigan

Dent, J. C.
Loughborough University of Technology
Leicestershire, England

Deris, J.
Factory Mutual Research Corporation
Norwood, Massachusetts

Dimick, D. L.
Engineering Staff, GMC
Warren, Michigan

Drexl, K.
Mercedes-Benz North America
Montvale, New Jersey

Dryer, F. L.
Princeton University
Princeton, New Jersey

Dyer, T. M.
Sandia National Laboratories
Livermore, California

Eckbreth, A. C.
United Technologies Research Center
East Hartford, Connecticut

Edelman, R.
Science Applications Inc.
Canoga Park, California

Elder, C. J.
Environmental Activities Staff, GMC
Warren, Michigan

Essenhigh, R. H.
Ohio State University
Columbus, Ohio

Faeth, G. M.
Pennsylvania State University
University Park, Pennsylvania

Farmer, R. C.
Science Applications Inc.
Canoga Park, California

Fenimore, C. P.
General Electric Company
Schnectady, New York

Ferguson, C. R.
Purdue University
West Lafayette, Indiana

Flower, W. L.
Sandia National Laboratories
Livermore, California

Frenklach, M.
Louisiana State University
Baton Rouge, Louisiana

Gallopoulos, N. E.
General Motors Research Laboratories
Warren, Michigan

Gann, R. G.
National Bureau of Standards
Washington, D.C.

Gardels, K. D.
General Motors Research Laboratories
Warren, Michigan

Gay, J. G.
General Motors Research Laboratories
Warren, Michigan

Glassman, I.
Princeton University
Princeton, New Jersey

Gordon, A. S.
University of California-San Diego
LaJolla, California

Groblicki, P. J.
General Motors Research Laboratories
Warren, Michigan

Groff, E. G.
General Motors Research Laboratories
Warren, Michigan

Hartley, D. L.
Sandia National Laboratories
Livermore, California

Haynes, B. S.
Massachusetts Institute of Technology
Cambridge, Massachusetts

Heap, M. P.
Energy and Environmental Research Corp.
Santa Ana, California

Heffner, F. E.
General Motors Research Laboratories
Warren, Michigan

Henein, N. A.
Wayne State University
Detroit

Heuss, J. M.
General Motors Research Laboratories
Warren, Michigan

Heywood, J. B.
Massachusetts Institute of Technology
Cambridge, Massachusetts

Hickling, R.
General Motors Research Laboratories
Warren, Michigan

Hilden, D. L.
General Motors Research Laboratories
Warren, Michigan

Hill, J. C.
General Motors Research Laboratories
Warren, Michigan

Hilliard, J. C.
University of Michigan
Ann Arbor, Michigan

Hills, F. J.
Mobil Research and Development Corp.
Paulsboro, New Jersey

Holmes, J.
California Air Resources Board
Sacramento, California

Homann, K. H.
Technische Hochschule Darmstadt
Darmstadt, West Germany

Howard, J. B.
Massachusetts Institute of Technology
Cambridge, Massachusetts

Huellmantel, L. W.
General Motors Research Laboratories
Warren, Michigan

Hunter, C.
Ford Motor Company
Dearborn, Michigan

Hutchinson, P.
AERE Harwell
Harwell, England

Jamerson, F. E.
General Motors Research Laboratories
Warren, Michigan

Japar, S.
Ford Motor Company
Dearborn, Michigan

Johnson, J. H.
Michigan Technological University
Houghton, Michigan

Kadota, T.
Hiroshima University
Hiroshima, Japan

Kamal, M.M.
General Motors Research Laboratories
Warren, Michigan

Keem, J. E.
General Motors Research Laboratories
Warren, Michigan

Kern, R. D.
University of New Orleans
New Orleans, Louisiana

Keski-Hynnica, D.
Detroit Diesel-Allison Div., GMC
Detroit, Michigan

Kittelson, D. B.
University of Minnesota
Minneapolis, Minnesota

Klimisch, R.L.
General Motors Research Laboratories
Warren, Michigan

Klomp, E. D.
General Motors Research Laboratories
Warren, Michigan

Kowalik, R. M.
Exxon Research and Engineering Company
Linden, New Jersey

Krieger, R. B.
General Motors Research Laboratories
Warren, Michigan

Lahaye, J.
C.N.R.S.
Mullhouse, France

Landis, J. R.
Chevrolet Division, GMC
Warren, Michigan

Larson, J. G.
General Motors Research Laboratories
Warren, Michigan

Lavoie, G.
Ford Motor Company
Dearborn, Michigan

Lee, M. L.
Brigham Young University
Provo, Utah

Levy, A.
Battelle-Columbus Laboratories
Columbus, Ohio

Lewis, L. L.
General Motors Research Laboratories
Warren, Michigan

Longwell, J. P.
Massachusetts Institute of Technology
Cambridge, Massachusetts

Ludecke, O.
Engineering Staff, GMC
Warren, Michigan

MacDonald, J. S.
General Motors Research Laboratories
Warren, Michigan

Magnussen, B.
University of Trondheim
Trondheim, Norway

Matekunas, F. A.
General Motors Research Laboratories
Warren, Michigan

Matsuoka, S.
Tokyo Institute of Technology
Tokyo, Japan

Mattavi, J. N.
General Motors Research Laboratories
Warren, Michigan

Matthews, C. C.
General Motors Research Laboratories
Warren, Michigan

Mayer, W. J.
General Motors Research Laboratories
Warren, Michigan

McDonald, R. J.
General Motors Research Laboratories
Warren, Michigan

Medalia, A. I.
Cabot Corporation
Billerica, Massachusetts

Mellor, A. M.
Purdue University
West Lafayette, Indiana

Modak, A. T.
Northern Research and Engineering Corp.
Norwood, Massachusetts

Muench, N. L.
General Motors Research Laboratories
Warren, Michigan

Muhlbaier, J L.
General Motors Research Laboratories
Warren, Michigan

Myers, J. P.
General Motors Research Laboratories
Warren, Michigan

Neoh, K. G.
Massachusetts Institute of Technology
Cambridge, Massachusetts

Newhall, H. K.
Chevron Research Company
Richmond, California

Novick, A.
Detroit Diesel Allison Division, GMC
Indianapolis, Indiana

Olson, D. B.
Aerochem Research Laboratories
Princeton, New Jersey

Otto, K.
Ford Motor Company
Dearborn, Michigan

Pagni, P. J.
University of California-Berkeley
Berkeley, California

Palmer, H. B.
Pennsylvania State University
University Park, Pennsylvania

Perez, J. M.
Caterpillar Tractor Company
Peoria, Illinoia

Peters, B. D.
General Motors Research Laboratories
Warren, Michigan

Plee, S. L.
General Motors Research Laboratories
Warren, Michigan

Prado, G.
Massachusetts Institute of Technology
Cambridge, Massachusetts

Pratt, D. T.
University of Michigan
Ann Arbor, Michigan

Ramohalli, K.
Jet Propulsion Laboratory
Pasadena, California

Rao, B.
General Motors Research Laboratories
Warren, Michigan

Richeson, D.
Chrysler Corporation
Detroit, Michigan

Rife, J. M.
Massachusetts Institute of Technology
Cambridge, Massachusetts

Rivin, D.
Cabot Corporation
Billerica, Massachusetts

Robinson, C. W.
Sandia National Laboratories
Livermore, California

Roessler, D. M.
General Motors Research Laboratories
Warren, Michigan

Rosebrock, T. L.
Engineering Staff, GMC
Warren, Michigan

Ruth, L. A.
Exxon Research and Engineering Company
Linden, New Jersey

Samuelsen, G. S.
University of California-Irvine
Irvine, California

Santoro, R. J.
National Bureau of Standards
Washington, D.C.

Sarofim, A. F.
Massachusetts Institute of Technology
Cambridge, Massachusetts

Schreck, R. M.
General Motors Research Laboratories
Warren, Michigan

Seery, D. J.
United Technologies Research Center
East Hartford, Connecticut

Shahed, S. M.
Cummins Engine Company, Inc.
Columbus, Indiana

Siegla, D.C.
General Motors Research Laboratories
Warren, Michigan

Sing, D. C.
General Motors Research Laboratories
Warren, Michigan

Skellenger, G. D.
General Motors Research Laboratories
Warren, Michigan

Skolnik, E. G.
Atlantic Research Corporation
Alexandria, Virginia

Sloane, C. S.
General Motors Research Laboratories
Warren, Michigan

Sloane, T. M.
General Motors Research Laboratories
Warren, Michigan

Smith, G. W.
General Motors Research Laboratories
Warren, Michigan

Smyth, K. C.
National Bureau of Standards
Washington, D.C.

Snelson, A.
  IIT Research Institute
  Chicago, Illinois

Sorensen, C.
  Kansas State University
  Manhattan, Kansas

Speck, C. E.
  General Motors Research Laboratories
  Warren, Michigan

Springer, G. S.
  University of Michigan
  Ann Arbor, Michigan

Stevenson, R.
  General Motors Research Laboratories
  Warren, Michigan

Stinton, H. C.
  Ricardo & Co. Engineers Ltd.
  Sussex, England

Stivender, D.L.
  General Motors Research Laboratories
  Warren, Michigan

Sturm, J. C.
  U.S. Department of Transportation
  Cambridge, Massachusetts

Talder, R. W.
  General Motors Research Laboratories
  Warren, Michigan

Taylor, K. C.
  General Motors Research Laboratories
  Warren, Michigan

Teets, R. E.
  General Motors Research Laboratories
  Warren, Michigan

Thurston, K. W.
  Oldsmobile Division, GMC
  Lansing, Michigan

Tracy, J. C.
  General Motors Research Laboratories
  Warren, Michigan

Ulrich, G. D.
  University of New Hampshire
  Durham, New Hampshire

Ultee, C. J.
  United Technologies Research Center
  Hartford, Connecticut

Vickers, P. T.
  General Motors Research Laboratories
  Warren, Michigan

Vostal, J. J.
  General Motors Research Laboratories
  Warren, Michigan

Wagner, H. Gg.
  Universität Göttingen
  Göttingen, West Germany

Williams, G. C.
  Massachusetts Institute of Technology
  Cambridge, Massachusetts

Williams, R. L.
  General Motors Research Laboratories
  Warren, Michigan

Wolff, G. T.
  General Motors Research Laboratories
  Warren, Michigan

Wolfson, B. T.
  U.S. Air Force
  Washington, D.C.

Wolgemuth, J. H.
  General Motors Research Laboratories
  Warren, Michigan

Yaccarino, P. A.
  General Motors Research Laboratories
  Warren, Michigan

Yang, L. C.
  Jet Propulsion Laboratory
  Pasadena, California

SUBJECT INDEX

Note: ff = following pages within chapter

---

nomenclature denoting similar structural concepts (see Preface, p. vi)